Characterization of Pharmaceutical Nano- and Microsystems

ADVANCES IN PHARMACEUTICAL TECHNOLOGY

A Wiley Book Series

Titles in the Series

Hot-Melt Extrusion: Pharmaceutical Applications
Edited by Dionysios Douroumis

Drug Delivery Strategies for Poorly Water-Soluble Drugs
Edited by Dionysios Douroumis and Alfred Fahr

Computational Pharmaceutics: Application of Molecular Modeling in Drug Delivery
Edited by Defang Ouyang and Sean C. Smith

Pulmonary Drug Delivery: Advances and Challenges
Edited by Ali Nokhodchi and Gary P. Martin

Novel Delivery Systems for Transdermal and Intradermal Drug Delivery
Edited by Ryan Donnelly and Raj Singh

Drug Delivery Systems for Tuberculosis Prevention and Treatment
Edited by Anthony J. Hickey

Continuous Manufacturing of Pharmaceuticals
Edited by Peter Kleinebudde, Johannes Khinast, and Jukka Rantanen

Pharmaceutical Quality by Design
Edited by Walkiria S Schlindwein and Mark Gibson

***In Vitro* Drug Release Testing of Special Dosage Forms**
Edited by Nikoletta Fotaki and Sandra Klein

Characterization of Pharmaceutical Nano- and Microsystems
Edited by Leena Peltonen

Forthcoming Titles:

Process Analytics for Pharmaceuticals
Edited by Jukka Rantanen, Clare Strachan and Thomas De Beer

Mucosal Drug Delivery
Edited by Rene Holm

Basic Biopharmaceutics
Edited by Hannah Batchelor

Characterization of Pharmaceutical Nano- and Microsystems

Edited by

LEENA PELTONEN
University of Helsinki, Finland

WILEY

This edition first published 2021

Registered Offices
John Wiley & Sons, Inc., 111 River Street, Hoboken, NJ 07030, USA
Wiley-VCH Verlag GmbH & Co. KGaA, Boschstr. 12, 69469 Weinheim, Germany

Editorial Office
Boschstr. 12, 69469 Weinheim, Germany

For details of our global editorial offices, customer services, and more information about Wiley products visit us at www.wiley.com.

Wiley also publishes its books in a variety of electronic formats and by print-on-demand. Some content that appears in standard print versions of this book may not be available in other formats.

Library of Congress Cataloging-in-Publication Data
Names: Peltonen, Leena Johanna, 1970- editor.
Title: Characterization of pharmaceutical nano- and microsystems / Leena Johanna Peltonen.
Description: First edition. | Hoboken : Wiley, 2020. | Series: Advances in pharmaceutical technology | Includes bibliographical references and index.
Identifiers: LCCN 2020027861 (print) | LCCN 2020027862 (ebook) | ISBN 9781119414049 (cloth) | ISBN 9781119414032 (adobe pdf) | ISBN 9781119414025 (epub)
Subjects: LCSH: Pharmaceutical technology. | Nanotechnology. | Drug development.
Classification: LCC RS192 .C522 2020 (print) | LCC RS192 (ebook) | DDC 615.1/9—dc23
LC record available at https://lccn.loc.gov/2020027861
LC ebook record available at https://lccn.loc.gov/2020027862

Cover Design: Wiley
Cover Image: © Inna Bigun/Shutterstock

Set in 10/12pt, TimesLTStd by SPi Global, Chennai, India.

MIX
Paper from responsible sources
FSC www.fsc.org FSC® C013604

Contents

List of Contributors

Malgorzata Iwona Adamczak, Department of Pharmacy, Faculty of Mathematics and Natural Sciences, University of Oslo, Norway; and GE Healthcare, Pharmaceutical Diagnostics, Oslo, Norway

Osmo Antikainen, Faculty of Pharmacy, University of Helsinki, Finland

Cláudia Azevedo, INEB – Instituto de Engenharia Biomédica; i3S – Instituto de Investigação e Inovação em Saúde; and Instituto de Ciências Biomédicas Abel Salazar, Universidade do Porto, Portugal

Erem Bilensoy, Department of Pharmaceutical Technology, Faculty of Pharmacy, Hacettepe University, Turkey

Preshita P. Desai, Western University of Health Sciences, Pomona, California, USA

Martin Dračínský, Czech Academy of Sciences, Prague, Czech Republic

Nazlı Erdoğar, Department of Pharmaceutical Technology, Faculty of Pharmacy, Hacettepe University, Turkey

Nazanin Zanjanizadeh Ezazi, Drug Research Program, University of Helsinki, Finland

Paulo J. Ferreira, CIEPQPF, Department of Chemical Engineering, University of Coimbra, Portugal

Margarida Figueiredo, CIEPQPF, Department of Chemical Engineering, University of Coimbra, Portugal

Flavia Fontana, Drug Research Program, University of Helsinki, Finland

Ellen Hagesaether, Department of Life Science and Health, Faculty of Health Sciences, Oslo Metropolitan University, Norway

Marianne Hiorth, Department of Pharmacy, Faculty of Mathematics and Natural Sciences, University of Oslo, Norway

Martti Kaasalainen, Laboratory of Industrial Physics, University of Turku, Finland, and Medicortex Finland Oy

Didem Şen Karaman, Pharmaceutical Sciences Laboratory, Faculty of Science and Engineering, Åbo Akademi University, Turku, Finland

Helene Kettiger, Pharmaceutical Sciences Laboratory, Faculty of Science and Engineering, Åbo Akademi University, Turku, Finland

Iiro Kiiski, Drug Research Program, University of Helsinki, Finland

Timo Laaksonen, Tampere University of Technology, Finland

Tiina Lipiäinen, Drug Research Program, University of Helsinki, Finland

M. José Moura, Department of Chemical and Biological Engineering, Polytechnic Institute of Coimbra, Portugal; and CIEPQPF, Department of Chemical Engineering, University of Coimbra, Portugal

Elisa Ollikainen, Drug Research Program, University of Helsinki, Finland

Vandana B. Patravale, Institute of Chemical Technology, Mumbai, Maharashtra, India

Inês Pereira, INEB – Instituto de Engenharia Biomédica; and i3S – Instituto de Investigação e Inovação em Saúde, Universidade do Porto, Portugal

Kaisa Rautaniemi, Tampere University of Technology, Finland

Jessica Rosenholm, Pharmaceutical Sciences Laboratory, Faculty of Science and Engineering, Åbo Akademi University, Turku, Finland

Jukka Saarinen, Drug Research Program, University of Helsinki, Finland

Helder A. Santos, Drug Research Program, University of Helsinki, Finland

Bruno Sarmento, INEB – Instituto de Engenharia Biomédica; and i3S – Instituto de Investigação e Inovação em Saúde, Universidade do Porto, Portugal; and CESPU, Instituto de Investigação e Formação Avançada em Ciências e Tecnologias da Saúde, Gandra, Portugal

Tiina M. Sikanen, Drug Research Program, and Helsinki Institute of Life Science, University of Helsinki, Finland

Marcin Skotnicki, Poznań University of Medical Sciences, Poland

Clare Strachan, Drug Research Program, University of Helsinki, Finland

Nayab Tahir, Drug Research Program, University of Helsinki, Finland; College of Pharmacy, University of Sargodha, Pakistan; and Faculty of Pharmacy and Alternative Medicine, The Islamia University of Bahawalpur, Pakistan

Ingunn Tho, Department of Pharmacy, Faculty of Mathematics and Natural Sciences, University of Oslo, Norway

Cem Varan, Department of Nanotechnology and Nanomedicine, Institute of Graduate Studies in Science, Hacettepe University, Turkey

Gamze Varan, Department of Nanotechnology and Nanomedicine, Institute of Graduate Studies in Science, Hacettepe University, Turkey

Elina Vuorimaa-Laukkanen, Tampere University of Technology, Finland

Series Preface

The series *Advances in Pharmaceutical Technology* covers the principles, methods and technologies that the pharmaceutical industry uses to turn a candidate molecule or new chemical entity into a final drug form and hence a new medicine. The series will explore means of optimizing the therapeutic performance of a drug molecule by designing and manufacturing the best and most innovative of new formulations. The processes associated with the testing of new drugs, the key steps involved in the clinical trials process and the most recent approaches utilized in the manufacture of new medicinal products will all be reported. The focus of the series will very much be on new and emerging technologies and the latest methods used in the drug development process.

The topics covered by the series include the following:

Formulation: The manufacture of tablets in all forms (caplets, dispersible, fast-melting) will be described, as will capsules, suppositories, solutions, suspensions and emulsions, aerosols and sprays, injections, powders, ointments and creams, sustained release and the latest transdermal products. The developments in engineering associated with fluid, powder and solids handling, solubility enhancement, colloidal systems including the stability of emulsions and suspensions will also be reported within the series. The influence of formulation design on the bioavailability of a drug will be discussed and the importance of formulation with respect to the development of an optimal final new medicinal product will be clearly illustrated.

Drug Delivery: The use of various excipients and their role in drug delivery will be reviewed. Among the topics to be reported and discussed will be a critical appraisal of the current range of modified-release dosage forms currently in use and also those under development. The design and mechanism(s) of controlled release systems including macromolecular drug delivery, microparticulate controlled drug delivery, the delivery of biopharmaceuticals, delivery vehicles created for gastrointestinal tract targeted delivery, transdermal delivery and systems designed specifically for drug delivery to the lung will all be reviewed and critically appraised. Further site-specific systems used for the delivery of drugs across the blood–brain barrier including dendrimers, hydrogels and new innovative biomaterials will be reported.

Manufacturing: The key elements of the manufacturing steps involved in the production of new medicines will be explored in this series. The importance of crystallization; batch and continuous processing, seeding; and mixing including a description of the key engineering principles relevant to the manufacture of new medicines will all be reviewed and reported. The fundamental processes of quality control including good laboratory practice, good manufacturing practice, Quality by Design, the Deming Cycle, regulatory requirements and the design of appropriate robust statistical sampling procedures for the control of raw materials will all be an integral part of this book series.

An evaluation of the current analytical methods used to determine drug stability, the quantitative identification of impurities, contaminants and adulterants in pharmaceutical materials will be described, as will the production of therapeutic bio-macromolecules, bacteria, viruses, yeasts, molds, prions and toxins through chemical synthesis and emerging synthetic/molecular biology techniques. The importance of packaging including the compatibility of materials in contact with drug products and their barrier properties will also be explored.

Advances in Pharmaceutical Technology is intended as a comprehensive one-stop shop for those interested in the development and manufacture of new medicines. The series will appeal to those working in the pharmaceutical and related industries, both large and small, and will also be valuable to those who are studying and learning about the drug development process and the translation of those drugs into new life-saving and life-enriching medicines.

Dennis Douroumis
Alfred Fahr
Jürgen Siepmann
Martin Snowden
Vladimir Torchilin

List of Abbreviations

μTAS	micro total analysis systems
AFM	atomic force microscopy
ALP	alkaline phosphatase
AMP	antimicrobial peptide
AOM	acousto-optic modulator
API	active pharmaceutical ingredient
APS	Aerodynamic Particle Sizer
AR	aspect ratio
ATR	attenuated total reflectance
AUC	area under the curve
BCS	biopharmaceutical classification system
CAD	computer-aided drawing
CARS	coherent anti-Stokes Raman scattering
CCD	charge-coupled device
CDF	continuous density function
CFM	confocal fluorescence microscopy
CLEA	crosslinked enzyme aggregate
CLEC	crosslinked enzyme crystal
CLS	classical least squares
CLSM	confocal laser scanning microscopy
CNS	central nervous system
CP	cross-polarization
CSA	chemical shift anisotropy
CT	contact time
CYP	cytochrome P450
DC	dendritic cell

DCLS	direct classical least squares
DDS	drug delivery system
DDM	derivative difference minimization
DE	direct excitation
DFT	density functional theory
DLS	dynamic light scattering
DLVO	Derjaguin, Landau, Verwey and Overbeek [theory]
DOE	design of experiments
DRIFTS	diffuse reflectance infrared Fourier transform spectroscopy
DSC	differential scanning calorimetry
DTGS	deuterated triglycine sulfate
ECM	extracellular matrix
EDFM	enhanced dark-field microscope
EDL	electrical double layer
EFG	electric field gradient
EM	electron microscopy
EPR	enhanced permeability and retention
ER	endoplastic reticulum
ESD	equivalent spherical diameter
FACS	fluorescence-activated cell sorting
FBRM	focused beam reflectance measurement
FD	Fraunhofer diffraction
FIR	far-infrared
FITC	fluorescein isothiocyanate
FLIM	fluorescence lifetime imaging microscopy
FRAP	fluorescent recovery after photobleaching
FRET	Förster (or fluorescence) resonance energy transfer
FSLN	furosemide solid lipid nanoparticle
FTIR	Fourier transform infrared
GFP	green fluorescent protein
GIT	gastrointestinal tract
GPC	gel permeation chromatography
HETCOR	heteronuclear correlation
HLM	human liver microsome
HMSN	hollow-type mesoporous silica nanoparticle
HPLC	high pressure liquid chromatography
HPPD	high-power proton decoupling
HPV	human papilloma virus
HIS	hyperspectral imaging
IEC	intestinal epithelial cell
IEP	isoelectric point
IgA	immunoglobulin A
IgG	immunoglobulin G
IONP	iron oxide nanoparticle
iPSC	induced pluripotent stem cell
iPSC-CM	induced pluripotent stem cell-derived cardiomyocytes

IR	infrared
IUPAC	International Union of Pure and Applied Chemistry
i.v.	intravenous
IVIVC	*in vitro–in vivo* correlation
LALLS	low-angle laser light scattering
LC	liquid chromatographic
LD	laser diffraction
LDA	linear discriminant analysis
LNC	lipid-core nanocapsule
LOD	limit of detection
LOQ	limit of quantification
MAE	mean absolute error
MAS	magic-angle spinning
MCR	multivariate curve resolution
MCR-ALS	multivariate curve resolution-alternating least squares
MCT	mercury cadmium telluride
MD	molecular dynamic
MDT	mean dissolution time
MIP	multiple image photography
MIR	mid-infrared
MPS	mononuclear phagocytic system
MRI	magnetic resonance imaging
MRT	mean residence time
MSN	mesoporous silica nanoparticle
MSP	mesoporous silica particle
NA	numerical aperture
NADPH-CPR	nicotinamide adenine dinucleotide phosphate-cytochrome P450 reductase
NCE	new chemical entity
NIR	near-infrared
NLC	nanostructured lipid carrier
NMR	nuclear magnetic resonance
NP	nanoparticle
OFAT	one factor at a time
OI	optical imaging
OSTE	off-stoichiometric thiol-ene
PAT	process analytical technology
PBMC	peripheral blood mononuclear cell
PBS	phosphate-buffered saline
PCA	principal component analysis
PCS	photon correlation spectroscopy
PCR	principal components regression
PDMS	poly(dimethyl siloxane)
PE	predictability evaluation
PEG	polyethylene glycol
PEI	poly(ethylene imine)
PET	positron emission tomography

P-gp	P-glycoprotein
pI	isoelectric point
PLGA	poly(lactic-co-glycolic) acid
PLS-DA	partial least-squares discriminant analysis
PMT	photomultiplier tube
PS	polystyrene
PSi	porous silicon
PTFE	polytetrafluoroethylene
PTX	paclitaxel
PZC	point of zero charge
QbD	Quality by Design
QCM	quartz crystal microbalance
QELS	quasi-elastic light scattering
RBC	red blood cell
RES	reticular endothelial system
RF	radio frequency
RGD	tripeptide Arg-Gly-Asp
RH	relative humidity
RMSE	root mean squared error
ROS	reactive oxygen species
SCXRD	single crystal X-ray diffraction
SEM	scanning electron microscopy
SEP	standard error of prediction
SFB	segmented filamentous bacteria
SFG	sum frequency generation
SGF	simulated gastric fluid
SHG	second harmonic generation
SIMCA	soft independent modeling of class analogy
SLN	solid lipid nanoparticle
SLS	static light scattering
SORS	spatially offset Raman spectroscopy
SPE	solid-phase extraction
SPR	surface plasmon resonance
SRG	stimulated Raman gain
SRL	stimulated Raman loss
SRS	stimulated Raman scattering
SSNMR	solid-state nuclear magnetic resonance
SUPAC	scale-up and post-approval change
SVM	support vector machine
TC	thermal carbonization
TEER	transepithelial electrical resistance
TEM	transmission electron microscopy
TERS	tip-enhanced Raman scattering
TGA	thermogravimetric analysis
THC	thermal hydrocarbonization
THG	third harmonic generation

TOF	time-of-flight
TOPSi	thermally oxidized porous silicon
TOSS	total suppression of spinning sidebands
TPEF	two-photon excited fluorescence
TPS	terahertz pulsed spectroscopy
UGT	uridine 5′-diphospho-glucuronosyltransferase
USFDA	United States Food and Drug Administration
UV	ultraviolet
VDOS	vibrational density of states
XPS	X-ray photoelectron spectroscopy
XRPD	X-ray powder diffraction
ZP	zeta potential

1

Selecting a Particle Sizer for the Pharmaceutical Industry

Margarida Figueiredo[1], M. José Moura[1,2] and Paulo J. Ferreira[1]

[1] *CIEPQPF, Department of Chemical Engineering, University of Coimbra, Portugal*
[2] *Department of Chemical and Biological Engineering, Polytechnic Institute of Coimbra, Portugal*

1.1 Introduction

1.1.1 Relevance of Particle Size in the Pharmaceutical Industry

Knowledge and understanding of particle size data is crucial in a wide range of industries, being vital for the pharmaceutical industry, with applications from drug development to production and quality control. The purpose of particle size analysis is to obtain quantitative data on the mean size, particle size distribution and, sometimes in addition, particle shape of the compounds used in pharmaceutical formulations. It is well known that particle size highly affects not only the final product performance (e.g., dissolution, bioavailability, stability and absorption rates), but also every step of the manufacturing process of both drug substances and excipients (like mixing, flowability, granulation, drying, milling, blending, coating and encapsulation) [1–8]. For example, particle size is often directly related to dissolution/solubility characteristics of solid or suspension delivery systems, which have a direct impact on the bioavailability of pharmaceutical products. Dissolution is directly proportional to particle surface area, which in turn is inversely proportional to particle size (i.e., finer particles promote faster drug dissolution). The same applies to the suspensions where precipitation is highly controlled by particle size (in practice, finer particles generally give more stable suspensions), equally affecting viscosity and flow (Stokes' law relates the settling velocity of particles to the square

Characterization of Pharmaceutical Nano- and Microsystems, First Edition. Edited by Leena Peltonen.

of particle diameter). Distribution of sizes is another key characteristic that influences, for instance, handling and processing (powder handling characteristics are profoundly affected by changes in flow properties and tendency to segregate, which are both highly dependent on powder size distribution). Ultimately, particle size also has a critical effect on the content uniformity of solid dosage forms. Size analysis also becomes of significant importance with new drug delivery formats such as liposomes and nanoparticles whose characterization requires sophisticated analytical techniques [9–12].

In brief, particle size simultaneously affects safety, efficacy and quality of the drug, and regulatory agencies are becoming increasingly aware of the importance of particle sizing, requiring developers to have greater control and understanding of this aspect of their drug products [3, 13–15].

1.1.2 Main Goals

This chapter intends to introduce the problem of particle sizing in the domain of the pharmaceutical industry, especially to those who are not very familiar with this topic. It is by no means an exhaustive description of particle sizing methods, but addresses the basic concepts associated with particle sizing, providing a basis to understand the most important details associated with particle sizing data and its interpretation. It was conceived not only to help the reader to select the most suitable techniques for your particle characterization needs, but also to be a valuable tool in daily work situations.

A considerable effort was made to condense in a single chapter topics that range from the interpretation of sizing data to the working principles, applications and limitations of some selected methods, including their selection criteria, subjects that are normally treated in separate publications/chapters. The idea was to provide the essential information to enable the reader to completely follow all the topics covered here. After discussing the reasons why choosing a particle sizer is not an easy task, some basic definitions of particle size, size distribution and their representations will be given in a concise manner, before addressing some of the most relevant parameters to be taken into consideration when selecting a particle sizing method. Finally, the underlying principles of some selected methods will be presented, together with their strengths and weaknesses. Naturally, the number of addressed methods had to be limited. Hence, this discussion will mainly be directed to sizing techniques normally available for *routine analyses* in the *pharmaceutical field*, from *nanoparticles* to some *hundred micrometer particles*. In order to encompass one of each class of particle sizing methods, the following techniques were selected: optical microscopy/image analysis and time-of-flight, representative of the counting techniques; static and dynamic light scattering, widely used ensemble techniques; and the cascade impactor, a separation technique frequently used for aerosol samples (nasal products). As mentioned, the ultimate goal will be to stimulate the reader's curiosity to consult other sources of information to complement this analysis.

1.1.3 Why it is So Difficult to Select a Particle Sizer

The apparent simplicity of particle size analysis is deceptive as particle sizing is a poorly posed problem. As is well known, only objects of simple geometry, namely spheres, can be unambiguously described by a single linear dimension. Non-spherical particles, as discussed below, are most conveniently described in terms of derived diameters calculated by measuring a size-dependent property of the particle and relating it to a linear dimension. As a result, different sizing methods, based on the measurement of different particle properties, might give different sizing data for the same sample. Moreover, the same measuring technique can also generate different sizing results as a consequence of distinct data processing algorithms used by the equipment manufacturers [2, 3, 16, 17]. Complicating this, a wide range of size distributions normally have to be analysed, being not uncommon that the size range of the particles is too wide to be measured with a single device. Besides, particles, namely pharmaceuticals, include dry powders, suspensions, aerosols, emulsions and nanoparticles, which in turn can be presented as primary (individual) particles, aggregates or agglomerates (in *aggregates* the primary particles are bound strongly by covalent bonds, whereas *agglomerates* are collections of aggregates loosely held together by weak forces). Also, the recent interest in measuring nanoparticles resulted in a burst of new techniques (or new applications of old techniques) for the nanometer range, being that the smaller the particles, the more difficult it is to characterize them. Accordingly, there has never been so much diversity of sizing equipment (hundreds of commercially available instruments), sample and data treatments.

Additionally, it should be pointed out that formal training in the field of particle technology is not often as widespread as in other fields. Further, the technical information available in particle technology, namely particle sizing vocabulary, is unique and complex, and a clear domain of fine particle technology terminology is indispensable for correct data interpretation.

As a final point, it should be highlighted that the determination of particle size distribution seldom is the ultimate objective: indeed, a particle size measurement is often carried out with the aim of relating particle sizing data to a particular property or behavior of the material, and this relationship should be taken into consideration when choosing a sizing instrument. For example, if we are studying the particles of an airborne aerosol and their deposition in the lungs, a sizing method based on the measurement of the aerodynamic diameter would be more appropriate; furthermore, if a drug product is to be administered as a dry powder, a particle characterization technique capable of measuring the sample as a dry powder dispersion should be used.

Sizing equipment is not often restricted to a specific application, being normally used for more general purposes. Nonetheless, it should be borne in mind that no single technique is superior in all applications. All these reasons render the selection of the most appropriate particle sizing method a challenging process.

1.2 Particle Size Distribution

1.2.1 Equivalent Diameter

It is not possible to rationally discuss the size of a particle without considering the three-dimensional characteristics of the particle itself (length, breadth, and height). In fact, only the sphere can be fully described by a single dimension, its radius or diameter. However, most real-world particles are far from round or uniform, and with regard to particle sizing, it is often most convenient to express particle size in terms of derived diameters such as equivalent spherical diameter (ESD). ESD is defined by ISO 9276-1:1998 [18] as the diameter of a sphere that produces a response by a given particle-sizing method that is equivalent to the response produced by the particle being measured. In many cases the equivalent sphere is the one with the same volume as the particle, the so-called volume-equivalent spherical diameter (a cube of length 1 μm has a volume-equivalent spherical diameter of 1.24 μm). However, the method of measurement and the property of interest of the particle can lead to the use of other diameters, such as, for instance, the surface-equivalent spherical diameter, which is the diameter of a sphere having the same surface area as the particle, or the projected area diameter, most used in image analysis, that is the diameter of a circle having the same area as the projected area of the particle. These and other frequently used particle-equivalent diameters are defined in Table 1.1 [16].

Table 1.1 *Definitions of equivalent spherical diameters (ESDs)*

ESD	Definition
Volume diameter	Diameter of a sphere having the same volume as the particle
Surface diameter	Diameter of a sphere having the same (external) surface area as the particle
Projected area diameter	Diameter of a circle having the same area as the projected area of the particle resting in a stable position
Surface volume diameter (Sauter diameter)	Diameter of a sphere having the same surface area-to-volume ratio as the particle
Sieve diameter	Diameter of a sphere passing through a sieve of defined mesh size (with square or circular apertures)
Stokes diameter	Diameter of a sphere with the same final settling velocity as the particle undergoing laminar flow in a fluid of the same density and viscosity
Hydrodynamic diameter	Diameter of a sphere with the same translational diffusion coefficient as the particle in the same fluid under the same conditions
Mobility diameter	Diameter of a sphere having the same mobility in an electric field as the particle
Fraunhofer diameter	Diameter of a sphere that will scatter light at the same intensity at the same angle as the particle (correspond to the projected area diameter of a particle in random orientation)
Optical diameter	Diameter of a sphere having the same optical cross-section as the particle
Aerodynamic diameter	Diameter of a unit density sphere that would have the identical settling velocity as the particle

Clearly, non-spherical particles can lead to very different equivalent diameters depending on the definition chosen, which in turn is related to the measured particle property and ultimately to the sizing instrument/technique used. The further away from spherical the actual particle shape is, the greater the difference in ESD (for non-spherical compact convex particles, the results will not differ greatly for the various size measurements, but for needles, disks or flakes, with one dimension significantly different from the others, the differences may be quite relevant). Moreover, ESD may not correlate with any single dimension of the particle. On the other hand, identical equivalent diameters may be obtained for different particle shapes. For that, particle size and particle size distribution results are frequently considered as relative measurements, and comparisons of size results from different instruments should be conducted with extreme caution.

Although the ESD approach is simplistic, it is very convenient and it is employed in almost all particle sizing techniques. However, it is absolutely essential to be clear and consistent as to which ESD is being used.

1.2.2 Reporting Particle Size

This section briefly addresses the representation of size distributions, focusing mainly on the types of curves used to express the distribution and some central tendencies. Nowadays, all particle sizers report the data in graphical form (some of which we can select) indicating some statistical parameters. However, a perfect understanding of the distributions and of the associated statistical parameters is absolutely essential for a correct interpretation of the sizing data [16, 19–22].

Almost all real-world samples exist as a distribution of particle sizes, normally expressed as a function of two coordinates: the size (mostly an ESD) plotted on the *x*-axis, and the amount of each size, plotted on the *y*-axis, as illustrated in Figure 1.1. The size distribution can be represented in the form of either a frequency (differential) distribution curve or a cumulative (normally undersize) distribution curve (typically with a sigmoidal shape), obtained by sequentially adding the percentage frequency values. Both types of plot are useful depending on the information we want from the graphical representation: the frequency distribution presents a clear description of the distribution spread and also shows if the distribution is monomodal or multimodal (i.e., with one or more peaks, respectively) and whether the peak is skewed from the centre; in a cumulative plot, multimodal peaks are not easily observed but the identification of the percentage of particles below a given diameter is much simpler.

However, we need to be aware that, depending on the sizing technique, the amount of each size can be weighted in different ways [23]. The more common weighted

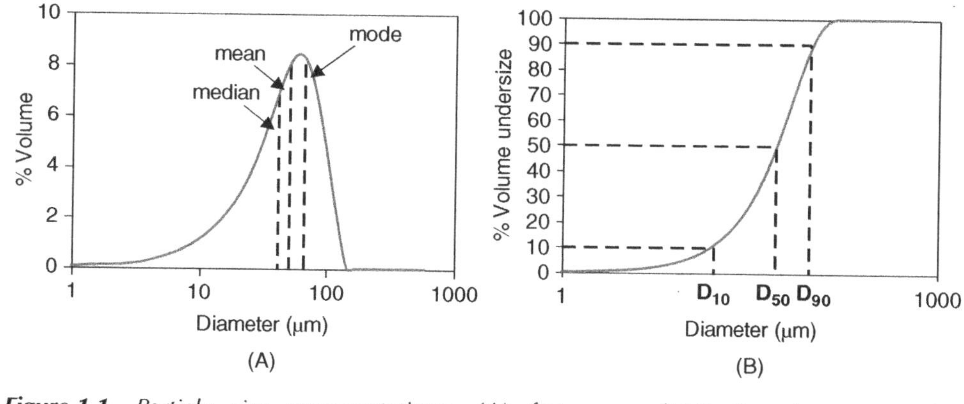

Figure 1.1 *Particle size representations: (A) frequency distribution (non-symmetric); (B) cumulative undersize distribution (with most common percentiles).*

distributions are: number-weighted distributions, resulting from counting techniques such as image analysis, where each particle is given equal weighting irrespective of its size; surface-weighted distribution (normally surface area) where each size is square weighted; volume-weighted distributions, common in static light-scattering techniques, in which the contribution of each particle in the distribution relates to its volume (being equivalent to a mass distribution if the density of the particles is uniform); and intensity-weighted distributions, where the contribution of each particle relates to the intensity of the light scattered by the particle, typical of dynamic light-scattering-based instruments. Number, surface and volume weightings vary as size raised to the zero, second and third powers, respectively. The case of intensity-weighted distributions this is not so simple, and depends on the type of light-scattering device and also on the size range [16] (for example, the intensity of the light scattered by very small particles (<50 nm) is proportional to $[\text{size}]^6$). Figure 1.2 clearly illustrates this point by showing the results of a size distribution of equal numbers of particles with diameters of 5 nm and 50 nm. As expected, the number-weighted distribution gives equal weighting to both types of particles, whereas the intensity-weighted distribution corresponds to a much stronger signal for the coarser 50 nm particles (one million times higher). The volume-weighted distribution is intermediate between the two. This example clearly shows how crucial it is, when reporting particle sizing data, to report not only the size measuring method but also the distribution base. It can then be concluded that different sizing techniques can generate different sizing results for the same sample, not only because different equivalent diameters are being measured, but also because different weighting factors are being used.

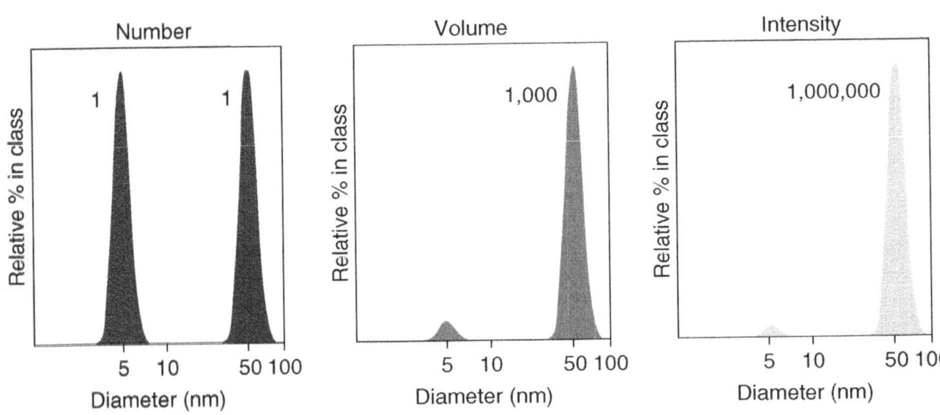

Figure 1.2 *Example of number-, volume- and intensity-weighted particle size distributions for the same sample [reproduced with permission of Malvern Panalytical].*

Volume-weighted (or mass-weighted) particle size distributions are common for most pharmaceutical materials; however, number-weighted representations are useful, for instance, for determining the size of primary particles in agglomerated systems or to detect impurities [3]. Although it is mathematically simple to convert from one type of weighting to another, the converted results are often erroneous [3]. In fact, additional information about particle characteristics (such as shape factors or optical properties (refractive index)) is normally required for a more reliable conversion, but in general these elements are not available in practice. Thus, whenever possible, a particle sizing technique that gives the desired weighting without transformation should be used.

1.2.3 Distribution Statistics

While a single number cannot describe the size distribution of the sample, sometimes it is tempting to report an "average size" or a central tendency of the distribution along with one or more values to describe the distribution width. A range of statistical parameters can be used for this purpose [16], as for example:

- **mean**: "average" size of a population;
- **median**: size where 50% of the population is below/above – this value is also called D_{50} and is one of the most meaningful parameters for particle size distributions;
- **mode**: size with the highest frequency (highest peak of the distribution), very useful if there is more than one peak in the distribution (multimodal).

For symmetric distributions (also called normal or Gaussian) all these values are numerically equivalent, but for asymmetric distributions with elongated tails, most common in real samples, these parameters correspond to different values, as illustrated in Figure 1.1.

Particular care should be taken with the "mean" values as there are multiple definitions for this parameter related to the basis of the distribution (e.g., number or volume). The various mean calculations are defined in standard documents [23]. Table 1.2 summarizes the most common.

The comparison between two or more particle size data is easier when using the cumulative distribution representations, in the same or separate graphs. Furthermore, in order to

Table 1.2 *Examples of mean diameters*

	Definition	Comment
Number-weighted mean ($D_{1,0}$) (also known as arithmetic mean)	$= \frac{\sum d}{n} = \frac{\sum d^1}{d^0}$	Most common in particle counting applications
Surface-weighted mean ($D_{3,2}$) (also called Sauter mean)	$= \frac{\sum d^3}{\sum d^2}$	Most relevant where specific surface area is important e.g., bioavailability, reactivity, dissolution
Volume-weighted mean ($D_{4,3}$)	$= \frac{\sum d^4}{\sum d^3}$	Most common in instruments where the result is displayed as a volume distribution, most sensitive to the presence of large particles
Intensity-weighted mean (also called Z-average diameter, D_Z, or harmonic mean ($D_{6,5}$))	$= \frac{\sum d_i^6}{\sum d_i^5}$	Most common in DLS for very small particles (Rayleigh scatterers)

quantify the width of the size distributions, it is common to use some parameters of the cumulative curve known as percentiles (D_x where x means the percentage of sample with sizes below this value), typically D_{10}, D_{50} and D_{90}. As mentioned before, D_{50} (the median) is the middle value of the cumulative distribution where the total frequency of values above and below is equal; D_{90} describes the diameter where 90% of the distribution has a smaller particle size than this value (and 10% has a larger particle size); and D_{10} means that 10% of the distribution have diameters lower than this value. These percentiles, easily recognized in a cumulative curve, as previously shown in Figure 1.1, are frequently used to quantify the width (or span) of the size distribution defined as:

$$\text{Span} = \frac{D_{90} - D_{10}}{D_{50}}$$

Span is normally defined as the distance between two points equally spaced from the median and thus other percentiles can also be used in this definition as, for example, D_{25} and D_{75} (also known as quartiles). Finally, it should be pointed out that cumulative distributions can be represented on linear and logarithmic axes for the particle size (the latter is especially suited for widely distributed data) [16, 18].

1.3 Selecting a Particle Sizer

1.3.1 Classification

As discussed above, the choice of a particle sizer is not an easy task due to several reasons, one of them being the arsenal of particle-sizing technologies and instrumentation currently available on the market, that range from the classical sieves to the more modern and sophisticated light-scattering instruments. As a result of this large variety of methodologies, it is difficult to classify the techniques used for particle size. Nonetheless, some attempts have been made to group them [14, 24, 25]. One criterion is to divide the sizing techniques into *imaging* and *non-imaging*. Imaging techniques obviously allow the direct visualization of the particles and thus can provide, besides size and size range, additional information on particle characteristics like shape, structure, degree of agglomeration and texture, which the non-imaging techniques are unable to give. These methods include optical microscopy/image analysis as well as electron microscopy (SEM and TEM), being mandatory whenever particle shape and structure information is required. They are normally slow and labour-intensive (especially manual microscopy) compared with the non-imaging methods that, on the other hand, are based on the measurement of a particle property related to its size through an equivalent spherical diameter.

Another type of classification is based on the measurement being made one particle at a time, accumulating counts of particles with similar sizes, as opposed to measuring an ensemble of particles at the same time and subsequently extracting the particle size distribution using an appropriate theory (model). The former are called *single particle techniques* (also referred to as *counting techniques*, as particles are individually counted), and typically exhibit high sensitivity and resolution but narrow dynamic size ranges. In contrast, the latter techniques, named *ensemble techniques*, normally have low resolution and low sensitivity but a broader dynamic size range and high statistical accuracy, being better suited for on-line and in-line applications. A high-resolution instrument can separate two close-together modes, while a low-resolution instrument can only detect one broad peak. Sensitivity in particle sizing can be viewed as a measure of the smallest amount of a given size particle that can be detected by the instrument.

Examples of counting methods are not only the microscopy-based techniques, as image analysis, but the electrozone counters (pioneered by the Coulter company and still known as the Coulter counter technique), the optical counters (optical equivalent to electrozone counters), and the time-of-flight counters targeted at aerosols. In these counting techniques, particles pass individually through the sensing zone (an electrical sensing zone in the case of the Coulter counter, or a photozone in the case of an optical counter) and so very low particle concentrations have to be used in order to avoid coincident effects (i.e., multiple particles being counted together). Another common feature of these methods is that they all need prior calibration, accomplished by using uniform particles of known sizes [16, 17].

As previously stated, ensemble methods rely on the measurement of a certain property of an ensemble of particles, being the raw detected signal "inverted" mathematically to estimate the particle size distribution of the entire population. For that, the results of these techniques normally depend on the mathematical algorithm used. Examples of ensemble averaging techniques are light-scattering techniques (static and dynamic) and acoustic spectroscopy [16, 17].

The ensemble methods can also involve fractionation of the samples prior to sizing, in which an outside separation force is applied to the particles, physically separating them according to size (*fractionating methods*). In order to provide a measurement of particle size distribution, the fractionation techniques must be combined with detection techniques, such as optical detection or light attenuation or scattering. Common fractionation techniques are sieves, gravitational sedimentation, differential centrifugal sedimentation, and various forms of particle chromatography [16, 17]. Details of some of these techniques, available on the market and commonly used to characterize pharmaceutical products, can be found in Table 1.3 together with the respective size ranges and the corresponding standards.

1.3.2 Selection Criteria

In making a decision on which technique to use, a number of criteria should be considered [24, 25]. First of all, it is necessary to consider the appropriateness of a particular sizing method to the problem at hand. As explained earlier, different techniques are likely to produce different size results for the same particle, and all of them are likely to be correct. The best instrument for a given application is probably the one that most closely relates particle size to the application of the particles. In other words, the particle property measured by the size analyser should be (whenever possible) related to the end use of the sizing data. Additionally, the measuring principle should be directly related to the weighting mechanism. As also discussed above, the sizing data can be expressed on a different basis (e.g., number, area, volume, light-scattering intensity) depending on the measuring principle. Conversions to another basis, although easily available from the software of most instruments, should be avoided. In the case of pharmaceutical products, a volume-weighted distribution is often the most relevant descriptor of the content of active ingredients as a function of particle size

Table 1.3 *Examples of sizing techniques commonly used in the pharmaceutical industry and respective measuring principles, measured equivalent spherical diameter (ESD) and primary distribution weight, size range and related standards*

Measurement technique	Method/physical principle	Technique layout	ESD/primary distribution weight	Size range (μm)	Standards
Light microscopy/ image analysis	The basic equipment consists of a microscope, a camera and a computer. The image of a dispersed sample is evaluated to assess the shape and size parameters of each particle. This process can be manual or automated.		Projected area diameter/ number weighted	Static: 1–5000 Dynamic: 30–30000	ISO 13322-1:2014 [26] ISO 13322-2:2006 [27]
Electrical zone sensing (Coulter counter)	Particles homogeneously suspended in an electrolyte solution are forced to flow through a small aperture that separates two electrodes of opposite potential. When a particle passes through the aperture, the resistance of the aperture increases, giving a voltage pulse proportional to the particle volume.	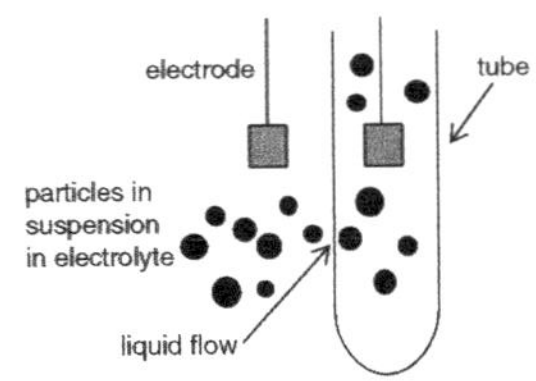	Volume diameter/ number and volume weighted	0.5–1500	ISO 13319:2007 [28]
Photo zone sensing (single particle optical sensing: SPOS)	Particles in a liquid suspension are forced between a light source and a detector, producing a shadow or blockage of light on the detector (light obscuration) related to the optical cross-section of the particle.	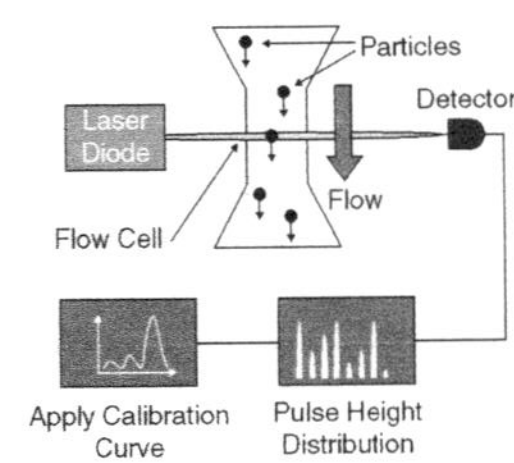 [Reproduced with permission of PSS – Particle Sizing Systems]	Optical diameter/ number weighted	0.5–5000	ISO 13099-2:2012 [29]
Time-of-flight	An air stream containing the particles is drawn through a fine nozzle into a partial vacuum producing a supersonic flow of air, causing particles to accelerate according to size.	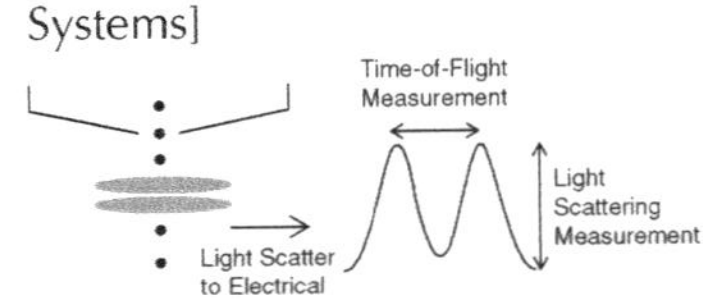 Adapted from [50]	Aerodynamic diameter/ number weighted	0.5–20	

Laser diffraction(LD)/ Fraunhofer diffraction (FD)/ low angle laser light scattering (LALLS)	A laser beam is passed through a sample of particles and the scattered light intensity is collected at low angles. The light scattering data is converted into a particle size distribution using Fraunhofer theory.	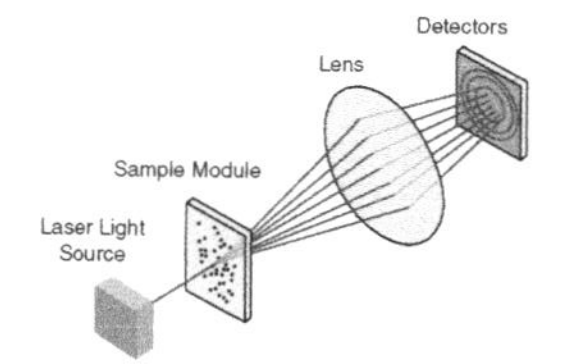	Fraunhofer diameter/ volume weighted	0.020–2000	ISO 13320:2009 [30] USP 34 NF 29 [31]
Dynamic light scattering (DLS)	The fluctuations of the light scattered by a suspension of submicron particles, due to Brownian motion, are collected over time at a given angle. From the autocorrelation function a diffusion coefficient and an average size is calculated.	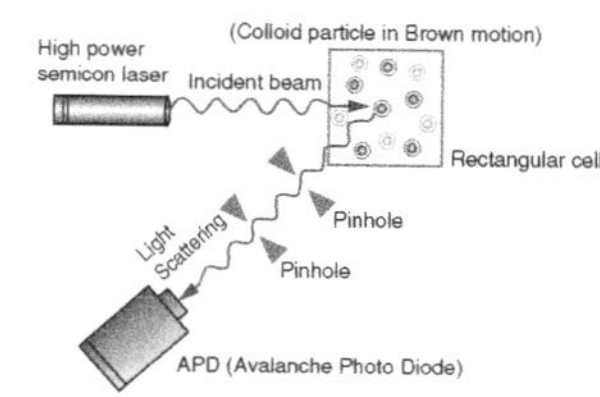	Hydrodynamic diameter/light intensity weighted	0.003–3	ISO 22412:2017 [32]
Sedimentation (gravitational and centrifugal)	A sample of particles is uniformly suspended in a fluid and allowed to settle due to gravity according to Stokes' law. Centrifugal sedimentation extends the range of analysis to much smaller particles.	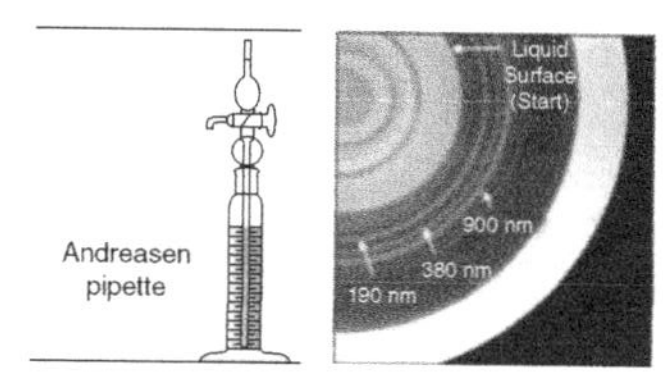	Stokes diameter	Gravitational: 1–250 Centrifugal: 0.01–100	ISO 13317-1:2001 [33] ISO 13318-1:2001 [34]
Cascade impactor	Consists of a number of impactor stages connected in series with smaller and smaller cutoff diameter. Particles having sufficient inertia will impact on a particular stage collection plate, whilst smaller particles with insufficient inertia will remain entrained in the air stream and pass to the next stage where the process is repeated.	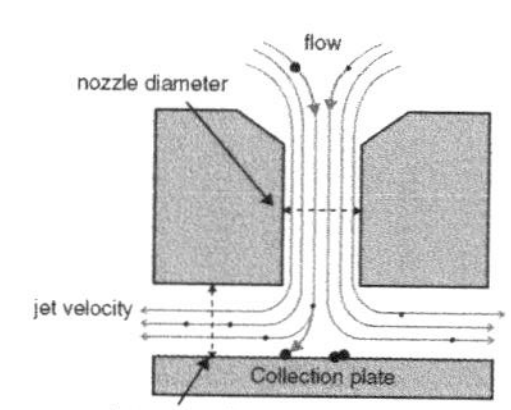	Aerodynamic diameter/ mass weighted	0.5–10	USP 26 NF 21 [35]

[1]. However, for studies like, for example, lung toxicity, surface-area-based measurements are indicated [6, 10]. On the other hand, number distributions are, as mentioned before, most suitable for an evaluation of particle contamination or for characterizing fiber-like particles [3].

Another important parameter is the particle size range. The size range of each instrument/technique is normally dictated by the assumptions/equations on which they are based. Many interesting applications involve particles around 1 micron. However this region is apparently a "natural frontier" for many sizing methods [24, 25]. In fact, this size region is considered the dividing line between sedimentation and centrifugation methods (in addition to particle density); Fraunhofer diffraction (for which the scattering pattern is independent of the material refractive index) is also recommended for particles larger than approximately 1 μm while, for submicron particles, dynamic light scattering is more appropriate. Additionally, this size region is regarded as a barrier between light and electronic microscopy. The same applies to electric conductivity and light obscuration based methods for which inconsistent data have been reported for the region 0.5–1 μm [3]. Nonetheless, many manufacturers claim to have increased the operational size range of their instruments below this value by adding some features in the original configuration. This normally results in the production of artefacts in the sizing data, especially at the extremes of the distribution. Thus, the user should be alerted for these details and should look for an instrument whose mid-range covers the size range of interest, that is, avoiding the extremes. Also, for the reasons indicated (namely, differences in the weights of the raw data), the combination of the results of two different sizing datasets in order to cover a larger size range should be avoided.

Other issues to be taken into account when selecting a particle sizer are: the ease of use, the speed, the flexibility, and the cost [24, 25]. Regarding the former, it should be kept in mind that behind the apparent simplicity of the measurement itself, the process of sampling and sample preparation may be complex/laborious. Thus the time of analysis may be just a (small) part of the total time needed. Also the possibility of measuring different types of samples, in a variety of conditions (liquid suspensions or particles in air), that is, the instrument flexibility, is another pertinent parameter, especially if different types of samples have to be measured with the same measuring device. Finally, subjects like accuracy, precision, resolution and sensitivity are also of major relevance in the decision-making process. Accuracy is a measure of how close an experimental result is to the "true" value. This concept is somehow problematic in particle sizing, especially for non-spherical particles for which the "true" values might not be known (but accepted as conventional true values). Precision is the closeness of agreement between a series of measurements from multiple sampling of the same sample. Measurement resolution, mentioned before, is normally defined as the minimum ratio of two monodisperse distributions that can be separated. Resolution and dynamic size range are often inversely related to each other [3]. There are currently many reference materials for checking these parameters.

As a final point, it should be stressed that sampling and dispersion, although of utmost importance in particle sizing, are beyond the scope of this evaluation. In fact, particle size data are only valid when the sample is representative and appropriate dispersion techniques have been used. It is well known that most of the variations in size distribution results are due to incorrect sampling and sample preparation [17, 36, 37].

1.4 Aspects of Some Selected Methods

This section encompasses the discussion of some selected particle sizing methods, probably the most widely applied in the pharmaceutical field. The light microscopy-based methods will be described first as they provide the most direct measurement of particle size and morphology. Next, the incontrovertible light-scattering methods (both static and dynamic), almost universally used, will be discussed. The time-of-flight method and cascade impaction, leading to the measurement of the aerodynamic diameter, of utmost importance to nasal drug formulations, close this selection of methods. It should be pointed out that, rather than describing in detail the operating principles of these methods (easily available in the open literature), this discussion is more focused on their weaknesses and strengths in order to better compare them.

1.4.1 Optical Microscopy-based Methods

Light microscopy-based methods are a powerful tool for particle characterization as they involve the direct observation of the particles and thus can provide information on surface features, besides shape and size, offering a better insight into the nature of the sample. Because of this, they are probably the first choice for investigation of unknown particles, from around 1 micron up to several millimeters. However, manual measurements are lengthy, require skilled operators, are labour-intensive and should be considered qualitative unless a statistically significant number of particles is inspected (statistically representative distributions can be constructed by measuring tens to hundreds of thousands of particles per sample).

Unlike other particle sizing techniques, these methods offer the advantage of providing the particle dimensions of each particle (though two-dimensional) in addition to a series of other shape parameters such as aspect ratio (ratio of the largest to the smallest dimension), circularity (ratio of the actual perimeter of the particle to the perimeter of a circle with the same area), convexity (ratio of the particle area to the total area), along with different geometric diameters, the most common being the projected area equivalent diameter (Table 1.1). Since the particle orientation on a substrate usually gives the maximum area, this leads to an apparently larger size than that measured by other techniques [3]. It is particularly useful for the characterization of particles with extreme shapes (e.g., acicular particles, platy particles, and fibres), as reporting these in terms of a spherical diameter can be, as discussed before, a gross approximation.

Nowadays microscopic methods are normally coupled to image analysis systems that record the data for many particles in very short intervals to be further processed, thus reducing the labour content and minimizing operator bias associated with manual inspection. Image analysis methods can be divided into *static image analysis* and *dynamic image analysis*. The main distinction is whether particles are presented in a static (stationary) orientation or dynamic, that is, flowing past the detector. These systems are currently applied to the characterization of many pharmaceutical products. The basic guidelines for particle size by microscopy image analysis are given in ISO 13322-1:2014, for static image analysis [26], and ISO 13322-2:2006, for dynamic image analysis [27].

Static image analysis uses an optical microscope to characterize particles dispersed on a slide that is moved by an automated stage. Each image is captured by a digital camera and a software routine performs various tasks to distinguish particles from background, separate touching particles, and assign size and shape parameters. The majority of static image analysis measurements are made on powders, typically used for solid oral dosage forms but they may also include suspensions, creams and even aerosols (actuating the device onto a slide) [38, 39]. The size range is typically between 1 μm and 1 mm.

Dynamic image analysis measures a stream of flowing particles using a rapid strobe light on one side of the stream and a digital camera on the other capturing the particle images used in the analysis. In this case, the sample is transported either by gravity, air pressure, or in a liquid which passes the camera, and pictures are taken in quick succession, giving tens of thousands of images per minute. Because of the size range measured by dynamic image analysis (30–30 000 μm) they have been related to sieve results with the bonus of shape information [38]. Modern systems use a hydrodynamic sheath flow mechanism to efficiently focus the particles [21], and others use two cameras with different magnifications to cover a wide measuring range, as illustrated in Figure 1.3.

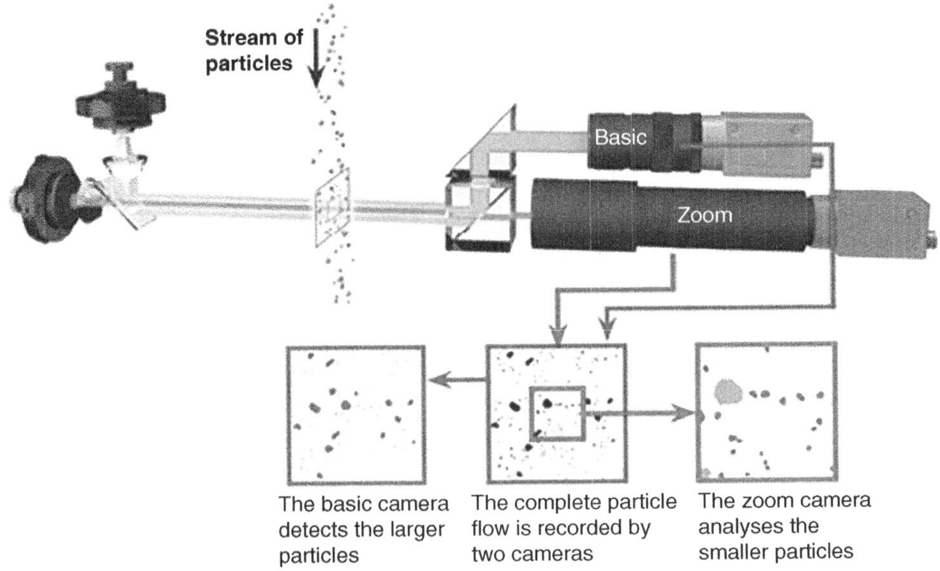

Figure 1.3 *Measuring principle of dynamic image analysis [reproduced with permission of Retsch Technology].*

Since image analysis is a high-resolution counting technique, it can effectively detect outlier populations (both larger and smaller than the main population). For active pharmaceutical ingredients this could be a great advantage for finding small amounts of particles that could negatively impact dose uniformity [38, 39].

It should be remembered that image analysis, as a counting method, generates number-based distributions. Volume-weighted distributions can be calculated assuming that the particles are spherical and of uniform density which, for particles deviating from sphericity, can lead to false results if adequate shape factors are not applied. Finally, it should be pointed out that this technique is often used in conjunction with ensemble-based particle sizing methods, such as laser diffraction, to gain a deeper understanding of the sample or to validate the ensemble-based measurements [7, 39, 40].

1.4.2 Laser Light-scattering Techniques

Laser scattering-based particle characterization techniques are probably the most popular for sizing particles as they are fast, easy to use, flexible (enabling the measurement of particles in liquids as well as in sprays, aerosols, emulsions and dry powders), highly repeatable, and have a wide dynamic size range (from a few tens of nanometers up to several millimeters). Additionally, these are well established and standardized techniques (ISO 13320:2009 [30] and USP 34 NF 29 [31] for laser diffraction, and ISO 22412:2017 [32] for dynamic light scattering).

In the pharmaceutical industry they are becoming a preferred standard method [41, 42]. However, they have been the subject of considerable criticism, mainly because some working parameters that are decisive for a correct size analysis are not always observed [3, 43]. In fact, although it is a very user-friendly technique, laser-scattering techniques can lead to erroneous data if not used properly. To avoid that, the users must have a sound knowledge of its basic principles and limitations. Additionally, the lack of definition (and misuse) of some terms and acronyms, indistinctly used in many situations, also entails some confusion. The next paragraphs will briefly mention the fundamentals of this technique, not only to clarify the meaning of those acronyms, but also to explain the reasons why some precautions should be taken to ensure a proper analysis and the consequences when they are not taken. Both methodologies of light scattering (static and dynamic) are included in this section.

Light scattering is the alteration of the direction and intensity of the light beam that strikes an object, due to the combined effects of reflection, refraction and diffraction (diffraction is the bending of light at the particle's boundary), as illustrated in Figure 1.4, creating a complex pattern known as "scattering pattern". More specifically, a scattering pattern is

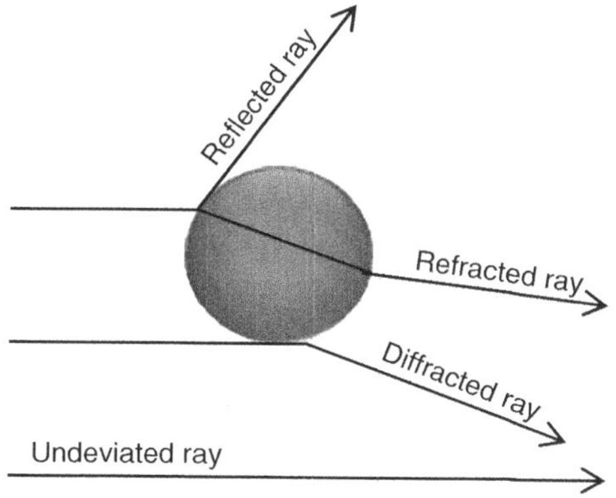

Figure 1.4 Interaction of light rays with a particle.

formed by light intensity as a function of the scattering angle (measured relative to the direction of the incident light) [16, 17, 44, 45].

The intensity of the scattered light is basically a function of the incident wavelength, the scattering angle, the particle size and shape, and the ratio of refractive indices of the particle to that of the medium. The shape of the scattering pattern is highly dependent on the ratio of the particle size to the incident wavelength, as illustrated in Figure 1.5. From this figure it is clear that particles larger than the wavelength of the incident light will scatter light essentially in a forwards direction, and as the particle size decreases the relative intensity of the backward scattered light increases compared to the forward light lobe, being approximately symmetric for particles much smaller than the wavelength (Rayleigh scatterers). This fact enables the measurement of particles of different sizes using different methodologies.

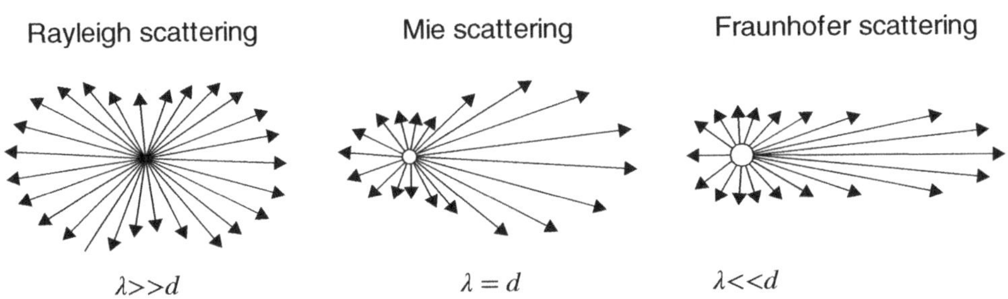

Figure 1.5 Effect of particle size (d) on the scattering pattern for a given light wavelength (λ).

Although generally called "scattering techniques", these techniques can be divided into two groups whose working principles are entirely different: static light scattering (SLS) and dynamic light scattering (DLS). In SLS, particle size information is extracted from the scattering pattern obtained when the sample material is illuminated with a light beam; the scattered light intensity averaged over a given time interval at various scattering angles (scattering pattern) is then used to derive particle size information, based on a suitable scattering theory (model). On the other hand, DLS records the fluctuations in the intensity of the light scattered by the particles as a function of time, measured at a fixed angle in order to calculate the particles' diffusion coefficient, which in turn can be used to determine their hydrodynamic radius. This methodology is recommended for submicron particles, including nanoparticles. Both of these techniques will be discussed in more detail below.

1.4.2.1 Laser Diffraction and Static Light Scattering

Included in the first group (SLS) is the popular laser diffraction (LD) technique. Although "laser diffraction" and "static light scattering" are often used interchangeably to refer to the same particle size determination, this is not correct as discussed next. As illustrated in Figure 1.5, large particles (i.e., with diameters considerably larger than the wavelength of the incident light) scatter light mainly in the forwards direction, being the scattering pattern dominated by diffraction phenomena. In a typical experiment of laser diffraction, a representative sample is passed through a collimated beam of monochromatic light (usually a laser) and the light scattered by the particles at various angles (scattering pattern) is measured in a multi-element detector (see technique layout in Table 1.3) and subsequently transformed into a particle size distribution. The method relies on the fact that diffraction angle is inversely proportional to particle size. However, it should be pointed out that the scattering pattern of an ensemble of particles is a composite pattern resulting from the contributions of all the individual particles (i.e., the resulting light at each point is the sum of the contribution of each individual pattern). These patterns are said to be *convoluted.* In the case of laser diffraction, the deconvolution of the scattering data into a volumetric size distribution is based on the Fraunhofer theory [16, 17, 44, 45]. This is accomplished by an iterative process of fitting the theoretical model to the data until the two converge within an acceptable amount defined by the manufacturer. In Fraunhofer theory, the refractive index of the particles is irrelevant, which is very convenient (namely when characterizing mixed or unknown powders) since normally this parameter is not easily obtained. Because of this, LD is also called Fraunhofer diffraction (FD). Furthermore, because information about larger particles is contained in low scattering angles (<35°), this technique is also named low-angle laser light scattering (LALLS). The variety of acronyms may be confusing to the less familiar user.

It should be kept in mind that Fraunhofer scattering is only valid for large particles, that is, particles that are at least 5–10 times larger than the incident wavelength, which limits the application of LD to particles larger than a few tens of micrometers. For smaller particle sizes (similar or smaller than the incident wavelength), the light is increasingly scattered with large angles (Figure 1.5), and the scattering pattern is not the result of diffraction phenomena exclusively. Thus Fraunhofer theory is no longer valid and a more complex theory, Mie theory, has to be used to account for all possible interactions between particles and light, yet it is only applicable to spheres [16, 17, 44, 45]. Figure 1.6 illustrates the discrepancy in the size distribution curves resulting from applying Fraunhofer and Mie theories to invert the scattering data obtained for a 6 μm latex standard (larger and/or more opaque particles lead to more convergent results).

Besides being much more complex than Fraunhofer approximation, from the mathematical point of view, the Mie formula requires information about the particle optical properties,

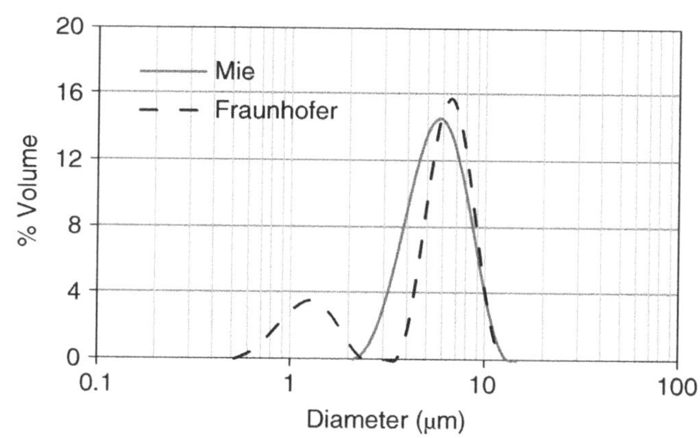

Figure 1.6 *Influence of the scattering model used in the deconvolution of the scattering data: Fraunhofer vs. Mie.*

such as refractive index (real refractive index and imaginary refractive index). This is a major drawback as these parameters are difficult to obtain in practice (especially the imaginary part), and errors in the estimation of the refractive index can give radically different particle size distributions mainly for particles smaller than approximately 10 µm [3, 40, 41, 46]. Most of the currently commercialized analyzers possess a databank with information on refractive indices of various materials and also provide ways to guess them based upon an iterative procedure to fit the modeled data to the actual data [21, 38].

Additionally, the intensity of the scattered light also decreases with particle size and thus the intensity of the signal collected by the detector also decreases (as mentioned earlier, scattered light intensity of very small particles (<50 nm) is proportional to the sixth power of the diameter). Depending on the incident wavelength, the minimum particle size theoretically possible would be around 500 nm. Nonetheless, the instruments in use nowadays claim to measure particles down to 20 nm. To extend the measuring range down to this level, additional techniques have to be combined with laser diffraction. These include the use of more than one laser (shorter wavelengths), supplementary detector arrays (located sideways and even backwards), different light polarizations (PIDS), additional lenses, and application of different inversion algorithms, among others [3, 43]. Figure 1.7 shows a configuration where, besides the ring-shaped detector located at small scattering angles (for larger particle measurement), additional large angle and backscatter detectors are used to allow the measurement of smaller particles.

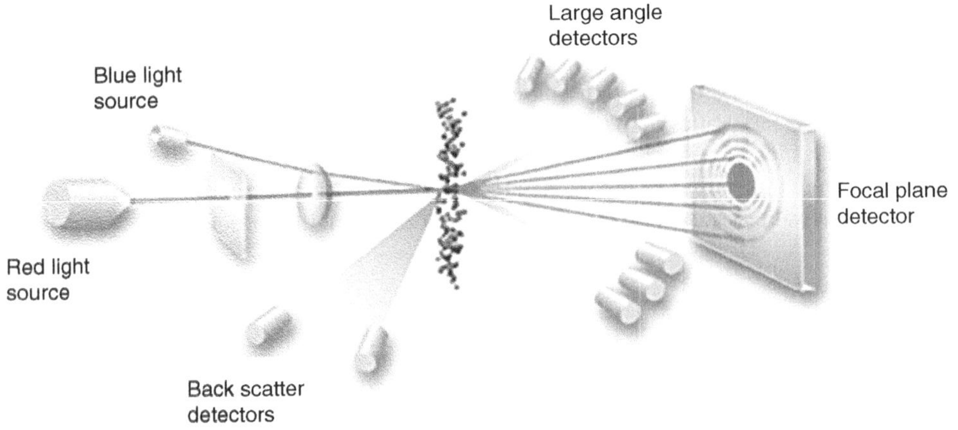

Figure 1.7 *Schematic representation of the static light scattering technique using different lights and rear detectors [reproduced with permission of Malvern Panalytical].*

These modifications, carried out in a diverse way by different equipment manufacturers, are responsible for many discrepancies in the sizing results that would not be expected when using "in principle" the same technique, and may justify some of the criticisms that have been directed to these techniques. In conclusion, modern laser "diffractometers" do not rely simply on laser diffractometry but include additional technologies to extend the measuring range to submicron particles. Therefore this technique, erroneously called "laser diffractometry", is indeed static light scattering [3, 43].

Major drawbacks of this technique are: the low resolving power (typical of ensemble methods), the fact that the result accuracy is also dependent on the accuracy of the optical parameters used, as explained, and not being recommended for particles with extreme aspect ratios [47]. Caution should also be taken when comparing laser diffraction data with other methods, as the Fraunhofer diameter is not a volume-equivalent diameter (it is associated with an average projected area diameter) [3, 20, 40]. Additionally, these techniques are normally applied to relatively low concentrations in order to avoid multiple scattering (where light scattered from one particle is scattered by a second particle before reaching the detector). Nonetheless, it continues to be very popular and one of the first options for general purposes.

1.4.2.2 Dynamic Light Scattering

Dynamic light scattering (DLS) is also known as photon correlation spectroscopy (PCS) or quasi-elastic light scattering (QELS) (because when photons are scattered by mobile particles, the process is quasi-elastic). It is a non-invasive, user-friendly method used since 1960 to routinely size submicron particles (from a few nanometers to a few micrometers, these limits being sample-dependent), yielding very accurate results especially for monomodal samples, in very short times, with low sample preparation requirements. It is an absolute technique as there are no adjustable parameters for a "calibration", in contrast to microscopes or particle counters. It has been widely used in the characterization of proteins, biomaterials, liposomes, micelles, colloidal dispersions and emulsions [32, 48].

This method is based on the determination of the hydrodynamic properties of submicron particles suspended in a liquid medium. If the particles are very small, their collisions with the solvent molecules result in an erratic motion called Brownian motion. When these particles are illuminated by a laser, the intensity of the scattered light measured by a single detector fluctuates as a result of the Brownian movements, which are indirectly proportional to size. The time dependence of the intensity fluctuations is analyzed with a very sensitive detector (as the intensity of the light scattered by very small particles is very low), normally a photomultiplier positioned at a suitable angle (either 90° or 175°) (see technique layout in Table 1.3). The signal of the photomultiplier is fed to a correlator to build the so-called "autocorrelation function" (which led to the acronym PCS), which is basically an exponential decay curve whose decay rate is related to the translational diffusion coefficient of the particles. This coefficient can be transformed into an average diameter (hydrodynamic diameter, Table 1.1), using the Stokes–Einstein equation. The hydrodynamic diameter is very similar to the geometric diameter, with the exception of very small (<300 nm) highly charged particles where the electric double layer gives a larger diameter than that measured by microscopy [40].

The autocorrelation function of polydisperse samples is a summation of single exponential functions (each corresponding to a different size) that, similarly to laser diffraction, needs to be deconvoluted to derive the corresponding particle size distribution. Different deconvolution algorithms have been applied for this purpose, depending on the type of information required. Normally a mean particle size (Z average diameter, Table 1.2) and the width of the distribution are sufficient for monomodal samples. The interpretation of data from polydisperse samples is considerably more difficult and requires the use of more complex algorithms that do not always have simple or unique solutions [48].

It should be pointed out that the size distribution generated by DLS is an intensity-based distribution. However, it can be converted to a volume-based distribution using Mie theory, which, as mentioned above, requires *a priori* knowledge of the particles' optical properties. One of the major drawbacks of this technique is the sensitivity to the presence of larger particles in the sample. As explained earlier, for this size range, light intensity is proportional to the sixth power of particle diameter, and thus the existence of a few larger particles can dominate the particle size result [40, 49]. In this case, complementary measures using laser diffraction are recommended.

The limitation of this technique to dilute suspensions has been partially overcome by new instrumentation configurations that have been developed (namely using backscattered detection). Finally, DLS is (as laser diffraction) a low-resolution method and, generally, can only resolve particle populations that differ by a factor larger than 2:1 [48]. Nonetheless, DLS is very good, quick and accurate for sizing narrow distributions of colloidal particles.

1.4.3 The Time-of-Flight Counter

This technique was developed in the late 1980s and is currently commercialized by TSI as the Aerodynamic Particle Sizer (APS), a third-generation instrument [50]. It is a fast stream scanning technique with high resolution, able to count and measure the size of aerosols between 0.5 and 20 μm in real time. Although the particle size distributions of pharmaceutical sprays and aerosols have traditionally been undertaken by methods involving multi-stage cascade impactors, these are slow and labour-intensive and other options have been adopted, namely the time-of-flight (TOF) method because of its ability to measure very small sample quantities and to count hundreds of thousands of particles in a very short time, and also because of its size-resolving capability [39]. The ability to determine the aerodynamic particle size (Table 1.1) is of particular importance, for instance to characterize orally inhaled dosage forms, as the knowledge of this parameter is quite useful for predicting, for example, the regional deposition of the particles in the human respiratory tract.

Size is measured as follows: after passing a dilution stage, incoming particles are accelerated singly through a well-defined flow field, in which particles experience ultra-Stokesian acceleration. Particles accelerate in the airflow according to size (smaller particles accelerating more rapidly than larger particles due to differences in mass). After the acceleration nozzle, particles pass into the sensing zone, where they are sized according to their transit time through two partially overlapping laser beams (see technique layout in Table 1.3). The first laser beam detects each particle and starts a TOF clock, while arrival at the second laser stops the clock. The transit time of the particles between two laser beams is a monotonic function of size. The aerodynamic diameter is calculated from calibration curves using spherical particles of known density. The equipment is easy to operate and maintain, but some problems related to the optimization of the analysis conditions remain to be solved [39, 51, 52].

The count-weighted size distribution can be transformed to a mass-weighted basis, assuming the particles are spherical and of uniform density. The size distribution can be biased when the drug-free droplets are mixed with the drug particles [53]. Recent versions of this equipment have minimized coincidence effects and extended the dynamic size range of the older models [3]. It has been considered adequate for aqueous droplet-based formulations, but for droplets larger than 5 µm, some distortion can occur. Also for non-spherical particles the results can be underestimated [3, 39]. The main limitation, compared with cascade impaction, is the lack of chemical assay of the active pharmaceutical ingredients, and for this reason it is not suitable for mixtures of active components and excipients.

1.4.4 Cascade Impactor

Inhalation products, namely nebulizers and dry and metered dose inhalers, represent a significant proportion of pharmaceutical products. The optimal design of dry powder inhalers is, in particular, challenging [14]. As mentioned above, aerodynamic diameter is a key parameter for these products, and thus its measurement is critical either during the product development cycle or for quality control.

From the various techniques available, multiple stage cascade impaction is the most widely used method, being the standard technique recommended by regulatory bodies [35] as it allows the analysis of mass-weighted aerodynamic particle size directly together with the capability for recovery and assay for active pharmaceutical ingredients in a traceable manner [54]. Alternative methods, such as TOF, that also measure aerodynamic particle diameters, do not provide (at least in their basic configuration) differentiation between active pharmaceutical ingredients and any other components in the formulation, and only deliver a number-based particle size distribution.

Cascade impactors are based on inertial separation (function of particle size and velocity) and allow direct measurement of mass-weighted aerodynamic particle size of drug substance in aerosols between 0.5 and 10 µm. These devices consist of a series of stages, each comprising a plate with a specific nozzle and a collection surface (see technique layout in Table 1.3). Sample-laden air is drawn into the impactor, flowing sequentially through the stages. As particles pass through the nozzle, they either remain in the air stream or impact on the collection surface. Particles with sufficient inertia are collected, while the rest pass on to the next stage. Each deposition stage provides a defined aerodynamic cutoff diameter (particles collected at 50% efficiency). As nozzle size decreases with increasing stage number, particle velocity increases as they proceed through the instrument, allowing the collection of increasingly small particles. The sample is thus separated into a series of size fractions, individually collected for subsequent analysis by HPLC (high pressure liquid chromatography) to determine the amount of active compound collected at each stage. The most widely used cascade impactors, the Andersen Cascade Impactor and the Next Generation Impactors, separate a sample into 7–8 discrete size fractions.

Nonetheless, cascade impactor determinations are rather time-consuming and complex to undertake, also being unsuitable for making size-resolved measurements of large liquid droplets as found in nasal sprays (20–200 µm). Thus real-time techniques, not only TOF instruments but also static light scattering and microscopy-automated image analysis (especially when combined with Raman chemical imaging), have been considered potential options for the rapid assessment of particle size in aerosols, especially in early-phase product development [39]. Additionally, these techniques are simple to perform and enable automatic data recording and processing. With regard to static light scattering, this technique is, as mentioned earlier, very versatile and enables the analysis of the aerosol by directly spraying it into the laser beam [41]. Equipment producers nowadays offer the possibility of *in situ* analysis of high concentration aerosols and sprays, using a so-called inhalation cell [55]. Several studies have been published comparing the multi-stage cascade impactor and laser diffraction systems [41, 56]. Despite the results not always being consistent with the cascade impactor, namely due to propellant evaporation and droplet break-up problems, and also because light scattering techniques deliver volume-equivalent diameters, the rapidity of measurement, the fact of being a non-invasive technique, the size-resolving capability and the much wider dynamic size range (up to 2 mm) are important points that make SLS a valuable tool for rapid screening [53]. However, in many instances, particularly with aerosols from aqueous solutions, aerodynamic and other physical diameters determined by microscopy and light scattering converge because the droplets are spherical and have unit density [39].

1.5 Conclusions

In the pharmaceutical field, particle size distribution should not only be known but should be controlled. However, the choice of a particle sizer is a more complex task than it seems. Indeed, from this chapter it is clear that there are no universal sizing techniques suitable for all samples, and that no technique can perform measurements from zero to infinity. Moreover, the wide diversity of instruments, based on different measuring principles, will generate different sizing data. Explanations for all these aspects were addressed above, including several points that need to be taken into account when choosing particle sizing equipment for a given application. Making the correct choice requires not only knowledge of the instrument working principle, but also of the type of equivalent diameter measured and the size distribution weight. This information has to be compatible with the analysis end-use and the particle properties.

A detailed discussion of some methods currently used for routine analysis of pharmaceutical products clearly indicated that they all exhibit advantages and limitations. Imaging techniques are most useful to visualize the particles, as well as to assess information about shape and structural parameters. Nonetheless, they suffer in general from statistical problems. On the other hand, ensemble techniques like the light-scattering techniques are more rapid and versatile, but involve deconvolution processes that limit resolution, requiring in most cases information about the particles' optical properties (not easily available). As for the measurement of the aerodynamic diameter, alternatives to cascade impaction have recently been considered.

The overall conclusion is, however, that a comparison of methods, yielding complementary information, is highly recommended.

Acknowledgements

The authors would like to thank Malvern Panalytical, Retsch Technology and PSS – Particle Sizing Systems for allowing the reproduction of Figures 1.2, 1.7, 1.3, and the photo zone sensing design for Table 1.3, respectively.

References

[1] Burgess, D.J., Duffy, E., Etzler, F., Hickey, A.J. (2004) Particle size analysis: AAPS workshop report, cosponsored by the Food and Drug Administration and the United States Pharmacopeia. *The AAPS Journal*, **6** (3), 23–34.
[2] Etzler, F.M., Deanne, R. (2006) Particle size analysis, in *Solid State Characterization of Pharmaceuticals* (eds Zakrzewski, A., Zakrzewski, M.), ASSA International, Inc., Danbury, CT, pp. 473–480.
[3] Shekunov, B.Y., Chattopadhyay, P., Tong, H.H.Y., Chow, A.H.L. (2006) Particle size analysis in pharmaceutics: principles, methods and applications. *Pharmaceutical Research*, **24** (2), 203–227.
[4] Snorek, S.M., Bauer, J.F., Chidambaram, N., *et al.* (2007) PQRI Recommendation on particle size analysis of drug substances used in oral dosage forms. *Journal of Pharmaceutical Science*, **96** (6), 1451–1467.
[5] Yu, L.X. (2008) Pharmaceutical quality by design: product and process development, understanding, and control. *Pharmaceutical Research*, **25** (4), 781–791.
[6] Iacocca, R.G. (2009) Physical characterization tests for API used in low dose formulations, in *Formulation and Analytical Development for Low Dose Oral Drug Products* (ed. Zheng, J.), Wiley & Sons, New York, pp. 309–324.
[7] Kippax, P. (2009) Particle size analysis. *Pharmaceutical Technology Europe PTE*, **21** (4), 1–4.
[8] Rawle, A.F. (2010) Analytical tools for suspension characterization, in *Pharmaceutical Suspensions From Formulation Development to Manufacturing* (eds Kulshreshtha, A.K., Singh, O.N., Wall, G.M.), Springer, New York, pp. 177–230.
[9] Möschwitzer, J. (2010) The role of particle size analysis in the development process of nanosized drug products. *American Pharmaceutical Review*, **13**, 98–101.
[10] Iacocca, R. (2011) Physical characterization of nano particulates used in the pharmaceutical industry. *American Pharmaceutical Review*, **14** (1), 30–34.
[11] Figueiredo, M. (2013) Sizing nanoparticles in liquids: an overview of methods, in *Drug Delivery Systems: Advanced Technologies Potentially Applicable in Personalised Treatment, Advances in Predictive, Preventive and Personalised Medicine* (ed. Coelho, J.), Springer Science C Business Media, Dordrecht, Vol. **4**, pp. 87–107.
[12] Figueiredo, M., Ferreira, P., Campos, E.J. (2015) Electrozone sensing goes nano, in *Encyclopedia of Analytical Chemistry* (ed. Meyers, R.), John Wiley & Sons, Ltd, Chichester, pp. 1–19.
[13] Sun, Z., Ya, N., Adams, R.C., Fang, F.C. (2010) Particle size specifications for solid oral dosage forms: a regulatory perspective. *American Pharmaceutical Review*, **13** (4), 68–73.
[14] Etzler, F.M. and Uddin, M.N. (2013) Powder technology and pharmaceutical development: particle size and particle adhesion. *KONA Powder and Particle Journal*, **30**, 125–143.
[15] Williams, H.D., Trevaskis, N.L., Charman, S.A., *et al.* (2013) Strategies to address low drug solubility in discovery and development. *Pharmacological Reviews*, **65** (1), 315–499.
[16] Allen, T. (1997) *Particle Size Measurement*, 5th edn, Chapman & Hall, London.
[17] Merkus, H.G. (2009) *Particle Size Measurements: Fundamentals, Practice Quality*, Springer Science, New York.
[18] ISO 9276-1:1998. (1998) *Representation of results of particle size analysis – Part I: Graphical representation*, International Standards Organization.
[19] Brittain, H.G. (2001) Particle size distribution, Part I: Representations of particle shape, size, and distribution. *Pharmaceutical Technology North America*, **25** (12), 38–44.
[20] Rawle, A. (2008) Basic principles of particle size analysis. *Malvern Instruments Technical Paper MRK034*, 1–8.
[21] Malvern Instruments, Ltd. (2015) *A Basic Guide to Particle Characterization*. https://www.cif.iastate.edu/sites/default/files/uploads/Other_Inst/Particle%20Size/Particle%20Characterization%20Guide.pdf

[22] Wolfrom, R.L. (2011) The language of particle size. *Journal of GXP Compliance*, **15** (2), 10–20.

[23] ISO 9276-2:2014. (2014) *Representation of results of particle size analysis – Part 2: Calculation of average particle sizes/diameters and moments from particle size distributions*, International Standards Organization.

[24] Weiner, B.B. (2009) Considerations in particle sizing. Part 1: Classification of the various sizing techniques. *Technical Brief* (Vol. **6**).

[25] Weiner, B.B. (2009) Considerations in particle sizing. Part 2: Specifying a particle size analyser. *Technical Brief* (Vol. **7**).

[26] ISO 13322-1:2014. (2014) *Particle size analysis – Image analysis methods – Part 1: Static image analysis methods*, International Standards Organization.

[27] ISO 13322-2:2006. (2006) *Particle size analysis – Image analysis methods – Part 2: Dynamic image analysis methods*, International Standards Organization.

[28] ISO 13319:2007. (2007) *Determination of particle size distributions – Electrical sensing zone method*, International Standards Organization.

[29] ISO 13099-2:2012. (2012) *Colloidal systems – Methods for zeta-potential determination – Part 2: Optical methods*, International Standards Organization.

[30] ISO 13320:2009. (2009) *Particle size analysis – Laser diffraction methods*, International Standards Organization.

[31] USP 34 NF 29. (2011) *Chapter <429> Light diffraction measurement of particle size*, United States Pharmacopeia National Formulary.

[32] ISO 22412:2017. (2017) *Particle size analysis – Dynamic light scattering (DLS)*, International Standards Organization.

[33] ISO 13317-1:2001. (2001) *Determination of particle size distribution by gravitational liquid sedimentation methods – Part 1: General principles and guidelines*, International Standards Organization.

[34] ISO 13318-1:2001. (2001) *Determination of particle size distribution by centrifugal liquid sedimentation methods – Part 1: General principles and guidelines*, International Standards Organization.

[35] USP 26 NF 21. (2003) *United States Pharmacopeia National Formulary*.

[36] Allen, T. (2003) *Powder Sampling and Particle Size Determination*, Elsevier, Amsterdam.

[37] Mingard, K., Morrell, R., Jackson, P., *et al.* (2009) Good practice guide for improving the consistency of particle size measurement, in *Measurement Good Practice Guide No. 111*, National Physical Laboratory, Middlesex, UK, pp. 1–66.

[38] HORIBA Instruments, Inc. (2012) *A Guidebook to Particle Size Analysis*. https://www.horiba.com/fileadmin/uploads/Scientific/Documents/PSA/PSA_Guidebook.pdf

[39] Mitchell, J., Bauer, R., Lyapustina, S., *et al.* (2011) Non-impactor-based methods for sizing of aerosols emitted from orally inhaled and nasal drug products (OINDPs). *AAPS PharmSciTech*, **12** (3), 965–988.

[40] Bowen, P. (2002) Particle size distribution measurement from millimeters to nanometers and from rods to platelets. *Journal of Dispersion Science and Technology*, **23** (5), 631–662.

[41] Mitchell, J.P., Nagel, M.W., Nichols, S., Nerbrink, O. (2006) Laser diffractometry as a technique for the rapid assessment of aerosol particle size from inhalers. *Journal of Aerosol Medicine*, **19** (4), 409–433.

[42] Kippax, P. (2005) Issues in the appraisal of laser diffraction particle sizing techniques. *Pharmaceutical Technology Europe*, 32–39.

[43] Cornelia, M.K. and Müller, R.H. (2008) Size of submicron particles by laser diffractometry – 90% of the published measurements are false. *International Journal of Pharmaceutics*, **355**, 150–163.

[44] Xu, R. (2002) *Particle Characterization: Light Scattering Methods*, Kluwer Academic Publishers, New York.

[45] Jillavenkatesa, A., Dapkunas, S.J., Lum, L-S H. (2001) *Particle Size Characterization*. NIST Recommended Practice Guide, Special publication 960-1, Washington, DC.

[46] Beekman, A., Shan D., Ali, A., *et al.* (2005) Micrometer-scale particle sizing by laser diffraction: critical impact of the imaginary component of refractive index. *Pharmaceutical Research*, **22** (4), 518–522.

[47] Kelly, R.N., Kazanjian, J. (2006) Commercial reference shape standards use in the study of particle shape effect on laser diffraction particle size analysis. *AAPS PharmSciTech*, **7** (2), E126–E137.

[48] Tscharnuter, W. (2000) Photon correlation spectroscopy in particle sizing, in *Encyclopedia of Analytical Chemistry* (ed. Meyers, R.A.), John Wiley & Sons Ltd, Chichester, pp. 5469–5485.

[49] Jamting, A.K., Cullen, J., Coleman, V.A., *et al.* (2011) Systematic study of bimodal suspensions of latex nanoparticles using dynamic light scattering. *Advanced Powder Technology*, **22** (2), 290–293.

[50] TSI Inc. (undated) https://www.tsi.com/products/particle-sizers/particle-size-spectrometers/aerodynamic-particle-sizer-(aps)-spectrometer-3321/#resources ().

[51] Peters, T.M., Leith, D. (2003) Concentration measurement and counting efficiency of the Aerodynamic Particle Sizer 3321. *Journal of Aerosol Science*, **34** (5) 627–634.

[52] Volckens, J., Peters, T.M. (2005) Counting and particle transmission efficiency of the aerodynamic particle sizer. *Journal of Aerosol Science*, **36** (12) 1400–1408.

[53] Pu, Y., Kline, L.C., Khawaja, N., *et al.* (2015) Comparison of optical particle sizing and cascade impaction for measuring the particle size of a suspension metered dose inhaler. *Drug Development and Industrial Pharmacy*, **41** (5) 737–743.

[54] Mitchell, J.P., Nagel, M.W. (2003) Cascade impactors for the size characterization of aerosols from medical inhalers: their uses and limitations. *Journal of Aerosol Science*, **16** (4), 314–377.

[55] Haynes, A., Sudhan, M., Krarup, H., Singh, M. (2004) Evaluation of the Malvern Spraytec with inhalation cell for the measurement of particle size distribution from metered dose inhalers. *Journal of Pharmaceutical Sciences*, **93** (2), 349–363.

[56] Sangolkar, S.S., Adhao, V.S., Mundhe, D.G., Sawarkar, H.S. (2012) Particle size determination of nasal drug delivery system: a review. *International Journal of Pharmaceutical Sciences Review and Research*, **17** (1), 66–73.

2

Spectroscopic Methods in Solid-state Characterization

Clare Strachan[1], Jukka Saarinen[1], Tiina Lipiäinen[2], Elina Vuorimaa-Laukkanen[2], Kaisa Rautaniemi[1], Timo Laaksonen[2], Marcin Skotnicki[3] and Martin Dračínský[4]

[1]Drug Research Program, University of Helsinki, Finland
[2]Tampere University of Technology, Finland
[3]Department of Pharmaceutical Technology, Poznań University of Medical Sciences, Poland
[4]Institute of Organic Chemistry and Biochemistry, Czech Academy of Sciences, Prague, Czech Republic

2.1 Solid-state Structure of Particulates

Most dosage forms, including tablets, capsules, oral and injectable suspensions, products for reconstitution and so forth, involve solid microparticles or nanoparticles. These may contain the active pharmaceutical ingredient (API or drug), excipient(s), or both. Most APIs on the market are small molecular weight molecules (<500 D), but increasingly commonly, a larger peptide or protein. Excipients may be of small or large (e.g. polymeric) molecular weight and, in some cases, inorganic solids. In the vast majority of pharmaceutical particles (including all microparticles), the molecules in the interior of the particles far outnumber those at the surface. The arrangement of these internal molecules with one another defines the solid-state structure. Solid-state structure is an intrinsic property that affects many critical quality attributes, such as (apparent) solubility (and by extension dissolution rate), flowability (through particle morphology), compactibility, as well as solid-state and chemical stability.

Characterization of Pharmaceutical Nano- and Microsystems, First Edition. Edited by Leena Peltonen.

A material can be broadly defined by its degree of order. In molecular crystalline solids, the molecules exhibit positional and conformational long-range order in three dimensions. In the "perfect" case (not usually observed in pharmaceutical solids), no defects exist, and the particle is a perfect single crystal. At the other extreme are amorphous solids, in which the molecules lack long-range orientational and positional order. Molecular arrangement is not normally completely random in amorphous solids, and as such, some orientational and/or positional order (e.g., directional hydrogen bonding) usually exists. Between these extremes, pharmaceutical solid particles occupy various positions on the order continuum. Most commonly the particles are classified as crystalline, but are composed of many small single crystals (known as crystallites) connected by less ordered regions. Liquid crystals, in which the molecules exhibit positional and/or long-range order in one or two dimensions, but not all three, are thus also of intermediate order. These have occasionally been observed in pharmaceutical particles [1].

Crystalline (and liquid crystalline) compounds are commonly capable of different molecular packing. This is known as polymorphism. One polymorph is thermodynamically stable at a given temperature and pressure, while the other polymorphs are metastable (occupy local Gibbs free energy minima, but not the global minimum). The amorphous form is unstable (not metastable since it does not occupy an energy minimum). While crystalline polymorphs have a defined structure regardless of their preparation method, the structure of the amorphous form can vary depending on the preparation method and storage conditions [2, 3].

Additional layers of complexity are added with multicomponent solid particles; these involve more than one chemical species, with the additional species varying from small molecular weight organic molecules, to polymeric structures, to inorganic materials (e.g., silicon, silica). These may be single-phase (with molecular-level mixing of the different components) or multi-phase. Multi-component single-phase crystalline systems include solvates (including hydrates), salts, and co-crystals [4], while those of the amorphous form include those with small molecular weight co-amorphous formers (often termed co-amorphous systems) [5, 6] and polymers (molecular solid dispersions) [3–5]. Multi-phase systems can involve either partial or complete separation of the components, both on the nano- and microscales [3].

In the majority of pharmaceutical formulations, the particles themselves are single phase (e.g., composed of pure API or molecularly dispersed within a polymer or other carrier), but these particles are dispersed within a matrix that may be liquid, semi-solid or solid. Thus, while analysis of particles before their incorporation in a dosage form tends to be more straightforward, analysis becomes much more challenging once the particles are dispersed within the dosage form. This is especially the case when the particles are present at low concentrations, for example in low dose formulations. Having said this, spectroscopic analysis, especially with recent developments, is also well suited for these more challenging cases.

2.2 Spectroscopy Overview

Spectroscopy – a broad discipline – is the study of the interaction between electromagnetic radiation and matter. It may be subcategorized according to the wavelength (or frequency

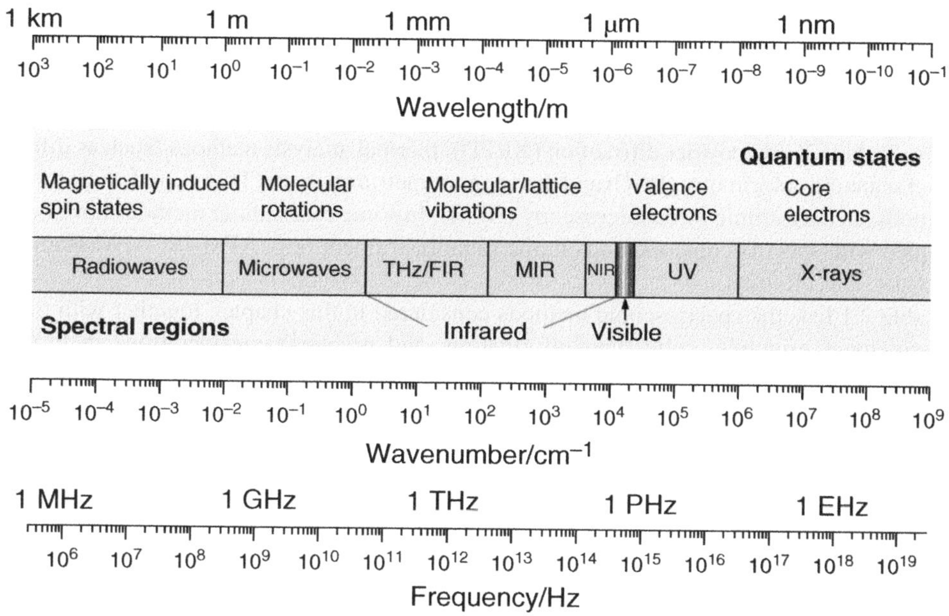

Figure 2.1 *Regions of the electromagnetic spectrum and associated quantum states.*

or wavenumber) of the radiation (Figure 2.1), the nature of interaction (e.g., absorption, scattering or emission), as well as the level being probed within the sample (e.g., nuclei, atoms, molecules, or crystals).

By definition and as introduced above, solid-state structure affects the nature of molecular (or atomic) interactions, including ionic bonding, hydrogen bonding, π–π stacking, and dipole–dipole interactions. In addition, conformations of the molecules themselves are also affected. Such differences are sufficient to affect the interaction of the solids with electromagnetic radiation and are the basis for using spectroscopy for probing solid-state structure.

Spectroscopic methods used to probe solid-state structure, and addressed in this chapter, include (mid-) infrared (MIR or IR), near-infrared (NIR), terahertz (similar to far-infrared (FIR)), Raman, fluorescence and solid-state nuclear magnetic resonance (SSNMR) spectroscopies. Nonlinear optics (encompassing several nonlinear phenomena such as second harmonic generation (SHG), coherent anti-Stokes Raman scattering (CARS) and two-photon excited fluorescence (TPEF)) has also been used in pharmaceutics more recently.

These methods are typically non-destructive and label-free, and, in most cases, can simultaneously yield data about both the chemical and solid-state nature of the sample. They each have different inherent sensitivities, as well as sampling setups, practical benefits and limitations. Thus, the most suitable technique to probe solid-state structure needs careful consideration and depends upon the nature of the sample (e.g., particle size, concentration of species of interest, and water content), type of information desired (e.g., bulk, surface or single particle, qualitative or quantitative), sampling environment (e.g., laboratory, processing environment, field), and other needs (e.g., speed, spatial resolution). In some cases, it can be beneficial or necessary to combine multiple methods for a more comprehensive understanding of the sample. Furthermore, spectroscopic analysis cannot yield all relevant solid-state information and it is often beneficial to combine it with orthogonal analytical methods. These include X-ray powder diffraction (XRPD), thermal analysis methods (such as differential scanning calorimetry (DSC) and thermogravimetric analysis (TGA)), and microscopy methods such as atomic force microscopy (AFM). In some cases, these methods have been coupled with spectroscopic methods in the same instrument (e.g. AFM-IR, AFM-Raman, TGA-IR, DSC-Raman).

Table 2.1 lists the spectroscopic methods considered in this chapter, together with their measurement principle, solid-state information, and practical considerations including sampling, and associated advantages and disadvantages. Figure 2.2 illustrates transitions involving absorption, emission and scattering of photons that are relevant to the spectroscopic methods addressed in this chapter (except NMR spectroscopy, since the energies involved are orders of magnitude lower).

2.3 Spectroscopic Data Analysis

2.3.1 Band Assignment

Spectral band assignment is often performed on some level, for example with group theory for MIR, NIR and Raman spectra. Such interpretation can be challenging, depending on the spectroscopic method and samples, and bands often remain unassigned or tentatively assigned, or worse still, wrongly assigned [8, 9]. In some cases interpretation may be aided by computational modeling and spectral prediction [10–12]. Although not trivial, such calculations have been performed in a solid-state context, including with MIR and Raman [11, 12], terahertz [13] and SSNMR [14] spectroscopies. Quantum mechanics-based calculations, which dominate vibrational spectra prediction especially for MIR and Raman spectroscopy, approximate the Schrödinger equation to varying degrees, and can be divided into three groups: *ab initio* (e.g., Hartree-Fock), density functional theory (DFT), and semi-empirical. For MIR and Raman spectral interpretation of small molecule pharmaceutics, DFT-based approaches with harmonic approximation of the vibrational modes exhibit relatively good accuracy with low computational cost. More details about the theory and methods of computational spectral prediction are available in the literature [10, 15–18]. While still not widespread, advances in computational power and availability, suitable algorithms and user-friendly software have increased their use and accuracy for solid-state spectroscopic analyses over the last 15 years [8, 9, 12, 19–22].

2.3.2 Statistical Analysis

Statistical analysis of spectral datasets can be either qualitative or quantitative, and univariate or multivariate. Univariate analysis involves the analysis of a single variable, such as peak height or area. Such analysis is popular in quantitative analysis, for example when determining concentration. Analysis of peak width may also be used as a measure of disorder (amorphousness) in the system. A challenge with univariate analysis is that it

Table 2.1 *Overview of spectroscopic methods used for solid-state characterization included in this chapter*

Spectroscopic method	Measurement principle	Solid-state information	Sampling options	Advantages	Disadvantages
Mid-infrared (MIR)	Absorption of MIR light upon dipole moment change during fundamental intramolecular vibrations.	Intermolecular interactions and conformation. Most sensitive to polar moieties and H-bonding. Peak shifts and relative intensity changes. Peak broadening with amorphous form.	Usually sample preparation involving dilution required (except ATR). Limited fiber optics. Transmission, reflection (ATR, DRIFTS) microscopy. Small sample volume needed.	Chemical and solid-state information. Sensitive to hydrogen bonding. Qualitative and quantitative. No sample preparation and very small amounts needed for ATR (surface analysis). Rapid. Directly complementary to mid/high-frequency Raman.	Sample preparation for DRIFTS, and especially transmission (compression) risks interactions with the diluent and solid-state conversions. Lack of convenient fiber optics and non-contact sampling probes. Water interference. Very limited sample penetration. Some particle size sensitivity.
Near-infrared (NIR)	Absorption of NIR light upon dipole moment change and anharmonicity during overtone and combination vibrations (of fundamentals observed in mid-IR region).	Intermolecular interactions and conformations, primarily with CH, OH and NH moieties. Peak shifts and relative intensity changes. Especially sensitive to H-bond and hydrate changes. Peak broadening with amorphous form.	No dilution or other sample preparation required. Transmission, reflection and microscopy. Larger sample volume needed. Flexible fiber optics.	Chemical (less than with MIR) and solid-state information. Sensitive to water mode of incorporation and hydrogen bonding. Qualitative and quantitative. No sample preparation or dilution needed. Deeper sample penetration. Flexible fiber optics and probes. Hand-held instruments available. Measurement through glass and plastic possible. Rapid.	Specific band identification more complex than with IR or Raman spectroscopy. Overlapping bands usually necessitate multivariate analysis. Low signal intensity limits spatial resolution in imaging. Water interference. Affected by particle size.
Terahertz (THz)	Absorption of THz (far-infrared) light upon dipole moment change during fundamental intermolecular/lattice (phonon) vibrations.	Intermolecular interactions and conformation. Very large spectral differences between different crystalline forms (phonon/lattice modes). Absence of distinct peaks for amorphous form (vibrational density of states). Molecular dynamics information (for amorphous form stability).	Sample preparation and dilution sometimes required (except with ATR and some dosage forms). Transmission and reflection (including with whole dosage forms). Imaging/mapping.	Chemical and solid-state information. Exquisitely sensitive to crystal structure. Direct determination of sample crystallinity/amorphousness (as with XRPD). Qualitative and quantitative. Largely insensitive to particle size. Sampling at depth in solid dosage forms. No thermal strain. Rapid. Measurement through some packaging possible. Directly complementary to low-frequency Raman. Molecular relaxation for amorphous form stability prediction below T_g.	Spectral bands difficult to interpret and predict. Water interference. More limited sampling options. Lower spatial resolution in imaging/mapping. Currently more expensive than most other vibrational spectroscopy methods.

(*continued*)

Table 2.1 (continued)

Spectroscopic method	Measurement principle	Solid-state information	Sampling options	Advantages	Disadvantages
Raman (mid- and high-frequency)	Raman scattering of (laser) light upon polarizability change during fundamental intramolecular vibrations.	Intermolecular interactions and conformation. Most sensitive to non-polar moieties. Peak broadening with amorphous form.	No dilution or other sample preparation required. Small or large volumes. Flexible fiber optics. Transmission, reflection and microscopy. Suitable for aqueous environments.	Chemical and solid-state information. Qualitative and quantitative. No sample preparation and very small amounts needed. Minimal water interference. Flexible fiber optics. Hand-held instruments available. Measurement through some glass and plastic possible. Rapid (except for mapping). Relatively insensitive to particle size. High spatial resolution with imaging setups. Directly complementary to MIR.	Fluorescence interference. Local heating of sample. Photodegradation. Ambient light often interferes with measurements. Imaging and especially mapping can be slow.
Low-frequency (terahertz) Raman	Raman scattering of (laser) light upon polarizability change during fundamental intermolecular/lattice (phonon) vibrations.	Intermolecular interactions and conformation. Very large spectral differences between different crystalline forms (phonon/lattice modes). Absence of distinct peaks with amorphous form (vibrational density of states).	Same as above (Raman).	Same as above (Raman). Additional advantage is that it probes phonon/lattice vibrations, with possibility for higher solid-state specificity. Adaptation can be made to standard Raman spectrometers to access the low frequency region (unlike terahertz IR spectroscopy). Directly complementary to low-frequency Raman.	Same as above (Raman).
Nonlinear optics	Nonlinear scattering of (laser) light upon interaction of multiple photons and nonlinear polarization of the sample. Stimulated coherent anti-Stokes Raman scattering (CARS) and stimulated Raman scattering (SRS) sensitive to molecular vibrations. Sum frequency generation (SFG)/second harmonic generation (SHG) sensitive to lack of inversion symmetry.	Intermolecular interactions and conformation. CARS: peak shifts and relative intensity changes. Most sensitive to non-polar moieties. SFG/SHG: signal for non-centrosymmetric crystals, and lack of signal for centrosymmetric crystals and amorphous form.	Usually small sample volumes. Most commonly with microscope for imaging, some fiber optics possible. Signal can be detected in forward (e.g., f-CARS) or backwards (e.g., epi-CARS) direction. Suitable for aqueous environments.	Chemical (CARS, SRS) and solid-state (CARS, SRS and SFG/SHG) information. Excellent 3D imaging capability (inherent confocality). (Sub)micron spatial resolution. Very rapid and label-free imaging. Minimal water and limited fluorescence interference. Simultaneous multi-modal nonlinear optical (e.g., CARS and SFG) imaging possible with some setups. Very low limits of detection based on imaging modality.	SFG/SHG: Limited value when analyzing solid-state forms with similar SFG/SHG activity, or complex-solid state mixtures. CARS: Non-resonant background. Nonlinear response complicates quantitative analysis. Generally less spectrally rich information than e.g. Raman mapping. For opaque samples, only surface information. Complexity and expense of instrument.

Fluorescence	Spontaneous emission of light by a molecule that, after absorbing a photon, is in its first excited singlet state.	Molecular interactions and conformation. Miscibility, anisotropy, spatial distribution.	Small or large sample volumes. Small sample thickness. Through sample or front face. Emission microscopy. Suitable for aqueous environments.	Sensitivity. Versatility. Rapid, generally non-destructive measurements. Little or no sample preparation. Spatial information with imaging setups. Multiple labeling allows simultaneous measurements of different properties.	Non-fluorescent samples need labeling. For opaque samples, gives only surface information. Sensitive to impurities and environmental changes.
Solid-state nuclear magnetic resonance (SSNMR)	Response of nuclei with non-zero spin quantum number to radio-frequency radiation in strong magnetic field.	Intramolecular and intermolecular interactions, conformations and molecular dynamics. Peak broadening with amorphous forms.	Typically, tens to hundreds of milligrams of sample is needed. No complex sample preparation required.	Can be applied to every solid physical state (crystalline, liquid crystalline, semicrystalline and amorphous), as well as to materials of very different chemical complexity, from pure APIs or excipients and their mixtures to formulated drug products (selective investigation of APIs does not usually require any chemical or physical treatment of the sample). Qualitative and quantitative. Complementary to X-ray crystallography. Can provide detailed information on the structure of amorphous materials, which is not possible to obtain by conventional X-ray diffraction techniques.	Low sensitivity for dilute nuclei (e.g., ^{13}C, ^{15}N, ^{17}O) and usually long experimental times. Complexity of experimental techniques and data analysis. High cost of instrument. The sample temperature can be strongly affected by frictional heating of the rotor (particularly at high MAS frequencies), which may trigger a phase transition during the experiment.

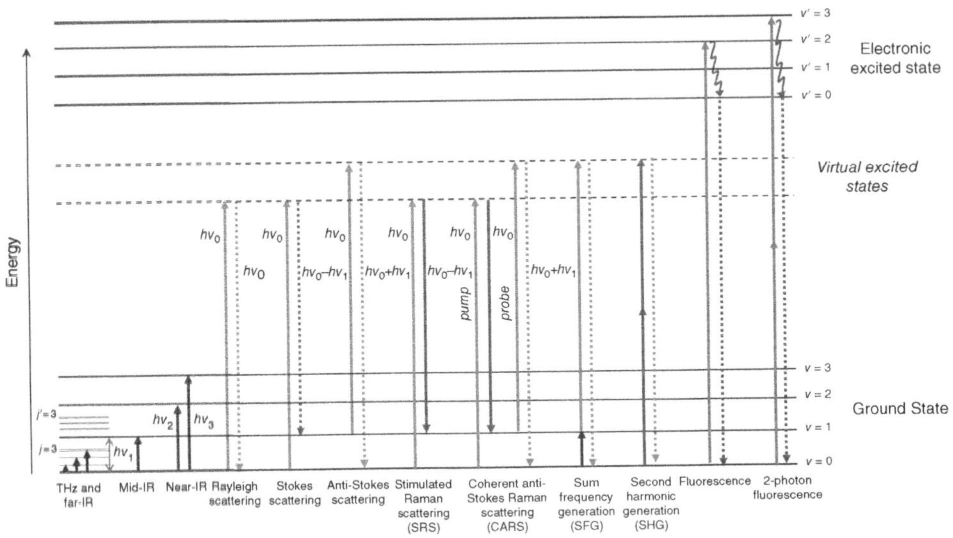

Figure 2.2 *Transitions associated with photon absorption, emission and scattering. The incident photon energy ($h\nu_0$), corresponding to the difference in energy between the ground state ($\nu = 0$) and the first vibrationally excited state ($\nu = 1$), is $h\nu_1$. The infrared (IR) and near-infrared (NIR) absorption processes involve fundamental and overtone (intra)molecular vibrational transitions, respectively. Terahertz absorption involves rotational and long-range order (phonon/lattice) modes. Scattering processes may be either elastic (Rayleigh) or inelastic (Raman). Inelastic processes either involve the incident photon providing energy to the molecule via a molecular vibrational transition ($h\nu_0 - h\nu_1$, Stokes scattering) or the molecule providing energy to the photon ($h\nu_0 + h\nu_1$, anti-Stokes scattering). Stimulated Raman scattering (SRS) involves a second laser with an energy of $h\nu_0 - h\nu_1$ to stimulate scattering. Coherent anti-Stokes Raman scattering (CARS) probes the anti-Stokes spectrum by stimulating population of the vibrationally excited state, which is then probed with another photon from the $h\nu_0$ laser. Sum frequency generation (SFG) is a two-photon excitation process (with different or equal energies) to induce anti-Stokes scattering at the sum frequency of the two incoming photons. Second harmonic generation (SHG) is a special case of SFG where the two photons have the same energy (and frequency). Fluorescence occurs upon absorption of a photon into an excited electronic state, internal relaxation, and emission from a lower-level excited electronic state back to the ground state. Two-photon excited fluorescence (TPEF) involves two photons to elevate the system into the electronic excited state followed by internal relaxation and emission back down to the ground state. Reprinted from Fraser et al. [7] with permission from Wiley.*

requires spectral bands that are resolved from one another. The spectral variation associated with solid-state differences can be subtle, with bands of different solid-state forms often overlapping (this is especially the case with NIR spectroscopy). In these cases, band-fitting or multivariate analysis may also be used. Multivariate analysis considers multiple spectral frequencies simultaneously to determine the relationship between the spectra. Principal component analysis (PCA) is the most commonly applied method for qualitative solid-state analysis. The method is a dimension reduction tool that extracts the largest systematic variation in the spectra and can be used to reject noise, while facilitating visualization of the relationships between the spectra. In addition, classification methods may be applied, such as soft independent modeling of class analogy (SIMCA), support vector machines (SVM), linear discriminant analysis (LDA), partial least-squares discriminant analysis (PLS-DA) and so forth. Multivariate analysis is also widely used for quantitative analysis. The most widely employed method in pharmaceutics is partial least squares or projection to latent structures (PLS) regression. This approach is similar to PCA, but instead of relying solely on the spectral data, the quantitative variable(s) of interest (e.g. concentration) is/are also taken into account, such that the covariance between the spectra and the quantitative variable(s) is maximized. Other (semi-)quantitative multivariate methods are also used for spectral analysis; these include principal components regression (PCR), multivariate curve resolution (MCR) and classical least squares (CLS). Typically, some form of spectral preprocessing (e.g., baseline correction, normalization, or noise reduction) is required prior to statistical analysis to remove sources of spectral variation unrelated to solid-state information (or other property of interest). The topic is vast – more detailed guidance on statistical analysis of spectra is available elsewhere [23–25].

2.4 Infrared Spectroscopy

2.4.1 Principle

Mid-infrared (MIR, also commonly referred to as infrared (IR)) spectroscopy is a vibrational spectroscopy method, along with near-infrared (NIR), far-infrared(FIR)/terahertz, and Raman spectroscopies (Figure 2.1). MIR spectroscopy (like NIR and terahertz, but not Raman) is based on the absorption of light. The MIR transition involves absorption of a single photon (Figure 2.2), and occurs when a photon of specific energy ($h\nu_I$) interacts with a molecular vibrational transition associated with a dipole moment (μ) change. The intensity of the absorbance is proportional to the size of dipole moment change. MIR probes transitions in the range $4000\,cm^{-1}$ to $400\,cm^{-1}$ (wavelengths of 2.5–25 μm) (Figure 2.1). This range corresponds to fundamental intramolecular vibrations (stretching and bending) in which a molecule is excited from the ground ($v = 0$) to first vibrational excited state ($v = 1$) as well as any associated rotational-vibrational transitions. In the absence of other interfering phenomena (e.g., nonlinear scattering effects), MIR (and NIR and terahertz) absorbance (but not transmittance) is linear to concentration, which facilitates quantitative analysis.

Experimentally, a MIR spectrometer includes an energy source (incandescent silicon carbide, or alternatively a synchrotron), a sample, a device to separate the different wavelengths (spectrometer or gratings) and a detector (deuterated triglycine sulfate (DTGS) or mercury cadmium telluride (MCT)). Nowadays, Fourier transform infrared (FTIR) setups are usually used, with a Michelson interferometer used as the spectrometer and the interferogram signal being converted to a spectrum using Fourier transformation. The instrument is usually purged with dry nitrogen, for example, to avoid spectral interference by water and, to a lesser extent, carbon dioxide.

Several sampling setups, involving either transmission or some form of reflection of the MIR light, are available. Due to the strong MIR absorption of most organic materials, transmission methods either require the sample to be extremely thin (10 μm or less) or diluted in, for example, a suitable mulling (e.g., Nujol) agent or alkali halide (usually potassium bromide, which is then compressed into a disc). Reflection methods include attenuated total reflectance (ATR) and diffuse reflectance (commonly referred to as diffuse reflectance infrared Fourier transform spectroscopy (DRIFTS)). For solid-state characterization, it is important to avoid transformations during sample preparation. These can be induced during compression (e.g., during alkali halide disc preparation) and grinding (to make the particles smaller than the wavelength of the MIR light to minimize scattering), or even upon chemical interactions with the alkali halide or mulling agent. In particular, the pressure applied during alkali halide disc preparation is substantial and is avoided when alternative gentler approaches are available. In the DRIFTS setup, the sample particles are usually ground and mixed with a diluent (e.g., KBr). ATR is now probably the most widely used MIR approach for solid-state characterization. Although it involves some degree of compression, it has the great advantage of simplicity: neither sample dilution nor grinding are required. With the sampling depth limited to approximately 1–2 μm (depending on crystal/sample refractive indices, as well as wavelength and incident angle of the light), the technique is highly surface-biased. This can be an advantage especially if solid-state properties at the surface of particles are of interest.

Spectroscopic microscopy and imaging are possible by coupling a spectrometer with a microscope. In this way, analysis of individual particles and their distribution in a matrix (e.g., tablet) becomes possible. On a general level, spectroscopic imaging can be divided into two broad categories based on the method of data acquisition: (global) imaging and (pixel-by-pixel) mapping (with a compromise being line-mapping). The difference between the two techniques is that in mapping, full spectra are sequentially acquired pixel-by-pixel, whereas in imaging, a full image is directly obtained over the whole imaging area with measurement only at one wavenumber (or a limited wavenumber range) using an array detector [26, 27]. In many cases, mapping pixel-by-pixel can achieve a more data-rich image; however, it is also generally more time-consuming. In this chapter, the term imaging is also used in its broader sense to encompass all spatially resolved spectroscopic analyses that generate images. In addition to these data acquisition variations, multiple MIR sampling orientations are available for imaging, including backscattering, transmission and ATR setups [7, 26, 27].

As with other optical imaging methods, the theoretical diffraction-limited minimum sampling spot diameter (d) depends on the wavelength (λ) of the light, the refractive index of the medium in which the sample and lens is in contact (n), and the maximal half-angle of the beam of light that enters the sample (θ), and can be estimated using Equation (2.1) [27]:

$$d \approx \frac{1.22\lambda}{n\ sin\theta} \tag{2.1}$$

The term $n\ sin\theta$ is the numerical aperture (NA) of the focusing objective. Thus, if the focusing objective has a NA of 0.90, the minimum spot diameter in the MIR region varies with extremes of 3.4 μm at $\lambda = 2.5$ μm ($4000\ cm^{-1}$) and 34 μm at $\lambda = 25$ μm ($4000\ cm^{-1}$). This means that sufficiently large individual microparticles, but not nanoparticles, can potentially be resolved.

The minimum theoretical possible depth of field (L) with diffraction-based imaging setups also depends on the refractive index (n) of the sample, and thus different materials with different refractive indices will give different sampling volumes, as in Equation (2.2):

$$L \approx \frac{4n\lambda}{\mathrm{NA}^2} \tag{2.2}$$

In practice, the scattering effects of particles increase the sampling volume and spot size above these theoretical limits. A more detailed treatment of resolution limits in imaging setups is provided elsewhere [27].

In ATR imaging, the high refractive index of the crystal in contact with the sample improves the theoretical diffraction-limited spatial resolution by a factor equal to the crystal refractive index (when compared to air), with the best increases obtained with germanium ($n = 4$) or silicon ($n = 3.4$) [27]. Furthermore, as with non-imaging ATR setups, the sampling depth is limited to the first approximately 1–2 μm. Intimate contact between the sample and crystal is needed with ATR imaging, which can present difficulties for particle imaging. Microtomed cross-sections can facilitate analysis of particles in solid dosage forms.

Near-field imaging allows spatial resolutions smaller than the diffraction-limited spot size. One example is atomic force microscopy (AFM) coupled with MIR spectroscopy (AFM-IR), with resolutions down to 30 nm [28–31].

2.4.2 MIR Applications

Vibrational modes in the MIR region provide molecular 'fingerprints' and appear as well defined and often resolved spectral bands, and this, together with its relatively widespread availability, helps to make MIR spectroscopy the most commonly employed vibrational spectroscopic method for chemical analysis. Since the band intensities and frequencies are also affected by molecular arrangement (intermolecular interactions and conformational changes) the method is widely used for solid-state characterization of particulate systems.

As an example, the experimental MIR spectra of the crystalline γ (commercially used) and amorphous forms of the drug indomethacin are presented in Figure 2.3, along with two theoretical predictions of the spectra using DFT. The calculations were used to perform a near-complete vibrational band assignment for the drug [32]. As expected, the experimental spectrum of the amorphous form (Figure 2.3d), prepared by quench-cooling the melt, features broader peaks than the crystalline form (Figure 2.3c). However, peak shifts are also visible, indicating specific differences in molecular arrangement between the two forms. For example, a carbonyl stretching mode in the carboxylic acid moiety (Figure 2.3, bottom) appears at 1716 cm^{-1} for the crystalline form, while the corresponding band appears at two different positions for the amorphous form: 1709 cm^{-1} and 1735 cm^{-1}.

The DFT spectral prediction was also used to help interpret these spectral differences on a structural level. From single crystal X-ray diffraction data, it is known that the γ polymorph contains molecular dimers with cyclic hydrogen bonding between the carboxylic acid moieties (Figure 2.3, bottom). Spectra were predicted with both a single indomethacin molecule, as well as this dimer motif, as starting points. The carboxylic acid C=O stretching mode is predicted by both the monomer and dimer predictions, but the dimer calculation predicts its position significantly closer (1735 cm^{-1}, Figure 2.3a) to

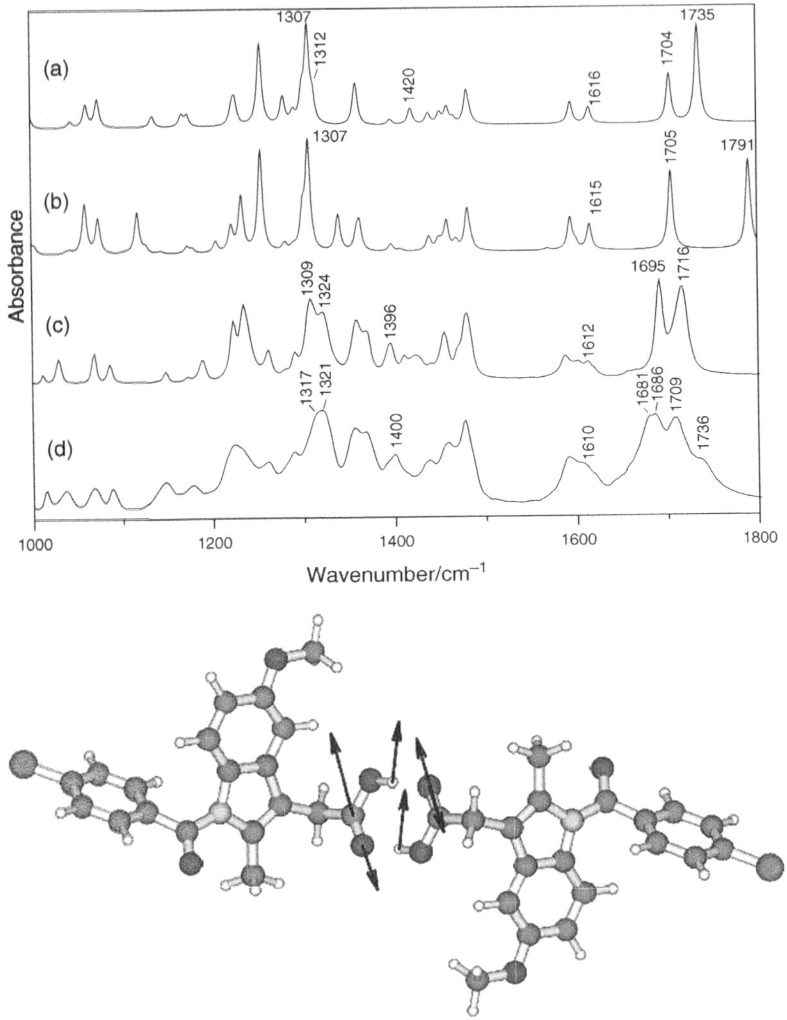

Figure 2.3 *MIR spectra of indomethacin: (a) calculated dimer; (b) calculated single molecule; (c) experimental γ polymorph; (d) experimental amorphous form (top), together with the predicted normal mode appearing at 1735 cm^{-1} for the indomethacin dimer (bottom). The main predicted atomic displacements of this mode are represented by eigenvectors. Calculations were performed with DFT with B3LYP functional and basis set 6-31G(d). Modified from Strachan et al. (2007) [32], with permission from Wiley.*

those in the experimental spectra of both the crystalline and amorphous forms than the monomer calculation (1791 cm^{-1}, Figure 2.3b). This, along with other spectral features, suggests that the quench-cooled amorphous form also exhibits this cyclic dimer motif. However, the additional peak in the experimental spectrum of the amorphous form at 1736 cm^{-1} suggests that not all the indomethacin molecules are involved in this dimer arrangement. For more details the reader is referred to the original publication [32].

Since MIR spectroscopy is most sensitive to vibrational modes involving polar moieties, the technique is exquisitely sensitive to water, limiting MIR analysis of particles in aqueous environments (e.g., suspensions). On the other hand, this sensitivity makes it excellent for probing hydrates and intermolecular hydrogen bonding. MIR spectroscopy is widely used in preformulation (e.g., solid-state screening) and formulation development to identify and quantify different solid-state forms including polymorphs, solvates, salts, co-crystals, and amorphous forms [33]. As introduced above with the example of indomethacin, the technique also allows a degree of solid-state structural analysis, especially with hydrogen bonding [34] and ionic interactions [35]. This can be useful when probing interactions and phases of multicomponent systems, such as polymeric dispersions, co-crystals and co-amorphous forms [36].

Although MIR spectroscopy has found more limited application in process analysis than NIR and Raman spectroscopies due to sampling challenges (dilution, limited fiber-optic cables, noncontact interfacing options, and sensitivity to water), it has been used extensively off-line to monitor solid-state transformations of particles, including from both kinetic and mechanistic perspectives [12, 33, 37, 38]. Recently, ATR FTIR spectroscopy was used to detect new polymorphs of indomethacin [39]. Crystallization of amorphous indomethacin in suspension was investigated as a function of pH and temperature, and multiple new metastable polymorphs (δ, ε and ζ) of the drug were discovered. The particles were separated from the aqueous medium by centrifugation (but not dried) before measurement. In this case, ATR FTIR spectroscopy was the only technique able to detect the ε form: it was too unstable for analysis with the available Raman, XRPD and DSC setups.

MIR spectroscopy combined with multivariate analysis has also been used to detect subtle variations between amorphous forms of indomethacin prepared by different methods (quench-cooling the melt, spray drying, milling, and cryo-milling, with the milling techniques prepared using both the γ and α forms of the drug) [40, 41]. Although the spectra were similar, PCA revealed reproducible differences between the preparation methods. The differently prepared forms have exhibited different crystallization rates and resulting polymorphs [41]. NIR and Raman spectroscopy with PCA have also been used for such analysis, although the analyses may be complicated by variation in water content and fluorescence for the two techniques respectively [40, 41].

While the sensitivity of FTIR spectroscopy to water, together with sampling restrictions, limits its in-line use for direct solid-state analysis during processing, MIR probes equipped with an ATR crystal can be used to monitor drug concentration changes in solution associated with solid-state changes (e.g., crystallization) [42]. Complimentary *in situ* methods can be used to directly probe the drug particles themselves. Schöll *et al.* (2006) combined *in situ* ATR IR spectroscopy to continuously monitor the concentration of dissolved L-glutamic acid, together with *in situ* Raman spectroscopy (to monitor polymorphic form) and focused beam reflectance measurement (FBRM) and particle vision measurement (to monitor crystal chord length and morphology), for a comprehensive analysis of the aqueous solution-mediated conversion of L-glutamic acid from its metastable α form to the stable β form [43]. The very limited ATR IR sampling depth (approximately 1–2 μm) allowed the solution concentration to be measured without interference from the suspended particles.

One example highlighting the importance of sampling geometries for solid-state analysis involved the crystallization of amorphous indomethacin particles [44]. The particles were prepared by quench-cooling the melt followed by grinding. Their crystallization was quantified during storage (30°C at 23% relative humidity) using three different analytical methods: ATR FTIR spectroscopy, XRPD, and DSC. According to the ATR FTIR spectra, crystallization was complete after 5 days, while with DSC, complete crystallization was observed at 140 days. XRPD analysis suggested crystallization was only about 60% complete after this time. The disparate results of the techniques resulted from their different sensitivities to disorder as well as sampling geometries. While DSC and XRPD (with the setups used) sample entire particles, ATR FTIR is limited to sampling depths of approximately 1–2 μm. Analysis of indomethacin particle cross-sections with scanning electron microscopy revealed the crystallization to be surface-specific, and thus the surface bias of ATR sampling is more sensitive to such crystallization behavior. This raises the question of what is the "best" technique to use. If increased dissolution rate is the aim of making particles amorphous, then the most appropriate analytical technique is the one that can correlate solid-state changes to changes in dissolution behavior (critical quality attribute). In this case, dissolution changes were best correlated to the crystallization as observed by ATR IR spectroscopy, which is logical considering dissolution occurs from the particle surfaces.

2.4.3 MIR Imaging

Imaging has been used to obtain spatially resolved information about particulate systems, with the potential to probe individual particles. In a series of publications using *in situ* ATR FTIR imaging, Kazarian *et al.* have demonstrated the technique's value in probing solid-state changes *in situ* during dissolution [35, 45–51]. Flow through dissolution setups were coupled with an ATR crystal, and the dosage form (e.g., tablet) is placed in the sample holder such that the dissolution medium flows around the tablet while the tablet is pressed against the ATR crystal. In one such study, Ewing *et al.* (2015) [35] investigated tablets of a proprietary antiviral drug. The drug is a weak acid and was formulated as the sodium salt, and the effect of dissolution in medium pH on its disproportionation was probed. In 0.1 M HCl solution, ATR FTIR imaging revealed the formation of a free acid salt shell around the tablet. The polymer and dissolution medium were also simultaneously imaged based on the IR spectra. The technique also has the advantage of being fast.

Near-field IR spectroscopy has been used to breach the diffraction-based resolution limit and obtain nanoscale resolution. Although the technique is not widely available, its benefit has been put to good use in probing nanoscale miscibility and phase separation analysis of drug-polymer blends [52, 53].

2.5 Near-infrared Spectroscopy

2.5.1 Principle

Near-infrared (NIR) spectroscopy includes wavenumbers between 12 500 and 4000 cm^{-1}, which corresponds to wavelengths of 0.8 to 2.5 μm (Figure 2.1). The transitions in this range include overtone ($v = 0$ to $v = 2, 3 \ldots$) and combination vibrations (simultaneous excitement of two or more fundamental modes). Figure 2.2 includes an example of a first-order overtone transition.

As is the case with MIR, the active modes are associated with a change in dipole moment. However, anharmonicity (deviation of the vibration from a simple harmonic oscillator) is also required for these transitions to be allowed. This occurs with a large difference in mass of the vibrating atoms. The occurrence of these higher-order transitions is much lower (10–100 fold weaker signal intensity) than that of their fundamental counterparts observed in the MIR range, which correspondingly makes NIR absorption weaker. Vibrational modes seen in the NIR range are typically X–H bonds, with X commonly being C, O, N or heavier atoms.

NIR spectrometers contain a broadband incandescent NIR light source for illuminating the sample, with the light separated into its constituent frequencies using a spectrograph or gratings, and then detected, typically with Si-based charge-coupled devices or InGaAs detectors [33]. Fourier transform (FT) setups are also common.

One of the great advantages of NIR spectroscopy is its sampling simplicity and flexibility. The inherently weak NIR modes make the technique suitable for non-contact sample analysis without any sample dilution, and the sample may be analyzed in transmission or reflectance modes, with sampling depth up to several millimeters into samples (depending on the sample interfacing setup, as well as absorbance and scattering effects). The NIR radiation is efficiently transmitted through readily-available fiber-optic materials, which can be hundreds of meters long, with a wide array of probes available. It is important to note that NIR spectroscopy is sensitive to particle size, due to particle size dependent scattering. This appears as a change in the baseline, a feature that can be harnessed for particle size characterization [54]. However, the resulting nonlinear baseline shifts are a challenge to interpret and quantitatively model in practice, and as such NIR spectroscopy is not considered a first-line method for accurate particle size determination [33].

NIR spectrometers may also be coupled with microscopes for imaging analysis [55]. The diffraction-limited spot size for NIR measurements is approximately 2–3 μm (smaller than for MIR microscopy due to the shorter wavelengths of NIR radiation). However, in practice, this spatial resolution is normally never achieved due to the inherently weak NIR signal, meaning that samples hundreds of micrometers thick are required, as well as scattering effects that increase the effective sampling volume. Spatial resolutions of tens to hundreds of micrometers are typical [27].

One of the challenges with NIR spectra is that they contain broad and complex overlapping peaks whose precise origins can be difficult to identify. This kept NIR spectroscopy out of favor for solid-state (and other) analyses until the introduction of multivariate analysis to the field of applied spectroscopy. This development, combined with the inherent advantages of speed, flexible non-contact sampling, and high specificity, led to its widespread implementation for in-line qualitative and quantitative analysis of pharmaceuticals during processing. The technique has been almost synonymous with process analytical technology (PAT) in pharmaceutics.

2.5.2 NIR Applications

Like MIR spectroscopy, NIR spectroscopy is especially sensitive to water. Although this generally precludes sampling in an aqueous environment (with certain exceptions involving specially designed flow through cells), it makes the technique ideal for probing the presence of moisture as well as solid-state hydrate changes. The two strongest water bands in the NIR region are the first overtone stretch of OH at around 1450 nm (6897 cm^{-1}), as well as the combination of OH stretching and bending at approximately 1940 nm (5155 cm^{-1}). NIR spectroscopy has been used to probe the moisture content of spray- and freeze-dried products, as well as other powders, with root mean squared errors (RMSEs) of detection of well under 1% [56, 57].

The technique has the further advantage in that it is often possible to differentiate between bulk/adsorbed/absorbed and crystal hydrate water. As well as peak position differences, the more uniform energetic state of crystalline hydrate water means that the hydrate water bands tend to be sharper. Different hydrates also exhibit band position and shape differences to one another. Blanco *et al.* exploited these aspects to characterize the transformations and solid-state stability of the antibiotic, azithromycin [58]. The drug can occur as an amorphous anhydrous form and two crystalline hydrates: a monohydrate and a dihydrate (most stable form in ambient conditions). The combination band of water appears at approximately 1920 nm for sorbed water, whereas the peak for the dihydrate appears at approximately 1955 nm. The anhydrous and monohydrate forms of the drug were stored at 60°C and 100% relative humidity (RH). An increase of the water band at approximately 1920 nm was associated with moisture sorption, while the appearance and growth of a new band at around 1955 nm revealed transformation to the dihydrate. In contrast, the monohydrate did not exhibit dihydrate transformation (Figure 2.4).

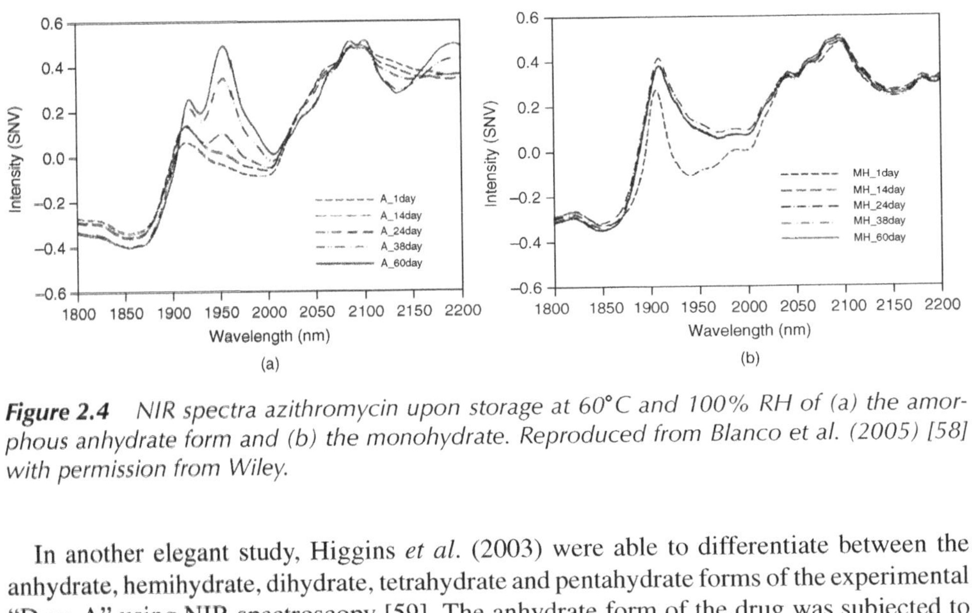

Figure 2.4 NIR spectra azithromycin upon storage at 60°C and 100% RH of (a) the amorphous anhydrate form and (b) the monohydrate. Reproduced from Blanco et al. (2005) [58] with permission from Wiley.

In another elegant study, Higgins *et al.* (2003) were able to differentiate between the anhydrate, hemihydrate, dihydrate, tetrahydrate and pentahydrate forms of the experimental "Drug A" using NIR spectroscopy [59]. The anhydrate form of the drug was subjected to increasing levels of RH (at 25°C and 40°C) and the combination band of water increased in a largely stepwise manner, indicative of the sequential appearance of increasingly higher hydrates up to the pentahydrate form at high humidities.

NIR spectroscopy is not only sensitive to water and hydrate forms — it can also be used to differentiate and quantify other solvates, co-crystals and salts, as well as different solid-state forms without any differences in chemical composition, including polymorphs [60, 61] and amorphous forms [33, 60]. The largest spectral differences are commonly associated with changes in intermolecular hydrogen bonding. Furthermore, transformations between these forms in different environments can be investigated [37]. Such analyses are usually aided by multivariate analysis of the NIR spectra, most commonly PCA (for qualitative analysis) and PLS regression (for quantitative analysis). One of the key challenges with NIR spectroscopy is the interference of sorbed water on the NIR spectra (when that is not the property of interest). Multivariate models that take into account such variation should be considered.

A case study demonstrating the different sensitivities of NIR and Raman spectroscopies to water and solid-state forms in particles is presented below, using the model drug theophylline (Figure 2.5). Theophylline exists as several solid-state forms including a monohydrate form (TPM), and anhydrate forms, one of which (TPA) is most commonly found in commercially available tablets. One of the metastable anhydrate forms of theophylline (TPA_M), is structurally similar to the TPM. TPM features channels within which the water molecules are accommodated, with TPA_M having a similar molecular arrangement of the theophylline molecules (without the water). TPA exhibits a substantially different structure and XRPD diffractogram to the other two forms (Figure 2.5a). The NIR and Raman spectra of the TPM, TPA and TPA_M are presented in Figure 2.5b–c. In the NIR spectra (Figure 2.5b), the OH bands associated with the crystal-bound water at 1490 nm and 1970 nm dominate the TPM spectrum, while these bands are absent for the two anhydrate forms. In contrast, the Raman spectra, in which water is weakly represented, are similar for TPM and TPA_M but quite different for TPA.

The TPM particles were subjected to isothermal dehydration (at 55°C) and NIR and Raman spectra were continuously recorded. PCA of these time-series spectra resulted in slightly different apparent transformation profiles with the two techniques, as evidenced by the PC1 score values (Figure 2.5d–e). For NIR, the decrease in PC1 score values (Figure 2.5d), was initially faster than for Raman (Figure 2.5e). The NIR change is due to water loss (loadings plot in Figure 2.5f) while Raman spectroscopy revealed that substantial rearrangement of the theophylline molecules lagged behind (loadings plot in Figure 2.5g). As such, TPM first formed TPA_M through loss of water molecules from the channels in TPM, before rearrangement of the theophylline molecules to the more stable TPA. Neither NIR or Raman could explain this transformation pathway alone. This same dehydration phenomenon has been quantified during fluid bed drying of TPM granules using NIR and Raman spectroscopies [62].

NIR has become widely utilized in the pharmaceutical industry for quality control of raw materials, including both APIs and excipients [33, 63]. With careful calibration model development, not only the solid-state form, but also potentially the chemical identity, water content and changes in particle/size and morphology may be checked [33, 63]. Hand-held NIR spectrometers have been introduced for this purpose [63].

NIR is also probably the most widely implemented in-line spectroscopic technique for monitoring pharmaceutical particles during processing, as part of the worldwide drive by regulatory authorities and industry alike for the adoption of process analytical technology (PAT) [54, 64, 65]. In terms of the solid-state form, NIR is used (in-line) qualitatively and quantitatively to monitor and control solid-state changes during, for example, wet granulation [66–68] and drying [62]. It is also used for solid-state analysis during polymorph screening [61].

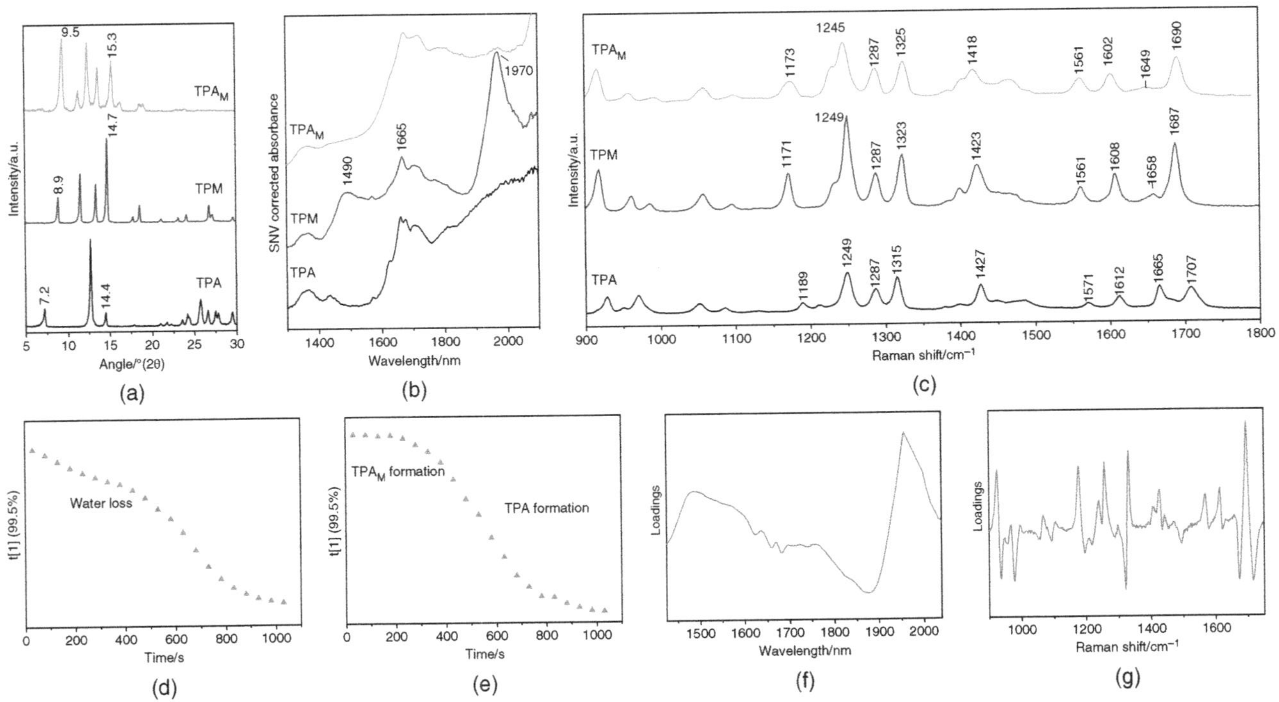

Figure 2.5 *(a) Experimental XRPD diffractograms and (b) NIR and (c) Raman spectra of metastable theophylline dehydrate (TPA_M), theophylline monohydrate (TPM) and theophylline anhydrate (TPA). Spectral changes upon isothermal dehydration of TPM at 55°C are represented by (d) NIR and (e) Raman PCA score plots for PC1, as well as the corresponding (f) NIR and (g) Raman PCA loadings plots.*

2.5.3 NIR Imaging

The majority of NIR imaging applications for pharmaceutical particle characterization have involved chemical analysis. They include drug and excipient distributions in powders, roller compacted ribbons, extrudates, tablets and printed products, as well as the detection of counterfeit products [26, 69–72]. NIR imaging has, however, also been used for solid-state form analysis. Schönbichler *et al.* (2013) performed a quantitative study on ternary blends of three polymorphs of furosemide [73]. They compared FT-NIR imaging with non-spatially resolved NIR, Raman and ATR FTIR spectroscopy. The ATR setup induced conversion between two polymorphs. The other three techniques performed similarly well quantitatively, but NIR imaging had the advantage of also providing distribution information (Figure 2.6).

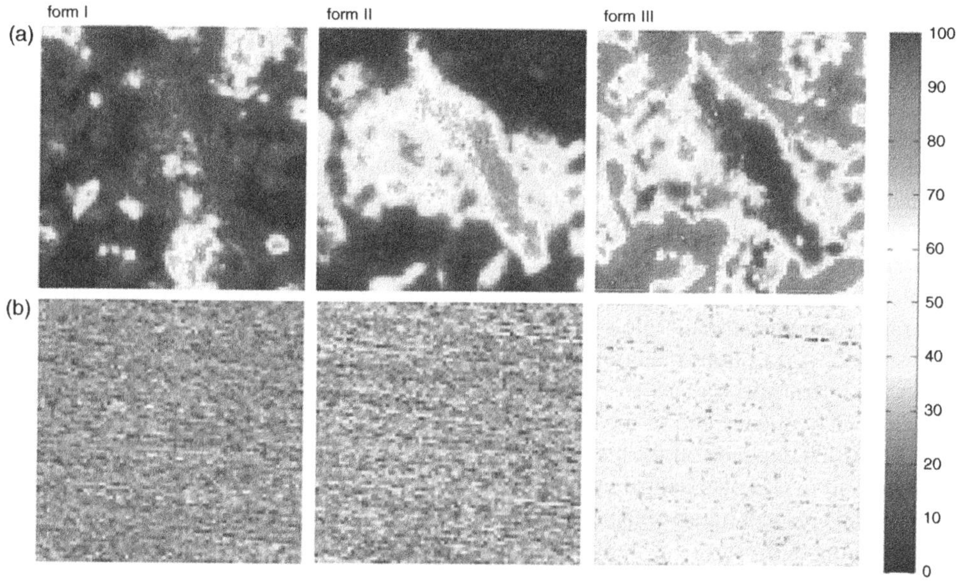

Figure 2.6 Quantitative FT-NIR imaging of two powder blends of three polymorphic forms (I, II, and III) of the drug furosemide, using PLS regression. The polymorph ratio in both blends is 25(I):25(II):50(III) with (a) an inhomogeneous blend and (b) a more homogeneous blend. The size range of the particles was similar for all three forms, with d50% values of 3–4 μm. The lateral resolution of 50 μm was chosen for the NIR imaging, with each image containing 80 × 80 pixels. Reprinted from Schönbichler et al. (2013) [73], with permission from Elsevier.

NIR has also been used to image solid-state transformations, most notably hydrate–anhydrate transitions [71, 74, 75]. Alexandrino *et al.* probed the dehydration of piroxicam and lactose hydrate particles in tablet formulations upon heating, together with the performance of different multivariate data analysis treatment methods including multivariate curve resolution-alternating least squares (MCR-ALS), and parallel factor analysis (PARAFAC and PARAFAC2) [74, 75]. Like MIR imaging, NIR imaging has also been coupled with specially adapted flow through cells to image disproportionation of an ionized drug *in situ* (together with Raman imaging mapping) [76]. One of the noted advantages of NIR imaging in many of these studies (relevant to all the imaging methods) is the detection of solid-state forms of low concentration species, when other majority species could otherwise overwhelm the spectral signal. However, it is important to select suitable model samples, spatial resolutions, and data preprocessing and analysis algorithms for reliable analysis. For example, the effects of covariance of concentration between different species on models should be considered.

2.6 Terahertz Spectroscopy

2.6.1 Principle

Terahertz radiation is sandwiched between the MIR and microwave regions of the electromagnetic spectrum (Figure 2.1). It has been defined as having a frequency range of 0.05 to 6 THz, which corresponds to 1.67 cm^{-1} to 200 cm^{-1}. It is somewhat synonymous with the far-infrared (FIR) region, which has a longer spectroscopic history. The FIR region is commonly defined as being between 10 cm^{-1} and 400 cm^{-1} (25–1000 μm). Terahertz radiation overlaps with a wide range of transitions, including molecular rotations (gases), hydrogen-bonding stretches and torsion vibrations (liquids), and in solids, low energy bending and torsional vibrations, crystalline phonon/lattice vibrations, as well as secondary dielectric relaxations [13, 77, 78].

Phonon or lattice vibrations can be thought of as collective vibrations that occur upon the interaction of many intramolecular vibrations. A prerequisite for their propagation is the long-range periodic structure present in crystals, and they appear as distinct peaks in terahertz spectra. Amorphous materials do not have the required long-range order for phonon modes, and their spectra lack distinct peaks. However, absorption still occurs, especially at the higher frequency end of the terahertz regime, due to the vibrational density of states (VDOS) which depends on frequency in a universal manner in this range [79]. Dielectric relaxation processes also contribute to the absorption by amorphous solids in this spectral range. In this respect, terahertz spectroscopy has recently shown much potential for better understanding of the physical stability of amorphous pharmaceuticals, especially below the glass transition temperature (T_g) [13].

FIR spectroscopy of (crystalline) solids was performed in a limited number of laboratories in the latter part of the twentieth century [77]. However, substantial challenges with the generation of sufficiently intense FIR radiation (with weak incoherent FIR sources such as globars or mercury arc lamps) and its detection over background thermal emission (requiring helium-cooled bolometers) severely limited its use. Terahertz radiation became much more accessible in the 1990s with the advent of ultra-short pulsed NIR lasers coupled with photoconductive switches that together allowed the generation and detection of coherent terahertz pulses [80, 81]. Although the terahertz beams generated with this technology have a low average power, their coherent nature combined with the time-gated detection provides high sensitivity without interference from background thermal emission. Spectrometers involving this technology were commercialized soon after the turn of the millennium, paving the way for the adoption of terahertz spectroscopy in several applications, including solid-state pharmaceutics. The technique is often referred to as terahertz pulsed spectroscopy (TPS) or terahertz time-domain spectroscopy (THz-TDS) to differentiate it from traditional FIR setups involving incoherent generation and detection of terahertz radiation. The most accessible frequency range with TPS technology is the terahertz range defined above.

Like MIR and NIR, terahertz spectroscopy is based on absorption involving transitions with dipole moment changes and is also a single photon event (Figure 2.2). Sampling is often performed in transmission mode; this may involve dilution of the particles in a non-absorbing matrix, such as polytetrafluoroethylene (PTFE) or polyethylene, but also more pharmaceutically relevant polymers, and then compression into a compact [82]. Many excipients, including most pharmaceutical polymers, are (partially) transparent to terahertz radiation. Thus, particles deep within pharmaceutical matrices such as tablets (up to approximately 3–4 mm deep) can also be directly probed nondestructively. Reflection measurements are also possible, including with ATR setups [83]. In ATR analysis, the longer wavelengths lead to a larger sampling depth than with MIR ATR. The coherent nature of TPS facilitates analyses in diverse temperatures without thermal interference. Furthermore, the method itself does not induce any thermal strain in the sample. Due to the longer wavelengths employed, the technique also does not suffer (micro)particle size-dependent scattering effects as with NIR and MIR spectroscopies. However, water strongly absorbs terahertz radiation, necessitating the samples to be dry and the sampling environment to be dry purged [13, 84].

Terahertz imaging of solid dosage forms has also been the subject of quite some research and development over the last 15 years or so. Most commonly this has been implemented through time-resolved detection and analysis of terahertz signal reflections in dosage forms, which occur when the terahertz pulse reaches a phase boundary associated with a change in refractive index [13, 85]. If the refractive indices of the different phases are known, the depth of the structural features can be determined, and, upon scanning the terahertz pulses across the sample surface, a 3D image of surfaces and buried structures can be generated. The axial (depth) resolution depends on the pulse emission; high bandwidth pulses allow axial resolutions of approximately 40 μm [86]. In pharmaceutics, this technology has been most successfully applied to imaging coatings of dosage forms, especially tablets [84, 87, 88]. However, extracting the solid-state sensitive spectral information from the time-domain waveforms in this setup is challenging.

An alternative approach to obtain spatially resolved terahertz spectral information is mapping in transmission mode. This setup is analogous to terahertz spectroscopy using transmission described above, but instead the terahertz beam is tightly focused onto the sample. Multiple terahertz signals are obtained as the focus is moved across the sample. The technique is sometimes referred to as terahertz spectroscopic imaging (despite, strictly speaking, being a mapping rather than imaging approach). The information with this approach is poorly resolved in the z-direction (e.g., over 2 mm) and thus analysis favors 2D imaging [89].

A further and major limitation of all imaging based on terahertz signals for microparticle analysis is its lateral (x–y) spatial resolution: the relatively long wavelengths of terahertz radiation lead to diffraction-limited lateral resolutions of approximately 100–250 μm. Most pharmaceutical grade microparticles are thus too small to image individually. Therefore, the principal potential benefit of solid-state imaging of particles based on terahertz radiation is the ability to probe solid-state forms of particle populations at depth (e.g., in tablets).

2.6.2 Terahertz Applications

Terahertz spectroscopy has been used to detect and quantify different crystalline solid-state forms including polymorphs, solvates (including hydrates), and co-crystals, as well as the amorphous form [13, 82, 87, 90–93]. It has also been used to probe transitions between these forms [13, 94–98]. Over the last few years, its potential to provide new insights into the stability of amorphous drugs has been increasingly realized [13, 99].

Proof-of-principle of TPS for pharmaceutical polymorph discrimination was illustrated by Taday *et al.* (2003) who demonstrated that two different polymorphs of ranitidine HCl (forms I and II) exhibited substantially different terahertz spectra [100]. The differences were also obvious in commercially available tablets. This drug had been the best-selling drug in the world, and these two solid-state forms had been the subject of numerous litigation battles when the patent for form I, but not form II, expired. One of the recurring themes in these cases was the sensitivity of established solid-state analysis methods to detecting polymorph contamination, including with XRPD and MIR spectroscopy. Since terahertz spectroscopy primarily probes intermolecular and long-range modes, it is potentially more sensitive than MIR, NIR, and (mid-frequency) Raman spectroscopy to solid-state differences of drugs. This has been one of the motivations behind the many pharmaceutical solid-state analysis investigations with terahertz spectroscopy.

Strachan *et al.* published terahertz spectra of two crystalline polymorphs of carbamazepine (forms I and III) and enalapril maleate (forms I and II), the crystalline (γ) and quench-cooled amorphous forms of indomethacin, and the crystalline and thermotropic liquid crystalline forms of fenoprofen calcium. Binary mixtures of these forms were also analyzed (Figure 2.7) [82, 101]. In all these cases, the powders were microparticulate, and diluted in a matrix of PTFE before compression into compacts and measurement in transmission mode. The terahertz spectra were completely different for the different solid-state forms, with the crystalline samples exhibiting multiple peaks associated with crystalline phonon modes. Both the amorphous and liquid crystalline forms exhibited the relatively featureless and diffuse absorption characteristic of disordered materials. In this way, terahertz spectroscopy is also suitable for direct determination of crystallinity, analogous to XRPD. The same authors also quantified binary solid-state mixtures of the materials using multivariate analysis (partial least squares regression), and were able to achieve limits of detection (LODs) of as low as approximately 0.5% and limits of quantification (LOQs) of less than 2% (similar to what has been achieved for other vibrational spectroscopic methods) [101].

In some cases, TPS can be used to resolve different polymorphs that are virtually identical structurally and very hard to resolve even with XRPD [96, 102]. This has been attributed to the fact that TPS probes the dynamics of intermolecular vibrational motions, in contrast to XRPD, which is purely structural in its analysis of atomic positions and lattice planes.

One example of phase transition analysis is where Zeitler *et al.* (2007) probed the dehydration of theophylline monohydrate *in situ* as a function of temperature (prepared as a matrix compact) [97]. In addition to the obvious spectral changes associated with the monohydrate to anhydrate transition, the evolution of gaseous water from the compacts upon dehydration was also detected. These rotational modes give narrower peaks in the terahertz

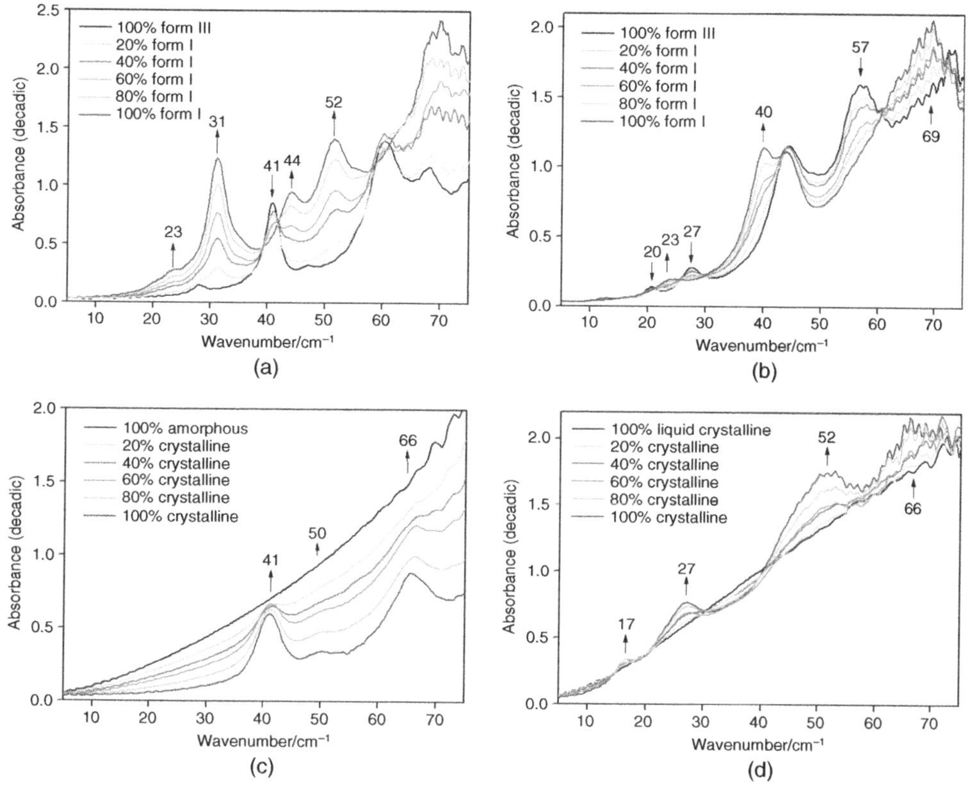

Figure 2.7 *Terahertz spectra of (a) carbamazepine (forms I and III), (b) enalapril maleate (forms I and II), (c) the crystalline (γ) and quench-cooled amorphous indomethacin, and (d) crystalline and thermotropic liquid crystalline fenoprofen calcium, as well their binary mixtures. Reprinted from Strachan et al. (2005) [101], with permission from Elsevier.*

spectra than the crystalline phonon modes. Disordering and loss of phonon modes during milling has been monitored (e.g., during the milling of β-lactose [92]), while co-crystal synthesis during milling has also been probed [94].

A drawback in the use of terahertz spectroscopy, when compared with its higher-frequency vibrational spectroscopy counterparts, has been the difficulty in interpreting the origins of the phonon modes. Due to the collective nature of the motions, group theory based on function groups is not generally applicable, nor is computational prediction involving single molecules or arguably even small clusters of molecules in vacuum (which is possible for MIR and Raman spectra). Computationally expensive calculations that consider order on a longer range (with periodic boundary conditions) are required. Overall, there are three groups of approach that have been successfully implemented to predict and assign vibrational modes in the terahertz region: atom–atom potential methods, periodic DFT methods, and molecular dynamics simulations [13, 103].

Recently, the potential of terahertz spectroscopy to provide new insights into and predict the stabilities of amorphous systems, especially below T_g, has been intensively investigated. This has involved analysis of the high-frequency dielectric relaxation and specifically β relaxation processes below T_g, with excellent reviews on the topic recently published [13, 99].

2.6.3 Terahertz Imaging

Transmission 2D terahertz solid-state spectroscopic mapping has been employed in a few studies involving pharmaceutical particles within matrices [89, 104, 105]. For example, Charron *et al.* (2013) used the technique (with a diffraction limited lateral resolution of 200 μm) to quantitatively map the distribution of caffeine-oxalic acid co-crystal and its co-crystal former in compacts using polyethylene as a matrix former [89]. Although the axial resolution was over 2 mm and sampling at each pixel traversed the entire compact thickness, the large sampling volumes helped ensure excellent quantitative agreement with the known concentration [89].

Phase transitions of particles have also been probed in tablets. Hisazumi *et al.* (2012) mapped the transformation of theophylline between one of its anhydrate and monohydrate forms in tablets containing microcrystalline cellulose and magnesium stearate as excipients, as a function of storage humidity [105]. They determined that changes in both directions followed the phase boundary mechanism, with transformation starting at the edges of the tablets and gradually progressing to the core.

Despite the relatively poor lateral and axial resolution of terahertz mapping when compared with the other vibrational spectroscopy methods, its potential advantage for solid-state particle characterization is probably for analysis of particle populations deep within solid dosage forms, as well as the resolution of structurally similar solid-state forms where other methods might fail to resolve them. Non-destructive analysis of different solid-state forms deep within tablets using reflection, rather than transmission, has also been investigated [106, 107].

2.7 Raman Spectroscopy

2.7.1 Principle

Raman spectroscopy can probe both fundamental intramolecular vibrations as well as intermolecular lattice vibrations, depending on the instrument setup. Detection of Raman scattering in the mid- and high-frequency region (~400–4000 cm^{-1}) allows the former, while detection in the low-frequency or terahertz region (below 100 cm^{-1}) allows the latter (Figure 2.1).

Different solid-state forms can usually be readily distinguished based on Raman spectra. In the mid-frequency spectral region (~400–2000 cm^{-1}), the solid-state-specific variation in intramolecular vibrations is normally presented as peak shifts or broadening, as with MIR. In contrast, the spectral features in the low-frequency region correspond to the transitions probed by terahertz spectroscopy, and can reveal distinct phonon peaks at unique positions for crystalline materials. Amorphous materials show inhomogeneous broadening of the phonon peaks, without sharp spectral features, similar to terahertz spectroscopy and XRPD. Assuming negligible absorption and additional scattering interactions of the Raman scattered photons with the analyte, Raman scattering intensity is directly proportional to the number of molecules sampled, giving the basis for quantitative Raman analysis. Particularly when combined with multivariate analysis methods, it is possible to quantify different solid-state forms of the same compound in a sample [108].

In contrast to MIR, NIR and terahertz spectroscopies described in the previous sections, which are based on absorption, Raman spectroscopy involves the detection of inelastically scattered light, originating from energy transfer between vibrational transitions in molecules and incident monochromatic light (energy of $h\nu_0$), where the energy exchange represents the gap between the ground and first vibrationally excited state (energy of $h\nu_1$). Raman scattering consists of Stokes Raman scattering ($h\nu_0 - h\nu_1$), where the incident photons have contributed energy to the molecules, and anti-Stokes scattering ($h\nu_0 + h\nu_1$), where the scattered photons have gained energy from the molecules (Figure 2.2). Anti-Stokes scattering is a much weaker event than Stokes scattering in normal conditions because it requires the molecules to be in a vibrationally excited state before interaction with the light.

For a vibrational transition to be Raman active, a polarizability change must occur during the vibration (in contrast to MIR, NIR and terahertz spectroscopies, which rely on a dipole moment change). Large polarizability changes are possible in systems where the electron density distribution surrounding the molecule or a bond is easily distorted during the vibration, and strong Raman bands are generally associated with symmetric and non-polar bonds (e.g. –C=C–), and for molecules with extended π systems (e.g., conjugated aromatic rings). In contrast, asymmetric and polar bonds (e.g. the O–H bond) give weak Raman scattering. This makes the technique highly complementary to MIR, NIR and terahertz spectroscopies.

The main components of Raman spectrometers are a source of monochromatic radiation (laser), optics for illuminating the sample and collecting the scattered photons, as well as rejecting elastically scattered Rayleigh photons (which far outnumber the rare Raman scattered photons), a device for spectrally discriminating the scattered light, and a detector. There are two types of common Raman spectrometers: Fourier-transform (FT) and dispersive Raman spectrometers. FT-Raman spectrometers are based on interferometers and are often coupled to liquid nitrogen-cooled germanium or indium-gallium-arsenide detectors. Dispersive spectrometers employ gratings to spatially separate the different wavelengths of light, which are then usually detected by a multichannel charge-coupled device (CCD) detector. FT-Raman instruments normally use near-infrared excitation at 1064 nm, while dispersive systems use shorter laser wavelengths. Dispersive Raman systems show higher sensitivities and allow shorter measurement times, but the beam diameter at the laser focus is smaller, which can cause sample heating and is more likely to induce fluorescence compared with FT-Raman. Encountering fluorescence is sample-dependent, but when it occurs, fluorescence can leave the Raman bands fully undetectable.

Rayleigh scattered light rejection is usually achieved with notch or edge filters, which efficiently block the laser frequency, but simultaneously prevent the detection of low-frequency Raman shifts. Spectrometers differ significantly in their ability to detect Raman shifts below 300 cm^{-1}. The best notch filters today allow detection of Raman shifts down to 50 cm^{-1}, but commonly spectral data below 150 cm^{-1} is not observed with a standard Raman instrument. Nevertheless, the detection of the full low-frequency spectrum requires more sophisticated instrumentation compared with the established technology. Recent technological advancements, particularly related to volumetric holographic grating notch filters, stable diode lasers and amplified spontaneous emission suppression filters, have made low-frequency Raman spectroscopy more practical and widely accessible than in previous decades. Though instrument-dependent, a single Raman measurement can provide low- and mid-frequency spectral data, both relevant for solid-state analysis (in contrast to MIR and terahertz spectroscopies, which depend on quite separate and different instrumental setups).

In general, non-invasive and non-destructive measurements with little or no sample preparation are possible. However, the use of laser powers that are too high can cause thermal damage or changes in the sample, such as dehydration. Orientation effects should be considered when analyzing single crystals, because the laser beam is polarized and if the direction of polarization is aligned with the direction of molecular vibrations, preferential enhancement of band intensities associated with these vibrations occurs. Both issues may be resolved by rotating the sample, by providing fresh sample under the laser beam to avoid heating, or a new angle for measuring a crystal. The sampled volume affects the Raman signal, and for comparative and reproducible signals between samples, it should be kept constant. Light compression of powder samples to ensure an even surface is recommended.

Raman spectroscopy is flexible regarding sampling options. The scattered photons can be detected using different collection geometries: 90° and 180° backscattering are common, and transmission techniques are also available. Additionally the spectrometer can be coupled with different types of optical fibers, allowing remote sampling. In the last decade, hand-held Raman spectrometers have also become a reality, dramatically increasing the potential for Raman spectroscopic analysis away from the laboratory.

Raman imaging is also possible and is carried out by coupling a Raman setup to an optical microscope, enabling signal collection from a small sample volume. Spontaneous Raman imaging can be achieved with direct imaging and mapping approaches (or line mapping) [7, 27, 109]. Raman mapping provides more data-rich spectra and allows 3D spatial resolution (with confocal setups where pinhole apertures are used to block out-of-focus light and suppress the background signal). However, on a practical level, times needed for Raman mapping can be substantial and frequently can take several hours depending on the Raman scattering efficiency of the sample, sampling size, and desired spatial resolution.

As with other optical imaging methods, the minimum theoretical sampling diameter is diffraction limited and can be estimated with Equation (2.1). Since Raman spectroscopy uses laser illumination, the diffraction limited resolution does not vary in the same way across the spectrum as with MIR microscopy. Small minimum spot diameters are obtained with shorter excitation laser wavelengths. As such, the minimum diffraction-limited resolution is lower than with MIR or even NIR microscopies (except where near-field optics are employed). For example, with a NA objective of 0.90, 785 nm and 532 nm laser illuminations will result in sampling diameters of 1.1 μm and 0.7 μm respectively (with sampling in air).

Based on Equation (2.2) the minimum sampling depth (depth of field) depends on the sample refractive index, with increasing refractive indices leading to larger depths of field. When using a 0.9 NA objective, a sample with an n of 1.5 (e.g., PMMA) has a theoretical depth of field of 3.9 μm when measured at 532 nm. However, the complex environment of particulate samples and mixtures in multiphase systems with varied refractive indexes will lead to alterations in sampling volume, which substantially complicates such determinations [110]. Furthermore, low levels of signal will also be detected from outside this sample volume; the extent of this depends on the relative Raman scattering intensities of the different components [110–112]. Making the situation even more complicated, elastic scattering effects and spherical aberration also affect the sampling volume and focal spot at depth. Thus, it can be very challenging to precisely define the sampling volume and especially focal spot at depth, and theoretical predictions need to be interpreted with caution [110–112].

Near-field optics with tip-enhanced Raman scattering (TERS) can allow spatial resolutions down to tens of nanometers (with tip-enhanced Raman microscopy). However, these measurements are technically demanding, and there is still a lack of publications on their use for pharmaceutical solid-state analysis.

2.7.2 Raman Applications

Raman spectroscopy is a routine solid-state characterization technique in pharmaceutical research and development, manufacturing, and quality control of finished products [113], because it allows qualitative and quantitative analysis of solid-state forms, including the possibility to simultaneously quantify multiple forms present in the sample [60, 108]. Many APIs give strong Raman signals compared with water and many excipients, enabling API analysis from aqueous mixtures and formulations even when the API is present at a relatively low concentration (approximately 1%). Non-spatially resolved Raman spectroscopy applications usually involve the analysis of bulk powders, suspensions, tablets or other solid dosage forms, where the Raman signal represents the average of a large number of particles.

Common applications include polymorph screening [114], crystalline content determination in amorphous systems [115–117] and analysis of interactions between the API and polymers in solid dispersions [34], *in situ* monitoring of solid-state conversions occurring during dissolution tests (e.g., solvent-mediated hydrate formation) [118], or induced by different temperature, humidity or pressure conditions [119–122]. Polymorph and formulation screening with Raman spectroscopy has become very appealing to the industry with the development of commercial high-throughput platforms.

The recent wider availability of Raman instruments with low-frequency capabilities has provided analytical benefits. Certain solid-state forms give very subtle differences in the mid-frequency spectral region [102, 123–125]. For example, in the case of the isostructural forms I and II of phenobarbital [124], as well as the five polymorphic forms of sulfathiazole [102], the different solid-state forms could not be differentiated by conventional mid-frequency Raman spectroscopy or even by XRPD. In contrast, in the low-frequency spectral region these forms had distinct spectra. Even in cases where differentiation is possible based on mid-frequency spectra, the use of low-frequency Raman spectral data may give better quantitative results. This was shown with ternary mixtures of piroxicam crystal forms (the β and $\alpha 2$ anhydrates, and the monohydrate) [126], as well as when detecting and quantifying early onset of crystallization of amorphous griseofulvin tablets [116]. Low-frequency Raman spectroscopy allowed better solid-state sensitivity because of larger spectral differences between the forms and higher Raman signal intensities compared to the

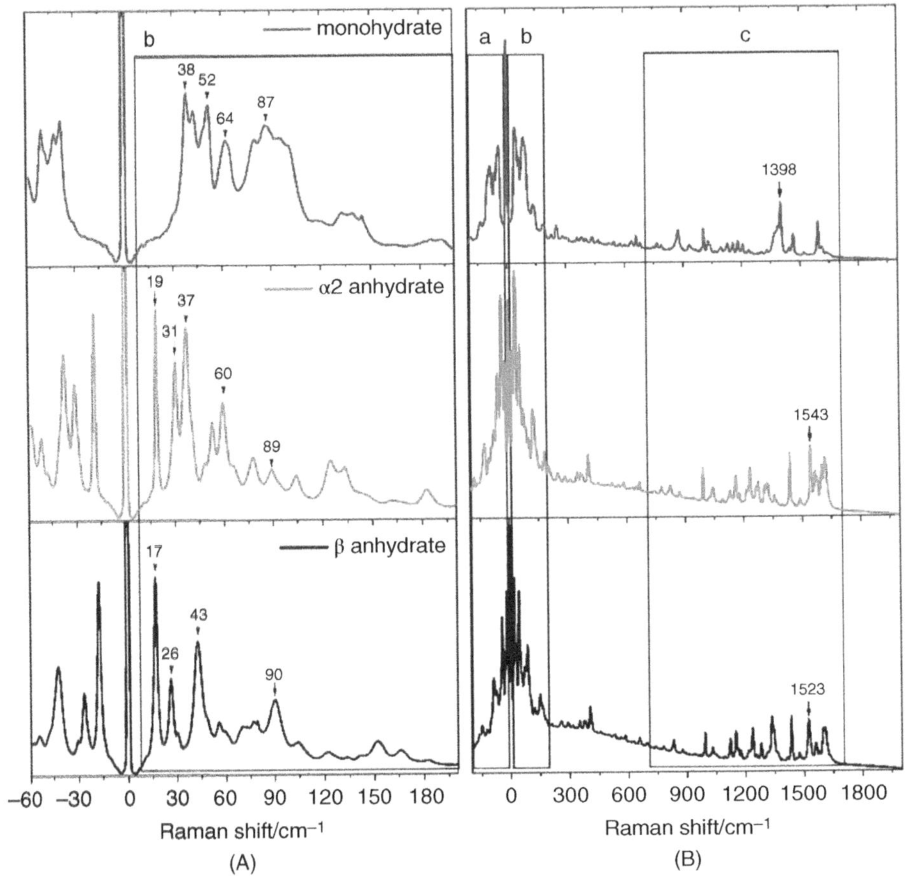

Figure 2.8 *(A) Low-frequency Raman spectra of three piroxicam crystal forms: β and α2 anhydrates and the monohydrate, measured with a low-frequency Raman instrument. (B) The full Raman spectra of the same crystal forms, showing both the low- (a = anti-Stokes and b = Stokes) and mid-frequency (c) spectral regions, measured with a Raman instrument capable of recording both spectral regions simultaneously. Adapted from Lipiäinen et al. [126], copyright 2018, with permission from Elsevier.*

mid-frequency region (Figure 2.8B). With the piroxicam crystal mixtures, use of the full low-frequency spectral region (down to 8 cm^{-1}) gave the best quantification accuracy.

The non-destructive measurements, possibility of measuring through transparent containers, and flexible coupling to fiber-optic probes, make Raman spectroscopy very well suited for process analytical technology (PAT) applications and quality control of finished products. It is used for real-time monitoring of solid-state changes (of the API or excipients) during unit operations in pharmaceutical production, such as granulation, drying, pelletization, and freeze-drying [127]. It has an important role in the transition from batch processes towards continuous pharmaceutical production processes. In one PAT example by De Beer *et al.* (2009), Raman spectroscopy was combined with NIR to monitor freeze-drying of formulations containing mannitol [128]. Unlike Raman spectroscopy, NIR spectroscopy is very sensitive to water, and it provided data for evaluating the drying steps and release of hydrate water. On the other hand, Raman spectroscopy gave more chemically and solid-state specific peaks, which revealed ice formation and crystallization of the solid-state form of mannitol (α, β, or δ-anhydrate, or hemi-hydrate), and the conversion from mannitol hemi-hydrate to δ-mannitol during the secondary drying step. When combined, these complementary techniques allowed mutual confirmation and continuous control of all critical process aspects, including the process end-point, as well as solid-state characterization of the end product.

A drawback of Raman spectroscopy is photoluminescence, including fluorescence, which occasionally hinders the applicability of Raman analysis of pharmaceuticals, because Raman scattering is a weaker phenomenon. As described in Section 2.9, APIs can be fluorescent in a solid-state specific manner, such as the amorphous and crystalline forms of indomethacin [22], or different crystal forms of piroxicam. In addition, other materials encountered with pharmaceuticals, including excipients (e.g., cellulose derivatives) or culture media components in bioprocesses can be the source of fluorescence, resulting in complications especially for process monitoring applications. There are several methods for rejecting fluorescence, including spectral processing by baseline removal and the use of a longer NIR wavelength (e.g., 785 nm or 1064 nm) for excitation [129]. An additional, more recently available option for the analysis of fluorescent materials is to utilize time-resolved methods, which exploit the different average lifetimes of Raman-scattered and fluorescence photons: Raman scattering occurs practically instantaneously during excitation, while fluorescence lifetimes are generally significantly longer (on the nanosecond scale) [130]. By using a pulsed laser for excitation, and by synchronizing the laser pulses with time-resolved detection, it is possible to temporally resolve the Raman and fluorescence signals, and reject a significant portion of the fluorescence photons. Time-resolved Raman measurements (using a picosecond pulsed 532-nm laser coupled with a CMOS-SPAD detector as one example) can be used to record fluorescence-suppressed Raman spectra of pharmaceuticals (Figure 2.9) [22, 131]. Such instrument-based fluorescence rejection can also provide better quantitative results compared with baseline-removal procedures, when analyzing solid-state form mixtures of fluorescent pharmaceuticals (piroxicam) [131]. Technological advances particularly in the field of CMOS-SPAD detectors have resulted in the recent availability of more affordable time-resolved Raman instruments, which are also more applicable for process environments (due to their compact size). The technique shows promise for diverse applications in the pharmaceutical and biopharmaceutical research, manufacturing and PAT settings.

When considering non-spatially resolved bulk analysis, such as monitoring solid-state transformations during manufacturing processes, it is preferable to have the entire product represented in the collected Raman signals. Sub-sampling can be problematic when analyzing mixtures – and mixtures of solid-state forms (as well as different chemical entities), rather than full transformations, are common. Also, for in-line quality control of tablets, it is desirable to detect the entire composition of the tablet before batch release. With traditional backscattering Raman setups, the signal originates mainly from the tablet surface, which might not be consistent in composition with the core of the tablet. Even though analysis of tablet surfaces is highly important in some settings, for determining the concentration and

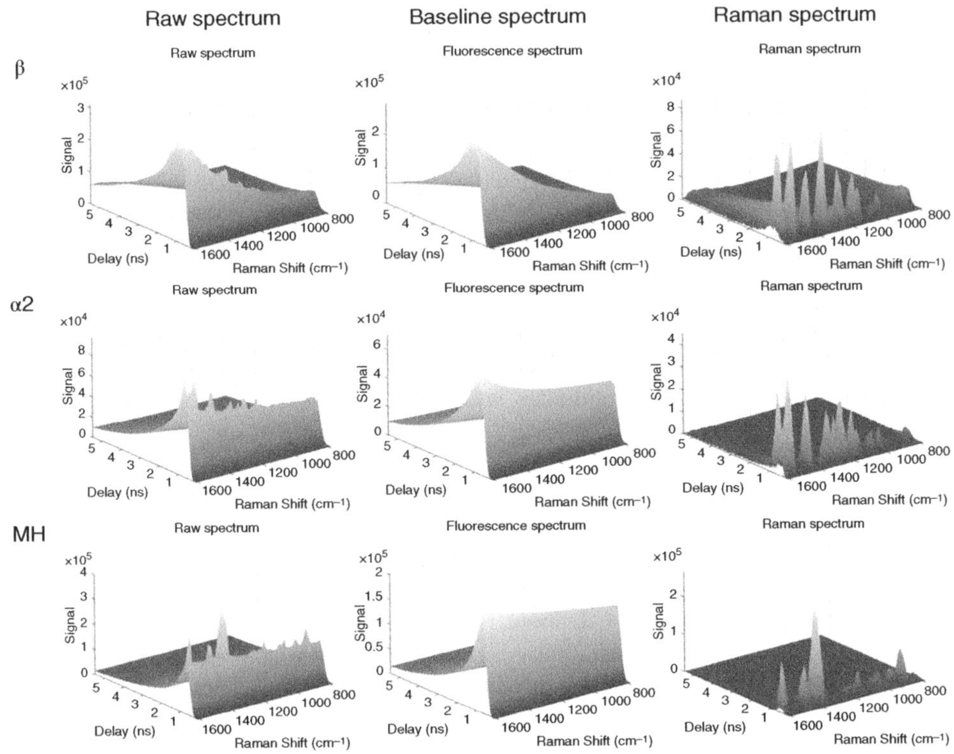

Figure 2.9 *Time-resolved spectra of piroxicam solid-state forms (β and α2 anhydrates, and the monohydrate, MH), measured with a time-gated Raman instrument using a picosecond pulsed 532-nm laser coupled with a CMOS-SPAD detector. The spectra show the detected photon intensities in the Raman shift dimension (between 800–1600 cm^{-1}) and in the time dimension (delay from laser pulse between 0–5.5 ns). The raw spectra contain all detected photons, including Raman scattered and fluorescence photons. The baseline spectra represent the fluorescence signals. The Raman spectra are the spectra after fluorescence rejection, presenting mainly Raman scattered photons. Adapted with permission from Lipiäinen et al. [131], copyright 2018, American Chemical Society.*

solid-state form of the API in a tablet or capsule, it is preferable to probe a larger volume of the tablet. With this aim, sampling probes with large sample spot sizes have been developed, by using several optical fibers in the probe head for illuminating and collection of the scattered light from a larger sample volume.

Another approach for bulk content analysis is transmission Raman spectroscopy, which is a spin-out from the development of spatially offset Raman spectroscopy (SORS) where the laser beam and the Raman collection zone are spatially separated [132]. In transmission Raman, the laser beam (with a large spot size) illuminates a large area and traverses the sample, and scattered light is collected from the opposite side. This suppresses subsampling especially in the *z*-direction (depth) and allows more representative signals of turbid samples such as tablets and capsules. A major pharmaceutical application of transmission

Raman spectroscopy is drug content uniformity determination in tablets and capsules [132, 133], and it is gaining regulatory approval for batch release testing [134]. However, transmission Raman spectroscopy is also suitable for quantification of different solid-state forms in solid dosage forms [135, 136].

2.7.3 Raman Imaging

Raman imaging (most commonly in mapping mode) has been used relatively extensively to investigate the spatial distribution of different components and solid-state forms in diverse particulate samples, including solid dispersions, powders, and tablets [27, 109]. Tres *et al.* (2014) demonstrated the potential of confocal Raman mapping for real-time monitoring of dissolution behavior of amorphous solid dispersions [137]. Their samples consisted of felodipine (5% and 50%) molecularly dispersed in a polymeric matrix of poly(vinylpyrrolidone-co-vinyl-acetate) (copovidone VA64). The extruded material was compressed into circular compacts that were analyzed during dissolution in a flow-through cell. First, eight Raman maps were acquired over a 70 min time-frame for the 5% felodipine sample and over 1705 min for the 50% felodipine sample. Each spectral map included 121 spectra (500 × 500 μm) and took 5 min to record. To obtain spatially resolved data with better resolution from the 50% extrudate compact, the dissolution experiment was repeated so that seven maps including 1156 spectra between 1100 and 1750 cm^{-1} (500 × 1000 μm) were recorded. This required 1.5 h. False color maps were formed from the spectral datasets using multivariate curve resolution (MCR). With the 50% felodipine compacts, the copovidone peak signal at 1426 cm^{-1} decreased after 65 min of dissolution, while the felodipine peak at 1497 cm^{-1} increased until the spectra of the compacts resembled that of amorphous felodipine. The authors attributed this to dissolution of the copovidone from the solid dispersion, leading to felodipine-rich amorphous material accumulating on the surface. However, after 522 min, the felodipine spectra matched that of crystalline felodipine. The Raman maps obtained at different time points revealed the surface area of crystalline felodipine increased drastically between 316 and 522 min. The distribution data showed that crystallization of felodipine did not occur homogeneously on the tablet surface. In addition, spectra were recorded in the phonon mode range of 30–400 cm^{-1}, which is extremely sensitive to different polymorphic forms and is especially useful to distinguish between amorphous and crystalline forms. The spectral data from this region was used to confirm the crystallization of felodipine after 522 min of dissolution.

Another example of Raman imaging of solid-state forms of particles on tablet surfaces is a study by Nakamoto *et al.* (2013) [138]. They mapped tablet surfaces containing ethosuximide (ESM) with different crystalline ratios, and excipients including lactose monohydrate, microcrystalline cellulose, hydroxyl-propyl-cellulose, and magnesium stearate. After recording spectra of individual components, they applied direct classical least squares (DCLS) to the baseline corrected and normalized data obtained from tablet surfaces, and crystalline and amorphous ESM were successfully mapped on the tablet surface (Figure 2.10). This study highlights that, in addition to differentiating solid-state forms, drug particles dispersed in excipients can be easily detected based on Raman scattering, since the drug molecules commonly contain strongly Raman scattering structures

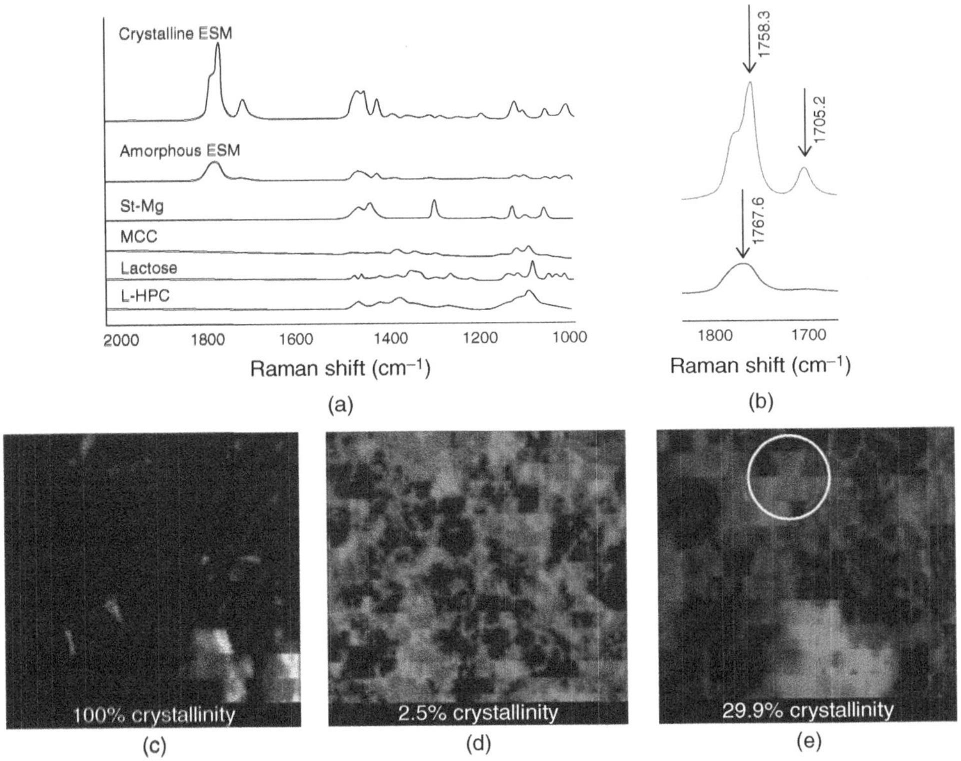

Figure 2.10 *(a) Raman spectra of crystalline ESM, amorphous ESM and excipients, (b) Raman spectra of crystalline and amorphous ESM from a shorter region exhibiting distinguishable peaks of both of the solid-state forms, and (c–e) Raman images of tablet surfaces containing different percentages of crystallinity. The circle in (e) indicates a region that maps crystalline ESM amongst amorphous ESM on the tablet surface. Adapted with permission from Nakamoto et al. [138], copyright 2013, Elsevier.*

(e.g., aromatic rings or double bonds), in contrast to most excipients. However, this discrepancy in Raman scattering efficiencies can lead to the more Raman-active molecules being over-represented in Raman mapping and imaging results.

One potential benefit of confocal Raman mapping is the capability to obtain chemical- and solid-state-specific images in 3D. For example, Ward *et al.* (2005) performed depth profiling with sorbitol [139]. They used a localized heat source to produce micron-sized amorphous domains on the surface of a sorbitol layer. These domains were imaged in 3D using peak intensity (850–900 cm^{-1}) and peak width. However, it is important to keep in mind the limitations of *z*-profiling due to the sources of error described in Section 2.7.1 above. This is an aspect that many publications involving *z*-profiling fail to acknowledge.

Despite it being comparatively time-consuming, Raman mapping is perhaps the best-established vibrational spectroscopic imaging method in solid-state analysis, perhaps due to its superior spatial resolution. As mentioned above, it also allows, to a certain degree, 3D imaging with confocal Raman microscopes. Probably the biggest drawback of Raman imaging/mapping based on spontaneous Raman scattering is the sometimes painfully slow data acquisition, since only approximately one out of 10^6 to 10^8 photons interacting with the sample undergoes Raman scattering. Improved imaging speed can be achieved using nonlinear coherent Raman techniques, such as stimulated Raman scattering (SRS) and coherent anti-Stokes Raman scattering (CARS), which are described below.

2.8 Nonlinear Optics

2.8.1 Principle

Nonlinear optical processes are multiphoton processes in which multiple photons simultaneously interact with each other in a sample to create a photon with a different energy that is then detected (Figure 2.2). Nonlinear processes only occur efficiently with sufficiently powerful irradiation. This can be achieved by using ultra-fast pulsed lasers with pulse durations of picoseconds or femtoseconds. Coherent nonlinear techniques include the Raman scattering-based techniques, coherent anti-Stokes Raman scattering (CARS) [140] and stimulated Raman scattering (SRS) [141], as well as sum-frequency generation (SFG), including second and third harmonic generation (SHG/THG) [142]. Two-photon excited fluorescence (TPEF) is a non-coherent second-order nonlinear optical phenomenon [143].

When electromagnetic radiation interacts with matter, the induced polarization can be described by the following power series expansion:

$$\boldsymbol{P} = \varepsilon_0\chi^{(1)}\boldsymbol{E} + \varepsilon_0\chi^{(2)}\boldsymbol{E}^2 + \varepsilon_0\chi^{(3)}\boldsymbol{E}^3 + \dots \tag{2.3}$$

where $\varepsilon_0$is the vacuum permittivity, $\chi^{(1)}$, $\chi^{(2)}$ and $\chi^{(3)}$ are the first-, second- and third-order electric susceptibilities, respectively, and $\boldsymbol{E}$ is the electric field. χ is a dimensionless proportionality unit (tensor) that describes the ease of polarization of a dielectric material in response to an applied electric field. In spontaneous Raman scattering it is the first term, $\varepsilon_0\chi^{(1)}\boldsymbol{E}$, that describes the polarization, and therefore spontaneous Raman is a linear method in this respect. When more energy is introduced to the system, the higher-order terms become significant. $\varepsilon_0\chi^{(2)}\boldsymbol{E}^2$ describes the second-order nonlinear interactions, including SHG, and $\varepsilon_0\chi^{(3)}\boldsymbol{E}^3$ the third-order nonlinear interactions, such as CARS and THG. These transitions are depicted in Figure 2.2.

CARS and SRS achieve chemically specific imaging by probing Raman-active molecular vibrations. CARS is a four-wave mixing process in which three ultra-fast laser pulses are tightly focused through a high NA objective and coherently drive molecular vibrations in the small focal volume [144–147]. In this way, resonant molecular vibrations are stimulated and imaging can be very fast (video-rate) [148]. Compared with spontaneous Raman imaging, CARS imaging is at least 100 times faster [149]. The photons (and lasers) involved in the CARS process are denoted Stokes (ω_S), pump (ω_p) and probe (ω_{pr}). In a typical narrow-band setup, the probe beam originates from the same source as the pump beam and has the same wavelength. The wavelength of the Stokes beam or pump/probe beam is tunable and a specific vibrational mode can be stimulated when the energy difference between the Stokes and pump beams corresponds to that specific vibrational frequency. This stimulated vibration is then probed with a probe photon. The resulting anti-Stokes signal (ω_{aS}) can be detected in the forward (f-CARS) or backward (epi-CARS) direction using a photomultiplier tube (PMT) or charge-coupled device (CCD) detector. Stokes and pump laser beams are also used in SRS, but the signal detection scheme differs from CARS. In SRS, when the energy difference between the Stokes beam and pump beam corresponds to some Raman-active molecular vibration, the pump beam loses intensity (stimulated Raman loss, SRL) and the Stokes beam increases intensity (stimulated Raman gain, SRG) [150]. Usually the Stokes beam is modulated, for example using an acousto-optic modulator (AOM), and the change in pump beam intensity can be detected using radio-frequency lock-in detection.

CARS imaging can be divided into two categories: narrowband (single-shift) [148] and broadband (multiplex) [151–153]. Narrowband CARS imaging utilizes laser beams with narrowband picosecond pulsed lasers. The energy difference between the pump and Stokes beams can be accurately tuned to match to a selected molecular vibrational frequency. This method is especially useful when spectral overlap between CARS active vibrations is not an issue and is very rapid: video-rate imaging is possible. Broadband CARS utilizes spectrally broader laser pulses that, for example, femtosecond lasers can offer. Such spectrally broad beams can be used to simultaneously excite many molecular vibrations that are then probed with a picosecond laser source. In this way, whole spectra are recorded pixel-by-pixel. Though more complex in its setup, and slower, this method is more suitable for resolving different species when there are spectrally overlapping bands from a sample. A third option is to use a narrowband CARS microscope for hyperspectral imaging. This is achieved when the wavelength of one of the lasers is systematically tuned to achieve a spectrum. If the number of wavelengths recorded is limited, this approach can be even faster than using broadband CARS; however, when the spectral region of interest is broad, broadband CARS is more efficient.

In contrast to CARS and SRS, second-order SFG (including SHG) signals do not originate from molecular vibrations. Structures that lack inversion symmetry (are non-centrosymmetric) can produce these signals with sufficiently intense laser irradiation. For example, in biomedical applications collagen can be detected using SFG or SHG [154]. In pharmaceutical applications, SFG is especially useful in solid-state analysis, since many pharmaceutical crystals are non-centrosymmetric and can generate these signals, whereas amorphous materials are not SFG/SHG active [155]. Nonlinear optical imaging based on SFG/SHG signals is thus useful when crystallinity and crystallization of particles is of interest, such as in powders [156]. SFG/SHG imaging is also useful for detecting very early stage crystallization and low-level crystallinity on particle and dosage form surfaces [156, 157].

In the SFG process, incoming photons are frequency-summed and the wavelength of the emitted SFG photon can be calculated as

$$\lambda_{\mathrm{SFG}} = \frac{1}{\frac{1}{\lambda_1} + \frac{1}{\lambda_2}} \tag{2.4}$$

where λ_1 and λ_2 are the irradiation wavelengths. In SHG, a special case of SFG where the two incident photons have the same wavelength, the wavelength of the emitted SHG photon is exactly half the wavelength of the incoming photons (double the energy) (Figure 2.2).

Another advantage of these nonlinear optical imaging methods, besides imaging speed [148, 158], chemical-selectivity [159] and sensitivity [156], is inherent confocality since the interacting photons must combine coherently in time and space [143]. Thus, signal generation is restricted to a small focal volume (lateral spatial resolution less than a micron) and the focus can be scanned across the sample in three dimensions to generate high-resolution 3D images [160]. The NIR wavelengths used in nonlinear imaging also penetrate deeper into many samples than visible and MIR light, and water has negligible or limited interference. Of importance to biomedical applications, including the imaging of nanoparticles in cells and tissues, the NIR wavelengths are also less phototoxic than shorter wavelengths [145]. A factor to consider in solid-state analysis is that the coherent non-linear signal is polarized, and thus the nonlinear signal can be affected by crystal orientation [161].

2.8.2 Nonlinear Optics Applications

While most recent nonlinear optical applications for pharmaceutical solid-state analysis have involved imaging, the first application of nonlinear optics for pharmaceutical analysis did not. Strachan *et al.* used non-spatially resolved SHG to analyze pharmaceutical microparticle mixtures [162–165]. A custom-built system involving a Nd:YAG laser (1064 nm, 60 mJ pulses, 10 ns duration, repetition rate of 10 Hz) was developed. In their first setup, the beam passed through a half-wave plate, thin-film polarizer and a dichroic filter prior to irradiating the powder sample, and the reflected light was directed through an imaging lens and a 10 nm bandpass filter centered on 532 nm (to reduce the TPEF signal) and onto a grating spectrometer [164]. The signal was then detected by a PMT detector and the fluorescence signal was further filtered using a box car integrator.

After considering the theoretical basis of optical nonlinearity of crystalline particles, the authors used SHG to quantify crystallinity and polymorphism in pharmaceutical powder mixtures. Proof-of-principle was demonstrated with mixtures of glass and quartz, and then more pharmaceutically relevant systems, including the API, enalapril maleate, and polymeric excipient, polyvinylpyrrolidone (PVP), were investigated [164]. Glass, being amorphous, is SHG-inactive, whereas crystalline quartz is SHG-active. The experimental data fit the theoretical treatment of bulk powder SHG well, with the effects of both particle size and the number of particles taken into account. Two crystalline polymorphs of enalapril maleate (forms I and II) were also analyzed. Both crystal forms are SHG-active, but form II is more SHG-active, allowing quantification. They also quantified enalapril maleate form II and PVP powder mixtures. The limits of detection and quantification were as low as 0.12% w/w and 0.41% w/w, respectively. The analytical setup was later optimized with computer-automated analysis, improved speed, and consideration of particle orientation effects [163]. Overall, these first forays into pharmaceutical solid-state analysis using nonlinear optics showed that SHG could be used reproducibly and sensitively to quantify crystallinity and polymorphism in powder mixtures and tablets. This suggests that pharmaceutical production could benefit from the technology as a PAT tool for solid-state control.

2.8.3 Nonlinear Optical Imaging

More recently, SFG and SHG have been exploited in imaging to obtain spatially resolved differentiation of crystalline and amorphous particles, as well as some polymorphic crystalline forms (based on the presence/absence of centrosymmetry) [156, 157, 161, 166–169]. Prof. Simpson and his group at Purdue University have been particularly active in using

SHG imaging in pharmaceutical applications, especially for characterizing powders [156, 161, 166]. The imaging setup consists of a femtosecond pulsed laser, which produces high peak powers for efficient generation of SHG signals. One important application has been the detection of the early stages of protein crystallization as demonstrated with green fluorescent protein (GFP) crystals [168]. SHG was used to detect crystalline GFP protein particles with lower background signals than with TPEF. Simpson *et al.* have also used SHG imaging to quantify crystallinity within nanosuspensions formulated with albumin-bound paclitaxel (PTX) [167]. SHG-active crystals were located, and subsequent confocal Raman analysis of these crystals allowed them to be identified as PTX. This method showed batch-to-batch variability and revealed water-insoluble PTX crystals up to 120 μm in size. SHG imaging was also successfully used to image trace crystallinity in griseofulvin powders with a detection limit of 4 ppm [156]. These studies demonstrate one of the key benefits of nonlinear optical imaging based on SHG compared with many bulk powder analysis methods (including XRPD and spectroscopy): extremely low levels of crystallinity can be efficiently detected due to the inherent confocality combined with video-rate imaging. This facilitates early-stage crystallization detection as well as analysis of fast crystallization kinetics.

As demonstrated, SHG is highly sensitive and promising for crystallinity and polymorphism detection. However, the univariate nature of the signal limits its potential for solid-state analysis of complex systems. Additional specificity can be achieved by coupling the method with vibrational spectroscopy methods such as Raman spectroscopy. Chowdhury *et al.* (2017) used SHG microscopy to visualize powder mixtures of clopidogrel bisulfate polymorphic forms I and II, and based on the differences in SHG activity (form I exhibiting ~250 times higher SHG than form II), the different polymorphic particles could be preliminarily classified [161]. The SHG images can guide the more solid-state specific but slower Raman and XRPD measurements to particles exhibiting strong or weak SHG activity, as demonstrated in Figure 2.11 with Raman spectroscopy.

An additional limitation of SHG arises from the fact that only non-centrosymmetric (chiral) crystals can produce SHG signals. Many pharmaceutical crystals are centrosymmetric. Coherent Raman imaging, including CARS and SRS, can be also be used to efficiently distinguish different solid-state forms of particles [149, 157, 170]. The most important benefit of coherent Raman imaging over spontaneous Raman imaging is speed. Hartshorn *et al.* (2013) used broadband CARS microscopy to detect different indomethacin solid-state forms on tablet surfaces [149]. Tablets consisted of the α and γ forms of indomethacin at different ratios, as well as lactose monohydrate, microcrystalline cellulose, croscarmellose sodium and magnesium stearate. The α and γ forms could be distinguished on the tablet surfaces. They also observed some unexpected spectral profiles on the tablet surfaces. These corresponded to amorphous indomethacin, presumably formed during the tableting process. The authors highlighted the benefit of using broadband CARS to rapidly record multiple spectra (in this case between 800 and 3400 cm^{-1}) pixel-by-pixel. A dwell time of 100 ms was enough to obtain images with sufficient signal-to-noise ratio. The imaging was approximately 100 times faster than with spontaneous Raman mapping. The measured broad spectral information facilitated detection of unexpected components – in this case amorphous indomethacin.

Novakovic *et al.* (2017) used simultaneous multimodal nonlinear imaging to visualize γ, α and amorphous indomethacin particles on tablet surfaces [157]. Very rapid narrowband CARS and simultaneous SFG/SHG imaging were used to visualize the different solid-state

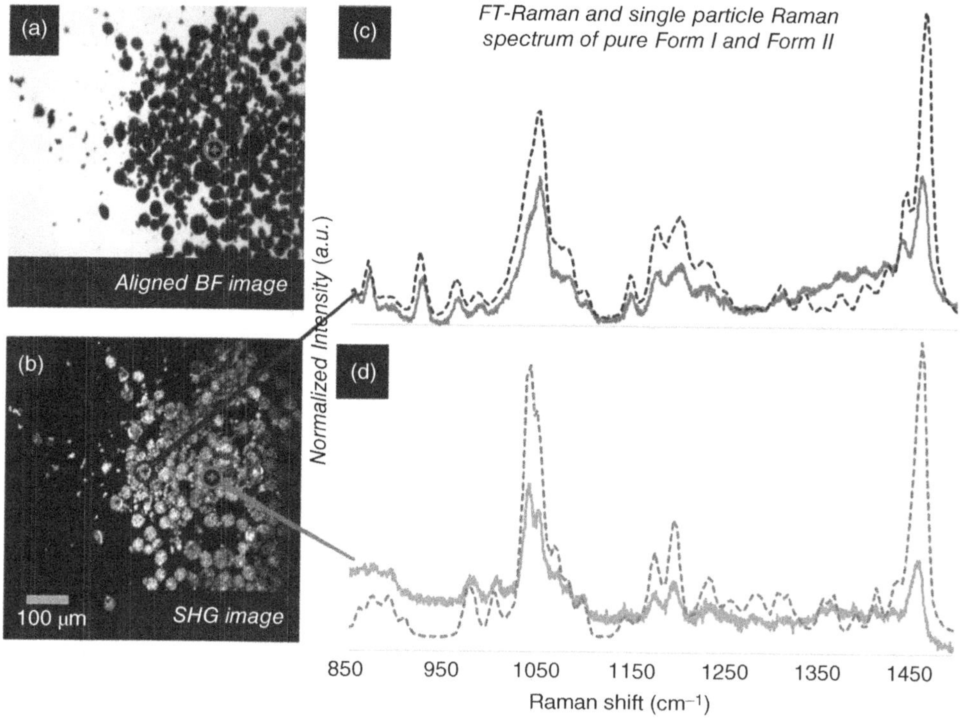

Figure 2.11 *Crystal form classification by independent SHG-guided Raman analysis on a per-particle basis. Parts a and b correspond to bright field and SHG images of a 10% mixture of Form II in Form I, respectively. Parts c and d correspond to single particle Raman spectrum and FT-Raman spectrum, respectively, of Form I and Form II. FT-Raman spectra of both forms are shifted a few wavenumbers to the right to match the spectra feature collected from single particle measurements. In both cases, Raman spectra support the preliminary classifications made by SHG microscopy. Reprinted with permission from Chowdhury et al. [161], copyright (2017), American Chemical Society.*

forms (the pixel dwell time was 1.2 μs) by exploiting the solid-state specific CARS shifts at 1701 cm^{-1} and 1676 cm^{-1} for the centrosymmetric γ and amorphous forms respectively, and the SFG/SHG signal for the non-centrosymmetric α form. Furthermore, surface crystallization could be probed using this approach, with amorphous indomethacin crystallizing to the γ form at 30°C/23% RH and to the α form at 30°C/75% RH. The nonlinear imaging allowed earlier detection of crystallization than more established solid-state methods (FTIR-ATR and spontaneous Raman microscopy). Single particles of minority solid-state species could also be detected.

In the same study, hyperspectral CARS imaging was also used to image the solid-state distribution on the tablet surfaces and the results were compared with those obtained with the narrowband CARS/SFG imaging, described above. This was achieved by systematically tuning one of the laser beams over the spectral region 1413 to 1800 cm^{-1}. PCA was applied to the spectral data and pixels were RGB colored based on their score values for the

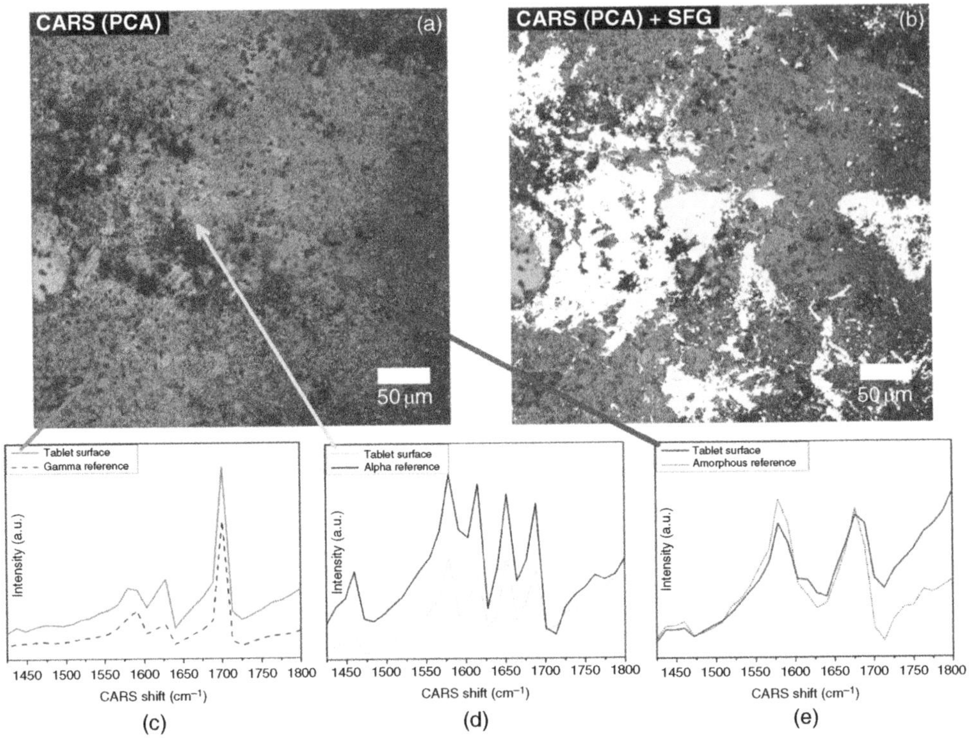

Figure 2.12 *(a) PCA-based CARS image and (b) overlaid PCA-based CARS image and SFG image of the surface of a tablet containing equal ratios of gamma, alpha, and amorphous indomethacin. CARS spectra extracted from the red, green, and blue regions are shown with arrows (c–e). The reference spectra of the different solid-state forms of indomethacin in powdered forms are shown for comparison (c–e). The RGB image is generated from a CARS spectral scan in the region 1413–1800 cm^{-1}, using the score values of the first three principal components. The SFG image is obtained from pixels generating SFG (480–490 nm) (light areas, b). Reprinted with permission from Novakovic et al. [157], copyright 2017, American Chemical Society.*

first three principal components (PCs). The three different solid-state forms of indomethacin were clearly distinguished using this approach (Figure 2.12). The area of CARS signal from α indomethacin matched that with SFG imaging. Overall, this study shows that multimodal nonlinear optical imaging is a sensitive and specific imaging approach for solid-state analysis. Narrowband CARS imaging with simultaneous SHG/SFG imaging is especially fast. Hyperspectral CARS imaging requires more time, but the spectral data can be especially beneficial if the component CARS signals overlap, making narrow-band CARS difficult or impossible. Synergistic multimodal nonlinear optical imaging can provide additional specificity and improve data interpretation confidence.

Solid-state changes of particles in tablets during dissolution have been visualized using CARS microscopy [170, 171]. The imaging modality has also been combined with *in situ* UV absorption spectroscopy for solution concentration analysis [170]. CARS microscopy revealed the theophylline anhydrate conversion to the monohydrate, and UV absorption spectroscopy revealed the dissolution rate decreased as a result of this conversion.

Hyperspectral 3D CARS microscopy has also been used to image drug distributions in mesoporous silica microparticles [172]. CARS microscopy has better 3D resolution than Raman mapping, and can also confirm or rule out the presence of drug material outside or on the surface of microparticles. Furthermore, the CARS spectra can be used to confirm the solid state form of the drug (inside and outside of the silica particles). Silica is optically transparent at the wavelengths commonly used for nonlinear optical imaging, which facilitates 3D analysis of the drug distribution within these particles.

In general, nonlinear optical imaging has much potential in solid-state analysis of particles in several environments such as powders, suspensions, and tablets. While SFG/SHG is sensitive to non-centrosymmetric crystals, CARS can provide additional chemical- and solid-state specificity by probing molecular vibrations that differ between solid-state forms. Potentially, the TPEF signal can also be harnessed for solid-state analysis. These features, combined with the potential for fast (3D) imaging, means that nonlinear optics is expected to gain attention rapidly for solid-state pharmaceutical analysis, as instruments are developed and commercialized.

2.9 Fluorescence Spectroscopy

2.9.1 Principle

Fluorescence is a phenomenon in which an atom or a molecule absorbs light at a particular wavelength and subsequently emits light of a longer wavelength after a brief interval, termed the fluorescence lifetime [173, 174]. As a molecule absorbs a photon of appropriate energy, a valence electron is promoted from the ground state to some vibrational level in one of the electronic excited singlet states (Figure 2.2). This process is extremely rapid, in the order of femtoseconds. After light absorption the excited state relaxes to the lowest vibrational level ($v' = 0$) of the first electronic excited state via internal conversion within picoseconds. This state is the starting point for fluorescence emission to the ground state and to the non-radiative decay routes (internal conversion and intersystem crossing). Fluorescence takes place on the nanosecond timescale. Fluorescence always originates from the lowest vibrational level of the first electronic excited state, irrespective of electronic energy level to which the molecules are initially excited. As a result, the emission spectrum is shifted to longer wavelengths than the excitation spectrum. The greater this shift, the Stokes shift, the easier it is to separate the excitation light from the emission light. In practice, the Stokes shift is measured as the difference between the maximum wavelengths in the absorption and emission spectra. The size of the shift varies with molecular structure, from just a few nanometers to over several hundred nanometers. In order to determine the emission spectrum of a particular fluorophore, it should be excited at a wavelength 10 nm shorter than the beginning of the fluorescence spectrum. Simultaneously, a sufficient number of excited molecules to enable the detection of the emitted light must be created. For fluorophores with a large Stokes shift, such as ethidium bromide (Figure 2.13a), the wavelength of maximum absorption can be utilized. However, for many fluorophores, such as fluorescein (Figure 2.13b), the Stokes shift is small and there is an overlap between the higher wavelength end of the absorption spectrum and the lower wavelength end of

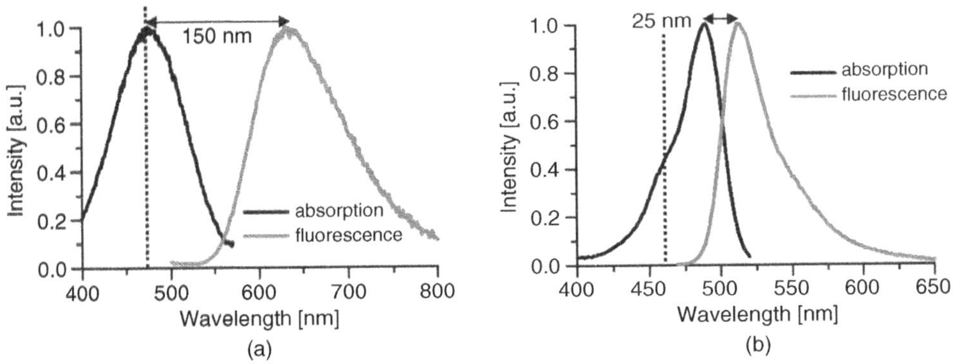

Figure 2.13 *Absorption and fluorescence spectra for (a) 20 µM ethidium bromide in MES-HEPES buffer with pH 7.4 and (b) 4.2 µM fluorescein in phosphate buffer with pH 8.0. The dashed line indicates the excitation wavelength used.*

the emission spectrum. Thus, the wavelength of maximum absorption cannot be used as the excitation wavelength. Excitation at a lower wavelength with lower absorbance must be utilized, leading to lower fluorescence intensity.

An excitation spectrum can be generated when the full absorption spectrum of a fluorophore is scanned, while recording the emission intensity at a single wavelength (usually the wavelength of maximum emission intensity). The result is a graph of the relative fluorescence intensities produced at each excitation wavelength and is thus another way of depicting the absorption spectrum. This is very useful in the case of opaque solid samples.

Fluorescence lifetime is the characteristic time that a molecule remains in an excited state before it returns to the ground state, and is defined as:

$$\tau = \frac{1}{k_f + k_{nr}} \tag{2.5}$$

where k_f is the rate constant of fluorescence and k_{nr} is sum of the rate constants for non-radiative processes (internal conversion and intersystem crossing) leading away from the excited state. During the lifetime of the excited state, a fluorophore can undergo conformational changes as well as interact with other molecules and move according to Brownian movement. The decay of fluorescence intensity for a uniform population of molecules is described by an exponential function:

$$I(t) = I_0 e^{-t/\tau} \tag{2.6}$$

where $I(t)$ is the fluorescence intensity measured at time t, I_0 is the initial intensity observed immediately after excitation, and τ is the fluorescence lifetime. Formally, the fluorescence lifetime is defined as the time in which the initial fluorescence intensity of a fluorophore decays to $1/e$ of the initial intensity. In a uniform solvent, fluorescence decay is usually a one-exponential function. In more complex systems, where the fluorophore is in a mixed set of different environments, multi-exponential behavior is often observed. The same applies to solid-state samples.

Fluorescence emission takes place in all directions, but is measured at a fixed angle and the detector collects only a part of the emission. Light scattering and reflection from the sample, which depend on the position of the sample and the sample itself, also affect the detected fluorescence intensity of the solid samples. Thus, although the measured intensity is proportional to the concentration of the excited fluorophores, the actual concentration cannot be determined from the intensity directly. In order to be able to do quantitative measurements utilizing fluorescence, calibration measurements with known amounts of the fluorophore must be performed. Another possibility for quantitative studies is to utilize a relative change in the fluorescence intensity obtained at equal experimental conditions, for example, as a function of time or an environmental factor such as temperature or relative humidity.

Photobleaching occurs when a fluorophore permanently loses its ability to fluoresce due to photon-induced chemical damage and covalent modification. This can be a serious problem for many pharmaceutically relevant molecules, such as indomethacin. To reduce the degree of photobleaching, the intensity of the excitation light must be reduced by adjusting the excitation slits and/or using a neutral density filter in the excitation light path before the illumination reaches the sample. Sampling time must also be minimized.

Each molecule has an absorption and emission transition dipole moment with defined orientations with respect to the molecular axes of the molecule, and which are separated from each other by an angle α [173, 174]. One condition for light absorption is that the electric component or vector of the electromagnetic wave must be parallel, or close to parallel, to the absorption transition moment of the molecule. The orientations of the absorption transition moments are random in solutions. When a solution sample is excited with linearly polarized light, the excitation will therefore only be efficient for the molecules who are oriented similarly to the direction of polarization. This leads to a very anisotropic distribution of orientations of the excited molecules. This ordering effect is called photoselection. After excitation, the molecules change their orientation according to Brownian rotational motion. In solution, this can be very fast and causes the initial orientational order of the excited molecules to vanish during the fluorescence lifetime. In the solid state, this rotational movement is hindered, giving an opportunity to determine the degree of order in the alignment of the fluorophores by detecting the anisotropy of their fluorescence.

2.9.2 Fluorescence from Solid-state Samples

Fluorescence is a rapid and non-invasive method, which can be applied both for material characterization and for following an ongoing process *in situ*. However, as a very sensitive method, care must be taken to avoid the effects of scattered and reflected light as well as possible background fluorescence. Fluorescence is usually collected at a 90° angle to the excitation direction in order to minimize the presence of scattered excitation light in the fluorescence channel (Figure 2.14). In measuring solid samples, the orientation of the sample towards the excitation light is important in this respect. For transparent samples, fluorescence can be monitored through the sample, thus avoiding most of the scattered light. Adjusting the angle between the excitation light and sample plane to about 60° usually gives the best result [175]. For opaque samples, most of the modern fluorescence spectrophotometers have a front face detection possibility (Figure 2.14). In this case, the angle

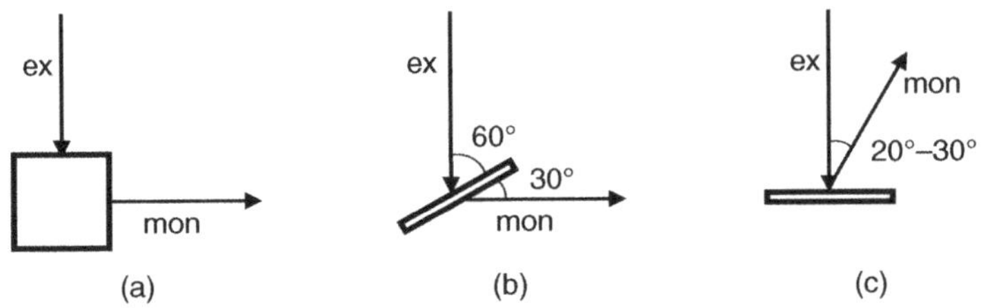

Figure 2.14 Sample orientations for collecting fluorescence (a) in solution measurement, (b) for transparent solid samples, and (c) for opaque solid samples.

between the sample plane and excitation light is 90° and the fluorescence is monitored at about a 20° angle towards the excitation light. It is also useful to apply cutoff filters in front of the monitoring light to avoid scattered excitation light reaching the detector.

In general, fluorescent systems are divided into two broad classes, termed intrinsic and extrinsic. Intrinsic fluorophores are those molecules of interest that also fluoresce naturally. These include aromatic amino acids, neurotransmitters, porphyrins, and some proteins such as the green fluorescent protein. Extrinsic fluorophores are synthetic dyes that are added to a studied sample in order to produce fluorescence with desired spectral properties. For instance, Purohit and Taylor [176] have used environmentally sensitive fluorophores to monitor the kinetics of phase transformations in drug-containing amorphous solid dispersions. In such cases, it is very important to be familiar with the fluorescence properties of the probe in solution to separate, for example, aggregation of the fluorophore from the environmental effects.

2.9.3 Intrinsic Fluorophores in Solid Samples

The close proximity and preferred orientation of molecules within molecular solids, both crystalline and amorphous, lead to intermolecular interactions, which affect the energetics of both the ground and excited states. The exciton theory predicts that the excited state levels of N identical fluorophore molecules form an exciton band of N discrete exciton states (Figure 2.15a) [177] due to dipole–dipole coupling of individual molecular transition moments. Whether the transition to a state is allowed or not depends on the net transition moment of the molecules: zero for forbidden states and >0 for allowed transitions. In 3D solids, in addition to purely allowed or forbidden states, multiple relative orientations between the transition dipole moments of neighboring molecules lead to partially allowed states and broader absorption and fluorescence spectra compared with those in solution.

Comparing the solid-state spectra to those in dilute solution gives an indication of the mutual orientation of the molecules and on the degree of order in the system. As an example, the excitation and fluorescence spectra for indomethacin in solution, and in crystalline and amorphous solid-state forms, are presented in Figure 2.15. In solution, both excitation and emission occur at shorter wavelengths and the excitation spectrum is narrow compared with solid indomethacin forms. Compared with crystalline forms, the amorphous form exhibits more intra- and intermolecular conformational variation and has more vibrational modes, which is seen as broader excitation and fluorescence spectra. More information on the organization of the molecules can be obtained by anisotropy measurements [173, 174].

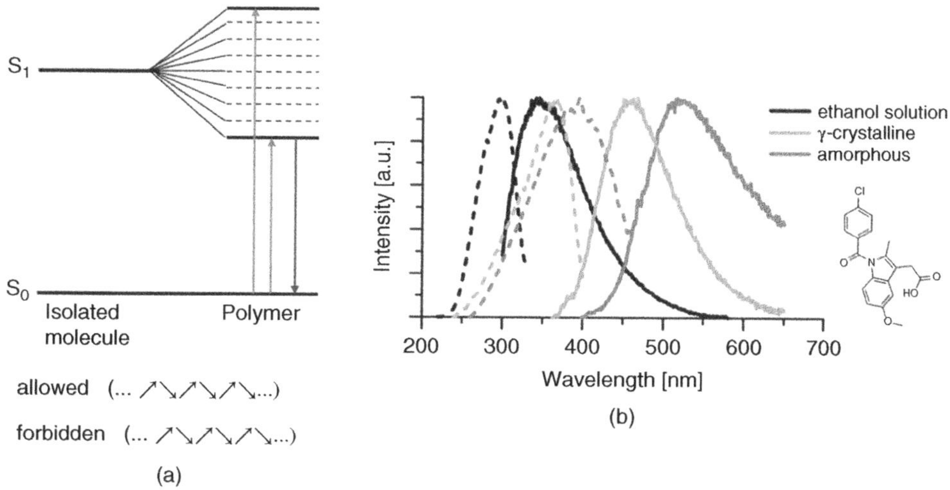

Figure 2.15 *(a) Exciton splitting in a linear molecular polymer with oblique transition dipole moments of the individual molecules. The solid lines indicate the allowed excited states, which can be reached by photoexcitation, and the dashed lines correspond to the forbidden states. Below the energy level diagram, the orientations of absorption transition dipole moments for individual fluorophores are shown for each case [177]. (b) Normalized excitation (dashed line) and emission (solid line) spectra for indomethacin ethanol solution, γ-crystalline and amorphous indomethacin. Adapted from Rautaniemi et al. (2018) [178] with permission from the American Chemical Society.*

The intrinsic fluorophores have also been used to probe protein tertiary structure in the solid state by using tryptophan fluorescence [179, 180], for real-time monitoring of the conversion of amorphous indomethacin to a crystalline form [178, 181], and for determining the amounts of pharmaceuticals in tablet formulations [182–184]. A solid-phase, fluorescence-based protein assay that quantifies proteins in the presence of detergents, urea and reducing agents, and can be applied to microplate readers, has been developed by Agnew *et al.* [185].

2.9.4 Fluorescence Imaging

Fluorescence microscopy is a non-invasive method for studying fluorescent materials that can be readily applied to systems exhibiting either intrinsic or extrinsic fluorescence. In addition, the fluorescence microscope can reveal the presence of fluorescing material with exquisite sensitivity. Although the fluorescence microscope is limited by its spatial resolution to the usual rules governed by numerical aperture and diffraction limits, the presence of fluorescing material can be detected even at sub-resolution quantities (such as a few molecules). Conventional confocal fluorescence microscopy (CFM) has not been able to provide spatial resolution below the diffraction limit, but the recently developed super-resolution fluorescence microscopy techniques have breached this limit [186, 187].

One of the most important applications of fluorescence microscopy is in the field of immunofluorescence. There, a covalent conjugate of a specific antibody and a fluorophore is first produced. The labeled antibody conjugate binds to its antigen, for example, on a test microscope slide or microarray, while the unbound conjugates are easily washed away. The presence of the chemically attached fluorescent conjugate and antigen is demonstrated by exciting the fluorophore, and the system can be easily studied visually or with various detectors and microscopes.

In pharmaceutical solid-state applications, CFM has been used to determine the miscibility of drug–polymer solid dispersions [188–190] and to visualize the distribution of immobilized biomolecules on different platforms, as well as to evaluate restrictions to diffusion within the support [191]. CFM has also been used to study the water-induced phase separation in amorphous solid dispersions, induced during preparation, upon storage at high relative humidity (RH) and during dissolution [192]. Different fluorescent dyes were added to the drug–polymer films and the location of the dyes was evaluated using CFM.

2.9.5 Fluorescence Lifetime Imaging Microscopy

In fluorescence lifetime imaging microscopy (FLIM), the confocal fluorescence microscope is combined with a fluorescence lifetime measurement setup. As a result, in addition to the distribution of fluorescence intensities, FLIM images show the distributions of fluorescence lifetimes in the sample as well. The intrinsic nature of the fluorescence lifetime allows quantitative analysis of the results, since fluorescence lifetimes and their amplitudes are proportional to each other, even in situations when the fluorescence intensities might not be. The signal arises from the nature of the energy transitions (Figure 2.2 and Figure 2.15), which are largely unrelated to the actual signal intensity. For example, if the sample is moved, or the fluorophore is unevenly distributed in a pharmaceutical product,

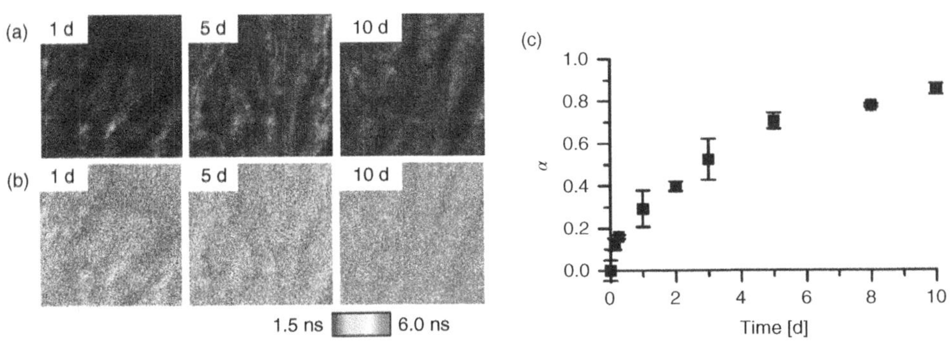

Figure 2.16 *FLIM images (80×80 μm) of the amorphous indomethacin stored at 60°C and 0% RH for 10 days. The same FLIM images are presented with (a) and without (b) intensity modulation. The images were taken at 40× magnification and air immersion. The excitation wavelength was 405 nm, and the fluorescence was monitored at 430–900 nm. The color scale for the fluorescence lifetimes is the same in all the images and is presented at the bottom right. (c) The degree of conversion, α. Adapted from Rautaniemi et al. (2018) [178] with permission from the American Chemical Society.*

the fluorescence intensity changes but the lifetime stays constant. In a recent study, the crystallization kinetics of amorphous indomethacin was monitored by FLIM [178]. All of the studied indomethacin forms (amorphous and γ- and α-crystalline) showed characteristic fluorescence lifetime profiles, making it possible to follow the crystallization process. The conversion to the crystalline form could be estimated from the changes in the fluorescence lifetime (Figure 2.16).

2.10 Solid-state Nuclear Magnetic Resonance

2.10.1 The Basic Theory of NMR Spectroscopy

Since the first observation of proton magnetic resonance in water [193], nuclear magnetic resonance (NMR) has grown from the study of an interesting physical phenomenon to be a most powerful analytical method, with applications ranging from solid-state physics to chemistry and biology. One of the most important applications of NMR spectroscopy is the characterization of pharmaceutical solids [194–199].

In quantum mechanical terms, an atomic nucleus is characterized by a nuclear spin quantum number *I*, referred to as nuclear *spin*. Nuclei can have integral, half-integral or zero spin quantum numbers. Nuclei with a non-zero spin quantum number possess a nuclear magnetic moment and can be observed by NMR spectroscopy. A nucleus can have $2I+1$ possible *spin states* defined by the nuclear magnetic quantum number m_I. Outside a magnetic field ($\boldsymbol{B}_0 = 0$) all magnetic quantum states of a nucleus are equivalent. In the presence of a magnetic field, the spin states have a different energy. The effect of the static field on the magnetic moment can be described in terms of classical mechanics. The magnetic moment has vector properties, that is, it has direction and magnitude. A static magnetic field applied to the nucleus causes its magnetic moment to precess at a rate dependent on the field strength ($\boldsymbol{B}_0$) and the magnetogyric ratio (γ) of the nucleus. This motion is referred to as *Larmor precession*. The rate of the precession as defined by the angular velocity (ω in rad s^{-1} or ν in Hz), known as a *Larmor frequency* of the nucleus, which can be expressed as:

$$\omega_{NMR} = -\gamma B_0 \tag{2.7}$$

or

$$\nu_{NMR} = \frac{-\gamma B_0}{2\pi} \tag{2.8}$$

The nuclear magnetic resonance occurs when the spin changes the state, driven by the absorption of a quantum of energy. This energy is applied as electromagnetic radiation, the frequency of which must match that of the Larmor precession for the resonance to occur. The energy required is:

$$\Delta E = h\nu_{NMR} = \frac{h\gamma B_0}{2\pi} \tag{2.9}$$

where *h* is Planck's constant. The difference in energy between two spins (ΔE) plays an important role in NMR spectroscopy. The energy difference is generally very small, leading to almost identical populations of all spin states and low sensitivity of NMR spectroscopy. As seen in Equation (2.9), increasing the applied field ($\boldsymbol{B}_0$) results in a greater ΔE and hence greater polarization of the spin system (larger differences between populations of the spin states), leading to a stronger NMR signal. Modern NMR spectrometers can currently use magnetic field strengths of up to 23.5 T, which correspond to resonant frequencies of 1000, 251 and 101 MHz for ^{1}H, ^{13}C and ^{15}N nuclei, respectively.

2.10.2 Solid-state NMR Technique

Solution-state NMR spectra usually contain very sharp, highly resolved signals due to rapid molecular tumbling, which averages anisotropic (orientation-dependent) NMR interactions. On the other hand, solid-state NMR (SSNMR) spectra contain broad resonances as the molecular motions are limited and anisotropic interactions are observed in the spectrum. The major causes of line broadening are dipolar couplings, chemical shift anisotropy (CSA) and quadrupolar coupling. Furthermore, the lack of molecular motion often results in long recovery back to an equilibrium of the spin system perturbed by excitation (long values of spin–lattice relaxation time T_1), increasing the time required for the experiment. However, a number of special techniques have been developed to overcome these difficulties.

2.10.2.1 Dipole–Dipole Interactions

Dipole–dipole interactions or dipolar couplings are interactions through space between two or more nuclear magnetic dipoles. The magnetic field of one nucleus will affect the magnetic moment of a neighboring nucleus. In solutions, dipolar couplings are not observed, because they are averaged to their isotropic value, zero, by rapid molecular reorientation. In the solid state, on the other hand, dipolar couplings are usually the major cause of line broadening. For a pair of two different spins I and S in a high magnetic field, the dipolar energy can be expressed as:

$$h^{-1}E_D = -D_{IS}m_I m_S(3\cos^2\theta - 1) \tag{2.10}$$

where θ is the angle between the internuclear vector and the magnetic field, m_I and m_S are magnetic quantum numbers of the nuclei, h is Planck's constant, and D_{IS} is the dipolar coupling constant between spins I and S, which is inversely proportional to the third power of the internuclear distance r_{IS}. It is seen from Equation (2.10) that the dipole–dipole interaction depends on the orientation of the dipoles. In a single crystal with the constant orientation of dipolar vectors, a single set of resonances will be observed. In the case of polycrystalline powders containing particles with random orientations of the internuclear vector, the spectra will be observed as the *Pake doublet*.

2.10.2.2 Chemical Shift Anisotropy

The nuclei of a molecule placed in a magnetic field are magnetically shielded by the surrounding electrons, leading to the well-known phenomenon of chemical shift. The shielding field $\boldsymbol{B}_S$ can be expressed by the equation:

$$\boldsymbol{B}_S = -\boldsymbol{\sigma}\boldsymbol{B}_0 \tag{2.11}$$

where $\boldsymbol{B}_0$ is the applied magnetic field and σ is the shielding tensor. This secondary field contributes to the total field felt by the nucleus, and therefore affects the resonance frequency of the nucleus. The magnitude of the shielding depends on the orientation of the molecule with respect to the magnetic field, and is referred to as *chemical shift anisotropy* (CSA). In a solution, the rapid motion of molecules averages the shielding tensor to a single isotropic chemical shift. The limited reorientation of molecules in the solid state and random orientations of molecules in powder samples leads to a distribution of chemical shifts known as a *powder pattern*. The shielding anisotropy has a similar orientation dependence as the dipolar interaction ($3\cos^2\theta - 1$), where θ is the angle between a molecular axis and the external magnetic field $\boldsymbol{B}_0$.

2.10.2.3 Quadrupolar Coupling

For nuclides with spin quantum numbers greater than ½ (for example, ^2H, ^{14}N, ^{17}O), the electric interaction between the nuclear electric quadrupole moment and the electric field gradient (EFG) at the nucleus leads to the observation of quadrupolar interaction. The EFG originates from the surrounding distribution of electrons and nuclei and can be described similarly as the dipolar coupling by a traceless tensor with an isotropic average of zero. Therefore, the quadrupolar coupling is not normally observable in solutions, although it still significantly contributes to relaxation and thus to line broadening. The quadrupolar effect, on the other hand, often dominates the appearance of solid-state NMR spectra since the quadrupole coupling constant may be hundreds of megahertz.

2.10.2.4 Indirect Coupling

Unlike the direct (through-space) dipole–dipole interaction, the indirect coupling (J-coupling) between the nuclear spins is mediated by electrons and therefore is usually classed as a through-bonds interaction. The determination of J-coupling constants allows for obtaining information on molecular-level connectivity and conformation. For example, the relation between dihedral angle and three-bond indirect coupling, known as the Karplus equation, has been widely used in structural studies of molecules in solution. Indirect coupling is usually less important in solid-state studies, because it often has the smallest magnitude of all NMR interactions in a solid sample.

2.10.2.5 Magic-angle Spinning and High-power Proton Decoupling

All NMR interactions are dependent on the orientation of the molecules. In solution, only the isotropic average of chemical shift and indirect coupling is observed; dipolar and quadrupolar interactions are averaged to zero. In solids, dipolar and quadrupolar coupling and chemical shift anisotropy give rise to line broadening. The magnitude of all NMR interactions depends on:

$$(3\cos^2\theta - 1) \tag{2.12}$$

Spinning the sample about an axis inclined at an angle β to the applied magnetic field (Figure 2.17) may mimic to some extent the molecular reorientation in solution, and therefore allows for significant reduction of the orientation dependence of solid-state NMR spectra [200, 201]. The average of ($3\cos^2\theta - 1$) about the conical path for the vector $\boldsymbol{r}$ is given as:

$$\langle(3\cos^2\theta - 1)\rangle = \frac{1}{2}(3\cos^2\beta - 1) \times (3\cos^2\Theta - 1) \tag{2.13}$$

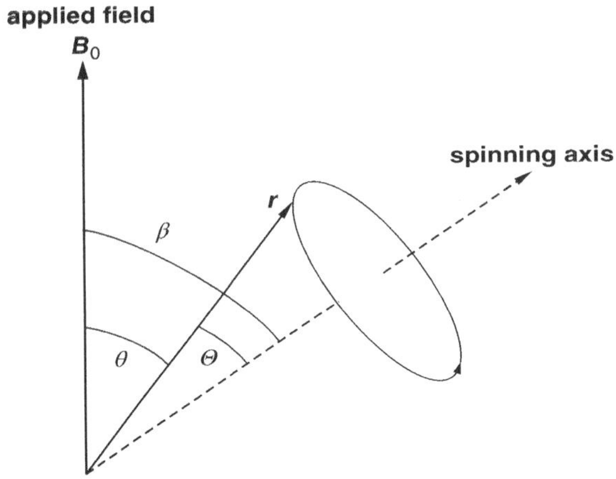

Figure 2.17 Schematic representation of the magic-angle experiment. The sample is spun in a rotor about a spinning axis orientated at the magic angle (β = 54.74°) with respect to the applied magnetic field (B_0) [196].

where β and Θ are defined in Figure 2.17. The contribution from the chemical shift anisotropy and dipolar coupling can be eliminated when (2.12) equals zero. This can be achieved when the value of a spinning angle β is 54.74°, the *magic angle*. Spinning the sample at 54.74° with respect to the applied magnetic field, $\boldsymbol{B}_0$, is termed *magic-angle spinning* (MAS).

The rate of MAS must be equal to or greater than the magnitude of the anisotropic interaction to average it to zero. If the spinning rate is lower than the magnitude of the interaction, spinning sidebands separated by the rate of the spinning become visible (Figure 2.18). To remove these spinning sideband artefacts, which can overlap with other signals and result in confusing spectra, different approaches have been developed. One simple way to remove spinning sidebands is to increase the spinning rate. Speeds of 111 kHz are achievable with rotors of 0.7 mm diameter [202]. However, the limited volume of these rotors leads to a loss of sensitivity. Alternatively, a pulse sequence developed by Dixon known as TOtal Suppression of Spinning Sidebands (TOSS) allows the effective suppression of the spinning sidebands [203].

The homonuclear (^1H–^1H) and heteronuclear dipolar interactions (for example ^{13}C–^1H or ^{15}N–^1H) can be very strong, up to tens of kHz. Thus, dipole–dipole interactions cannot usually be fully removed by the MAS technique. Homonuclear ^1H–^1H dipolar couplings cause extensive line broadening when observing the ^1H nucleus. The most important nuclei observed by SSNMR in pharmaceuticals are ^{13}C and ^{15}N. Homonuclear ^{13}C–^{13}C and ^{15}N–^{15}N interactions are not an issue because of the low probability of two nuclei being in close proximity, given the low natural abundance of these nuclides. However, the problem of heteronuclear $^{13}C(^{15}N)$–^1H dipolar couplings remains. To overcome this interaction, high-power proton decoupling (HPPD) is used [204, 205]. Radio-frequency (RF) pulses with a nutation frequency greater than the dipolar interaction are applied in the proton channel while acquiring in the carbon (nitrogen) channel. The effect of the RF field decouples the proton from the carbon (nitrogen).

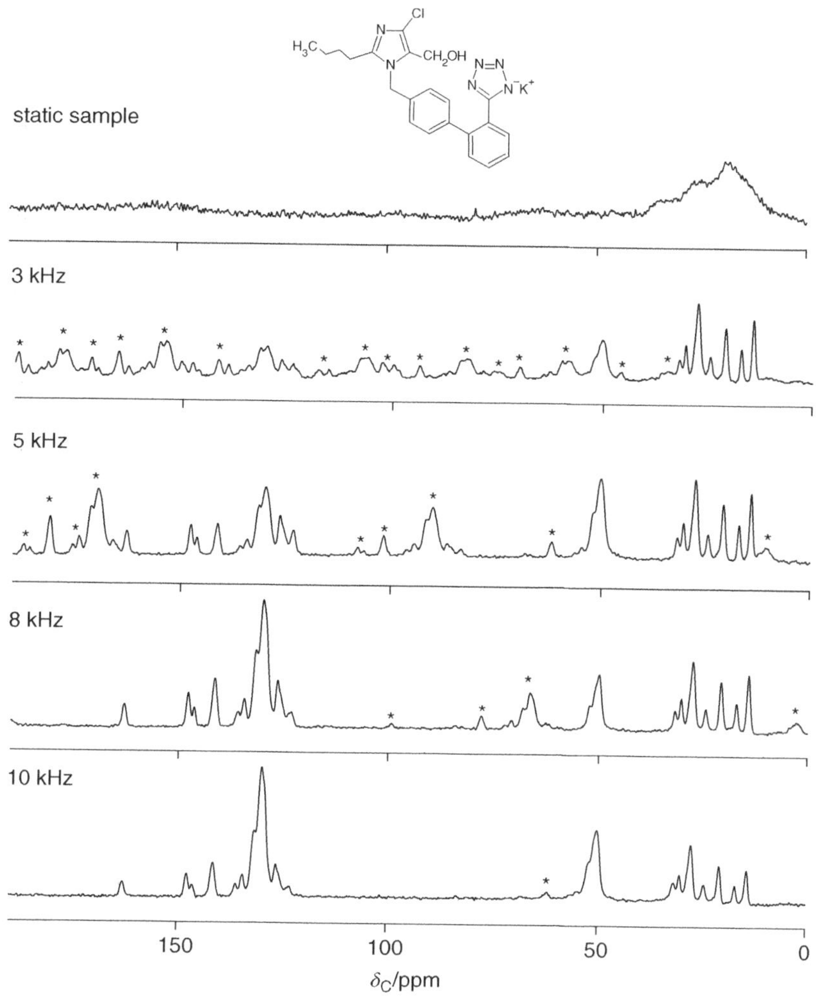

Figure 2.18 Solid-state ^{13}C NMR spectra of crystalline losartan potassium recorded with MAS at different spinning rates from 0 to 10 kHz. All spectra were obtained at 125 MHz with cross-polarization and high-power proton decoupling. 128 scans were acquired for each spectrum. Asterisks (*) denote spinning sidebands.

2.10.3 Solid-state NMR Experiments

2.10.3.1 Sample Preparation

SSNMR does not usually require complex sample preparation. Typically, tens to hundreds of milligrams of a powdered sample is packed into the rotor with outer diameters ranging from 0.7 to 7 mm. The rotors are made from zirconia (ZrO_2). Both powdered samples and entire formulations (tablets) can be packed into the rotor. The drawback of the technique is that the sample's temperature can be strongly affected by frictional heating of the rotor (particularly at high MAS frequencies), which may trigger a phase transition during the experiment.

2.10.3.2 Cross-polarization

Cross-polarization (CP) is a technique that allows the transfer of polarization (magnetization) from abundant nuclei (^{1}H or occasionally ^{19}F, ^{31}P) to dilute nuclei S (e.g., ^{13}C or ^{15}N) [196, 206, 207]. The natural abundance of the most commonly measured nuclides is presented in Table 2.2.

Table 2.2 *The spin properties of commonly used spin-½ nuclides in the pharmaceutical analysis*

Isotope	Natural abundance (x) [%]	Magnetogyric ratio (γ) [10^7 rad s^{-1} T^{-1}]	Relative receptivity vs. ^{1}H (D^p)
^{1}H	99.9885	26.7522128	1.000
^{13}C	1.07	6.728284	1.70×10^{-4}
^{15}N	0.368	−2.71261804	3.84×10^{-6}
^{19}F	100	25.18148	0.834
^{31}P	100	10.8394	6.65×10^{-2}

The relative receptivity gives measures of the intensities to be expected, relative to those for the proton. It is proportional to $\gamma^3 I(I+1)x$ [196].

CP is used to improve the sensitivity and to reduce the waiting times between successive experiments. The signal is enhanced by a factor equal to a magnetogyric factors ratio (γ^H/γ^S), which is 4:1 and 10:1 for ^{1}H and ^{13}C, and ^{1}H and ^{15}N nuclei, respectively. It allows a 16- or 100-fold reduction in the experiment time, respectively, assuming 100% magnetization transfer. The efficacy of cross-polarization depends on the sample nature and the experimental conditions. Furthermore, because the magnetization originates from the ^{1}H nucleus, the recycle delay is limited by the recovery of the ^{1}H magnetization, and not that of the S spin (usually ^{13}C or ^{15}N). Therefore, the pulse delay depends on spin–lattice relaxation $T_1{}^H$ and not $T_1{}^S$. Usually $T_1{}^H \ll T_1{}^S$, and thus the pulse sequence can be repeated faster than in a direct excitation (DE) experiment, leading to a further increase of the signal-to-noise ratio in the spectrum.

In a typical CP experiment, the initial 90° RF pulse applied at the proton frequency rotates the ^{1}H magnetization from the equilibrium (z-axis) into the y-axis. Then the proton magnetization is spin-locked in the y-direction by RF irradiation. During the spin-lock period, another RF irradiation is applied simultaneously at the S frequency. The time when both protons and S nuclei are irradiated is termed *contact time* (CT). To achieve the cross-polarization, ^{1}H and S RF fields must fulfil the Hartmann-Hahn match condition:

$$\gamma^H B_1^H = \gamma^S B_1^S \tag{2.14}$$

where $B_1{}^H$ and $B_1{}^S$ are magnetic field strengths generated by the RF irradiation applied to ^{1}H and the S nuclei and γ^H and γ^S magnetogyric ratios of ^{1}H nucleus and the dilute spin, respectively.

The CP pulse sequence is generally used with MAS and HPPD, which allows high-resolution spectra to be recorded.

2.10.3.3 Heteronuclear Correlation Experiments

To assign complex 1D spectra and/or to obtain further information about the structure of the system, 2D experiments can be performed. For example, heteronuclear correlation (HETCOR) correlates the carbon or nitrogen signals with proton signals [196, 208]. The correlation is obtained when ^{1}H and S nuclei are dipolar coupled; it is, therefore, a through-space correlation. Using HETCOR with a short contact time, a 2D ^{1}H–S dipolar correlation map is obtained with correlations present between S resonances and the protons, to which they are most strongly dipolar coupled, i.e., typically those separated by one covalent bond. If a longer contact time is used, more correlations will appear resulting from longer-range dipolar couplings and ^{1}H spin diffusion. As ^{1}H spins are strongly coupled, obtaining useful resolution in the ^{1}H dimension may be difficult. Usually, homonuclear decoupling is required in order to narrow the abundant spin linewidths.

2.10.4 Pharmaceutical Applications of Solid-state NMR

SSNMR is a non-destructive, powerful technique for obtaining structural and dynamic information on solid pharmaceutical systems [194–199]. This technique can be employed to characterize active pharmaceutical ingredients (APIs) and excipients alone. One of the advantages of SSNMR is that the API usually has different signals than the excipients. Therefore, APIs in the formulated drug products [209, 210], drug carriers such as polymer matrices [211, 212] or mesoporous silica materials [213–216] can also be successfully characterized. SSNMR enables information to be obtained on both crystalline [217, 218] and amorphous materials [195, 199]. In pharmaceutical sciences, SSNMR is mainly used for polymorphic form detection, identification and quantitation [194, 219], structural characterization and molecular dynamics information in amorphous or crystalline phases [199, 220–223], as well as API–API/excipient interactions [211, 213, 214, 224].

Structural characterization of nanoparticles, especially disordered nanocrystalline and amorphous phases, can be challenging [225, 226]. Traditional crystallographic techniques such as X-ray powder diffraction (XRPD), widely and effectively used for characterization of crystalline materials, may be of limited utility for aforementioned nanomaterials characterization, as conventional XRPD measures a crystalline property (i.e., long-range order), which leads to intense and sharp diffraction peaks from the crystalline phase of the materials [225–227]. Also, X-ray beams are diffracted from the surface of the material, which may be an issue for nanoparticles embedded in the matrices [215]. In contrast, SSNMR does not have this limitation and can provide detailed information on the short-range order, hence can be successfully used to characterize nanocrystalline and disordered materials [198, 199, 215, 220, 223]. For example, Yang *et al.* (2014) studied four different model drugs (ibuprofen – IBP, griseofulvin – GSF, fenofibrate – FEN, indomethacin – IMC) embedded into a porous matrix (silica with pore size ~40 nm) [215]. The aim of the work was to develop a novel formulation method to produce nano-sized API nanocrystals while controlling the polymorphic outcome under rigid confinement. ^{13}C CPMAS NMR was used as the main analytical tool for the materials characterization, together with differential scanning calorimetry (DSC) and XRPD as supporting techniques. XRPD analysis showed no crystalline diffraction peaks, suggesting an amorphous nature of APIs embedded in the matrices in all four cases. SSNMR provided information regarding the phase of all APIs embedded in the silica. Spectra of the system with IBP clearly showed resonances arising from the polymorphic form I of the API (79%) and some additional signals (1' and 4', see Figure 2.19b), which, the authors speculated, could have arisen from an unknown IBP polymorphic form (not confirmed by DSC) or from an interaction with silica.

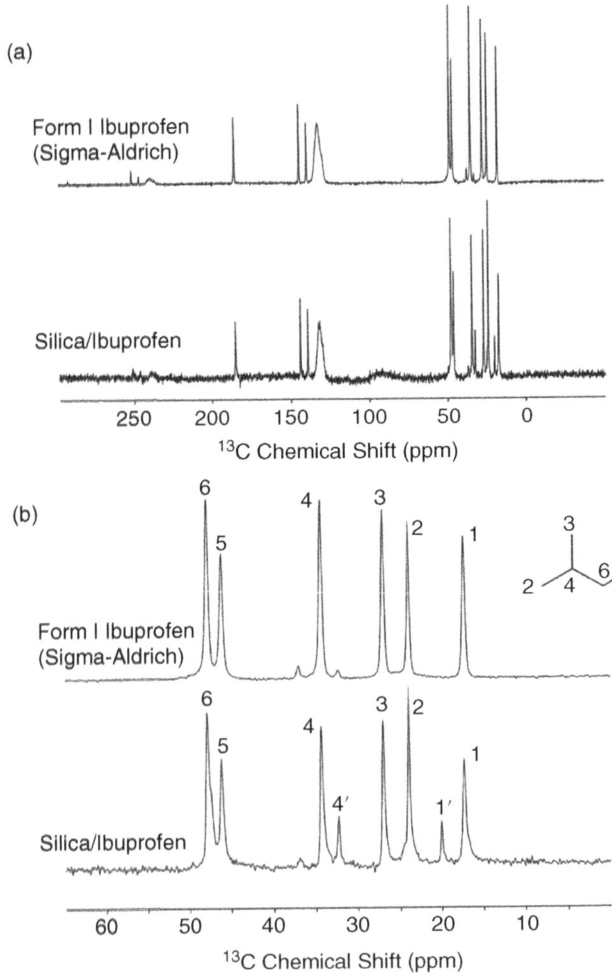

Figure 2.19 *^{13}C CPMAS spectra of form I IBP and silica-IBP system: (a) full spectra and (b) expanded spectra of the aliphatic carbon region, which clearly shows the onset of additional peaks for silica-IBP (peak 1' and 4'). Reproduced from Yang et al. [215] with permission of the Royal Society of Chemistry.*

Furthermore, with the FEN-silica system, it was possible to detect the polymorphic form I of FEN in the matrices, and in the case of GSF-silica, a mixture of two polymorphic forms and an amorphous form of GSF. In these cases, the amorphous "halo" in the XRPD pattern was probably due to X-ray diffraction from the surface of the silica. In the case of IMC, only the amorphous phase was detected by both SSNMR and XRPD.

A change of the solid form of drug nanoparticles can arise during all steps of the manufacturing process or storage of the final product. Usually, the transformation between unstable amorphous and crystalline form is an issue [228]. As the phase transition may be induced by the presence of water, the issue of physical instability is crucial in the case of nanosuspensions (drug nanoparticles dispersed in a liquid medium) [229]. The unique technique, *in situ* solid-state NMR, can be employed to characterize solids in nanosuspensions. Zhang *et al.* (2012) characterized nanoparticles of probucol obtained by dispersion of a probucol, polyvinylpyrrolidone (PVP) and sodium dodecyl sulfate (SDS) ternary co-ground mixture in water by *in situ* and conventional solid-state NMR, solution-state NMR, XRPD and cryogenic-scanning electron microscopy [230]. The authors prepared and characterized three ternary probucol/PVP/SDS mixtures: a physical mixture (PM), a ground mixture (GM) and a freeze-dried mixture (FD). The ^{13}C CPMAS NMR spectrum of the ternary GM showed, in comparison to PM, broadened signals resulting from the co-existence of distribution of local environments and hence dispersion of chemical shifts for each resonance, suggesting that the probucol was amorphous. The amorphousness of the API was confirmed by XRPD. The *in situ* ^{13}C CPMAS of GM suspended in water and stored at 25°C for 24 hours showed sharp peaks corresponding to the crystalline form I of probucol. It was speculated that a transformation of the amorphous form or very small disordered crystals into nanocrystals was induced by water. Furthermore, no signals arising from SDS and PVP molecules were detected, which indicated that PVP and SDS molecules were either freely mobile or soluble in water. SSNMR spectra of the FD sample were quite similar to those of the PM sample, indicating both probucol and SDS were crystalline; however, some broadening of the probucol crystals in the FD sample was observed, which was attributed to nanocrystal formation. The solution-state NMR suggested that PVP and SDS interact with probucol nanocrystals, providing stability to the nanoparticles.

Another example of using SSNMR for solid form detection and interaction between components in pharmaceutical systems was reported by Vogt and Williams [212]. The authors analyzed a nanocrystalline drug–polymer dispersion consisting of ebselen (organoselenium drug molecule) and polyvinylpyrrolidone-vinyl acetate (PVP-VA) by SSNMR, Raman microscopy, DSC, and XRPD. The ^{1}H, ^{13}C and the less commonly employed ^{77}Se 1D SSNMR techniques confirmed that ebselen was predominantly present in the crystalline state in the dispersion. Furthermore, ^{77}Se CPMAS NMR analysis of the ebselen and the dispersion were compared, and a line broadening of resonances arising from ebselen was observed for the dispersion (Figure 2.20). This broadening is likely to result from defects, particularly at the surface of the nanocrystalline ebselen in the dispersion. A ^{1}H 2D spin diffusion experiment and a modified ^{1}H–^{13}C CP-HETCOR experiment were used to analyze the interaction between the API and the polymer, and it was found that the aromatic rings of ebselen are in close contact with PVP-VA at the nanocrystalline interface.

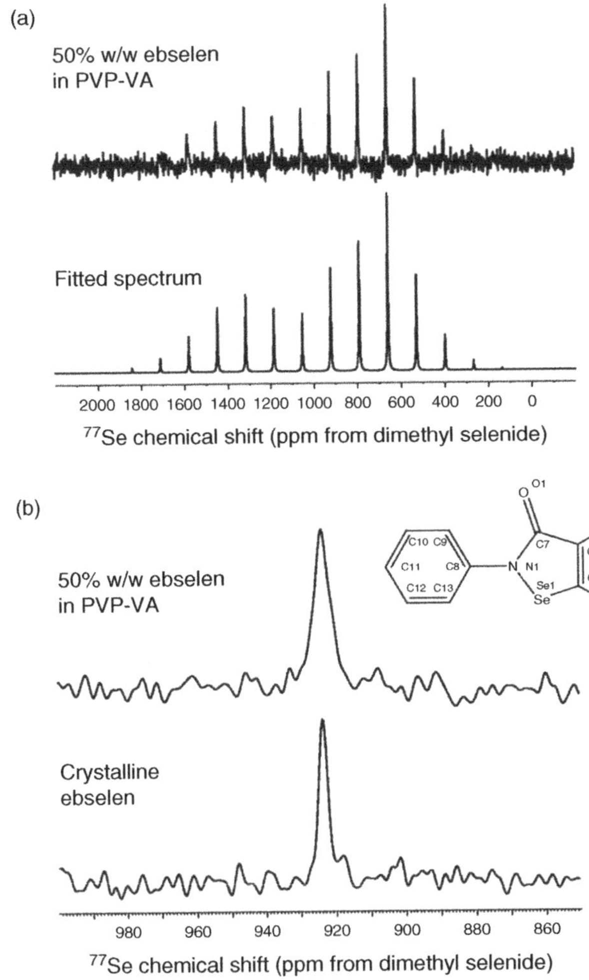

Figure 2.20 *(a) ^{77}Se CPMAS spectrum of the 50% w/w dispersion of ebselen in PVP-VA. (b) ^{77}Se CP-MAS spectra comparing the center band of the 50% w/w dispersion of ebselen in PVP-VA with that of crystalline ebselen. Adapted by permission from Springer, Pharmaceutical Research, Vogt and Williams, copyright 2012 [212].*

The pharmaceutical multicomponent systems (solvates, co-crystals, co-amorphous forms) are used to improve the physicochemical properties of APIs [4, 231, 232]. The nanoscaling of multicomponent solids may further enhance the pharmaceutical performance due to the larger surface-to-volume ratio of nanoparticles [233].

SSNMR is commonly employed to characterize pharmaceutical multicomponent systems, which is of paramount importance. For instance, Othman *et al.* (2007) characterized solvates of finasteride [219]. The crystal structures of three finasteride solvates were found to be isomorphous, as determined by single crystal X-ray diffraction (SCXRD), and their XRPD patterns were almost indistinguishable. The solvates were distinguished by the characteristic ^{13}C CPMAS NMR signals from the solvent, clearly resolved from the finasteride peaks.

A novel approach to drug delivery systems containing an API in multicomponent materials was reported by Skorupska *et al.* [216]. The authors obtained co-crystals and incorporated them into mesoporous silica particles (MSPs). The obtained co-crystals formed by racemic (*R*/*S*) and enantiomeric (*S*) ibuprofen (IBU) and nicotinamide (NA) as co-former, were embedded into the mesoporous material (MCM-41) and fully characterized by advanced 1D and 2D SSNMR as the main analytical tool. Firstly, the (*R*/*S*)-IBU:NA and (*S*)-IBU:NA co-crystals obtained by two different methods (grinding and melting) were fully characterized by ^{13}C, ^{15}N and ^{1}H SSNMR. Samples obtained by both methods resulted in the same spectral patterns, revealing that both methods produced the same co-crystal forms of (*R*/*S*)-IBU:NA and (*S*)-IBU:NA. However, distinct features of ^{13}C and ^{15}N CP MAS spectra of (*S*)-IBU:NA and (*R*/*S*)-IBU:NA co-crystals were observed, indicating a difference in molecular packing. Signal splitting of the carbonyl group and some aromatic and aliphatic carbons in ^{13}C spectra, as well as of nitrogen signals in ^{15}N (*S*)-IBU:NA spectra, were observed. These splittings revealed the difference in the crystal structure with one independent molecule of the (*R*/*S*)-IBU:NA and two independent molecules in the asymmetric unit of the (*S*)-IBU:NA sample, and/or distinction in hydrogen-bonding interactions involving the carboxylic groups. Furthermore, advanced 1D and 2D techniques were used to obtain further insight into the structure of both co-crystals. A 1H very fast (60 kHz) MAS NMR suggested that the carboxylic group of IBU contributes to hydrogen bonding in both co-crystal forms, and these bonds are stronger than in pure IBU. In the next step, 2D experiments with proton detection (^{1}H–^{13}C and ^{1}H–^{15}N inv-HETCOR NMR) with high spinning rate (60 kHz) were employed to provide more information about the nature of the hydrogen bonding. The results of these experiments suggested the mode of hydrogen bonding in both co-crystals. Finally, a ^{1}H–^{1}H 2D single quantum–double quantum (SQ-DQ) experiment was employed to reveal further intermolecular interactions (Figure 2.21). The proposed hydrogen bonding pattern was consistent with X-ray data.

Additionally, co-crystals embedded into the mesoporous silica were characterized. The ^{1}H very fast MAS NMR provided information on whether the IBU:NA co-crystal is embedded into the pores and/or located outside on the walls of MSPs, as this is crucial for the pharmaceutical performance of the drug delivery system. The authors used the signal of the proton of the carboxylic group of IBU as a diagnostic tool for estimation of the co-crystals embedding into MCM-41. Spectra with different IBU:NA/MCM-41 ratios (1:3, 1:1 and 2:1) were recorded (Figure 2.22). The acidic proton of IBU was not observed in the spectrum with the 1:3 concentration. This indicated that IBU:NA was trapped in the pores as the proton was transferred to the MCM-41 wall.

Nanoparticles based on biopolymers, such as the cationic polysaccharide chitosan, have excellent biocompatibility and biodegradability, but a molecular-level understanding of their morphology, interactions and dynamics is difficult to obtain. Detailed characterization of chitosan–poly(*N*-3-acrylamidophenylboronic acid) nanoparticles was achieved by employing a combination of several SSNMR experiments. For example, ^{11}B quadrupolar coupling parameters provided information on the coordination geometry of boron atoms,

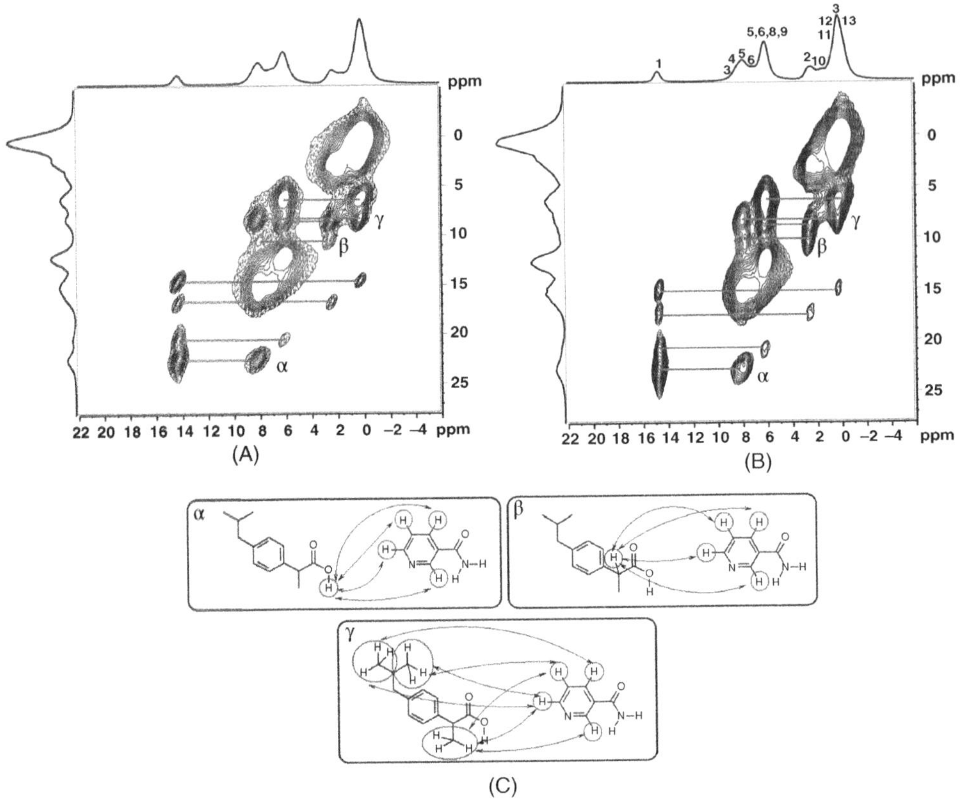

Figure 2.21 $^1H–^1H$ SQ-DQ Back-to-Back (BaBa) experiment of (A) (R/S)-IBU:NA and (B) (S)-IBU:NA. (C) Correlations labelled as α, β, γ represent inter-molecular interactions. Reprinted with permission from Skorupska et al. [216], copyright 2017, American Chemical Society.

and using a ^1H spin-diffusion experiment with ^{13}C detection allowed for determination of the internal morphology and nanostructure size [234].

In summary, SSNMR is a useful technique in the characterization of pharmaceutical solids, providing detailed information at the atomic level. It can be applied at both preformulation and formulation stages of drug development. Despite its potential, SSNMR is still not widely used in the pharmaceutical industry, mostly due to several factors such as high cost of the instrument, relatively long analysis time, and the expertise needed to acquire and interpret the data.

2.11 Conclusions

The solid-state structure of nano- and microparticles ranges from the relative simplicity of perfect single crystals to complex structures including multiple compounds, phases and solid-state forms. The structures can substantially affect many pharmaceutically relevant

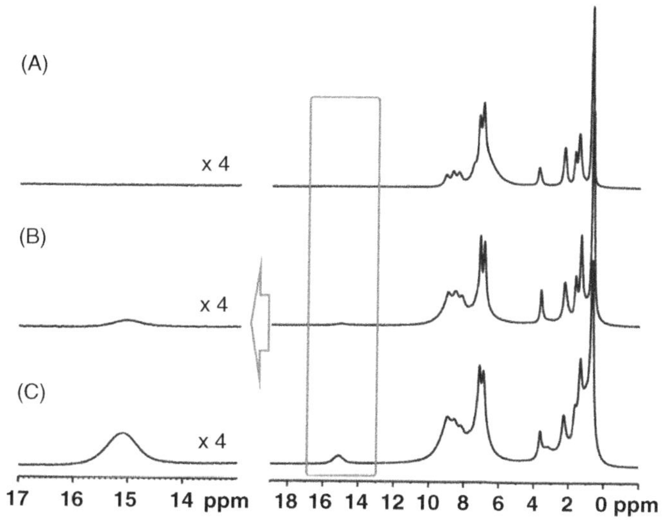

Figure 2.22 *^{1}H MAS spectra of IBU:NA in MCM-41 with different weight ratios: (A) 1:3; (B) 1:1; (C) 2:1, recorded with spinning rate of 60 kHz at ambient temperature. Reprinted with permission from Skorupska et al. [216], copyright 2017c American Chemical Society.*

properties including solubility, dissolution, stability (chemical and physical), and processability. Furthermore, regulatory, commercial, and patent infringement implications also need to be considered. Therefore, understanding, controlling and optimizing the solid-state structure during the entire lifecycle of the particles from synthesis to drug delivery is critical. Spectroscopy has a key role to play.

Solid-state spectroscopic analysis is wide-ranging and includes the more established techniques described in this chapter: (mid)-infrared (MIR), near-infrared (NIR), Raman, and solid-state nuclear magnetic resonance (SSNMR) spectroscopies. More recently, terahertz (both IR and Raman versions), fluorescence and nonlinear optical spectroscopies have been demonstrated to have much to offer in this context. Most of the techniques may be used in imaging modalities, facilitating analysis of single particles and sample heterogeneity. Collectively they have been used for solid-state identification and quantification, crystallinity and stability analysis, as well as solid-state transformation mechanisms and interpretation of molecular interactions. While the techniques are typically non-destructive and information-rich, each has its own particular strengths and weakness in different solid-state analysis contexts.

Spectroscopy has much potential to help solve a number of "hot topic" analytical challenges in the pharmaceutical industry today, including, but not limited to, in-line PAT applications, analysis of API particles in low-dose formulations, the detection of minority solid-state forms, as well as prediction of stability of the amorphous form. No doubt further technical and commercial developments, as well as recognition and support by regulatory authorities, will increase their application. While recognizing these opportunities, it is equally important to take into account the limitations of the techniques to avoid erroneous interpretations.

References

[1] Bunjes, H., Rades, T. (2005) Thermotropic liquid crystalline drugs. *Journal of Pharmacy and Pharmacology*, **57**, 807–816.

[2] Bhugra, C., Pikal, M.J. (2008) Role of thermodynamic, molecular, and kinetic factors in crystallization from the amorphous state. *Journal of Pharmaceutical Sciences*, **97** (4), 1329–1349.

[3] Laitinen, R., *et al.* (2014) Theoretical considerations in developing amorphous solid dispersions, in *Amorphous Solid Dispersions* (eds Shah, N., *et al.*), Springer, New York, pp. 35–90.

[4] Berry, D.J., Steed, J.W. (2017) Pharmaceutical cocrystals, salts and multicomponent systems; intermolecular interactions and property based design. *Advanced Drug Delivery Reviews*, **117**, 3–24.

[5] Laitinen, R., *et al.* (2012) Emerging trends in the stabilization of amorphous drugs. *International Journal of Pharmaceutics*, **453** (1), 65–79.

[6] Löbmann, K., *et al.* (2014) Stabilized amorphous solid dispersions with small molecule excipients, in *Amorphous Solid Dispersions* (eds Shah, N., *et al.*), Springer, New York, pp. 613–636.

[7] Fraser, S.J., Saarinen, J., Strachan, C.J. (2016) Vibrational spectroscopic imaging, in *Analytical Techniques in the Pharmaceutical Sciences* (eds Mullertz, A., Perrie, Y., Rades, T.). Controlled Release Society.

[8] Amado, A.M., Azevedo, C., Ribeiro-Claro, P.J.A. (2017) Conformational and vibrational reassessment of solid paracetamol. *Spectrochimica Acta – Part A: Molecular and Biomolecular Spectroscopy*, **183**, 431–438.

[9] Nolasco, M.N., Amado, A.M., Ribeiro-Claro, P.J.A. (2006) Computationally-assisted approach to the vibrational spectra of molecular crystals: study of hydrogen-bonding and pseudo-polymorphism. *ChemPhysChem*, **7**, 2150–2161.

[10] Young, D.C. (2001) *Computational Chemistry: A Practical Guide for Applying Techniques to Real World Problems*, 1st edn, John Wiley & Sons, Inc., New York.

[11] Gordon, K.C., *et al.* (2007) The use of quantum chemistry in pharmaceutical research as illustrated by case studies of indometacin and carbamazepine. *Journal of Pharmacy and Pharmacology*, **59** (2), 271–277.

[12] Aaltonen, J., *et al.* (2008) Perspectives in the use of spectroscopy to characterise pharmaceutical solids. *International Journal of Pharmaceutics*, **364** (2), 159–169.

[13] Zeitler, J.A. (2016) Pharmaceutical terahertz spectroscopy and imaging, in *Analytical Techniques in the Pharmaceutical Sciences* (eds Mullertz, A., Rades, T., Perrie, Y.), Controlled Release Society, pp. 171–222.

[14] Bērziņš, A., Hodgkinson, P. (2015) Solid-state NMR and computational investigation of solvent molecule arrangement and dynamics in isostructural solvates of droperidol. *Solid State Nuclear Magnetic Resonance*, **65**, 12–20.

[15] Foresman, J., Frisch, Æ. (1995–96) *Exploring Chemistry with Electronic Structure Methods*, 2nd edn, Gaussian, Inc., Pittsburgh.

[16] Harris, R.K., *et al.* (2007) Chemical shift computations on a crystallographic basis: some reflections and comments. *Magnetic Resonance in Chemistry*, **45** (S1), S174–S186.

[17] Bonhomme, C., *et al.* (2012) First-principles calculation of NMR parameters using the gauge including projector augmented wave method: A chemist's point of view. *Chemical Reviews*, **112** (11), 5733–5779.

[18] Dracinsky, M., Hodgkinson, P. (2013) A molecular dynamics study of the effects of fast molecular motions on solid-state NMR parameters. *CrystEngComm*, **15** (43), 8705–8712.

[19] Meier, R.J. (2007) Calculating the vibrational spectra of molecules: an introduction for experimentalists with contemporary examples. *Vibrational Spectroscopy*, **43**, 26–37.

[20] Wysokinski, R., Kuduk-Jaworska, J., Michalska, D. (2006) Electronic structure, Raman and infrared spectra, and vibrational assignment of carboplatin. *Density functional theory studies. Journal of Molecular Structure-Theochem*, **758** (2–3), 169–179.

[21] Cipriani, P., Smith, C. (2008) Characterisation of thalidomide using Raman spectroscopy. *Spectrochimica Acta – Part A*, **69**, 333–337.

[22] Rojalin, T., *et al.* (2016) Fluorescence-suppressed time-resolved Raman spectroscopy of pharmaceuticals using complementary metal-oxide semiconductor (CMOS) single-photon avalanche diode (SPAD) detector. *Analytical and Bioanalytical Chemistry*, **408** (3), 761–774.

[23] Forina, M., Casolino, M., de la Pezuela Martinez, C. (1998) Multivariate calibration: applications to pharmaceutical analysis. *Journal of Pharmaceutical and Biomedical Analysis*, **18**, 21–33.

[24] Gabrielsson, J., Lindberg, N.-O., Lundstedt, T. (2002) Multivariate methods in pharmaceutical applications. *Journal of Chemometrics*, **16**, 141–160.

[25] Rajalahti, T., Kvalheim, O.M. (2011) Multivariate data analysis in pharmaceutics: A tutorial review. *International Journal of Pharmaceutics*, **417** (1–2), 280–290.

[26] Amigo, J.M. (2010) Practical issues of hyperspectral imaging analysis of solid dosage forms. *Analytical and Bioanalytical Chemistry*, **398** (1), 93–109.

[27] Griffiths, P.R., Miseo, E.V. (2014) Infrared and Raman Instrumentation for mapping and imaging, in *Infrared and Raman Spectroscopic Imaging*, Wiley-VCH Verlag GmbH & Co., pp. 1–56.

[28] Higgins, D.A., *et al.* (1996) Polarization-modulation near-field scanning optical microscopy of mesostructured materials. *Journal of Physical Chemistry*, **100** (32), 13794–13803.

[29] Samson, J.-S., *et al.* (2006) Setup of a scanning near field infrared microscope (SNIM): Imaging of sub-surface nano-structures in gallium-doped silicon. *Physical Chemistry Chemical Physics*, **8** (6), 753–758.

[30] Dazzi, A., *et al.* (2012) AFM-IR: combining atomic force microscopy and infrared spectroscopy for nanoscale chemical characterization. *Applied Spectroscopy*, **66** (12), 1365–1384.

[31] Sonntag, M.D., *et al.* (2014) Recent advances in tip-enhanced Raman spectroscopy. *Journal of Physical Chemistry Letters*, **5** (18), 3125–3130.

[32] Strachan, C.J., Rades, T., Gordon, K.C. (2007) A theoretical and spectroscopic study of gamma-crystalline and amorphous indometacin. *Journal of Pharmacy and Pharmacology*, **59** (2), 261–269.

[33] Reich, G. (2016) Mid and near infrared spectroscopy, in *Analytical Techniques in the Pharmaceutical Sciences* (eds Müllertz, A., Perrie, Y., Rades, T.), Springer, New York, pp. 61–138.

[34] Taylor, L.S., Zografi, G. (1997) Spectroscopic characterization of interactions between PVP and indomethacin in amorphous molecular dispersions. *Pharmaceutical Research*, **14** (12), 1691–1697.

[35] Ewing, A.V., Wray, P.S., Clarke, G.S., Kazarian, S.G. (2015) Evaluating drug delivery with salt formation: Drug disproportionation studied in situ by ATR-FTIR imaging and Raman mapping. *Journal of Pharmaceutical and Biomedical Analysis*, **111**, 248–256.

[36] Löbmann, K., *et al.* (2013) A theoretical and spectroscopic study of co-amorphous naproxen and indomethacin. *International Journal of Pharmaceutics*, **453** (1), 80–87.

[37] Heinz, A., *et al.* (2009) Analysis of solid-state transformations of pharmaceutical compounds using vibrational spectroscopy. *Journal of Pharmacy and Pharmacology*, **61**, 971–988.

[38] Pindelska, E., Sokal, A., Kolodziejski, W. (2017) Pharmaceutical cocrystals, salts and polymorphs: Advanced characterization techniques. *Advanced Drug Delivery Reviews*, **117**, 111–146.

[39] Surwase, S.A., *et al.* (2013) Indomethacin: New polymorphs of an old drug. *Molecular Pharmaceutics*, **10** (12), 4472–4480.

[40] Savolainen, M., *et al.* (2007) Screening for differences in the amorphous state of indomethacin using multivariate visualization. *European Journal of Pharmaceutical Sciences*, **30** (2), 113–123.

[41] Karmwar, P., *et al.* (2011) Investigation of properties and recrystallisation behaviour of amorphous indomethacin samples prepared by different methods. *International Journal of Pharmaceutics*, **417** (1–2), 94–100.

[42] Groen, H., Roberts, K.J. (2001) Nucleation, growth, and pseudo-polymorphic behavior of citric acid as monitored in situ by attenuated total reflection Fourier transform infrared spectroscopy. *Journal of Physical Chemistry B*, **105** (43), 10723–10730.

[43] Schöll, J., *et al.* (2006) In situ monitoring and modeling of the solvent-mediated polymorphic transformation of L-glutamic acid. *Crystal Growth & Design*, **6** (4), 881–891.

[44] Priemel, P.A., *et al.* (2012) The impact of surface- and nano-crystallisation on the detected amorphous content and the dissolution behaviour of amorphous indomethacin. *European Journal of Pharmaceutics and Biopharmaceutics*, **82** (1), 187–193.

[45] Chan, K.L.A., Kazarian, S.G. (2006) Dissolution of solid dispersions of ibuprofen studied by Fourier transform infrared imaging, in *Polymeric Drug Delivery II: Polymeric Matrices and Drug Particle Engineering*, American Chemical Society, Washington, pp. 203–214.

[46] Ewing, A.V., *et al.* (2015) Comparison of pharmaceutical formulations: ATR-FTIR spectroscopic imaging to study drug–carrier interactions. *International Journal of Pharmaceutics*, **495** (1), 112–121.

[47] Kazarian, S.G., *et al.* (2005) Spectroscopic imaging applied to drug release. *Food and Bioproducts Processing*, **83** (C2), 127–135.

[48] Kazarian, S.G., van der Weerd, J. (2008) Simultaneous FTIR spectroscopic imaging and visible photography to monitor tablet dissolution and drug release. *Pharmaceutical Research*, **25** (4), 853–860.

[49] van der Weerd, J., Chan, K.L.A., Kazarian, S.G. (2004) An innovative design of compaction cell for in situ FT-IR imaging of tablet dissolution. *Vibrational Spectroscopy*, **35** (1–2), 9–13.

[50] van der Weerd, J., Kazarian, S.G. (2004) Validation of macroscopic attenuated total reflection-Fourier transform infrared imaging to study dissolution of swelling pharmaceutical tablets. *Applied Spectroscopy*, **58** (12), 1413–1419.

[51] van der Weerd, J., Kazarian, S.G. (2005) Release of poorly soluble drugs from HPMC tablets studied by FTIR imaging and flow-through dissolution tests. *Journal of Pharmaceutical Sciences*, **94** (9), 2096–2109.

[52] van Eerdenbrugh, B., *et al.* (2012) Nanoscale mid-infrared imaging of phase separation in a drug-polymer blend. *Journal of Pharmaceutical Sciences*, **101** (6), 2066–2073.

[53] van Eerdenbrugh, B., *et al.* (2012) Nanoscale mid-infrared evaluation of the miscibility behavior of blends of dextran or maltodextrin with poly(vinylpyrrolidone). *Molecular Pharmaceutics*, **9** (5), 1459–1469.

[54] Räsänen, E., Sandler, N. (2007) Near infrared spectroscopy in the development of solid dosage forms. *Journal of Pharmacy and Pharmacology*, **59**, 147–159.

[55] Reich, G. (2005) Near-infrared spectroscopy and imaging: basic principles and pharmaceutical applications. *Advanced Drug Delivery Reviews*, **57**, 1109–1143.

[56] Grohganz, H., *et al.* (2010) Classification of lyophilised mixtures using multivariate analysis of NIR spectra. *European Journal of Pharmaceutics and Biopharmaceutics*, **74** (2), 406–412.

[57] Grohganz, H., *et al.* (2010) *Towards a robust water content determination of freeze-dried samples by near-infrared spectroscopy. Analytica Chimica Acta*, **676** (1–2), 34–40.

[58] Blanco, M., *et al.* (2005) Application of NIR spectroscopy in polymorphic analysis: study of pseudo-polymorphs stability. *Journal of Pharmaceutical Sciences*, **94** (6), 1336–1342.

[59] Higgins, J.P., Arrivo, S.M., Reed, R.A. (2003) Approach to the determination of hydrate form conversions of drug compounds and solid dosage forms by near-infrared spectroscopy. *Journal of Pharmaceutical Sciences*, **92** (11), 2303–2316.

[60] Heinz, A., *et al.* (2007) Quantifying ternary mixtures of different solid-state forms of indomethacin by Raman and near-infrared spectroscopy. *European Journal of Pharmaceutical Sciences*, **32** (3), 182–192.

[61] Aaltonen, J., *et al.* (2003) Polymorph screening using near-infrared spectroscopy. *Analytical Chemistry*, **75** (19), 5267–5273.

[62] Aaltonen, J., *et al.* (2007) In-line monitoring of solid-state transitions during fluidisation. *Chemical Engineering Science*, **62** (1–2), 408–415.

[63] da Silva, V.H., da Silva, J.J., Pereira, C.F. (2017) Portable near-infrared instruments: Application for quality control of polymorphs in pharmaceutical raw materials and calibration transfer. *Journal of Pharmaceutical and Biomedical Analysis*, **134**, 287–294.

[64] US Department of Health and Human Services, Food and Drug Administration (2004) *Guidance for Industry: PAT – A Framework for Innovative Pharmaceutical Development, Manufacturing, and Quality Assurance*, Food and Drug Administration.

[65] Simon, L.L., *et al.* (2015) Assessment of recent process analytical technology (PAT) trends: A multiauthor review. *Organic Process Research & Development*, **19** (1), 3–62.

[66] Jorgensen, A., *et al.* (2002) Hydrate formation during wet granulation studied by spectroscopic methods and multivariate analysis. *Pharmaceutical Research*, **19** (9), 1285–1291.

[67] Jorgensen, A.C., *et al.* (2004) Visualization of a pharmaceutical unit operation: Wet granulation. *Analytical Chemistry*, **76** (18), 5331–5338.

[68] Rasanen, E., *et al.* (2001) Novel identification of pseudopolymorphic changes of theophylline during wet granulation using near infrared spectroscopy. *Journal of Pharmaceutical Sciences*, **90** (3), 389–396.

[69] Ewing, A.V., Kazarian, S.G. (2018) Recent advances in the applications of vibrational spectroscopic imaging and mapping to pharmaceutical formulations. *Spectrochimica Acta Part A: Molecular and Biomolecular Spectroscopy*, **197**, 10–29.

[70] Khorasani, M., *et al.* (2015) Near-infrared chemical imaging (NIR-CI) as a process monitoring solution for a production line of roll compaction and tableting. *European Journal Pharmaceutics Biopharmaceutics*, **93**, 293–302.

[71] Khorasani, M., *et al.* (2016) Near-infrared chemical imaging (NIR-CI) of 3D printed pharmaceuticals. *International Journal of Pharmaceutics*, **515** (1–2), 324–330.

[72] Vakili, H., *et al.* (2015) Hyperspectral imaging in quality control of inkjet printed personalised dosage forms. *International Journal of Pharmaceutics*, **483** (1–2), 244–249.

[73] Schönbichler, S.A., *et al.* (2013) Comparison of NIR chemical imaging with conventional NIR, Raman and ATR-IR spectroscopy for quantification of furosemide crystal polymorphs in ternary powder mixtures. *European Journal of Pharmaceutics and Biopharmaceutics*, **84** (3), 616–625.

[74] Alexandrino, G.L., *et al.* (2017) Unveiling multiple solid-state transitions in pharmaceutical solid dosage forms using multi-series hyperspectral imaging and different curve resolution approaches. *Chemometrics and Intelligent Laboratory Systems*, **161**, 136–146.

[75] Alexandrino, G.L., *et al.* (2015) *Monitoring of multiple solid-state transformations at tablet surfaces using multi-series near-infrared hyperspectral imaging and multivariate curve resolution. European Journal of Pharmaceutics and Biopharmaceutics*, **93**, 224–230.

[76] Wray, P.S., *et al.* (2015) The use of in situ near infrared imaging and Raman mapping to study the disproportionation of a drug HCl salt during dissolution. *International Journal of Pharmaceutics*, **493** (1–2), 198–207.

[77] Chantry, G.W. (1971) *Submillimetre Spectroscopy: A Guide to the Theoretical and Experimental Physics of the Far Infrared*, Academic Press, London, p. 385.

[78] Möller, K.D., Rotschild, W.G. (1971) *Far-Infrared Spectroscopy*. Wiley Series in Pure and Applied Optics (ed. Ballard, S.S), John Wiley & Sons, Inc., New York, p. 797.

[79] Taraskin, S.N., *et al.* (2006) Universal features of terahertz absorption in disordered materials. *Physical Review Letters*, **97** (5), 055504.

[80] Peiponen, K.-E., Zeitler, A., Kuwata-Gonokami, M. (eds) (2013) *Terahertz Spectroscopy and Imaging*, 1st edn, Springer Series in Optical Sciences, Vol. **171**, Springer-Verlag, Berlin, p. 644.

[81] Jepsen, P.U., Cooke, D.G., Koch, M. (2011) Terahertz spectroscopy and imaging – Modern techniques and applications. *Laser & Photonics Reviews*, **5** (1), 124–166.

[82] Strachan, C.J., *et al.* (2004) Using terahertz pulsed spectroscopy to study crystallinity of pharmaceutical materials. *Chemical Physics Letters*, **390** (1–3), 20–24.

[83] Newnham, D.A., Taday, P.F. (2008) Pulsed terahertz attenuated total reflection spectroscopy. *Applied Spectroscopy*, **62** (4), 394–398.

[84] Zeitler, J.A., *et al.* (2007) Terahertz pulsed spectroscopy and imaging in the pharmaceutical setting – a review. *Journal of Pharmacy and Pharmacology*, **59** (2), 209–223.

[85] Zeitler, J.A., Gladden, L.F. (2009) In-vitro tomography and non-destructive imaging at depth of pharmaceutical solid dosage forms. *European Journal of Pharmaceutics and Biopharmaceutics*, **71**, 2–22.

[86] Shen, Y.C., *et al.* (2005) 3D chemical mapping using terahertz pulsed imaging, in *Terahertz and Gigahertz Electronics and Photonics IV* (eds Hwu, R.J., Linden, K.J.), SPIE, San Jose, CA.

[87] Shen, Y.C. (2011) Terahertz pulsed spectroscopy and imaging for pharmaceutical applications: A review. *International Journal of Pharmaceutics*, **417** (1–2), 48–60.

[88] Ulmschneider, M. (2014) Terahertz imaging of drug products, in *Infrared and Raman Spectroscopic Imaging*, 2nd edn (eds Salzer, R., Siesler, H.W.), Wiley.

[89] Charron, D.M., *et al.* (2013) Chemical mapping of pharmaceutical cocrystals using terahertz spectroscopic imaging. *Analytical Chemistry*, **85** (4), 1980–1984.

[90] Korter, T.M., *et al.* (2006) Terahertz spectroscopy of solid serine and cysteine. *Chemical Physics Letters*, **418**, 65–70.

[91] Parrott, E.P.J. *et al.* (2007) Using terahertz time-domain spectroscopy to identify pharmaceutical cocrystals, in *2007 Joint 32nd International Conference on Infrared and Millimeter Waves and 15th International Conference on Terahertz Electronics, Vols 1 and 2*, pp. 645–646.

[92] Smith, G., *et al.* (2015) Quantification of residual crystallinity of ball-milled, commercially available, anhydrous β-lactose by differential scanning calorimetry and terahertz spectroscopy. *Journal of Thermal Analysis and Calorimetry*, **121** (1), 327–333.

[93] Strachan, C., *et al.* (2005) Using terahertz pulsed spectroscopy to quantify pharmaceutical polymorphism and crystallinity. *Journal of Pharmaceutical Sciences*, **94** (4), 837–846.

[94] Nguyen, K.L., *et al.* (2007) Terahertz time-domain spectroscopy and the quantitative monitoring of mechanochemical cocrystal formation. *Nature Materials*, **6** (3), 206–209.

[95] Parrott, E.P.J., *et al.* (2008) Using terahertz time-domain-spectroscopy to follow the kinetics and mechanism of cocrystal formation, in *2008 33rd International Conference on Infrared, Millimeter and Terahertz Waves, Vols 1 and 2*, pp. 653–654.

[96] Parrott, E.P.J., *et al.* (2009) Testing the sensitivity of terahertz spectroscopy to changes in molecular and supramolecular structure: A study of structurally similar cocrystals. *Crystal Growth & Design*, **9** (3), 1452–1460.

[97] Zeitler, J.A., *et al.* (2007) Drug hydrate systems and dehydration processes studied by terahertz pulsed spectroscopy. *International Journal of Pharmaceutics*, **334** (1–2), 78–84.

[98] Zeitler, J.A., *et al.* (2006) Characterization of temperature induced phase transitions in the five polymorphic forms of sulfathiazole by terahertz pulsed spectroscopy and differential scanning calorimetry. *Journal of Pharmaceutical Sciences*, **95** (11), 2486–2498.

[99] Sibik, J., Zeitler, J.A. (2016) Direct measurement of molecular mobility and crystallisation of amorphous pharmaceuticals using terahertz spectroscopy. *Advanced Drug Delivery Reviews*, **100**, 147–157.

[100] Taday, P.F., *et al.* (2003) Using terahertz pulse spectroscopy to study the crystalline structure of a drug: a case study of the polymorphs of ranitidine hydrochloride. *Journal of Pharmaceutical Sciences*, **92** (4), 831–838.

[101] Strachan, C.J., *et al.* (2005) Using terahertz pulsed spectroscopy to quantify pharmaceutical polymorphism and crystallinity. *Journal of Pharmaceutical Sciences*, **94** (4), 837–846.

[102] Zeitler, J.A., *et al.* (2006) Characterization of temperature-induced phase transitions in five polymorphic forms of sulfathiazole by terahertz pulsed spectroscopy and differential scanning calorimetry. *Journal of Pharmaceutical Sciences*, **95** (11), 2486–2498.

[103] Tomerini, D., Day, G.M. (2013) Computational methods for the assignment of vibrational modes in crystalline materials, in *Terahertz Spectroscopy and Imaging* (eds Peiponen, K.-E., Zeitler, A., Kuwata-Gonokami, M.), Springer, Berlin, pp. 151–190.

[104] Ajito, K., *et al.* (2014) Nondestructive multicomponent terahertz chemical imaging of medicine in tablets. *Journal of the Electrochemical Society*, **161** (9), B171–B175.

[105] Hisazumi, J., *et al.* (2012) Chemical mapping of hydration and dehydration process of theophylline in tablets using terahertz pulsed imaging. *Chemical and Pharmeutical Bulletin (Tokyo)*, **60** (7), 831–836.

[106] Issei, T., *et al.* (2014) Nondestructive analysis of structure and components of tablet coated with film by the usage of terahertz time-domain reflection spectroscopy. *Journal of Pharmaceutical Sciences*, **103** (1), 256–261.

[107] Cogdill, R.P., *et al.* (2006) An efficient method-development strategy for quantitative chemical imaging using terahertz pulse spectroscopy. *Journal of Pharmaceutical Innovation*, **1** (1), 63–75.

[108] Strachan, C.J., *et al.* (2007) Raman spectroscopy for quantitative analysis of pharmaceutical solids. *Journal of Pharmacy and Pharmacology*, **59** (2), 179–192.

[109] Smith, G.P.S., *et al.* (2015) Raman imaging of drug delivery systems. *Advanced Drug Delivery Reviews*, **89**, 21–41.

[110] Everall, N.J. (2010) Confocal Raman microscopy: common errors and artefacts. *Analyst*, **135** (10), 2512–2522.

[111] Everall, N.J. (2009) Confocal Raman Microscopy: Performance, Pitfalls, and Best Practice: Invited Lecture at the Symposium "50 Years of SAS: Looking to the Future with Vibrational Spectroscopy" at Pittcon 2008, New Orleans, Louisiana. *Applied Spectroscopy*, **63** (9), 245A–262A.

[112] Everall, N.J. (2000) Confocal Raman microscopy: Why the depth resolution and spatial accuracy can be much worse than you think. *Applied Spectroscopy*, **54** (10), 1515–1520.

[113] Paudel, A., Raijada, D., Rantanen, J. (2015) Raman spectroscopy in pharmaceutical product design. *Advanced Drug Delivery Reviews*, **89**, 3–20.

[114] Aaltonen, J., *et al.* (2009) Solid form screening – A review. *European Journal of Pharmaceutics and Biopharmaceutics*, **71** (1), 23–37.

[115] Sinclair, W., *et al.* (2011) Physical stability and recrystallization kinetics of amorphous ibipinabant drug product by fourier transform raman spectroscopy. *Journal of Pharmaceutical Sciences*, **100** (11), 4687–4699.

[116] Mah, P.T., *et al.* (2015) Use of low-frequency Raman spectroscopy and chemometrics for the quantification of crystallinity in amorphous griseofulvin tablets. *Vibrational Spectroscopy*, **77**, 10–16.

[117] Netchacovitch, L., *et al.* (2017) Development of an analytical method for crystalline content determination in amorphous solid dispersions produced by hot-melt extrusion

using transmission Raman spectroscopy: A feasibility study. *International Journal of Pharmaceutics*, **530** (1), 249–255.

[118] Paaver, U., *et al.* (2012) Insight into the solubility and dissolution behavior of piroxicam anhydrate and monohydrate forms. *International Journal of Pharmaceutics*, **431** (1–2), 111–119.

[119] Kogermann, K., *et al.* (2008) Establishing quantitative in-line analysis of multiple solid-state transformations during dehydration. *Journal of Pharmaceutical Sciences*, **97** (11), 4983–4999.

[120] Koradia, V., *et al.* (2010) Integrated approach to study the dehydration kinetics of nitrofurantoin monohydrate. *Journal of Pharmaceutical Sciences*, **99** (9), 3966–3976.

[121] Helmdach, L., Feth, M.P., Ulrich, J. (2012) Online analytical investigations on solvent-, temperature- and water vapour-induced phase transformations of citric acid. *Crystal Research and Technology*, **47** (9), 967–984.

[122] Hedoux, A., Guinet, Y., Descamps, M. (2011) The contribution of Raman spectroscopy to the analysis of phase transformations in pharmaceutical compounds. *International Journal of Pharmaceutics*, **417** (1–2), 17–31.

[123] Hédoux, A., *et al.* (2011) Low- and high-frequency Raman investigations on caffeine: Polymorphism, disorder and phase transformation. *Journal of Physical Chemistry B*, **115** (19), 5746–5753.

[124] Roy, S., Chamberlin, B., Matzger, A.J. (2013) Polymorph discrimination using low wavenumber Raman spectroscopy. *Organic Process Research and Development*, **17** (7), 976–980.

[125] Ayala, A.P., *et al.* (2012) Conformational polymorphism of the antidiabetic drug chlorpropamide. *Journal of Raman Spectroscopy*, **43** (2), 263–272.

[126] Lipiäinen, T., *et al.* (2018) Direct comparison of low- and mid-frequency Raman spectroscopy for quantitative solid-state pharmaceutical analysis. *Journal of Pharmaceutical and Biomedical Analysis*, **149**, 343–350.

[127] De Beer, T., *et al.* (2011) Near infrared and Raman spectroscopy for the in-process monitoring of pharmaceutical production processes. *International Journal of Pharmaceutics*, **417** (1), 32–47.

[128] De Beer, T.R., *et al.* (2009) In-line and real-time process monitoring of a freeze drying process using Raman and NIR spectroscopy as complementary process analytical technology (PAT) tools. *Journal of Pharmaceutical Sciences*, **98** (9), 3430–3446.

[129] Wei, D., Chen, S., Liu, Q. (2015) Review of fluorescence suppression techniques in Raman spectroscopy. *Applied Spectroscopy Reviews*, **50** (5), 387–406.

[130] Kostamovaara, J., *et al.* (2013) Fluorescence suppression in Raman spectroscopy using a time-gated CMOS SPAD. *Optics Express*, **21** (25), 31632–31645.

[131] Lipiäinen, T., *et al.* (2018) Time-gated Raman spectroscopy for quantitative determination of solid-state forms of fluorescent pharmaceuticals. *Analytical Chemistry*, **90** (7), 4832–4839.

[132] Buckley, K., Matousek, P. (2011) Recent advances in the application of transmission Raman spectroscopy to pharmaceutical analysis. *Journal of Pharmaceutical and Biomedical Analysis*, **55** (4), 645–652.

[133] Hargreaves, M.D., *et al.* (2011) Characterisation of transmission Raman spectroscopy for rapid quantitative analysis of intact multi-component pharmaceutical capsules. *Journal of Pharmaceutical and Biomedical Analysis*, **54** (3), 463–468.

[134] Villaumie, J., Jeffreys, H. (2015) Revolutionising Raman with the transmission technique. *European Pharmaceutical Review*, **20** (3), 41–45.

[135] McGoverin, C.M., *et al.* (2012) Pharmaceutical polymorphs quantified with transmission Raman spectroscopy. *Journal of Raman Spectroscopy*, **43** (2), 280–285.

[136] Griffen, J.A., *et al.* (2016) Rapid quantification of low level polymorph content in a solid dose form using transmission Raman spectroscopy. *Journal of Pharmaceutical and Biomedical Analysis*, **128**, 35–45.

[137] Tres, F., *et al.* (2014) Real time Raman imaging to understand dissolution performance of amorphous solid dispersions. *Journal of Controlled Release*, **188**, 53–60.

[138] Nakamoto, K., *et al.* (2013) Evaluation of the crystalline and amorphous states of drug products by nanothermal analysis and Raman imaging. *Journal of Pharmaceutical and Biomedical Analysis*, **75**, 105–111.

[139] Ward, S., *et al.* (2005) Identifying and mapping surface amorphous domains. *Pharmaceutical Research*, **22** (7), 1195–1202.

[140] Evans, C.L., Xie, X.S. (2008) Coherent anti-Stokes Raman scattering microscopy: chemical imaging for biology and medicine. *Annual Review of Analytical Chemistry*, **1**, 883–909.

[141] Prince, R.C., Frontiera, R.R., Potma, E.O. (2017) Stimulated Raman scattering: From bulk to nano. *Chemical Reviews*, **117** (7), 5070–5094.

[142] Campagnola, P.J., *et al.* (1999) High-resolution nonlinear optical imaging of live cells by second harmonic generation. *Biophysical Journal*, **77** (6), 3341–3349.

[143] Helmchen, F., Denk, W. (2005) Deep tissue two-photon microscopy. *Nature Methods*, **2** (12), 932–940.

[144] Cheng, J.-X., Volkmer, A., Xie, X.S. (2002) Theoretical and experimental characterization of coherent anti-Stokes Raman scattering microscopy. *Journal of the Optical Society of America B*, **19** (6), 1363–1363.

[145] Zumbusch, A., Holtom, G., Xie, X.S. (1999) Three-dimensional vibrational imaging by coherent anti-Stokes Raman scattering. *Physical Review Letters*, **82** (20), 4142–4145.

[146] Cheng, J.-X., Xie, X.S. (2004) Coherent anti-Stokes Raman scattering microscopy: instrumentation, theory, and applications. *Journal of Physical Chemistry B*, **108**, 827–840.

[147] Cheng, J.-X., Xie, X.S. (2015) Vibrational spectroscopic imaging of living systems: An emerging platform for biology and medicine. *Science*, **350** (6264), aaa8870-1–aaa8870-9.

[148] Evans, C.L., *et al.* (2005) Chemical imaging of tissue in vivo with video-rate coherent anti-Stokes Raman scattering microscopy. *Proceedings of the National Academy of Sciences of the United States of America*, **102** (46), 16807–16812.

[149] Hartshorn, C.M., *et al.* (2013) Multicomponent chemical imaging of pharmaceutical solid dosage forms with broadband CARS microscopy. *Analytical Chemistry*, **85** (17), 8102–8111.

[150] Tipping, W.J., *et al.* (2016) Stimulated Raman scattering microscopy: an emerging tool for drug discovery. *Chemical Society Reviews*, **45** (8), 2075–2089.

[151] Müller, M., Schins, J.M. (2002) Imaging the thermodynamic state of lipid membranes with multiplex CARS microscopy. *Journal of Physical Chemistry B*, **106** (14), 3715–3723.

[152] Cheng, J.-X., *et al.* (2002) Multiplex coherent anti-Stokes Raman scattering microspectroscopy and study of lipid vesicles. *Journal of Physical Chemistry B*, **106** (34), 8493–8498.

[153] Camp Jr, C.H., *et al.* (2014) High-speed coherent Raman fingerprint imaging of biological tissues. *Nature Photonics*, **8** (8), 627–634.

[154] Campagnola, P.J., Loew, L.M. (2003) Second-harmonic imaging microscopy for visualizing biomolecular arrays in cells, tissues and organisms. *Nature Biotechnology*, **21** (11), 1356–1360.

[155] Strachan, C.J., Windbergs, M., Offerhaus, H.L. (2011) Pharmaceutical applications of non-linear imaging. *International Journal of Pharmaceutics*, **417** (1–2), 163–172.

[156] Wanapun, D., *et al.* (2011) Single particle nonlinear optical imaging of trace crystallinity in an organic powder. *Analytical Chemistry*, **83** (12), 4745–4751.

[157] Novakovic, D., *et al.* (2017) Multimodal nonlinear optical imaging for sensitive detection of multiple pharmaceutical solid-state forms and surface transformations. *Analytical Chemistry*, **89** (21), 11460–11467.

[158] Hashimoto, K., *et al.* (2016) Broadband coherent Raman spectroscopy running at 24,000 spectra per second. *Nature Scientific Reports*, February, 21036–21036.

[159] Evans, C.L., *et al.* (2007) Chemically-selective imaging of brain structures with CARS microscopy. *Optics Express*, **15** (19), 12076–12087.

[160] Wang, H., *et al.* (2005) Coherent anti-stokes Raman scattering imaging of axonal myelin in live spinal tissues. *Biophysical Journal*, **89** (1), 581–591.

[161] Chowdhury, A.U., *et al.* (2017) Second harmonic generation guided Raman spectroscopy for sensitive detection of polymorph transitions. *Analytical Chemistry*, **89** (11), 5958–5965.

[162] Lee, C.J., *et al.* (2007) Characterization of the bulk properties of pharmaceutical solids using nonlinear optics – a review. *Journal of Pharmacy and Pharmacology*, **59** (2), 241–250.

[163] Rawle, C.B., *et al.* (2006) Towards characterization and identification of solid state pharmaceutical mixtures through second harmonic generation. *Journal of Pharmaceutical Sciences*, **95** (4), 761–768.

[164] Strachan, C.J., Lee, C.J., Rades, T. (2004) Partial characterization of different mixtures of solids by measuring the optical nonlinear response. *Journal of Pharmaceutical Sciences*, **93** (3), 733–742.

[165] Strachan, C.J., Rades, T., Lee, C.J. (2005) Determination of the optical second harmonic response of pharmaceutical solid-solid mixtures. *Optics and Lasers in Engineering*, **43** (2), 209–220.

[166] Kestur, U.S., *et al.* (2012) Nonlinear optical imaging for sensitive detection of crystals in bulk amorphous powders. *Journal of Pharmaceutical Sciences*, **101** (11), 4201–4213.

[167] Schmitt, P.D., *et al.* (2015) Finding the needle in the haystack: Characterization of trace crystallinity in a commercial formulation of paclitaxel protein-bound particles by Raman spectroscopy enabled by second harmonic generation microscopy. *Molecular Pharmaceutics*, **12** (7), 2378–2383.

[168] Wampler, R.D., *et al.* (2008) Selective detection of protein crystals by second harmonic microscopy. *Journal of the American Chemical Society*, **130** (43), 14076–14077.

[169] Mah, P.T., *et al.* (2017) Elucidation of compression-induced surface crystallization in amorphous tablets using sum frequency generation (SFG) microscopy. *Pharmaceutical Research*, **34** (5), 957–970.

[170] Fussell, A., *et al.* (2013) In situ dissolution analysis using coherent anti-Stokes Raman scattering (CARS) and hyperspectral CARS microscopy. *European Journal of Pharmaceutics and Biopharmaceutics*, **85** (3 Part B), 1141–1147.

[171] Windbergs, M., *et al.* (2009) Chemical imaging of oral solid dosage forms and changes upon dissolution using coherent anti-Stokes Raman scattering microscopy. *Analytical Chemistry*, **81** (6), 2085–2091.

[172] Fussell, A.L., *et al.* (2014) Coherent anti-Stokes Raman scattering microscopy driving the future of loaded mesoporous silica imaging. *Acta Biomaterialia*, **10** (11), 4870–4877.

[173] Lakowicz, J.R. (2006) *Principles of Fluorescence Spectroscopy*, Springer, Boston, MA.

[174] Valeur, B., Berberan-Santos, M.R.N. (2013) *Molecular Fluorescence: Principles and Applications*, 2nd edn, Wiley-VCH.

[175] Tkachenko, N.V. (2006) *Optical Spectroscopy: Methods and Instrumentation*, Elsevier, Amsterdam.

[176] Purohit, H.S., Taylor, L.S. (2015) Phase separation kinetics in amorphous solid dispersions upon exposure to water. *Molecular Pharmaceutics*, **12** (5), 1623–1635.

[177] Kasha, M. (1963) Energy transfer mechanisms and the molecular exciton model for molecular aggregates. *Radiation Research*, **20** (1), 55.

[178] Rautaniemi, K., *et al.* (2018) Crystallization kinetics of an amorphous pharmaceutical compound using fluorescence-lifetime-imaging microscopy. *Molecular Pharmaceutics*, **15** (5), 1964–1971.

[179] Ramachander, R., *et al.* (2008) Solid state fluorescence of lyophilized proteins. *Analytical Biochemistry*, **376** (2), 173–182.

[180] Sharma, V.K., Kalonia, D.S. (2003) Steady-state tryptophan fluorescence spectroscopy study to probe tertiary structure of proteins in solid powders. *Journal of Pharmaceutical Sciences*, **92** (4), 890–899.

[181] Frenette, M., Cosa, G., Friščić, T. (2013) Characterisation of organic solid forms and real-time in situ monitoring of their transformations using solid-state fluorescence. *CrystEngComm*, **15** (25), 5100.

[182] Alves, J.C.L., Poppi, R.J. (2009) Simultaneous determination of acetylsalicylic acid, paracetamol and caffeine using solid-phase molecular fluorescence and parallel factor analysis. *Analytica Chimica Acta*, **642** (1–2), 212–216.

[183] Moreira, A.B., *et al.* (2004) Solid-phase fluorescence spectroscopy for the determination of acetylsalicylic acid in powdered pharmaceutical samples. *Analytica Chimica Acta*, **523** (1), 49–52.

[184] Woltmann, E., *et al.* (2014) Applicability of UV laser-induced solid-state fluorescence spectroscopy for characterization of solid dosage forms. *Analytical and Bioanalytical Chemistry*, **406** (25), 6347–6362.

[185] Agnew, B.J., Murray, D., Patton, W.F. (2004) A rapid solid-phase fluorescence-based protein assay for quantitation of protein electrophoresis samples containing detergents, chaotropes, dyes, and reducing agents. *Electrophoresis*, **25** (15), 2478–2485.

[186] Leung, B.O., Chou, K.C. (2011) Review of super-resolution fluorescence microscopy for biology. *Applied Spectroscopy*, **65** (9), 967–980.

[187] Blom, H., Brismar, H. (2014) STED microscopy: increased resolution for medical research? *Journal of Internal Medicine*, **276** (6), 560–578.

[188] Purohit, H.S., *et al.* (2017) Insights into nano- and micron-scale phase separation in amorphous solid dispersions using fluorescence-based techniques in combination with solid state nuclear magnetic resonance spectroscopy. *Pharmaceutical Research*, **34** (7), 1364–1377.

[189] Purohit, H.S., Taylor, L.S. (2015) Miscibility of itraconazole–hydroxypropyl methylcellulose blends: Insights with high resolution analytical methodologies. *Molecular Pharmaceutics*, **12** (12), 4542–4553.

[190] Tian, B., Tang, X., Taylor, L.S. (2016) Investigating the correlation between miscibility and physical stability of amorphous solid dispersions using fluorescence-based techniques. *Molecular Pharmaceutics*, **13** (11), 3988–4000.

[191] Torres-Salas, P., *et al.* (2011) Immobilized biocatalysts: Novel approaches and tools for binding enzymes to supports. *Advanced Materials*, **23** (44), 5275–5282.

[192] Saboo, S., Taylor, L.S. (2017) Water-induced phase separation of miconazole-poly (vinylpyrrolidone-co-vinyl acetate) amorphous solid dispersions: Insights with confocal fluorescence microscopy. *International Journal of Pharmaceutics*, **529** (1–2), 654–666.

[193] Bloch, F., Hansen, W.W., Packard, M. (1946) Nuclear induction. *Physical Review*, **69** (3–4), 127.

[194] Harris, R.K. (2007) Applications of solid-state NMR to pharmaceutical polymorphism and related matters. *Journal of Pharmacy and Pharmacology*, **59** (2), 225–239.

[195] Geppi, M., *et al.* (2008) Solid-state NMR studies of pharmaceutical systems. *Applied Spectroscopy Reviews*, **43** (3), 202–302.

[196] Apperley, D.C., Harris, R.K., Hodgkinson, P. (2012) *Solid-State NMR Basic Principles & Practice*, Momentum Press, New York.

[197] Paradowska, K., Wawer, I. (2014) Solid-state NMR in the analysis of drugs and naturally occurring materials. *Journal of Pharmaceutical and Biomedical Analysis*, **93**, 27–42.

[198] Skorupska, E., *et al.* (2014) Recent progress in solid-state NMR studies of drugs confined within drug delivery systems. *Solid State Nuclear Magnetic Resonance*, **57–58**, 2–16.

[199] Vogt, F.G. (2015) Characterization of pharmaceutical compounds by solid-state NMR. *eMagRes*, **4** (2).
[200] Lowe, I.J. (1959) Free induction decays of rotating solids. *Physical Review Letters*, **2** (7), 285–287.
[201] Andrew, E.R., Bradbury, A., Eades, R.G. (1959) Removal of dipolar broadening of nuclear magnetic resonance spectra of solids by specimen rotation. *Nature*, **183** (4678), 1802–1803.
[202] Bruker (2018) 111 KHz probe for ultra-fast magic angle spinning. https://www.bruker.com/products/mr/nmr/probes/probes/solids/very-fast-mas/07-mm/overview.html (accessed 21 April 2018).
[203] Dixon, W.T. (1982) Spinning-sideband-free and spinning-sideband-only NMR spectra in spinning samples. *Journal of Chemical Physics*, **77** (4), 1800–1809.
[204] Sarles, L.R., Cotts, R.M. (1958) Double nuclear magnetic resonance and the dipole interaction in solids. *Physical Review*, **111** (3), 853–859.
[205] Bloch, F. (1958) Theory of line narrowing by double-frequency irradiation. *Physical Review*, **111** (3), 841–853.
[206] Hartmann, S.R., Hahn, E.L. (1962) Nuclear double resonance in the rotating frame. *Physical Review*, **128** (5), 2042–2053.
[207] Duer, M.J. (2002) *Solid State NMR Spectroscopy: Principles and Applications*. Blackwell Science, Oxford.
[208] Burum, D.P., Bielecki, A. (1969) An improved experiment for heteronuclear-correlation 2D NMR in solids. *Journal of Magnetic Resonance*, **94** (3), 645–652.
[209] Harris, R.K., *et al.* (2005) Quantification of bambuterol hydrochloride in a formulated product using solid-state NMR. *Journal of Pharmaceutical and Biomedical Analysis*, **38** (5), 858–864.
[210] Zielińska-Pisklak, M., Pisklak, D.M., Wawer, I. (2012) Application of ^{13}C CPMAS NMR for qualitative and quantitative characterization of carvedilol and its commercial formulations. *Journal of Pharmaceutical Sciences*, **101** (5), 1763–1772.
[211] Pham, T.N., *et al.* (2010) Analysis of amorphous solid dispersions using 2D solid-state NMR and 1H T1 relaxation measurements. *Molecular Pharmaceutics*, **7** (5), 1667–1691.
[212] Vogt, F.G., Williams, G.R. (2012) Analysis of a nanocrystalline polymer dispersion of ebselen using solid-state NMR, Raman microscopy, and powder x-ray diffraction. *Pharmaceutical Research*, **29** (7), 1866–1881.
[213] Babonneau, F., *et al.* (2004) Solid state NMR characterisation of encapsulated molecules in mesoporous silica. *Journal of Sol-Gel Science and Technology*, **31** (1), 219–223.
[214] Azaïs, T., *et al.* (2006) Solid-state NMR study of ibuprofen confined in MCM-41 material. *Chemistry of Materials*, **18** (26), 6382–6390.
[215] Yang, X., *et al.* (2014) Formation of organic molecular nanocrystals under rigid confinement with analysis by solid state NMR. *CrystEngComm*, **16** (39), 9345–9352.
[216] Skorupska, E., Kaźmierski, S., Potrzebowski, M.J. (2017) Solid state NMR characterization of ibuprofen:nicotinamide cocrystals and new idea for controlling release of drugs embedded into mesoporous silica particles. *Molecular Pharmaceutics*, **14** (5), 1800–1810.
[217] Dračínský, M., Hodgkinson, P. (2015) Solid-state NMR studies of nucleic acid components. *RSC Advances*, **5** (16), 12300–12310.
[218] Xu, Y., *et al.* (2016) The role of solid-state nuclear magnetic resonance in crystal engineering. *CrystEngComm*, **18** (28), 5236–5252.
[219] Othman, A., *et al.* (2007) Structural study of polymorphs and solvates of finasteride. *Journal of Pharmaceutical Sciences*, **96** (5), 1380–1397.
[220] Apperley, D.C., *et al.* (2005) Characterisation of indomethacin and nifedipine using variable-temperature solid-state NMR. *Magnetic Resonance in Chemistry*, **43** (11), 881–892.

[221] Brus, J., *et al.* (2011) New perspectives of ^{19}F MAS NMR in the characterization of amorphous forms of atorvastatin in dosage formulations. *International Journal of Pharmaceutics*, **409** (1), 62–74.

[222] Yuan, X., Sperger, D., Munson, E.J. Investigating miscibility and molecular mobility of nifedipine-PVP amorphous solid dispersions using solid-state NMR spectroscopy. *Molecular Pharmaceutics*, **11** (1), 329–337.

[223] Skotnicki, M., et al. (2016) Characterization of two distinct amorphous forms of valsartan by solid-state NMR. *Molecular Pharmaceutics*, **13** (1), 211–222.

[224] Skotnicki, M., *et al.* (2015) Bisoprolol and bisoprolol-valsartan compatibility studied by differential scanning calorimetry, nuclear magnetic resonance and X-ray powder diffractometry. *Pharmaceutical Research*, **32**, 414–429.

[225] Billinge, S.J., Levin, I. (2007) The problem with determining atomic structure at the nanoscale. *Science*, **316** (5824), 561–565.

[226] Billinge, S.J., *et al.* (2010) Characterisation of amorphous and nanocrystalline molecular materials by total scattering. *CrystEngComm*, **12** (5), 1366–1368.

[227] Bates, S., *et al.* (2006) Analysis of amorphous and nanocrsytalline solids from their x-ray diffraction patterns. *Pharmaceutical Research*, **23** (10), 2333–2349.

[228] Sun, Y., *et al.* (2012) Stability of amorphous pharmaceutical solids: Crystal growth mechanisms and effect of polymer additives. *AAPS Journal*, **14** (3), 380–388.

[229] Wu, L., Zhang, J., Watanabe, W. (2011) Physical and chemical stability of drug nanoparticles. *Advanced Drug Delivery Reviews*, **63** (6), 456–469.

[230] Zhang, J., *et al.* (2012) Molecular-level characterization of probucol nanocrystal in water by in situ solid-state NMR spectroscopy. *International Journal of Pharmaceutics*, **423** (2), 571–576.

[231] Healy, A.M., *et al.* (2017) Pharmaceutical solvates, hydrates and amorphous forms: A special emphasis on cocrystals. *Advanced Drug Delivery Reviews*, **117**, 25–46.

[232] Laitinen, R., *et al.* (2017) Supersaturating drug delivery systems: The potential of co-amorphous drug formulations. *International Journal of Pharmaceutics*, **532** (1), 1–12.

[233] Yan, D. (2015) Micro-/nanostructured multicomponent molecular materials: design, assembly, and functionality. *Chemistry – A European Journal*, **21** (13), 4880–4896.

[234] Wang, F., *et al.* (2014) Probing the nanostructure, interfacial interaction, and dynamics of chitosan-based nanoparticles by multiscale solid-state NMR. *ACS Applied Materials & Interfaces*, **6** (23), 21397–21407.

3

Microfluidic Analysis Techniques for Safety Assessment of Pharmaceutical Nano- and Microsystems

Tiina M. Sikanen, Iiro Kiiski and Elisa Ollikainen

Faculty of Pharmacy, Drug Research Program, University of Helsinki, Finland

3.1 Microfluidic Bioanalytical Platforms

Since the 1990s, microfluidics has been increasingly applied to chemical and biological research, including drug discovery and development [1, 2]. Microfluidics is by definition a field of study that encompasses the physics of fluid behavior in micrometer structures, and the engineering aspects of design and fabrication of miniaturized devices for controlling the flow of small amounts of fluids (typically in the range of pico–nanoliters). Using microfabrication, all critical operations can be combined on a single microfluidic chip, including but not limited to fluid manipulation and mixing, sample purification and enrichment, (bio)chemical reactions, cell manipulation and culturing, as well as separation and detection of the chemical and biological sample components [3–5]. This field of research is often referred to as "micro total analysis systems" (μTAS) or lab(oratory)-on-a-chip. From the perspective of (bio)chemical analysis, the high degree of integration results in negligible dead volumes (no time lag) between the different units, which substantially decreases the total analysis time per sample. The analytical throughput may be further increased through microfabrication of multiple integrated assays in parallel. From the perspective of cell culturing, introduction of microfluidic flow enables, for example, efficient supply of nutrients to, and removal of metabolic waste from, the cell cultures. In addition, microfluidic

Characterization of Pharmaceutical Nano- and Microsystems, First Edition. Edited by Leena Peltonen.

actuation enables creation of spatial chemical gradients and their precise control over time, which benefits especially mechanism-based studies on both organ and (single) cell levels. Overall, miniaturization also reduces the consumption of expensive and/or toxic chemicals, thus saving resources and producing less chemical waste.

Besides (bio)analytical techniques, microfluidics has been extensively applied to custom manufacturing, such as synthesis and functionalization, of pharmaceutical nano- and microsystems, but these are categorically out of the scope of this book. For an overview of the state-of-the-art in this field, the reader is advised to familiarize themselves with other recent reviews focusing on microfluidic techniques in drug delivery [6, 7]. Here, the aim is to provide the reader with an idea of the rationale design of microfluidic (bio)analytical devices, paying particular attention to the possibilities and limitations associated with the applicable microfabrication methods and materials.

3.2 Microfabrication Methods and Materials

In this section the evolution of microfabrication methods and materials, applicable to manufacturing of micro total analysis systems (or lab-on-a-chip), is reviewed from a general perspective. The material properties critical to each reviewed application field are further emphasized in later sections.

The progressive development of silicon microfabrication processes in the early 1980s enabled the explosive growth of semiconductor technology so that electrical components (transistors, diodes, integrated circuits, solar cells, etc.) could be fabricated at practically zero cost, and mobile electronic devices became available to average users. In the 1990s, the same microfabrication processes were harnessed for manufacturing of miniaturized chemical analysis devices that incorporated integrated micrometer-scale channels for sample loading and electrophoretic separation on a single chip [8, 9]. Although the first silicon-based gas chromatographic separation chip was reported by the end of the 1970s [10], the greater breakthrough emerged via the introduction of glass micromachining techniques to fabricate microfluidic electrophoresis devices [8, 9]. This was because the concept of microchip electrophoresis, on an electrically insulating material such as glass, facilitates a substantial decrease in the duration of analysis down to around 1 min per sample (for more details, see Section 3.5). However, microfabrication of glass by wet or plasma etching requires specialty cleanroom equipment and relatively harsh chemicals, such as hydrogen fluoride [11], which significantly limits the use of glass-based microfluidics in regular research laboratories. Therefore, the introduction of poly(dimethyl siloxane) (PDMS) soft lithography in 1998 [12, 13] was the critical next step to wider exploitation of microfluidics in chemical and biological research, by enabling rapid replication of microfabricated structures in normal laboratory conditions.

In PDMS soft lithography, a master mold is typically prepared (in cleanroom conditions) by transferring the microstructures from a photomask to a photoresist (such as the negative tone epoxy polymer SU-8 [14, 15]), spincoated on a silicon wafer, by ultraviolet (UV) lithography (Figure 3.1). The photomask is typically based on a computer-aided drawing (CAD) translated into a chromium-coated glass plate, although nowadays the quality and feature resolution of printed plastic masks (less expensive) are often sufficient for most microfluidic applications. The silicon wafer (typically 4-inch in diameter) supporting the photoresist pattern is the standard borrowed from the semiconductor industry, as is the UV lithographic patterning of photoresists to reproduce the photomask design onto the silicon wafer. Several commercial vendors exist for both types of photomasks and silicon (and glass) wafers, as well as for both negative and positive tone photoresists. In the case of negative tone photoresists, the areas exposed to UV will crosslink, whereas the masked areas do not and can be dissolved using appropriate developer solution, analogous to photography (Figure 3.1). In a subsequent step, the microstructures of the master mold will be replicated to PDMS by pouring the elastomer solution onto the master, crosslinking the PDMS by heating, and detaching the PDMS replica from the master mold [12, 13], as illustrated in Figure 3.1. Although in this protocol the master fabrication typically still relies on cleanroom processes, the subsequent PDMS crosslinking can be performed in regular laboratory conditions and the master typically tolerates reproduction of multiple PDMS replicas, which significantly reduces the cost and complexity of microfabrication. Equally important

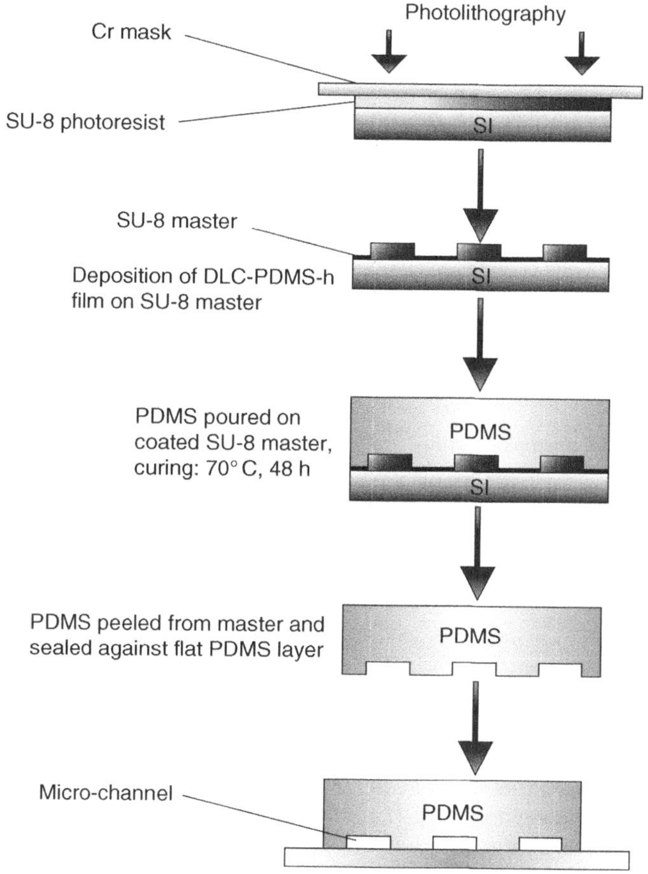

Figure 3.1 Schematic representation of the typical PDMS soft lithography process. Republished with permission of the Royal Society of Chemistry, from Huikko et al. (2003) [17]; permission conveyed through Copyright Clearance Center, Inc.

is the fact that PDMS is self-adhesive to itself and to most other materials, which enables straightforward sealing of the PDMS microstructures [12, 13]. The PDMS–PDMS bonding is by default reversible in nature, but permanent bonding of two PDMS layers for applications requiring higher pressures (>1 bar) may be achieved by treating the surfaces with, for example, oxygen plasma prior to their bonding [16]. For comparison, sealing of glass or silicon-based microstructures with another glass/silicon wafer is much more complicated and typically requires high temperatures or specialist equipment [11].

Besides manufacturing of PDMS-based microfluidic devices, the PDMS replica itself can also be used as a stamp or a mold for further replication steps, as illustrated in Figure 3.2. In principle, the PDMS microstructures can be replicated to any heat- or UV-curable polymer by casting the monomer solution onto the PDMS mold, crosslinking the cast polymer when in contact with PDMS, and detaching the new polymer replica from the PDMS mold. The lifetime of the PDMS mold is typically less than that of the cleanroom master molds,

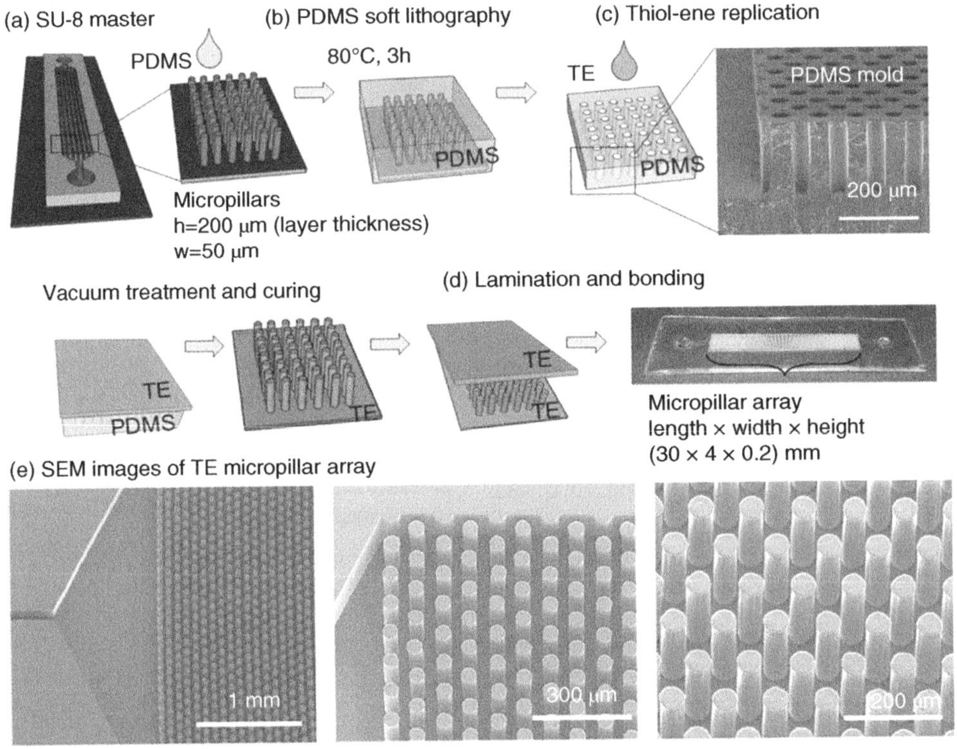

Figure 3.2 *Schematic presentation of the fabrication steps of micropillar arrays reproduced in UV-curable, off-stoichiometric thiol-ene (TE) polymers: (a) SU-8 master fabrication in cleanroom; (b) PDMS soft lithography; (c) TE replica-molding: pouring TE monomer solution onto the PDMS mold, removal of trapped air in vacuum and crosslinking by UV; and (d) sealing of the cured micropillar layer by another (flat) TE layer by lamination and further UV curing of the bonded interface. (e) Scanning electron microscopy (SEM) images of the micropillar arrays. Republished with permission of Springer Nature, from Tähkä et al. (2019) [18]; under the terms of the Creative Commons Attribution 4.0 International License.*

because PDMS tends to absorb the monomers of the cast solution, but it easily lasts for at least a handful of repeated replication cycles [18]. In comparison with direct replication of heat- or UV-curable polymers (other than PDMS) with the help of cleanroom masters (e.g., SU-8), the elasticity of PDMS plays a critical role in facilitating straightforward detachment of the mold and the replica from each other (after curing) without damaging the reproduced microstructures. If both the mold and the replica are mechanically rigid materials, the detachment is much more complicated. Thereby, the introduction of PDMS soft lithography at the end of 1990s significantly increased the freedom in selection of the microfabrication material on a best-fit-for-the-purpose basis so that the materials (properties) could be chosen from the end application viewpoint. However, not all polymers can be sealed in an equally straightforward manner as PDMS devices, since most other polymers do not feature self-adhesive properties. One exception is UV-curable off-stoichiometric thiol-ene (OSTE) polymers, which – depending on the bulk monomer composition and the applied crosslinking conditions – may be sealed by one another following a fairly simple lamination protocol [19, 20]. In addition to straightforward sealing, the OSTE chemistry provides the possibility of adjusting both the bulk properties and the surface chemistry (type and number of functional groups available for further chemical conjugation reactions) on the basis of the end-application needs [18, 20–22]. These possibilities are briefly reviewed in the context of relevant applications in the next sections.

To date, a wealth of polymers, primarily UV-curable or thermoplastic, have been introduced for prototyping of microfluidic devices by replication (e.g., soft lithography, embossing) or direct microfabrication methods (e.g., lithography, micromachining) [23]. Thermoplastic materials, on one hand, can be cost-efficiently processed by a variety of techniques, including injection molding, thermoforming, hot embossing, laser machining, and precision mechanical machining [24]. On the other hand, direct UV lithography of negative tone photoresists, such as SU-8 [14, 15], and organically modified ceramics [25, 26], typically provides superior feature resolution compared with other microfabrication techniques. In recent years, the progressive development and the low-cost of 3D printing has also significantly increased the use of fused deposition modeling and stereolithography for additive manufacturing of both master molds and micro-millifluidic devices [27, 28]. All of these techniques together currently provide versatile possibilities for rapid prototyping of microfluidic devices in both cleanroom and regular laboratory settings. The selection of the proper material and method can in most cases be made from the perspective of the critical materials properties required in the end application. However, a technology barrier still exists in terms of translating the prototypes into commercially viable microfluidic products, since not all materials are feasible for mass manufacturing. Commercial vendors primarily exist for glass-based microfluidic devices as well as for thermoplastics. Lately, roll-to-roll manufacturing has also been introduced to mass production of PDMS [29] and thiol-ene [30] microdevices, which may pave the way for their commercial use as well.

3.3 Microfluidic Cell Cultures

In this section, the possibilities and limitations associated with microfluidic cell culturing, or so called organ-on-a-chip technology, are briefly reviewed together with selected examples of their exploitation to characterization of pharmaceutical nano- and microsystems. Since comprehensive reviews of the previous literature in this field have been published elsewhere [31], the emphasis in this chapter will be on the design of an organ-on-a-chip from the perspective of the microfabrication methods and materials (selection).

3.3.1 Selection of the Microfabrication Material by Design

Compared with conventional static cell cultures, microfluidic cell culturing benefits from efficient nutrient supply and improved elimination of metabolic waste thanks to the through-flow applied over the cells seeded onto the microchannel bottom (Figure 3.3). However, differences in proliferation, glucose metabolism, signaling pathway activation and protein expression levels between cells cultured in traditional macroscale cultures and in microfluidic cultures have also been reported [33]. Much of the original work on organ-on-a-chip development has centered around fully PDMS-based or PDMS–glass hybrid devices, not only because of the straightforward prototyping, but also because of the inherent bio- and cell compatibility and superior optical transparency (down at the UV range) of PDMS [34]. Moreover, PDMS also has high oxygen permeability [35], which is critical to ensuring

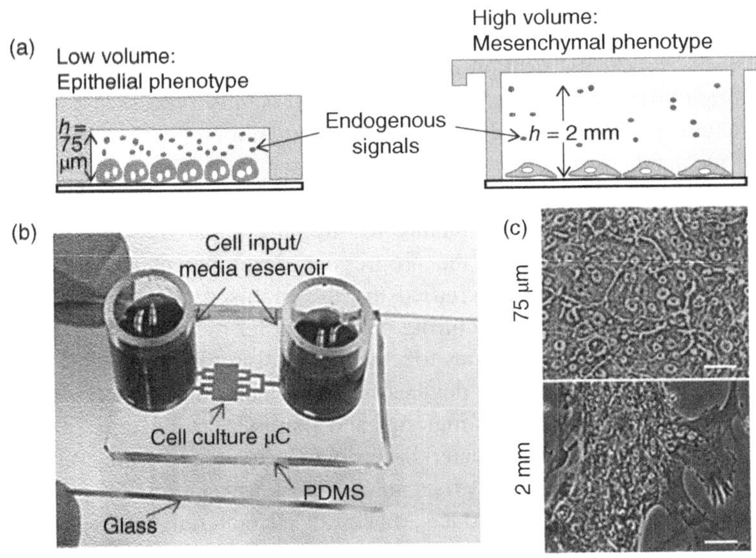

Figure 3.3 (a) Illustration of the difference in phenotype of primary hepatocytes in small vs. large volumes, suggested to arise from the accumulation of endogenous factors in small volumes (h is the height of cell culture chamber). (b) A PDMS-based microdevice perfused with food dye. (c) Bright-field images of primary hepatocytes after seven days in a 12-well plate (h = 2 mm) or inside the microfluidic device (h = 75 μm). Scale bar is 25 μm. Republished with permission of Springer Nature, from Haque et al. (2016) [32]; under the terms of the Creative Commons Attribution 4.0 International License.

sufficient oxygen transfer to the cell culture through the cover (PDMS) layer, especially during the static (no-flow) cell seeding step. On their own, neither PDMS nor glass readily support cell adhesion (at the bottom of the microchannel), and therefore the microchannel is typically coated with proteins promoting cell adhesion prior to cell seeding [34]. However, readily cell-adherent and biocompatible microfabrication materials also exist, such as the commercially available organically modified ceramics [36–38]. Besides material properties, the spatial confinement and shear force will impact upon cell adhesion and proliferation (Figure 3.3). On the one hand, care should be taken to adjust the microfluidic flow rate so as to ensure that the shear stress does not exceed the threshold for cell adhesion [39]. On the other hand, manipulation of the shear force is a prerequisite to ensure physiologically relevant conditions in, for example, vascular cell cultures [40]. The cell-compatibility of the chosen microfabrication material may also be affected by chemicals (e.g., uncrosslinked monomers) leaching from the bulk to the cell culture medium, which limits especially the use of many polymers (other than PDMS) in organ-on-a-chip applications. For example, the leaching monomers of methacrylate-based cell culturing platforms have been reported to induce apoptosis of human osteoblasts [41]. Similar material-induced cell death has also been associated with certain 3D printing [42] as well as thiol-rich OSTE polymers [43]. To eliminate the impacts of leaching chemicals, particular attention thus needs to be paid to post-processing of the cell culture platforms prior to cell seeding. Typically, relatively simple heat treatment, possibly combined with a pre-incubation in the culturing medium, has sufficed to eliminate the leaching monomers [41, 44].

The optical properties (clarity and transparency) of the chosen microfabrication materials mainly determine the suitability of the different cell stains for visualization of cell viability and/or apoptotic biomarkers. Apart from PDMS, many polymers exhibit fairly high autofluorescence in the near-UV range and below [45], although organically modified ceramics and OSTE polymers too are optically clear and transparent down to ca. 300 nm [46, 47], almost the same as glass. To a certain degree, the optical clarity also depends on the chosen microfabrication method. For example, 3D printing by default results in a rough and thus somewhat opaque surface compared with lithographically defined microstructures or the replicated patterns thereof [48]. Because of these reasons, hybrid devices made from PDMS (top layer) and glass (bottom layer) are probably the most used prototype design for an organ-on-a-chip, as this configuration enables high-quality optical monitoring of the on-chip cell culture by both inverted and upright microscopy (see, e.g., Figure 3.3). Besides optical detection, electrical impedance spectroscopy can be applied as a label-free, non-invasive method for monitoring the growth rate (or disruption) of the cell monolayers on microfluidic devices [49]. This approach, however, requires integration of conductive (metal) electrodes at the bottom of the microchannel, which sets certain limitations on the material selection. Namely, the electrode material has to be cell-compatible (such as gold or indium tin oxide) [50, 51] and the substrate material (microchannel bottom) has to support metal adhesion. Metallization processes (sputtering or evaporation) typically rely on cleanroom equipment, but are fairly well established for glass, for example. Instead, metallization of polymers typically suffers from poor adhesion (especially on PDMS [52]) and necessitates development of custom, material-specific processes (see, e.g., metallization of organically modified ceramics [53]). At best, however, a combination of optical and impedance detection allows for long-term culturing with continuous parallel and mutually independent monitoring of the cell growth rates by means of impedance measurements and of specific other cellular events by means of optical or fluorescence microscopy [49].

Besides facilitating cell adhesion and proliferation, and their uninterrupted monitoring, careful selection of the microfabrication materials enables the design of fairly complex organ-on-a-chip platforms. For example, it is well known that the atmospheric oxygen levels (21%) are suboptimal to cell culturing from the perspective of *in vitro–in vivo* correlation [54]: in physiological conditions (oxygen levels in the body range from 1–12%), or under hypoxia, the cells grow faster, live longer, and show lower stress. In addition to conventional approaches (adding nitrogen gas to the cell incubator), hypoxic conditions can also be reproduced on chip, for example by adding oxygen-scavenging chemicals to the liquid feed [55]. In this case, however, the microfluidic oxygen sink often has to be physically separated from the cell culture compartment, so as to avoid the toxic oxygen scavengers from attacking the cells. Figure 3.4 describes a schematic overview of a chip design, where a PDMS membrane was exploited to prevent chemical diffusion into the cell culture (on top of the membrane), while allowing for oxygen transfer from the cell culture to the underlying chamber where oxygen scavenging was chemically induced. In this case a vertical oxygen gradient was created across the atmospheric cell culture (Figure 3.4), but to achieve uniform hypoxia, attention has to be paid to the use of gas-impermeable materials only, so as to prevent oxygen transfer to the cell culture through the microchannel cover layer. The oxygen permeability of most other microfabrication materials is insignificantly small compared with PDMS [56] and does not thereby set limitations to the material selection from the design perspective. Besides chemical additives, it has been reported that certain thiol-rich OSTE polymers feature inherent oxygen-scavenging properties, which can be controlled by simply changing the surface-to-volume ratio of the microfluidic channel or the temperature and the curing time during the fabrication process [44]. Although the inherent oxygen-scavenging capability of thiol-rich OSTE polymers is long-lasting, and thus feasible for long-term cell culturing under hypoxic conditions, it is associated with an excess of uncrosslinked thiol monomers and thus disappears during the heat treatment required to avoid the material-induced cell death of OSTE polymers.

3.3.2 Additional Design Considerations

Materials selection plays a pivotal role in terms of ensuring the cell adhesion and viability as well as defining the prevailing culture conditions (especially oxygen level) inside the microfluidic channels. Besides the surface chemistry and wetting properties, inducing cell alignment via engineered topographical cues may elicit a cellular phenotype similar to aligned tissues *in vivo* [57]. The topography of the microchannel surface may be easily manipulated by microfabrication means. A common approach is to make use of microfabricated (on axis) grooves, which resemble the groove-like topographical features of the extracellular matrix. The impacts of topographical cues, in combination with and in the absence of microfluidic flow, have been particularly studied with a view to maturation of induced pluripotent stem cell-derived cardiomyocytes (iPSC-CM) [58, 59], although they are also beneficial for cell types of skeletal and neuronal lineages [60, 61]. The microgrooves alone have been shown to bring about cellular alignment ($p < 0.0001$) and

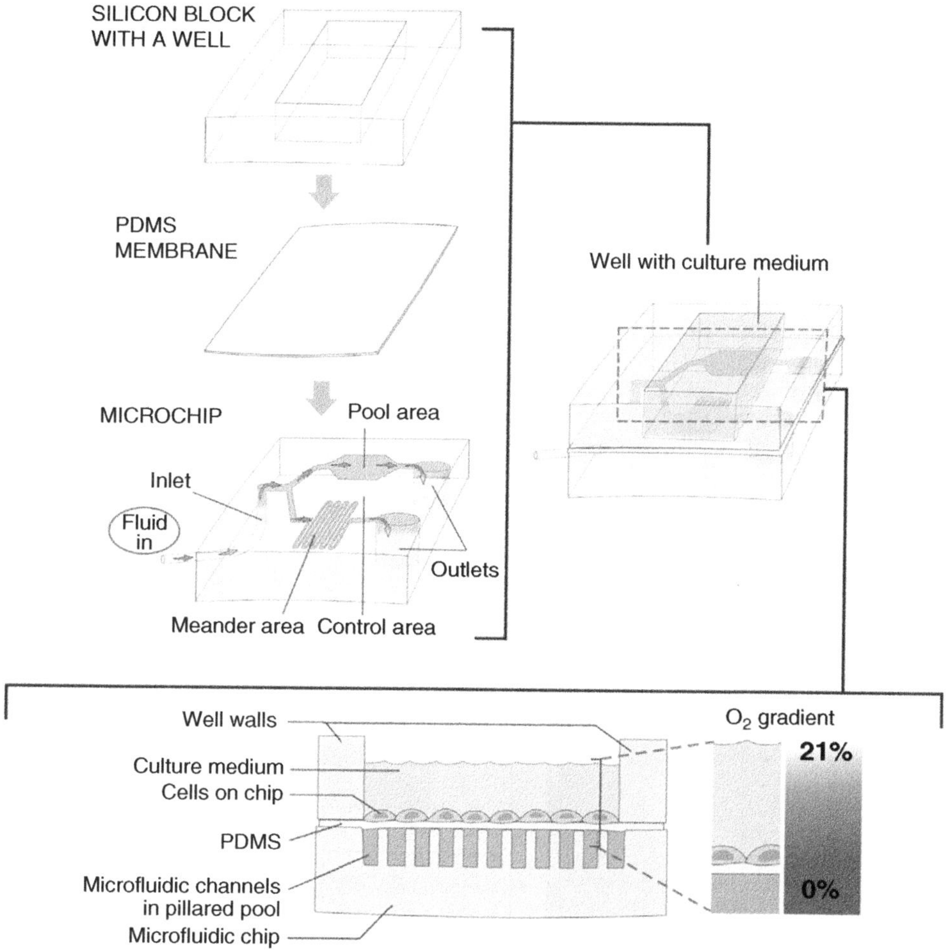

Figure 3.4 *A schematic presentation of the microfluidic organ-on-a-chip used for creating hypoxic conditions on chip, while simultaneously isolating the toxic oxygen-scavenging chemicals from the cell culture compartment with the help of a PDMS membrane. Republished with permission of Elsevier, from Barmaki et al. (2018) [55]; permission conveyed through Copyright Clearance Center, Inc.*

more organized sarcomeres (Figure 3.5), and thus improved Ca^{2+} cycling not associated with modifications in gene expression of iPSC-CMs [58]. The synergistic impact of topographical cues and sustained release of biochemical growth factors has been reported to improve even differentiation of human mesenchymal stem cells toward myogenic lineage [61].

Along with cell monolayers, microfluidic three-dimensional (3D) cell culturing techniques have received increased recognition by both researchers and administrators [62]. In 3D, the cells are known to retain tissue-specific architecture better than in monolayers [63, 64], because the flat and rigid surface requires cytoskeleton to establish contact

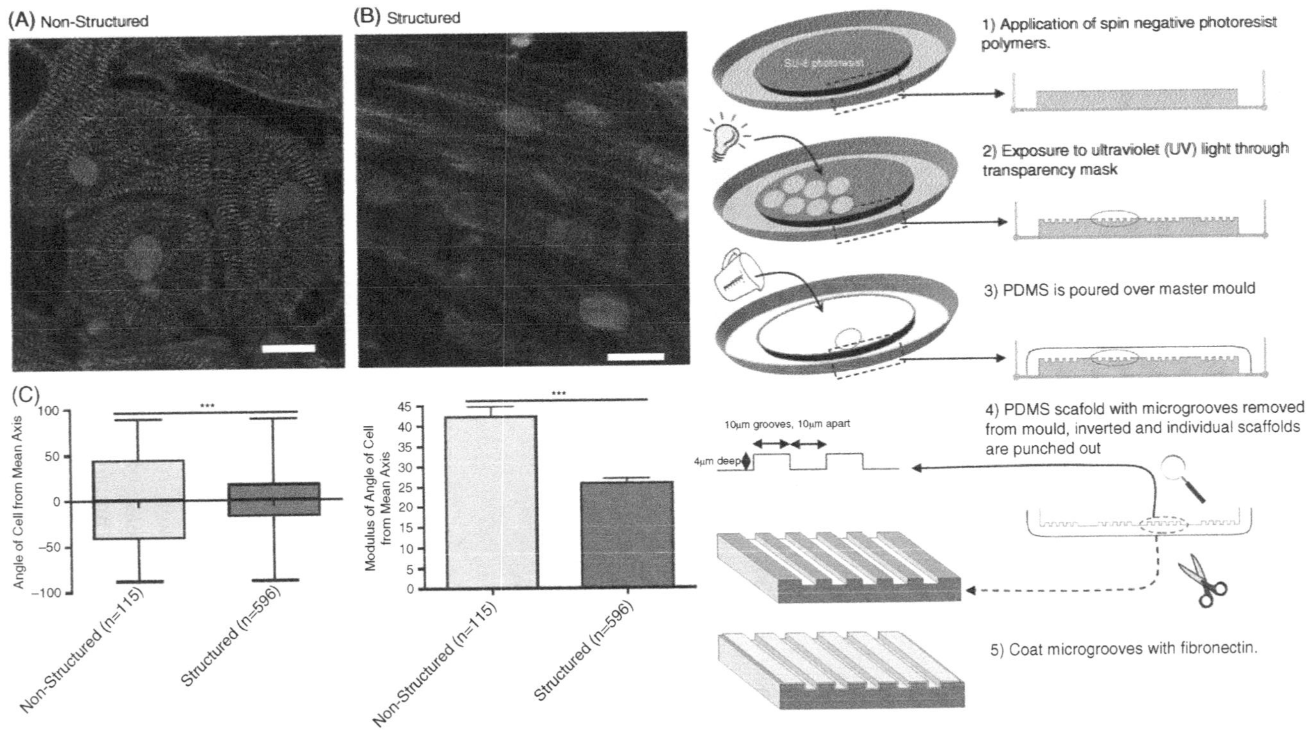

Figure 3.5 *Left: Representative immunofluorescence of iPSC-CM cultured on (A) unstructured PDMS and (B) microgrooved PDMS; dark grey – sarcomeric α-actin, light grey – DAPI, scale bar 20 μm. (C) Quantification of cell alignment iPSC-CM on structured and unstructured constructs. Right: Schematic demonstrating the fabrication of microgrooved tissue culture substrates (not drawn to scale). Republished with permission of Elsevier, from Rao et al. (2013) [58]; under the terms of the Creative Commons Attribution 4.0 International License.*

between neighboring cells and exert artificial polarity [65], which prevents the formation of adequate extracellular matrix (ECM) and the cell–cell and cell–matrix interactions [66]. Moreover, the lack of certain cell-specific functionalities in monolayer cultures (e.g., subphysiological expression of metabolizing enzymes or cell polarization) favor the use of 3D cell models in modern drug discovery and development [67]. Overall, culturing cells in 3D often improves the accuracy of *in vitro–in vivo* correlation in cell-based safety and efficacy assessment. For example, 3D tumor models are often more resistant to drug therapies compared with monolayer cultures [68]. The microfluidic 3D cell cultures can be established by a variety of techniques, including both scaffold-based (e.g., encapsulation of cells in 3D hydrogels) and scaffold-free strategies, such as forced floating in hanging drops or in cell-repellant microwells that result in the formation of spherical cell aggregates [62, 69]. However, the increasing complexity does not automatically mean better results. For example, the diffusion-limited penetration of antibodies and other cell stains are prevailing challenges commonly associated with 3D cell cultures, especially those exploiting scaffold-based strategies. Scaffold-based strategies also raise issues of biocompatibility and cell-material biorecognition, whereas biodegradable scaffolds substitute a large amount of ECM and result in 3D cultures composed of less densely packed cells [70]. In contrast, scaffold-free approaches initiate interactions between cells, which aids the formation of self-generated ECM. Compared with conventional scaffold-free methods, often featuring limited reproducibility and size uniformity, microfluidic 3D cell culturing devices hold the promise of high throughput in both generation and handling of a multitude of uniform-sized spheroids on a single platform [69]. Fabrication of microwells, and particularly their integration with microfluidic channels, is generally more straightforward than that of hanging drop devices. However, achieving cell-repellant surfaces to induce forced floating in microwells, as illustrated in Figure 3.6, typically requires additional post-processing steps. The common approaches taken to reduce cell adhesion include PEGylation [72] and nanostructuring of hydrophobized surfaces [73, 74]. Moreover, most microfabrication methods and materials yield vertical or near-vertical walls, i.e., cylindrical microwells, which are suboptimal for controlling the spheroid growth. Although substantial effort has gone on development of feasible methods for reproducing rounded microstructures (cross-section profiles), their fabrication is generally challenging and in most cases relies on nonstandard techniques, such as overexposure of organically modified ceramics [26] or replica-molding of microdroplet-shaped features obtained via photoresist reflow [75]. Many of the other available techniques, such as isotropic etching [76], laser ablation [77], and milling [78], are limited in terms of achievable aspect ratio (height/width), which may set limitations on total culturing time due to spheroids outgrowing the microwells. Nevertheless, U-shaped microwells are preferred for on-chip spheroid culturing, because in cylindrical microwells, the cells are initially located far apart from each other, which may impair the cell–cell interactions and subsequently reduce the reproducibility and size uniformity from well to well. Increasing the initial cell count or decreasing the well dimensions often improve reproducibility, but simultaneously limit the total culturing time as the spheroids outgrow the wells faster. In contrast, U-shaped wells enable single-pointed gravitational force, which brings the cells closer to each other even in relatively large wells, resulting in superior reproducibility [71].

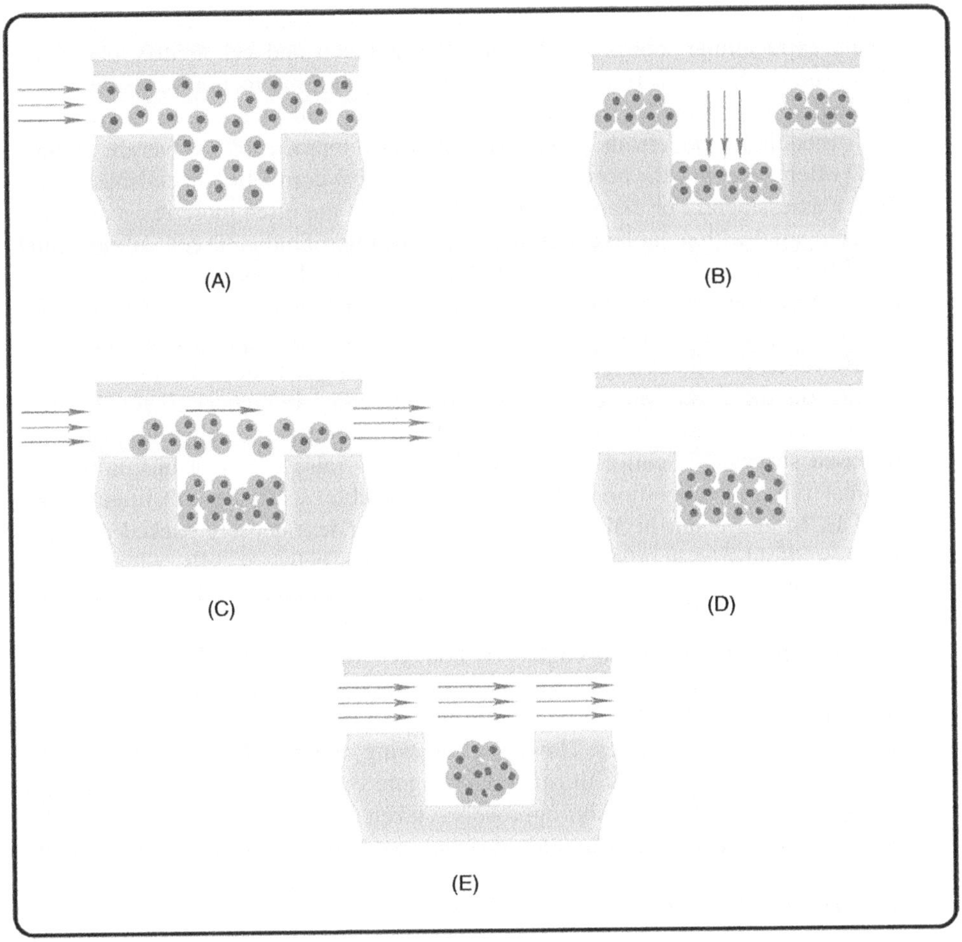

Figure 3.6 Spheroid formation process in a microwell-based organ-on-a-chip. (A) Introduction of a cell suspension to the chip inlet. The cell suspension fills all the microchannels and microwells rapidly due to the capillary effect. (B) Cells start depositing on the bottom of the microchannels and microwells. (C) Pure culture medium flows through the chip to rinse the excess cells without disturbing the cells lying on the microwell bottom. (D) Cell secretions and signaling lead to establishment of cell–cell interactions on the non-adherent microwell bottom. (E) Driving spheroid formation under a perfusing flow of culture medium. Republished with permission of Elsevier, from Moshksayan et al. (2018) [71]; permission conveyed through Copyright Clearance Center, Inc.

3.3.3 Characterization of Pharmaceutical Nano- and Microsystems Using Organ-on-a-chip

By enabling simulation of dynamic fluid flows, chemical and oxygen gradients, and partitioning between organs, a microfluidic organ-on-a-chip can offer a cost-efficient approach to rapid *in vitro* screening of pharmaceutical nanocarriers in physiologically relevant conditions. As most pharmaceutical nanocarriers are intravenously administered into the blood, the preliminary work on microfluidic devices has addressed optimal nanoparticle (NP) design, for example via NP margination, effect of vessel geometry and shear stress on NP accumulation, interactions between red blood cells/platelets and NPs, and vessel permeability on NP translocation [31]. Microfluidic organ-on-a-chip devices have also been developed to assess the NP haemocompatibility [79] and NP transport across the blood–brain barrier [80] and the alveolar–capillary barrier in the lungs [81]. The classical example of a microfluidic lung-on-a-chip device, capable of mimicking the mechanical distortion of the alveolar–capillary interface during normal breathing, is illustrated in Figure 3.7. Besides these selected examples, a variety of organ-on-a-chip models, including single cell traps, have been developed to enable assessment of the cellular uptake and cytotoxicity of NPs under microfluidic flow, so as to avoid inaccuracies associated with, for example, NP sedimentation (common to static cell cultures). For a more comprehensive overview, the reader is advised to familiarize themselves with references [6, 31].

3.4 Immobilized Enzyme Microreactors for Hepatic Safety Assessment

The possibility of targeted identification and treatment of tumors remains one of the main drivers of nanomedicine development [82]. However, the challenges associated with the selectivity of NP targeting to tumors [82] often results in substantial accumulation and sequestering of NPs to the liver, by estimates of up to 30–99% of the administered dose [83]. NP accumulation to the liver inevitably results in reduced NP delivery to the tumor and potentially leads to increased hepatic toxicity at the cellular level, thus creating a major barrier to clinical translation of NP-based therapies. As a result, the clinical benefits of NPs so far have mainly arisen from formulations that improve the pharmacokinetics and toxicity profiles of chemotherapeutic agents [84]. This section focuses on the hepatic safety assessment of nanoparticles (NP) and gives an overview of the development of microfluidic immobilized enzyme reactors that could facilitate examination of the (adverse) hepatic effects of nanomedicines under physiologically relevant conditions. In comparison to cell-based *in vitro* models, the subcellular enzyme assays likely enable more detailed assessment of the mechanistic basis of possible NP interactions once taken up by cells. Similar to cell-based assays, conducting these experiments under microfluidic flow will help to avoid inaccuracies associated with NP sedimentation. Microfluidic actuation also allows for precise control of the NP exposure time so as to determine whether the possible impacts are reversible or permanent in nature.

3.4.1 Nanoparticle Impacts on the Hepatic Clearance of Xenobiotics

The nonspecific accumulation of NPs within the liver can be explained by its normal physiology. Since the metabolism and clearance of foreign materials are the primary functions of the liver, engineered NPs that cannot be cleared by the renal system (threshold ca. >6 nm) will eventually be processed in the liver [85]. Moreover, the primary mechanism for the passive accumulation of NPs in tumors is considered to result from the combination of fenestrations in the vasculature and poor lymphatic drainage from tumors [86], known as the

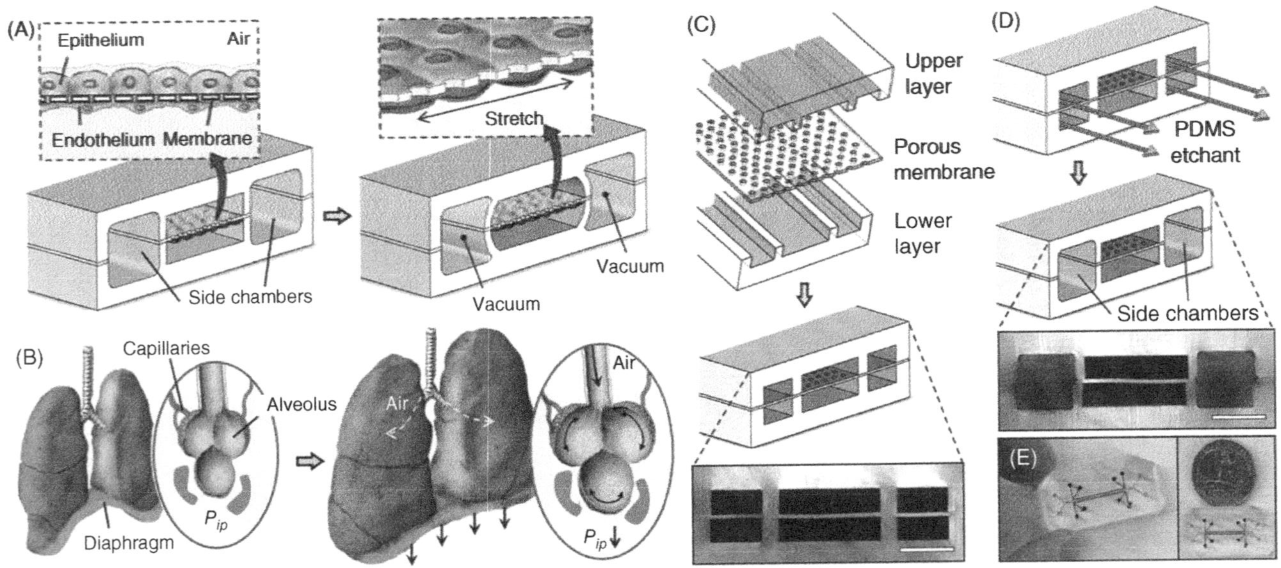

__Figure 3.7__ Biologically inspired design of a human breathing lung-on-a-chip microdevice. (A) The microfabricated lung-mimicking device uses compartmentalized PDMS microchannels to form an alveolar–capillary barrier on a thin, porous, flexible PDMS membrane coated with ECM. The device recreates physiological breathing movements by applying a vacuum to the side-chambers and causing mechanical stretching of the PDMS membrane forming the alveolar–capillary barrier. (B) During inhalation in the living lung, contraction of the diaphragm causes a reduction in intrapleural pressure (P_{ip}), leading to distension of the alveoli and physical stretching of the alveolar–capillary interface. (C) Three PDMS layers are aligned and irreversibly bonded to form two sets of three parallel microchannels separated by a 10-mm thick PDMS membrane containing an array of through-holes with an effective diameter of 10 mm. Scale bar, 200 mm. (D) After permanent bonding, PDMS etchant flows through the side-channels. Selective etching of the membrane layers in these channels produces two large side-chambers to which a vacuum is applied to cause mechanical stretching. Scale bar, 200 mm. (E) Images of an actual lung-on-a-chip microfluidic device viewed from above. Republished with permission of the American Association for the Advancement of Science, from Huh et al. (2010) [81]; permission conveyed through Copyright Clearance Center, Inc.

enhanced permeability and retention (EPR) effect [87]. However, similar fenestrations in the blood vessels are a part of the normal physiology of the liver and provide the pathway to passive accumulation of NPs in the hepatic tissue [88]. Once trapped in the liver, NPs have been reported to interact with hepatocytes, liver sinusoidal endothelial cells, B cells and Kupffer cells [89]. Even NP uptake in the hepatocytes has been reported and is typically impacted by the NP surface charge and other chemical modifications [90, 91]. As hepatocytes make up 60% of all the cells in the liver and are responsible for the metabolism and elimination of most foreign compounds, including pharmaceuticals, *in vivo*, understanding the NP interactions in the liver is critical to the clinical translation of nanomedicine for diagnosing and treating diseases.

If taken up by hepatocytes, NPs may interact with the cellular content, such as the cytochrome P450 (CYP) system, prior to their elimination via the hepatobiliary route [85]. The CYP system is in charge of the phase I metabolism of the majority of drugs in clinical use [92]. Although further conjugation reactions (e.g., glucuronidation) may follow CYP metabolism, the substrate specificity associated with CYPs forms a bottleneck and renders the CYP-mediated metabolic clearance most critical in respect to the safety and efficacy of medical therapies, especially in the case of highly polymorphic isoforms such as CYP2D6 [93]. Moreover, CYP enzymes are membrane-bound proteins that reside on the cytosolic side of the endoplastic reticulum (ER) and are thus more liable to NP interactions compared with, for example, the phase II uridine 5′-diphospho-glucuronosyltransferases (UGTs), which face the lumen side of the ER. Currently, several *in vitro* models exist for studying the hepatic drug metabolism via the CYP system, including transgenic cell lines and primary hepatocytes, liver microsomal fractions and recombinant (single) CYP isoforms expressed in insect cell (referred to as baculosome or supersome in the commercial contexts) [94, 95]. Of the different available *in vitro* models, the subcellular preparations (microsomes, recombinant enzymes) provide the most straightforward approach to mechanism-based interaction screening. Human liver microsomes (HLMs) have become the industry standard for pre-clinical metabolic profiling and drug interaction studies because of their relative affordability and acceptable *in vivo* resemblance [95, 96]. HLMs consist of vesicles of the hepatocyte ER, prepared by differential centrifugation, so as to ensure that the microsomal preparation contains all the CYP isoforms as well as the redox-partners. The recombinant microsomal human CYPs produced in the insect cells are primarily used for preliminary screening of isoform-specific interaction risks owing to their limited *in vivo* correlation with respect to predicting the hepatic clearance. Via protein engineering, it is also possible to manufacture soluble (not membrane-bound) human CYP isoforms by truncating the hydrophobic N-terminal membrane bound domain of CYPs [97]. However, this neither omits the need for the auxiliary proteins or an electron source, nor provides a viable model with a view to *in vivo* predictions.

The research regarding CYP interactions has mainly centered around metallic NPs (silver, gold), which have been shown to alter the CYP metabolism both *in vitro* and *in vivo* [98]. In most *in vitro* studies, the impact of silver NPs on human and rat CYP systems (assessed based on liver microsomes or recombinant enzymes) has been weak or insignificant [99–101], although moderate inhibition of CYP2C9, CYP2C19, and CYP3A4 activities in human liver microsomes has also been reported [99]. The impacts of gold NPs *in vitro* are often somewhat similar to those of silver NPs [102–104]. For example, silver NPs have been reported to cause moderate to strong inhibition of recombinant CYP1A2,

CYP2C9, and CYP2C19 activities *in vitro* [104], whereas the gold NPs moderately inhibited recombinant CYP2C9, CYP2C19, and CYP3A4 activities, but not so much the CYP1A2 [104]. However, the *in vivo* interactions vary greatly depending on the NP type, size and charge, which are the critical NP properties in terms of cellular uptake. For example, an almost 28-fold increase in the CYP1A1 expression in rats *in vivo* was reported after a single intravenous injection of gold NPs [102]. In the case of silver NPs, the strong inhibition of CYP2C and 2D activities observed in rat liver microsomes *in vitro* was not reproduced *in vivo* [101].

Generally, the risk for CYP inhibition tends to increase as a function of decreasing particle size, as has been confirmed by both gold and carboxyl polystyrene NPs [103, 105]. In addition to size, the surface charge is known to alter the NP uptake by hepatic cell types due to differences in protein adsorption to the NP surface and the electrostatic interactions between the NPs and the cell membrane [106, 107]. However, the limited data available on NP uptake in hepatocytes results in intricate *in vitro–in vivo* correlation with respect to the CYP inhibition. Moreover, the subcellular fractions (such as liver microsomes) commonly used in the enzyme inhibition studies *in vitro* are not capable of predicting the possible NP impacts on gene expression (i.e., CYP induction). Nevertheless, the determination of the NP impacts on the *in vitro* clearance of drugs via the CYP system could provide the preliminary tool for identification of possibly hazardous NP–drug interactions. For example, the pioneering work with porous silicon-based NPs reported significantly reduced activity (ca. 50–60% of the control, $p < 0.001$) for the highly polymorphic CYP2D6 in human liver microsomes in the presence of NPs independent of their surface chemistry or charge [108]. Besides enzyme activity (V_{MAX}), the NPs were shown to alter the enzyme affinity (K_M) so that the total intrinsic clearance (CL_{int}) was most impacted by the alkyne-modified porous silicon NPs. Owing to the relatively large size of the porous silicon NPs used in the study (160–180 nm), competitive binding to the enzyme's active size (typically below 3 nm^3 [109]) was not likely. Instead, the inhibition mechanism was concluded to be a combination of noncompetitive and uncompetitive binding [108]. Considering the highly variable expression of CYP2D6 between individuals (rapid vs. slow metabolizer genotypes) [92], these preliminary *in vitro* findings suggest that NP interactions with the CYP system should not be neglected. Currently, the "assay cascade" for the characterization of nanomaterials, developed by the European Nanomedicine Characterisation Laboratory (EUNCL), does not include characterization of the CYP interactions, but focuses on determination of the physical and chemical properties of NPs, as well as screening of their immunological, hematological, and toxicological properties *in vitro* [110]. The microfluidic subcellular characterization techniques reviewed in the next section could provide the means for high-throughput preliminary screening of the impacts of NPs on the CYP system *in vitro*. Besides NP effects, the same technology readily fits for screening of drug–drug interactions under microfluidic flow, for example, in combination with controlled release studies.

3.4.2 Cytochrome P450 Interaction Studies in Through-flow Conditions

In microfluidic cell-based assays, the cells readily adhere to the microchannel bottom strongly enough that they remain "immobilized" upon application of the through-flow, if the flow-induced shear stress is tolerable. In contrast to intact cells, the use of cellular components, such as microsomes or recombinant enzymes, requires custom methodology for incorporating the enzymes into the microchannel in a permanent fashion. Once immobilized, the enzymes are compartmentalized in a confined space, which facilitates both spatial and temporal control of, for example, NP exposure with the help of microfluidic through-flow. In enzyme kinetic characterization, the microfluidic actuation enables the creation of concentration gradients of the substrate of interest and thus determination of the kinetic constants with a single experiment [111–113]. From the analysis viewpoint, enzyme immobilization omits the need for sample purification (separation of the protein and lipid content from the analytes of interest), which facilitates online coupling of the reactor to a mass spectrometer, for example [114]. But this approach is applicable to NP characterization only to a limited extent. Sometimes, immobilization also enhances the stability of the enzymes toward denaturation during both storage and use. Section 3.4.2.1 gives an overview of the commonly used strategies feasible for enzyme immobilization on microfluidic devices and the special precautions that should be accounted for when establishing immobilized CYP microreactors. Section 3.4.2.2 provides additional considerations regarding the design of the microreactors and the selection of the microfabrication materials.

3.4.2.1 Immobilization Strategies for Cytochrome P450 Enzymes

The CYP system as a whole comprises multiple membrane-bound cooperative enzymes, which places an extra challenge to the drafting of biologically relevant immobilization strategies. Besides CYP isoforms, various redox partner enzymes are required to supply electrons for the metabolism reactions, most importantly nicotinamide adenine dinucleotide phosphate-cytochrome P450 reductase (NADPH-CPR) [115, 116]. In addition, the reactions require both oxygen and NADPH as cofactors and the enzymes need to remain in their natural environment, embedded in the lipid bilayer, to maintain their activity [117]. The immobilization as well as operational conditions should also be gentle enough to ensure that the enzymes are not denatured during the process. For example, elevated temperatures [118], organic solvents [119] and oxygen radicals generated in CYP-mediated reactions [120, 121] have all been reported to promote CYP inactivation. To be able to draw clinically relevant conclusions based on the *in vitro* data of the immobilized enzymes, the kinetic parameters should remain unchanged compared with non-immobilized enzymes, which complicates the use of scaffold-based immobilization techniques and incorporation of chemical modifications to the protein structures [113]. The pros and cons of the common enzyme immobilization techniques with respect to CYP immobilization are briefly discussed below. These techniques are generally divided into three main categories: enzymes can either be (i) entrapped inside a porous matrix; (ii) crosslinked with each other to form enzyme aggregates or crystals; or (iii) bound to an inert support material via physical (adsorption) or chemical interactions (covalent or affinity-based binding) (Figure 3.8).

With the entrapment techniques, enzymes are physically retained inside a natural or synthetic polymer matrix, which is synthetized in the presence of the enzyme. The matrix forms physical constraints that enable mechanical enzyme immobilization in high volumetric concentrations and relatively mild immobilization conditions [123, 124]. Entrapment is thus readily feasible for immobilization of both soluble and membrane-bound enzymes, as it does not require any specific chemistry. However, the density of the matrix plays a critical

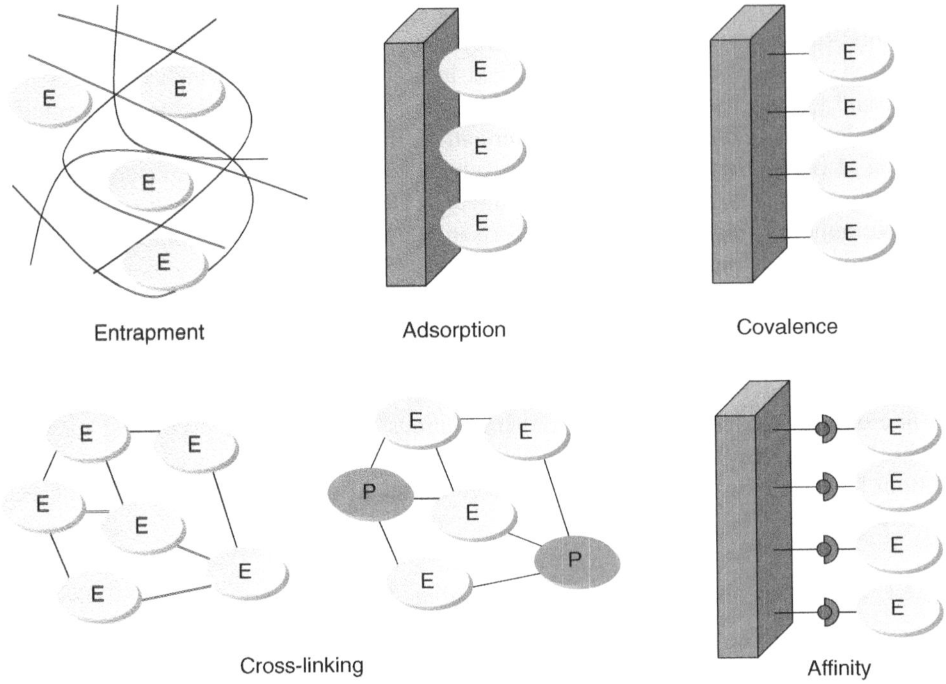

Figure 3.8 *Schematic representation of the main different methods of enzyme immobilization. E: enzyme, P: inert protein. Republished with permission of Elsevier, from Sassolas et al. (2012) [122]; permission conveyed through Copyright Clearance Center, Inc.*

role and has to be carefully optimized to avoid enzyme leaching (too loose matrix) and diffusion-limited kinetics resulting from retarded mass transfer (too dense matrix) [125]. On microfluidic devices, both sol-gels [126] and hydrogels [127] have been used as matrices for enzyme entrapment, but hydrogels are often preferred because of their better transparency and biocompatibility (no toxic reagents needed for polymerization) [128]. Commonly used hydrogels include synthetic polyethylene glycol and polyacrylamide as well as chitosan, agarose and other polymeric biomolecules [129]. In a pioneering work, the shelf-life of HLM (during storage) was substantially improved (weeks vs. days) as the result of sol-gel entrapment, but the enzyme affinity (testosterone to CYP3A4) was simultaneously increased compared with that of non-immobilized enzymes [126]. The CYP activities of entrapped HLM are generally similar to those of nonimmobilized enzymes, but long-term stabilities during use are seldom reported. Besides gel-based matrices, HLMs have also been entrapped in microfluidic devices with the help of porous (pore size 0.4 μm) polycarbonate membranes [130].

In contrast to scaffold-based enzyme immobilization, scaffold-free enzyme macroparticles can be prepared via crosslinking, including crosslinked enzyme crystals (CLECs) and crosslinked enzyme aggregates (CLEAs) [125, 128]. CLECs are typically prepared from crystallized enzyme by the addition of a bifunctional reagent, such as glutaraldehyde, which results in stable and highly active enzyme particulates of controlled size. Compared with

CLECs, crosslinking of aggregates (CLEAs) is somewhat easier and does not necessarily require such high enzyme purity. However, neither of the two techniques is very feasible for immobilization of membrane proteins, such as mammalian CYPs, and crosslinking approaches have thus been primarily used for soluble bacterial CYPs only [131, 132].

The enzyme binding on support materials can be either covalent or non-covalent. Covalent binding typically relies on functional groups ($-NH_2$ and $-COOH$) naturally occurring in enzymes. This often results in non-uniform orientation with respect to the active site of the immobilized enzymes, as it is difficult to control where in the amino acid chain the covalent bond will form. If the enzyme's active site is "face down" toward the support material, the overall activity is inevitably decreased because of apparent steric hindrance. Therefore, spacer molecules, attached to the support material prior to enzyme binding, are often employed to distance the enzyme from the surface [133]. Covalent binding usually allows for a relatively stable anchorage of the enzyme to the carrier surface. However, if the charged amino acid residues are extensively employed by the covalent bonds, the enzyme's surface charge, and thus its natural conformity may change and activity decrease upon immobilization [134], although increased activities have also been reported [135]. In a prior work, CYP2C9 baculosomes were covalently bound on magnetic particles with the carbodiimide method to create a packed bed capillary reactor compatible with online capillary electrophoresis analysis [136]. Although the setup facilitated enzyme kinetic determinations in through-flow conditions, rapid activity loss of the immobilized enzyme during the measurements was reported. In most other reports, covalent binding has mainly been applied to soluble recombinant CYPs (lacking the hydrophobic N-terminus sequence) [137–139]. However, the disadvantage of soluble enzyme systems is the lack of auxiliary proteins (redox partners), which need to be separately introduced along with the through-flow [137] or immobilized onto the same carrier. Adjusting the stoichiometric ratios and the spatial relationships of the different redox proteins is, however, technically and economically challenging. Moreover, with respect to *in vitro–in vivo* extrapolation, the immobilized enzymes should ideally maintain their physiology unchanged, which reduces the biological significance of the *in vitro* data derived from soluble (not membrane-bound) enzymes.

The non-covalent methods (Figure 3.8) can be divided into physical (hydrophobic and van der Waals interactions), ionic (electrostatic interactions), and affinity-based (e.g., antibody-mediated binding) methods [125, 134]. Weak non-covalent interactions do not usually alter the tertiary structure of the enzymes as much as covalent binding, which helps the enzymes to retain their natural activity [140]. However, the non-covalently bound enzymes are easily washed away from the carrier surface, especially at elevated temperatures, owing to their low binding energies compared with covalent binding [125, 134]. An exception is affinity-based biotin–avidin binding, which has been exploited to immobilize recombinant CYPs by incorporating the biotin tag to the protein structure and binding the biotinylated CYPs on an avidin-coated chromatography column [141]. However, this approach also required chemical modification of the protein structure, which resulted in loss of the enzyme activity at physiological temperatures. In another work [142], biotin-containing, unilamellar fusogenic liposomes were employed to transfer the biotin tag to the HLM membrane (Figure 3.9), which is a substantially less laborious and more universal protocol compared with biotin labeling of the protein structure [141]. By modifying the lipid membrane instead of the protein structure, the overall immobilization process was gentle enough that the enzyme kinetic parameters remained similar to those

(a)

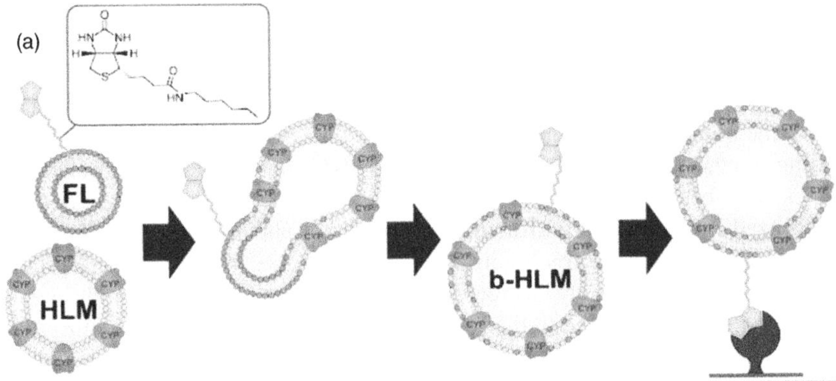

(b)

		Isoenzyme			
		CYP3A4	CYP2C9	CYP2D6	CYP2A6
K_M (μM)	Soluble	5.2 ± 0.5	75.1 ± 5.3	21.0 ± 2.4	2.9 ± 0.3
	Immobilized	15.8 ± 1.2	116.4 ± 37.3	32.0 ± 5.2	2.2 ± 0.5

Figure 3.9 *Immobilization of human liver microsomes (HLMs) on streptavidin-functionalized magnetic particles with the help of fusogenic liposomes. (a) A schematic describing the membrane fusion between biotin-containing fusogenic liposomes and HLM resulting in biotinylated HLM. (b) Isoenzyme-specific K_M values (± standard error) of four main CYPs for immobilized and non-immobilized (soluble) HLMs. Republished with permission of John Wiley and Sons, from Kiiski et al. (2019) [142]; permission conveyed through Copyright Clearance Center, Inc.*

of non-immobilized enzymes [142]. Besides fusogenic liposomes, artificial lipid bilayers prepared on solid supports have also been employed to immobilize microsomal vesicles via membrane fusion [143].

3.4.2.2 *Microfabrication Materials and Design Considerations*

In addition to enzyme activity and stability, implementation of microfluidic immobilized enzyme reactors necessitates careful selection of proper microfabrication materials as well as design that is mindful of the through-flow reaction chamber. Integration of scaffold-based immobilization matrices with a microfluidic network is relatively straightforward. For example, polyethylene glycol based hydrogels can be microstructured *in situ* (through the microchannel cover layer) by UV lithography, provided the covering layer is transparent to the chosen wavelengths [127]. But as the scaffold-based systems suffer from retarded diffusion, direct binding of enzymes on a solid support material is practically the better choice for enzyme kinetic determinations. However, functionalization of mere microchannel walls rarely enables immobilization of high enough amounts of enzymes to produce detectable amounts of metabolites for deriving the kinetic parameters. The total surface area (per volume) available for enzyme binding is typically increased by "packing" the microreactor with microbeads [136, 137] or porous polymer monoliths [144, 145] (Figure 3.10). Microstructured pillar arrays similar to those in Figure 3.2 can also be used to increase the total surface area [142, 147, 148].

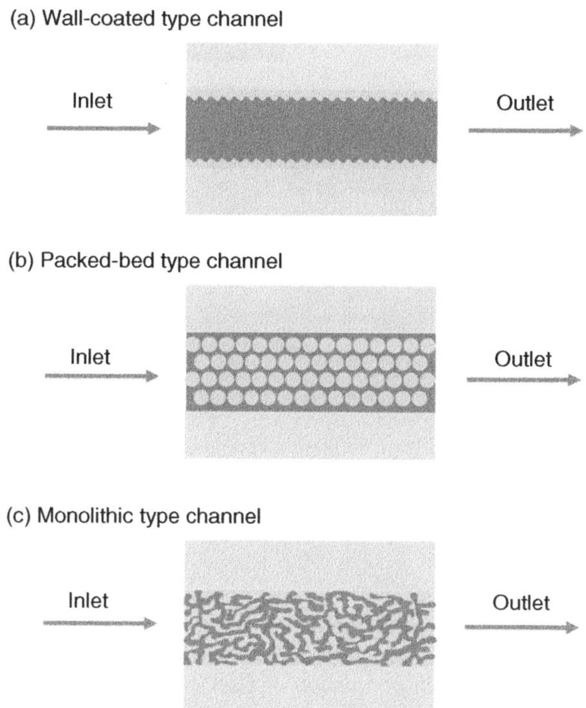

Figure 3.10 *Typical designs for increasing the surface-to-volume ratio of microfluidic channels: (a) wall-coated type channel; (b) packed-bed type channel; and (c) monolithic type channel. Republished with permission of the Royal Society of Chemistry, from Zhu et al. (2020) [146]; permission conveyed through Copyright Clearance Center, Inc.*

Microchannel packing with beads is relatively straightforward: the beads can be introduced into the channel by injecting, and retained with the help of magnetic field (in case of magnetic microbeads [149]) or microfabricated frit structures (Figure 3.11a). There is also a wide range of commercially available microbeads with different surface chemistries for both affinity-based and covalent binding of enzymes. However, tight microbead packings easily increase the back pressure in the microfluidic system and thus necessitate strong bonding between the microchannel bottom and cover layers, rendering, for instance, PDMS hardly feasible for packed bed applications. *In situ* curing of porous polymer monoliths by UV (through the microchannel cover layer) or heating, in the presence of porogens, is also a relatively straightforward process, which only requires UV transparent

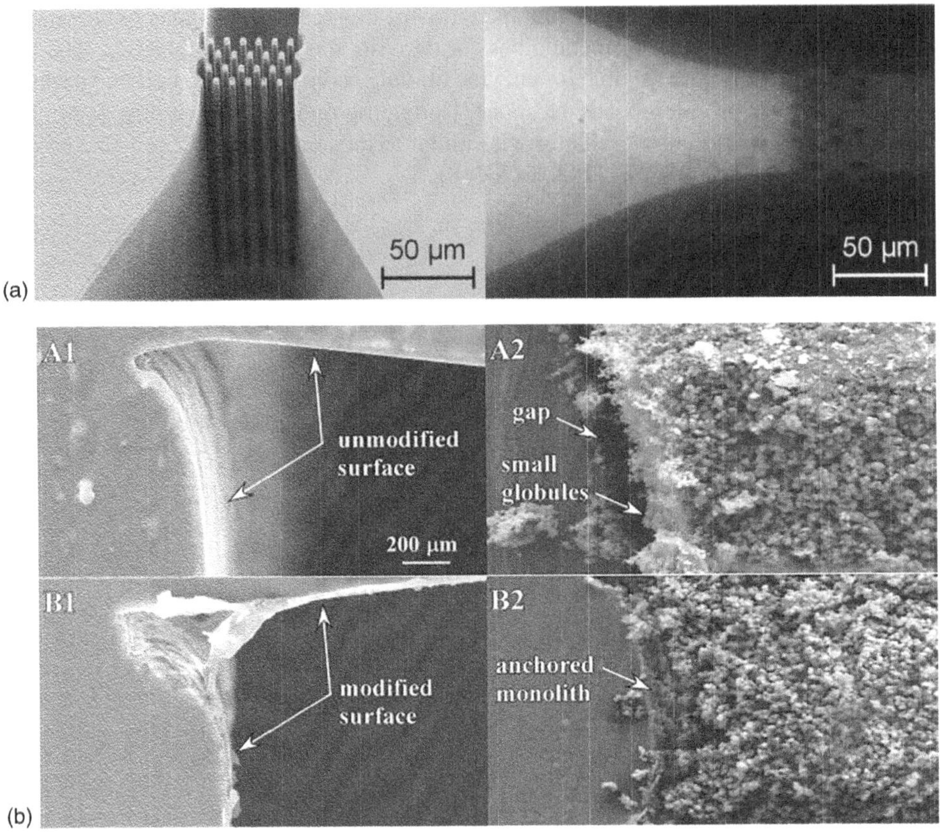

Figure 3.11 (a) Scanning electron micrograph of microfabricated frit structures (left) acting as a mechanical barrier to retain fluorescent microbeads injected into a microfluidic channel (right, top view). Republished with permission of Elsevier, from Haapala et al. (2010) [150]; permission conveyed through Copyright Clearance Center, Inc. (b) Methacrylate-based porous polymer monoliths cured in situ in an unmodified (A1, A2) and modified (B1, B2) PDMS microchannel, illustrating the void volume next to the microchannel wall (A2) if anchoring of the polymer monolith is deficient. Republished with permission of AIP Publishing, from Burke and Smela (2012) [151]; permission conveyed through Copyright Clearance Center, Inc.

or thermally stable microfabrication materials. Compared with microbead-based packings, the porous polymer monoliths typically show lower back pressures and do not require mechanical frit structures as there is no risk of solid phase escape similar to that of packed beads [152, 153]. However, controlling the pore size uniformity and the homogeneity of the packed bed may be challenging. Seamless anchoring of the monolith to the microchannel surfaces is often particularly difficult and gives rise to void volumes at the sharp corners and microchannel edges, unless special anchoring chemistry is applied (Figure 3.11b) [151]. More importantly, both microbead-based packings and porous polymer monoliths may retard the mass transfer into and out of the immobilized enzymes, similar to the scaffold-based immobilization matrices. As a result, the micropillar arrays often appear as the most feasible "packing" strategy for enzyme kinetic determinations, especially in the context of NP characterization.

Ordered micropillar arrays can be made with high precision from silicon by anisotropic etching [148, 154] or from polymer photoresists by direct UV lithography [155]. The lithographically defined micropillar arrays may also act as master molds for replicating the structures into other UV and heat-curable polymers or thermoplasts, as described in Section 3.1, although the achievable aspect ratios of replicated pillars may be limited compared with direct microfabrication techniques [18]. Although silicon (oxide) readily facilitates a multitude of surface reactions, the possibility of using replication techniques increases the freedom in material selection with a view to enzyme immobilization on micropillar surfaces. Thus, from the material perspective, many different chemistries are available for enzyme immobilization, but they all have their specific requirements in terms of pH, temperature, and applicable solvents, which need to match with the properties of the fabrication material [156]. For example, PDMS has very limited solvent compatibility as it tends to absorb the solvent molecules and swell upon exposure [157]. PDMS is also inherently hydrophobic and relatively inert, which complicates its chemical surface modification [158, 159]. Typically, some kind of chemical surface modification, such as covalent coupling of appropriate linker molecule, is a prerequisite for enzyme immobilization directly on micropillar surfaces, although some polymer surfaces readily contain functional groups. For example, the OSTE polymers feature either free surface thiols or allyl groups, the density of which can be tuned by simply manipulating the molar ratio of the thiol and allyl monomers in the bulk composition [20]. Especially the free thiol groups of the OSTEs have been extensively employed in a variety of microfluidic immobilized enzyme reactors, making use of photopolymerization [142, 160] or gold nanoparticle interaction with the free thiol residues on both the chip surface and in the protein structure [18]. Besides surface chemistry, attention should be paid to the possible nonspecific interactions between the polymers and the biomolecules to ensure that the catalytic activity is not lost because of protein fouling. Many commonly used polymers, including PDMS, tend to adsorb biomolecules [159]. A common solution is to functionalize the surface with polyethylene glycol, which not only reduces the non-specific binding [161], but may also increase the enzyme activity via the reduction of steric constraints and stabilization of the protein structure [162]. To a certain degree, the non-specific binding may also be prevented by oxygen plasma treatment [163, 164], although the impact is typically reversible. However, sometimes the plasma treatment may give rise to newly-formed functional groups, which may then be used for further chemical coupling reactions [165].

From the design perspective, the primary benefits of microstructured pillar arrays over other packing strategies include the possibility to adjust and control the total surface area in a reproducible manner, which is critical to be able to quantitate the amount of the bound enzyme. Furthermore, by adjusting the geometry and dimensions of the micropillars (diameter, spacing), the surface-to-volume ratio may also be optimized so as to avoid diffusion-limited kinetics and to maximize the amount of the immobilized enzyme. The ordered micropillar arrays also support capillary filling [154] and reduce pressure differences within the microchannel. In addition to the amount of immobilized enzyme, the microchannel design impacts the enzymatic reaction time, which is equal to the residence time of the target components inside the microfluidic device. Besides flow rate (linear velocity), the residence time can be adjusted by manipulating the microchannel

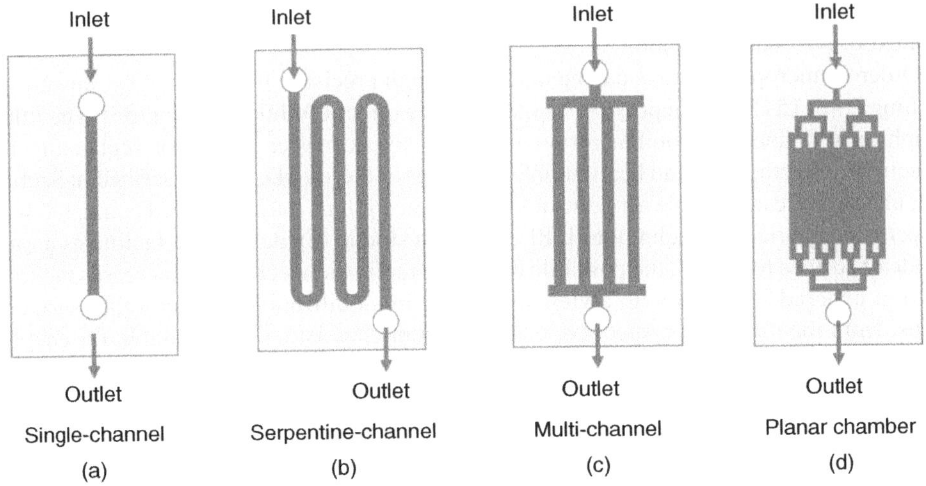

Figure 3.12 Representative configuration designs of microfluidic chips that can be used for adjusting the residence time: (a) single-channel chip; (b) serpentine-channel chip; (c) multi-channel chip; and (d) planar chamber chip. Republished with permission of the Royal Society of Chemistry, from Zhu et al. (2020) [146]; permission conveyed through Copyright Clearance Center, Inc.

dimensions (width, length) as illustrated in Figure 3.12. Under optimized conditions, the metabolic activity should be stable over time and, in the case of CYP enzymes, often dependent on the reaction temperature, as illustrated in Figure 3.13. To maintain physiological temperature, the microreactors may be placed in an incubator, on a hot plate or on an external resistive heater element. In addition, microfabricated resistive heater elements may also be integrated with the microfluidic channels via thin-film metallization. Besides temperature control, mixing of the reagents may be necessary for creating the concentration gradients over time. For a comprehensive review of the heating and mixing possibilities on microfluidic devices, the reader is advised to familiarize themselves with relevant prior reviews [166, 167].

3.5 Microfluidic Total Analysis Systems

This section provides a brief overview of the future prospects regarding system-level integration possibilities facilitated by microfabrication of miniaturized separation and sample preparation systems as integral parts of microfluidic *in vitro* models. Here, the focus will be on microfluidic total chemical analysis systems feasible for characterization of the impacts of nanomedicine on, for example, cell metabolism or drug elimination on a molecular level, rather than characterization of the properties of pharmaceutical nano- and microsystems. Besides molecular chemical analysis, a wealth of microfluidic techniques have also been developed for various particle sorting needs [168], but these techniques are typically applicable to micrometer-sized particles only. However, somewhat similar approaches and techniques can be used for cell separation and sorting on microfluidic devices [169], which

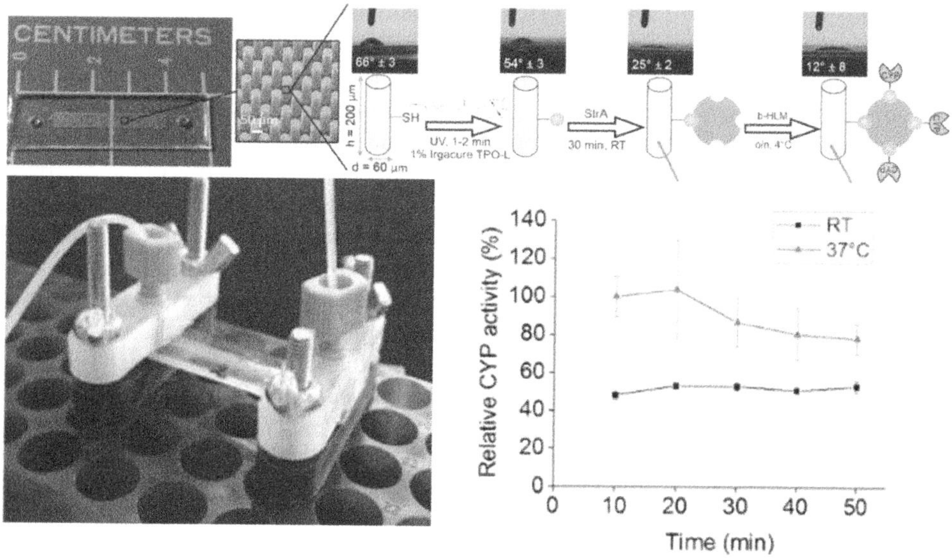

Figure 3.13 *Top: A micropillar-based immobilized enzyme reactor made from OSTE polymers by replication techniques and functionalized with biotin-alkyne by photopolymerization and streptavidin via biotin–avidin binding, prior to immobilization of biotinylated HLM to the streptavidin surface. Left: Photograph of the microfluidic setup. Right: Stability of the CYP activity over time on the OSTE reactor. Republished with permission of John Wiley and Sons, from Kiiski et al. (2019) [142]; permission conveyed through Copyright Clearance Center, Inc.*

may appear useful for the characterization of the impacts of nanomedicine in selected cases. Regarding chemical analysis with the help of microfluidic devices, the separation systems reviewed below represent the gold-standard techniques originally established to speed up the analysis of small organic molecules, peptides, and proteins in biological and synthetic matrices. To date, these techniques have hardly been applied to characterization of nanomedicine, and therefore this section provides only a brief discussion of the main available techniques and their current technology readiness level, which sets the framework for their possible future use in the context of nanomedicine characterization. Comprehensive reviews of the feasibility of these techniques to metabolomics, proteomics, and other clinical and forensic applications can be found elsewhere [2, 170, 171].

3.5.1 Microfluidic Separation Systems

The entire evolution of microfluidic total analysis systems (lab-on-a-chip) originates from the pioneering work in the early 1990s, which set the theoretical frame and experimental evidence for rapid chemical sensing on a microfabricated, planar glass chip by a technique called microchip capillary (zone) electrophoresis [8, 9]. This technique has since become the gold standard for microfluidic separations and has also transferred to practically any available microfabrication material. In microchip electrophoresis, the sample solution is introduced into a microfabricated separation channel via an intersecting side (injection) channel (Figure 3.14a) [176]. Compared with conventional capillary-based electrophoresis, the initial sample plug is much narrower [177], which enables significantly shorter separation times (at best <1 min) under a high electric field. As a result, the length of the separation column can be decreased (from ca. 1 m to only a few centimeters) and much lower separation voltages can be used on a microfluidic device to reach the same electric field strength (volts per meter), which is the critical parameter in terms of the resolving power. From the materials perspective, the only requirements are that the microchannel surfaces are charged and do not non-specifically adsorb the target analytes. In practice, most microfabrication materials maintain cathodic electro-osmosis, even if the bulk composition does not contain any functional groups. According to a theory, the apparent negative surface charge may originate from adsorption of hydroxyl ions to the microchannel surfaces [178]. Instead, non-specific adsorption (biomolecule fouling) is a pronounced problem with many polymer materials, especially PDMS, which often needs to be coated to avoid surface interactions [179]. However, inherently biofouling-resistant materials also exist, such as the lithographically defined organically modified ceramics [180].

In a classical setup, the electrophoresis chip is interfaced with a standard epifluorescence microscope to facilitate monitoring of the fluorescence signal over time, perpendicular to the separation channel (Figure 3.14c). As the microchannel heights are typically in the range of few tens of micrometers only, the optical path length on microfluidic devices is significantly short and optical absorbance detection is thus often compromised. Via thin-film metallization, it is also possible to integrate electrodes to the microchannel bottom layer [181] so as to enable electrochemical detection of the separated sample components at the end of the separation channel (Figure 3.14d). However, metal adhesion to the bottom layer, especially in the case of polymer devices as discussed in Section 3.3.1, as well as sealing of the metallized microstructures, may pose additional challenges to microfabrication of electrochemical sensor chips [53]. Alternatively, the separation channel outlet can be left open to allow transfer of the separated components (in liquid phase) into gas phase ions with the help of electrospray ionization (Figure 3.14e). In this manner, the microfluidic separation devices may also interface with mass spectrometric (MS) detection. The stability of the electrospray ionization (gas-phase transition), and thus the quality of the MS signal, benefits from a sharp-pointed, overhanging emitter tip structure, which is challenging to achieve by most microfabrication methods [182]. An alternative approach is to attach an external emitter needle or capillary tip at the microchannel outlet. Besides being laborious, this technique is also prone to dead volume generation at the junction point, which may result in loss of the achieved separation. For a more comprehensive overview of the possibilities and limitations of the microfluidics interfacing to MS, the reader is advised to consult the indicated reviews [182, 183]. Although microchip electrophoresis appears as a very powerful separation technique, a considerable technological barrier exists in the translation of the laboratory-scale methodologies from engineers into commercial or clinical benefit by biologists and pharmacologists. To a large degree, this is because of lack of standardization of both the components and the techniques, which inevitably decreases the level of automation and thus usability of the technology in routine analysis. As a result, most of the previous literature centers around proof-of-concept research, and only in limited cases are the technologies thoroughly validated for their intended purpose with authentic samples and datasets [174].

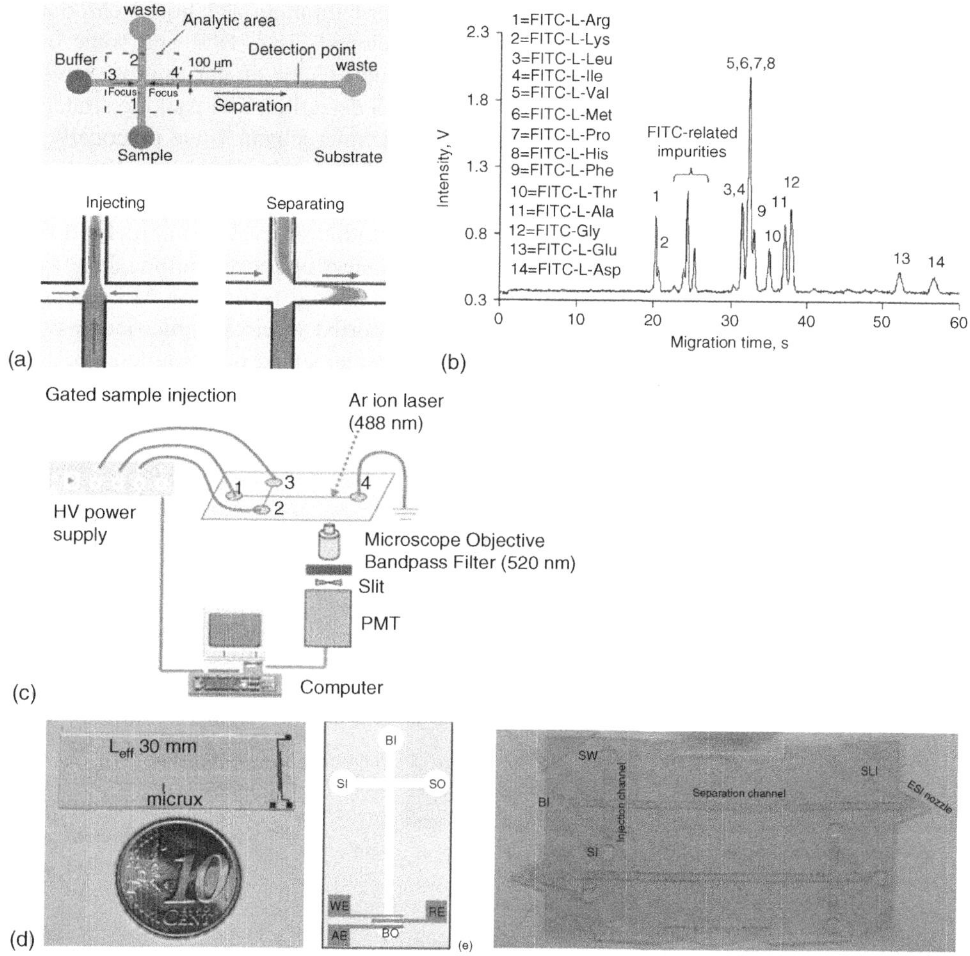

Figure 3.14 *(a) Sample loading and dispensing steps on a microchip electrophoresis device. Republished with permission of Springer Nature, from Fu and Tsai (2008) [172]; permission conveyed through Copyright Clearance Center, Inc. (b) An example of electrophoretic separation of fluorescent derivatized amino acids on a microchip made of organically modified ceramics. Republished with permission of the American Chemical Society, from Sikanen et al. (2010) [25]; permission conveyed through Copyright Clearance Center, Inc. (c) Experimental arrangement of microchip electrophoresis in combination with fluorescence detection. Republished with permission of Elsevier, from Kim et al. (2005) [173]; permission conveyed through Copyright Clearance Center, Inc. (d) Photograph and schematic view of a commercial electrophoresis chip featuring platinum electrodes for amperometric detection. BI/BO, buffer inlet/outlet; SI/SO, sample inlet/outlet; WE/RE/AE, working/reference/auxiliary electrode. Republished with permission of Springer Nature, from Ollikainen et al. (2019) [174], under the terms of the Creative Commons Attribution 4.0 International License. (e) An SU-8-based electrophoresis chip featuring an integrated, sharp opening at the microchannel outlet for electrospray ionization mass spectrometry. BI, buffer inlet; SI/SO, sample inlet/outlet; SLI, sheath liquid inlet. Republished with permission of John Wiley and Sons, from Nordman et al. (2010) [175]; permission conveyed through Copyright Clearance Center, Inc.*

In addition to microchip electrophoresis, a variety of miniaturized liquid chromatographic (LC) separation devices have also been developed [184, 185] and some even commercialized. The detection possibilities in microchip LC are similar to those of electrophoresis chips, although most are interfaced with MS detection. The packing strategies applicable to incorporation of the stationary phase in the microchannels are practically the same as those used for immobilized enzyme reactors, including microbead-based packings, porous polymer monoliths, and microfabricated pillar arrays, typically functionalized with C18 solid phase [184, 185]. Overall, the technology readiness level of microchip LC is closer to the commercial level than that of microchip electrophoresis, although in most designs only the chromatographic column and the ion source facilitating MS detection are integrated on-chip. Very few designs have been reported where the micropumps (for actuation) and sample loading would also have been implemented on a single chip, as in Figure 3.15.

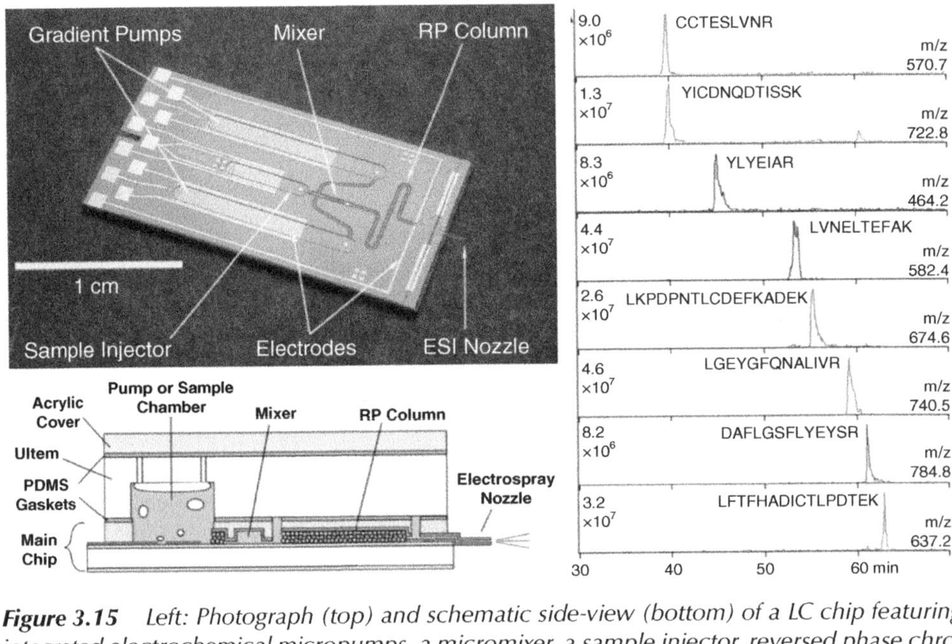

Figure 3.15 Left: Photograph (top) and schematic side-view (bottom) of a LC chip featuring integrated electrochemical micropumps, a micromixer, a sample injector, reversed phase chromatographic column, and an electrospray ionization nozzle. Right: Separation of eight tryptic peptides of bovine serum albumin performed using the LC chip. Republished with permission of the American Chemical Society, from Xie et al. (2005) [186]; permission conveyed through Copyright Clearance Center, Inc.

3.5.2 Toward n-in-one Analytical Platforms

Since the 1990s, the feasibility of microfluidic separation devices has been well established for a variety of pharmaceutically and biologically relevant analytical tasks, including analysis of cell metabolites, drugs and their metabolites, peptides and proteins [2, 170, 171].

Besides separation systems, a range of electrokinetic [187], solid phase extraction [188] and affinity-based sample preparation techniques have also been transferred on microfluidic devices to facilitate online enrichment and purification of the target analytes prior to their analysis. Although complete integration of all possible operations on a single chip is not yet a reality (in routine use), one of the key drivers of biologically inspired microfluidics is nevertheless the possibility of combining two or more microfluidic operations in sequence or in parallel – so as to improve the efficiency of analysis or to simulate the interplay between selected model organs in a systemic fashion *in vitro*. A classic example of an integrated microfluidic device, applied to drug cytotoxicity screening and featuring multiple parallel units for sequential cell culturing, sample preparation, and MS detection, is presented in Figure 3.16. Another somewhat similar concept combined immobilized enzyme reactors (HLM entrapped in polyethylene glycol), microfluidic culture of human hepatocellular carcinoma cells, and an integrated solid-phase extraction (SPE) unit with online MS analysis to study the cytotoxicity associated with drug metabolites [190]. Overall, the majority of the previous literature centers around n-in-one pressure-driven analytical platforms, whereas integration of molecular separation systems with *in vitro* models is still rare. At least in part this is due to certain technical requirements of microchip electrophoresis and LC configurations (i.e., high voltages and high pressures, respectively), which complicate their seamless

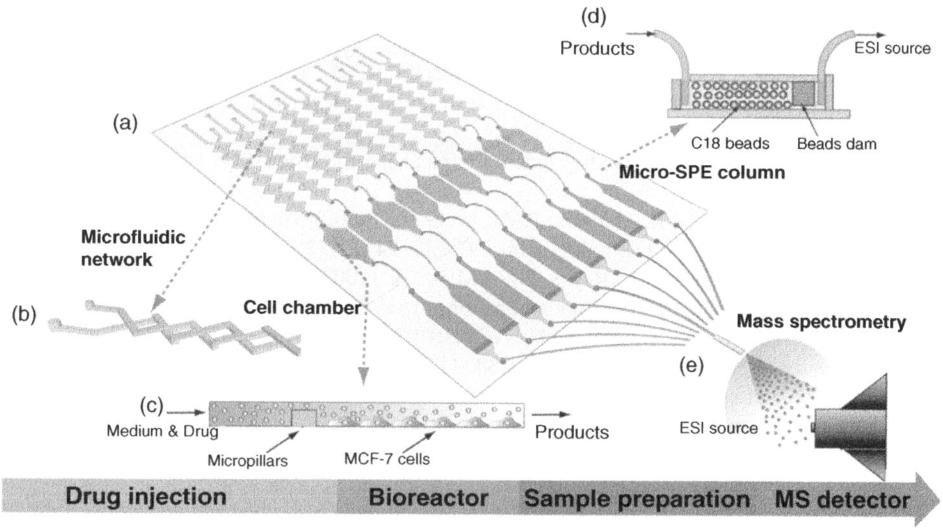

Figure 3.16 *Schematic diagram of the chip–ESI-MS system. (a) The system consists of a microfluidic network for culture medium and drug injections, cell culture chambers, on-chip micro-SPE columns for sample desalting and purifying, and an ESI-Q-TOF-MS. (b) Microfluidic network design for the concentration gradient generator during cell culture and drug screening. (c) Magnified view of the MCF-7 cells culture chamber for cell metabolism on the microfluidic chip. (d) Magnified view of an integrated micro-SPE column on the microdevice for sample pretreatment prior to ESI-MS detection. (e) The ESI source was coupled with the micro-SPE column by capillaries. Republished with permission of the American Chemical Society, from Chen et al. (2012) [189]; permission conveyed through Copyright Clearance Center, Inc.*

integration with cell-based assays. From this perspective, increasing interest in modular microfluidics may pave the way for a true lab-on-a-chip.

3.6 Epilogue

To date, a wealth of microfluidic platforms have been developed that facilitate detailed *in vitro* analysis of biological pathways on organ, cell and molecular levels. As the technology readiness level and automation increases, the integrated "body-on-a-chip" platforms will become a viable and biologically relevant *in vitro* model for preclinical drug testing, thus reducing the need for laboratory animals and improving the quality of *in vitro–in vivo* predictions. By using a microfluidic platform that connects multiple engineered tissues from different organs, it is possible to replicate human organ interactions for weeks at a time, allowing us to measure the systemic drug effects on different parts of the body. Thereby, microfluidics may also pave the way to increased translation of nanomedicine into clinical use, for instance, by improving the prediction of NP targeting to the desired tissue already in the *in vitro* phase. The aim of this chapter was to bring the reader near to the relevant literature by highlighting three application areas most relevant to characterization of pharmaceutical nano- and microsystems.

References

[1] Kang, L., Chung, B.G., Langer, R., Khademhosseini, A. (2008) Microfluidics for drug discovery and development: From target selection to product lifecycle management. *Drug Discov. Today*, **13** (1), 1–13.
[2] Verpoorte, E. (2002) Microfluidic chips for clinical and forensic analysis. *Electrophoresis*, **23** (5), 677–712.
[3] Reyes, D.R., Iossifidis, D., Auroux, P.-A., Manz, A. (2002) Micro total analysis systems. 1. Introduction, theory, and technology. *Anal. Chem.*, **74** (12), 2623–2636.
[4] Auroux, P.-A., Iossifidis, D., Reyes, D.R., Manz, A. (2002) Micro total analysis systems. 2. Analytical standard operations and applications. *Anal. Chem.*, **74** (12), 2637–2652.
[5] Patabadige, D.E.W., Jia, S., Sibbitts, J., *et al.* (2016) Micro total analysis systems: Fundamental advances and applications. *Anal. Chem.*, **88** (1), 320–338.
[6] Ahn, J., Ko, J., Lee, S., *et al.* (2018) Microfluidics in nanoparticle drug delivery; From synthesis to pre-clinical screening. *Adv. Drug Deliv. Rev.*, **128**, 29–53.
[7] Santos, H.A., Liu, D., Zhang, H. (eds) (2018) *Microfluidics for Pharmaceutical Applications*, 1st edn, Elsevier.
[8] Manz, A., Graber, N., Widmer, H.M. (1990) Miniaturized total chemical analysis systems: A novel concept for chemical sensing. *Sens. Actuators B*, **1** (1), 244–248.
[9] Harrison, D.J., Manz, A., Fan, Z., *et al.* (1992) Capillary electrophoresis and sample injection systems integrated on a planar glass chip. *Anal. Chem.*, **64** (17), 1926–1932.
[10] Terry, S.C., Jerman, J.H., Angell, J.B. (1979) A gas chromatographic air analyzer fabricated on a silicon wafer. *IEEE Trans. Electron Dev.*, **26** (12), 1880–1886.
[11] Iliescu, C., Taylor, H., Avram, M., *et al.* (2012) A practical guide for the fabrication of microfluidic devices using glass and silicon. *Biomicrofluidics*, **6** (1), 16505–16516.
[12] Duffy, D.C., McDonald, J.C., Schueller, O.J.A., Whitesides, G.M. (1998) Rapid prototyping of microfluidic systems in poly(dimethylsiloxane). *Anal. Chem.*, **70** (23), 4974–4984.
[13] Xia, Y., Whitesides, G.M. (1998) Soft lithography. *Annual Rev. Mater. Sci*, **28** (1), 153–184.

[14] Tuomikoski, S., Franssila, S. (2005) Free-standing SU-8 microfluidic chips by adhesive bonding and release etching. *Sensors Act. A*, **120** (2), 408–415.
[15] Sikanen, T., Heikkilä, L., Tuomikoski, S., *et al.* (2007) Performance of SU-8 microchips as separation devices and comparison with glass microchips. *Anal. Chem.*, **79** (16), 6255–6263.
[16] Eddings, M.A., Johnson, M.A., Gale, B.K. (2008) Determining the optimal PDMS–PDMS bonding technique for microfluidic devices. *J. Micromech. Microeng.*, **18** (6), 067001.
[17] Huikko, K., Östman, P., Grigoras, K., *et al.* (2003) Poly(dimethylsiloxane) electrospray devices fabricated with diamond-like carbon–poly(dimethylsiloxane) coated SU-8 masters. *Lab Chip*, **3** (2), 67–72.
[18] Tähkä, S., Sarfraz, J., Urvas, L., *et al.* (2019) Immobilization of proteolytic enzymes on replica-molded thiol-ene micropillar reactors via thiol-gold interaction. *Anal Bioanal Chem*, **411** (11), 2339–2349.
[19] Sikanen, T.M., Lafleur, J.P., Moilanen, M.-E., *et al.* (2013) Fabrication and bonding of thiol-ene-based microfluidic devices. *J Micromech. Microeng.*, **23** (3), 037002.
[20] Carlborg, C.F., Haraldsson, T., Öberg, K., *et al.* (2011) Beyond PDMS: off-stoichiometry thiol-ene (OSTE) based soft lithography for rapid prototyping of microfluidic devices. *Lab Chip*, **11** (18), 3136–3147.
[21] Tähkä, S.M., Bonabi, A., Nordberg, M.-E., *et al.* (2015) Thiol-ene microfluidic devices for microchip electrophoresis: Effects of curing conditions and monomer composition on surface properties. *J. Chromatogr. A*, **1426**, 233–240.
[22] Tähkä, S.M., Bonabi, A., Jokinen, V.P., Sikanen, T.M. (2017) Aqueous and non-aqueous microchip electrophoresis with on-chip electrospray ionization mass spectrometry on replica-molded thiol-ene microfluidic devices. *J. Chromatogr. A*, **1496**, 150–156.
[23] Becker, H., Gärtner, C. (2008) Polymer microfabrication technologies for microfluidic systems. *Anal Bioanal Chem*, **390** (1), 89–111.
[24] Tsao, C.-W., DeVoe, D.L. (2009) Bonding of thermoplastic polymer microfluidics. *Microfluid Nanofluid*, **6** (1), 1–16.
[25] Sikanen, T., Aura, S., Heikkilä, L., *et al.* (2010) Hybrid ceramic polymers: new, nonbiofouling, and optically transparent materials for microfluidics. *Anal. Chem.*, **82** (9), 3874–3882.
[26] Bonabi, A., Cito, S., Tammela, P., *et al.* (2017) Fabrication of concave micromirrors for single cell imaging via controlled over-exposure of organically modified ceramics in single step lithography. *Biomicrofluidics*, **11** (3).
[27] Weisgrab, G., Ovsianikov, A., Costa, P.F. (2019) Functional 3D printing for microfluidic chips. *Adv. Mater. Technol.*, **4** (10), 1900275.
[28] Bazaz, S.R., Kashaninejad, N., Azadi, S., *et al.* (2019) Rapid softlithography using 3D-printed molds. *Adv. Mater. Technol*, **4** (10), 1900425.
[29] Hiltunen, J., Liedert, C., Hiltunen, M., *et al.* (2018) Roll-to-roll fabrication of integrated PDMS-paper microfluidics for nucleic acid amplification. *Lab Chip*, **18** (11), 1552–1559.
[30] Senkbeil, S., Aho, J., Yde, L., *et al.* (2016) Roll-to-plate fabrication of microfluidic devices with rheology-modified thiol-ene resins. *J. Micromech. Microeng.*, **26** (7), 075014.
[31] Zhu, D., Long, Q., Xu, Y., Xing, J. (2019) Evaluating nanoparticles in preclinical research using microfluidic systems. *Micromachines (Basel)*, **10** (6).
[32] Haque, A., Gheibi, P., Gao, Y., *et al.* (2016) Cell biology is different in small volumes: endogenous signals shape phenotype of primary hepatocytes cultured in microfluidic channels. *Sci Rep.* **6**, 33980.
[33] Paguirigan, A.L., Beebe, D.J. (2009) From the cellular perspective: exploring differences in the cellular baseline in macroscale and microfluidic cultures. *Int Bio (Cam)*, **1** (2), 182–195.
[34] Halldorsson, S., Lucumi, E., Gómez-Sjöberg, R., Fleming, R.M.T. (2015) Advantages and challenges of microfluidic cell culture in polydimethylsiloxane devices. *Biosens. Bioelectron.*, **63**, 218–231.

[35] Firpo, G., Angeli, E., Repetto, L., Valbusa, U. (2015) Permeability thickness dependence of polydimethylsiloxane (PDMS) membranes. *J. Membrane Sci*, **481**, 1–8.
[36] Käpylä, E., Sorkio, A., Teymouri, S., *et al.* (2014) Ormocomp-modified glass increases collagen binding and promotes the adherence and maturation of human embryonic stem cell-derived retinal pigment epithelial cells. *Langmuir*, **30** (48), 14555–14565.
[37] Klein, F., Richter, B., Striebel, T., *et al.* (2011) Two-component polymer scaffolds for controlled three-dimensional cell culture. *Adv. Mater.*, **23** (11), 1341–1345.
[38] Schlie, S., Ngezahayo, A., Ovsianikov, A., *et al.* (2007) Three-dimensional cell growth on structures fabricated from ORMOCER® by two-photon polymerization technique. *J Biomater Appl*, **22** (3), 275–287.
[39] Lu, H., Koo, L.Y., Wang, W.M., *et al.* (2004) Microfluidic shear devices for quantitative analysis of cell adhesion. *Anal. Chem.*, **76** (18), 5257–5264.
[40] van der Meer, A.D., Poot, A.A., Duits, M.H.G., *et al.* (2009) Microfluidic technology in vascular research. *J. Biomed. Biotechnol.*, 823148.
[41] Gough, J.E., Downes, S. (2001) Osteoblast cell death on methacrylate polymers involves apoptosis. *J. Biomed. Mater. Res.*, **57** (4), 497–505.
[42] Piironen, K., Haapala, M., Järvinen, P., Sikanen, T. (2018) Cell proliferation on common 3D printing materials used in stereolithographic patterning of microfluidic devices. *Proceedings of the 22nd International Conference on Miniaturized Systems for Chemistry and Life Sciences (MicroTAS 2018)*, 1469–1472.
[43] Piironen, K., Järvinen, P., Kiiski, I., Sikanen, T. (2019) Understanding cell proliferation and material-induced cell death on microfluidic devices made of off-stoichiometric thiol-enes. *Proceedings of the 23rd International Conference on Miniaturized Systems for Chemistry and Life Sciences (MicroTAS 2019)*, 316–317.
[44] Sticker, D., Rothbauer, M., Ehgartner, J., *et al.* (2019) Oxygen management at the microscale: A functional biochip material with long-lasting and tunable oxygen scavenging properties for cell culture applications. *ACS Appl. Mater. Interfaces*, **11** (10), 9730–9739.
[45] Piruska, A., Nikcevic, I., Lee, S.H., *et al.* (2005) The autofluorescence of plastic materials and chips measured under laser irradiation. *Lab Chip*, **5** (12), 1348–1354.
[46] Aura, S., Sikanen, T., Kotiaho, T., Franssila, S. (2008) Novel hybrid material for microfluidic devices. *Sensors Act. B*, **132** (2), 397–403.
[47] Feidenhans'l, N.A., Lafleur, J.P., Jensen, T.G., Kutter, J.P. (2014) Surface functionalized thiol-ene waveguides for fluorescence biosensing in microfluidic devices. *Electrophoresis*, **35** (2–3), 282–288.
[48] Hartcher-O'Brien, J., Evers, J., Tempelman, E. (2019) Surface roughness of 3D printed materials: Comparing physical measurements and human perception. *Mater. Today Commun.*, **19**, 300–305.
[49] Chawla, K., Bürgel, S.C., Schmidt, G.W., *et al.* (2018) Integrating impedance-based growth-rate monitoring into a microfluidic cell culture platform for live-cell microscopy. *Microsyst. Nanoeng.*, **4** (1), 1–12.
[50] Reiss, B., Wegener, J. (2015) Impedance analysis of different cell monolayers grown on gold-film electrodes. *2015 37th Annual International Conference of the IEEE Engineering in Medicine and Biology Society (EMBC)*, 7079–7082.
[51] Choi, Y.H., Min, J., Cho, S. (2015) Indium tin oxide based chip for optical and electrochemical characterization of protein–cell interaction. *Jpn. J. Appl. Phys.*, **54** (6S1), 06FN03.
[52] Ryspayeva, A., Jones, T.D.A., Esfahani, M.N., *et al.* (2019) A rapid technique for the direct metallization of PDMS substrates for flexible and stretchable electronics applications. *Microelectron. Eng.*, **209**, 35–40.

[53] Bonabi, A., Tähkä, S., Ollikainen, E., *et al.* (2019) Metallization of organically modified ceramics for microfluidic electrochemical assays. *Micromachines*, **10** (9).

[54] Wenger, R.H., Kurtcuoglu, V. Scholz, C.C., *et al.* (2015) Frequently asked questions in hypoxia research. *Hypoxia (Auckl).* **3**, 35–43.

[55] Barmaki, S., Jokinen, V., Obermaier, D., *et al.* (2018) A microfluidic oxygen sink to create a targeted cellular hypoxic microenvironment under ambient atmospheric conditions. *Acta Biomater.*, **73**, 167–179.

[56] Brennan, M.D., Rexius-Hall, M.L., Elgass, L.J., Eddington, D.T. (2014) Oxygen control with microfluidics. *Lab Chip*, **14** (22), 4305–4318.

[57] Kim, D.-H., Lipke, E.A., Kim, P., *et al.* (2010) Nanoscale cues regulate the structure and function of macroscopic cardiac tissue constructs. *PNAS*, **107** (2), 565–570.

[58] Rao, C., Prodromakis, T., Kolker, L., *et al.* (2013) The effect of microgrooved culture substrates on calcium cycling of cardiac myocytes derived from human induced pluripotent stem cells. *Biomaterials*, **34** (10), 2399–2411.

[59] Kim, J.Y., Choi, Y.-S., Lee, B.-K., Lee, D.-W. (2016) Surface-patterned SU-8 cantilever arrays for preliminary screening of cardiac toxicity. *Biosens. Bioelectron.*, **80**, 456–462.

[60] Anene-Nzelu, C.G., Peh, K.Y., Fraiszudeen, A., *et al.* (2013) Scalable alignment of three-dimensional cellular constructs in a microfluidic chip. *Lab Chip*, **13** (20), 4124–4133.

[61] Boroujeni, S.M., Mashayekhan, S., Vakilian, S., *et al.* (2016) The synergistic effect of surface topography and sustained release of TGF-β1 on myogenic differentiation of human mesenchymal stem cells. *J. Biomed. Mater Res. Part A*, **104** (7), 1610–1621.

[62] Castiaux, A.D., Spence, D.M., Martin, R.S. (2019) Review of 3D cell culture with analysis in microfluidic systems. *Anal. Methods*, **11** (33), 4220–4232.

[63] Duval, K., Grover, H., Han, L.-H., *et al.* (2017) Modeling physiological events in 2D vs. 3D cell culture. *Physiology*, **32** (4), 266–277.

[64] Pampaloni, F., Reynaud, E.G., Stelzer, E.H.K. (2007) The third dimension bridges the gap between cell culture and live tissue. *Nature Rev. Mol. Cell Biol.*, **8** (10), 839–845.

[65] Cukierman, E., Pankov, R., Stevens, D.R., Yamada, K.M. (2001) Taking cell-matrix adhesions to the third dimension. *Science*, **294** (5547), 1708–1712.

[66] Nath, S., Devi, G.R. (2016) Three-dimensional culture systems in cancer research: Focus on tumor spheroid model. *Pharmacol. Ther.*, **163**, 94–108.

[67] Godoy, P., Hewitt, N.J., Albrecht, U., *et al.* (2013) Recent advances in 2D and 3D in vitro systems using primary hepatocytes, alternative hepatocyte sources and non-parenchymal liver cells and their use in investigating mechanisms of hepatotoxicity, cell signaling and ADME. *Arch Toxicol*, **87** (8), 1315–1530.

[68] Verjans, E.-T., Doijen, J., Luyten, W., *et al.* (2018) Three-dimensional cell culture models for anticancer drug screening: Worth the effort? *J. Cellular Physiol.*, **233** (4), 2993–3003.

[69] Vadivelu, R.K., Kamble, H., Shiddiky, M.J.A., Nguyen, N.-T. (2017) Microfluidic technology for the generation of cell spheroids and their applications. *Micromachines*, **8** (4).

[70] Liao, J., Guo, X., Grande-Allen, K.J., *et al.* (2010) Bioactive polymer/extracellular matrix scaffolds fabricated with a flow perfusion bioreactor for cartilage tissue engineering. *Biomaterials*, **31** (34), 8911–8920.

[71] Moshksayan, K., Kashaninejad, N., Warkiani, M.E., *et al.* (2018) Spheroids-on-a-chip: Recent advances and design considerations in microfluidic platforms for spheroid formation and culture. *Sensors Act. B*, **263**, 151–176.

[72] Otsuka, H., Nagamura, M., Kaneko, A., *et al.* (2012) Chondrocyte spheroids on microfabricated PEG hydrogel surface and their noninvasive functional monitoring. *Sci Technol Adv Mater*, **13** (6).

[73] Jokinen, V., Kankuri, E., Hoshian, S., *et al.* (2018) Superhydrophobic blood-repellent surfaces. *Adv. Mater.*, **30** (24), 1705104.

[74] Oliveira, S.M., Song, W., Alves, N.M., Mano, J.F. (2011) Chemical modification of bioinspired superhydrophobic polystyrene surfaces to control cell attachment/proliferation. *Soft Matter*, **7** (19), 8932–8941.

[75] Wang, G.-J., Ho, K.-H., Hsu, S., Wang, K.-P. (2007) Microvessel scaffold with circular microchannels by photoresist melting. *Biomed Microdevices*, **9** (5), 657–663.

[76] Camp, J.P., Stokol, T., Shuler, M.L. (2008) Fabrication of a multiple-diameter branched network of microvascular channels with semi-circular cross-sections using xenon difluoride etching. *Biomed Microdevices*, **10** (2), 179–186.

[77] Tu, T.-Y., Wang, Z., Bai, J., *et al.* (2014) Rapid prototyping of concave microwells for the formation of 3D multicellular cancer aggregates for drug screening. *Adv Healthc Mater*, **3** (4), 609–616.

[78] Ma, L.-D., Wang, Y.-T., Wang, J.-R., *et al.* (2018) Design and fabrication of a liver-on-a-chip platform for convenient, highly efficient, and safe in situ perfusion culture of 3D hepatic spheroids. *Lab Chip*, **18** (17), 2547–2562.

[79] Rodrigues, R.O., Bañobre-López, M., Gallo, J., *et al.* (2016) Haemocompatibility of iron oxide nanoparticles synthesized for theranostic applications: a high-sensitivity microfluidic tool. *J Nanopart Res*, **18** (7), 194.

[80] Falanga, A.P., Pitingolo, G., Celentano, M., *et al.* (2017) Shuttle-mediated nanoparticle transport across an in vitro brain endothelium under flow conditions. *Biotechnol. Bioeng.*, **114** (5), 1087–1095.

[81] Huh, D., Matthews, B.D., Mammoto, A., *et al.* (2010) Reconstituting organ-level lung functions on a chip. *Science*, **328** (5986), 1662–1668.

[82] Stylianopoulos, T., Jain, R.K. (2015) Design considerations for nanotherapeutics in oncology. *Nanomed.: Nanotech. Biol. Med.*, **11** (8), 1893–1907.

[83] Zhang, Y.-N., Poon, W., Tavares, A.J., *et al.* (2016) Nanoparticle–liver interactions: Cellular uptake and hepatobiliary elimination. *Journal of Controlled Rel.*, **240**, 332–348.

[84] Ragelle, H., Danhier, F., Préat, V., *et al.* (2017) Nanoparticle-based drug delivery systems: a commercial and regulatory outlook as the field matures. *Expert Opin. Drug Deliv.*, **14** (7), 851–864.

[85] Yu, M., Zheng, J. (2015) Clearance pathways and tumor targeting of imaging nanoparticles. *ACS Nano*, **9** (7), 6655–6674.

[86] Maeda, H. (2012) Macromolecular therapeutics in cancer treatment: The EPR effect and beyond. *J. Controlled Rel.*, **164** (2), 138–144.

[87] Matsumura, Y., Maeda, H. (1986) A new concept for macromolecular therapeutics in cancer chemotherapy: Mechanism of tumoritropic accumulation of proteins and the antitumor agent smancs. *Cancer Res*, **46** (12 Part 1), 6387–6392.

[88] Haute, D.V., Berlin, J.M. (2017) Challenges in realizing selectivity for nanoparticle biodistribution and clearance: lessons from gold nanoparticles. *Ther. Deliv.*, **8** (9), 763–774.

[89] Tsoi, K.M., MacParland, S.A., Ma, X.-Z., *et al.* (2016) Mechanism of hard-nanomaterial clearance by the liver. *Nature Mater.*, **15** (11), 1212–1221.

[90] Cormode, D.P., Skajaa, G.O., Delshad, A., *et al.* (2011) A versatile and tunable coating strategy allows control of nanocrystal delivery to cell types in the liver. *Bioconjugate Chem.*, **22** (3), 353–361.

[91] Cheng, S.-H., Li, F.-C., Souris, J.S., *et al.* (2012) Visualizing dynamics of sub-hepatic distribution of nanoparticles using intravital multiphoton fluorescence microscopy. *ACS Nano*, **6** (5), 4122–4131.

[92] Zanger, U.M., Schwab, M. (2013) Cytochrome P450 enzymes in drug metabolism: Regulation of gene expression, enzyme activities, and impact of genetic variation. *Pharmacol. Ther.*, **138** (1), 103–141.
[93] Zhou, S.F., Liu, J.P., Lai, X.S. (2009) Substrate specificity, inhibitors and regulation of human cytochrome P450 2D6 and implications in drug development. *Curr. Med. Chem.* **16** (21), 2661–2805.
[94] Brandon, E.F.A., Raap, C.D., Meijerman, I., *et al.* (2003) An update on in vitro test methods in human hepatic drug biotransformation research: pros and cons. *Toxicol. Appl. Pharmacol.*, **189** (3), 233–246.
[95] Fasinu, P., Bouic, P.J., Rosenkranz, B. (2012) Liver-based in vitro technologies for drug biotransformation studies – A review. *Current Drug Metab.*, **13** (2), 215–224.
[96] Fowler, S., Zhang, H. (2008) In vitro evaluation of reversible and irreversible cytochrome P450 inhibition: Current status on methodologies and their utility for predicting drug–drug interactions. *AAPS J*, **10** (2), 410–424.
[97] Kumar, S. (2010) Engineering cytochrome P450 biocatalysts for biotechnology, medicine and bioremediation. *Expert Opin. Drug Metab. Toxicol.*, **6** (2), 115–131.
[98] Midde, N.M., Kumar, S. (2015) Development of NanoART for HIV treatment: minding the cytochrome P450 (CYP) enzymes. *J Pers Nanomed*, **1**, 24–32.
[99] Lamb, J.G., Hathaway, L.B., Munger, M.A., *et al.* (2010) Nano-silver particle effects on drug metabolism in vitro. *Drug Metab Dispos*, **8** (12), 2246–2251.
[100] Warisnoicharoen, W., Hongpiticharoen, P., Lawanprasert, S. (2011) Alteration in enzymatic function of human cytochrome P450 by silver nanoparticles. *Res. J. Environ. Toxicol.*, **5** (1), 58–64.
[101] Kulthong, K., Maniratanachote, R., Kobayashi, Y., *et al.* (2012) Effects of silver nanoparticles on rat hepatic cytochrome P450 enzyme activity. *Xenobiotica*, **42** (9), 854–862.
[102] Balasubramanian, S.K., Jittiwat, J., Manikandan, J., *et al.* (2010) Biodistribution of gold nanoparticles and gene expression changes in the liver and spleen after intravenous administration in rats. *Biomaterials*, **31** (8), 2034–2042.
[103] Ye, M., Tang, L., Luo, M., *et al.* (2014) Size- and time-dependent alteration in metabolic activities of human hepatic cytochrome P450 isozymes by gold nanoparticles via microsomal coincubations. *Nanoscale Res. Lett.*, **9** (1), 642.
[104] Sereemaspun, A., Hongpiticharoen, P., Rojanathanes, R., *et al.* (2008) Inhibition of human cytochrome P450 enzymes by metallic nanoparticles: A preliminary to nanogenomics. *Int. J. Pharmacol.*, **4** (6), 492–495.
[105] Fröhlich, E., Kueznik, T., Samberger, C., *et al.* (2010) Size-dependent effects of nanoparticles on the activity of cytochrome P450 isoenzymes. *Toxicol. Applied Pharmacol.*, **242** (3), 326–332.
[106] Lee, M.-Y., Yang, J.-A., Jung, H.S., *et al.* (2012) Hyaluronic acid–gold nanoparticle/interferon α complex for targeted treatment of hepatitis C virus infection. *ACS Nano*, **6** (11), 9522–9531.
[107] Yang, Y., Yuan, S.-X., Zhao, L.-H., *et al.* (2015) Ligand-directed stearic acid grafted chitosan micelles to increase therapeutic efficacy in hepatic cancer. *Mol. Pharmaceutics*, **12** (2), 644–652.
[108] Ollikainen, E., Liu, D., Kallio, A., *et al.* (2017) The impact of porous silicon nanoparticles on human cytochrome P450 metabolism in human liver microsomes in vitro. *Eur. J. Pharm. Sci*, **104**, 124–132.
[109] Gay, S.C., Roberts, A.G., Halpert, J.R. (2010) Structural features of cytochromes P450 and ligands that affect drug metabolism as revealed by X-ray crystallography and NMR. *Future Med Chem*, **2** (9), 1451–1468.

[110] EUNCL – Nanomedicine Characterisation Laboratory. http://www.euncl.eu/
[111] Hadd, A.G., Raymond, D.E., Halliwell, J.W., *et al.* (1997) Microchip device for performing enzyme assays. *Anal. Chem.*, **69** (17), 3407–3412.
[112] Cohen, C.B., Chin-Dixon, E., Jeong, S., Nikiforov, T.T. (1999) A microchip-based enzyme assay for protein kinase A. *Anal. Biochem.*, **273** (1), 89–97.
[113] Mao, H., Yang, T., Cremer, P.S. (2002) Design and characterization of immobilized enzymes in microfluidic systems. *Anal. Chem.*, **74** (2), 379–385.
[114] Křenková, J., Foret, F. (2004) Immobilized microfluidic enzymatic reactors. *Electrophoresis*, **25** (21–22), 3550–3563.
[115] Munro, A.W., Girvan, H.M., McLean, K.J. (2007) Variations on a (t)heme – novel mechanisms, redox partners and catalytic functions in the cytochrome P450 superfamily. *Nat. Prod. Rep.*, **24** (3), 585–609.
[116] Munro, A.W., Girvan, H.M., Mason, A.E., *et al.* (2013) What makes a P450 tick? *Trends Biochem. Sci*, **38** (3), 140–150.
[117] Imaoka, S., Imai, Y., Shimada, T., Funae, Y. (1992) Role of phospholipids in reconstituted cytochrome P450 3A form and mechanism of their activation of catalytic activity. *Biochemistry*, **31** (26), 6063–6069.
[118] Pearce, R.E., McIntyre, C.J., Madan, A., *et al.* (1996) Effects of freezing, thawing, and storing human liver microsomes on cytochrome P450 activity. *Arch. Biochem. Biophys.*, **331** (2), 145–169.
[119] Busby, W.F., Ackermann, J.M., Crespi, C.L. (1999) Effect of methanol, ethanol, dimethyl sulfoxide, and acetonitrile on in vitro activities of cDNA-expressed human cytochromes P-450. *Drug Metab. Dispos.*, **27** (2), 246–249.
[120] Foti, R.S., Fisher, M.B. (2004) Impact of incubation conditions on bufuralol human clearance predictions: Enzyme lability and nonspecific binding. *Drug Metab Dispos*, **32** (3), 295–304.
[121] Karuzina, I.I., Archakov, A.I. (1994) Hydrogen peroxide-mediated inactivation of microsomal cytochrome P450 during monooxygenase reactions. *Free Radical Biol. Med.*, **17** (6), 557–567.
[122] Sassolas, A., Blum, L.J., Leca-Bouvier, B.D. (2012) Immobilization strategies to develop enzymatic biosensors. *Biotechnol. Adv.*, **30** (3), 489–511.
[123] Kim, D., Herr, A.E. (2013) Protein immobilization techniques for microfluidic assays. *Biomicrofluidics*, **7** (4), 041501.
[124] Datta, S., Christena, L.R., Rajaram, Y.R.S. (2013) Enzyme immobilization: an overview on techniques and support materials. *3 Biotech*, **3** (1), 1–9.
[125] Sheldon, R.A. (2007) Enzyme immobilization: The quest for optimum performance. *Adv. Synthesis Cat.*, **349** (8–9), 1289–1307.
[126] Sakai-Kato, K., Kato, M., Homma, H., *et al.* (2005) Creation of a P450 array toward high-throughput analysis. *Anal. Chem.*, **77** (21), 7080–7083.
[127] Zguris, J.C., Itle, L.J., Hayes, D., Pishko, M.V. (2005) Microreactor microfluidic systems with human microsomes and hepatocytes for use in metabolite studies. *Biomed Microdevices*, **7** (2), 117–125.
[128] Sheldon, R., Pelt, S. van (2013) Enzyme immobilisation in biocatalysis: why, what and how. *Chem. Soc. Rev.*, **42** (15), 6223–6235.
[129] Honiger, J., Balladur, P., Mariani, P., *et al.* (1995) Permeability and biocompatibility of a new hydrogel used for encapsulation of hepatocytes. *Biomaterials*, **16** (10), 753–759.
[130] Wu, Q., Gao, D., Wei, J., *et al.* (2014) Development of a novel multi-layer microfluidic device towards characterization of drug metabolism and cytotoxicity for drug screening. *Chem. Commun.*, **50** (21), 2762–2764.
[131] Tan, C.Y., Hirakawa, H., Nagamune, T. (2015) Supramolecular protein assembly supports immobilization of a cytochrome P450 monooxygenase system as water-insoluble gel. *Sci Rep.*, **5** (1), 1–8.

[132] Do, M.Q., Henry, E., Kato, M., Cheruzel, L. (2018) Cross-linked cytochrome P450 BM3 aggregates promoted by Ru(II)-diimine complexes bearing aldehyde groups. *J. Inorg. Biochem.*, **186**, 130–134.

[133] Luo, Q., Mao, X., Kong, L., *et al.* (2002) High-performance affinity chromatography for characterization of human immunoglobulin G digestion with papain. *J. Chromatogr. B*, **776** (2), 139–147.

[134] Wong, L.S., Khan, F., Micklefield, J. (2009) Selective covalent protein immobilization: Strategies and applications. *Chem. Rev.*, **109** (9), 4025–4053.

[135] Tudorache, M., Mahalu, D., Teodorescu, C., *et al.* (2011) Biocatalytic microreactor incorporating HRP anchored on micro-/nano-lithographic patterns for flow oxidation of phenols. *J. Mol. Cat. B*, **69** (3), 133–139.

[136] Schejbal, J., Řemínek, R., Zeman, L., *et al.* (2016) On-line coupling of immobilized cytochrome P450 microreactor and capillary electrophoresis: A promising tool for drug development. *J. Chromatogr. A*, **1437**, 234–240.

[137] Wollenberg, L.A., Kabulski, J.L., Powell, M.J., *et al.* (2014) The use of immobilized cytochrome P4502C9 in PMMA-based plug flow bioreactors for the production of drug metabolites. *Appl Biochem Biotechnol*, **172** (3), 1293–1306.

[138] Ménard, A., Huang, Y., Karam, P., *et al.* (2012) Site-specific fluorescent labeling and oriented immobilization of a triple mutant of CYP3A4 via C64. *Bioconjugate Chem.*, **23** (4), 826–836.

[139] Gannett, P.M., Kabulski, J., Perez, F.A., *et al.* (2006) Preparation, characterization, and substrate metabolism of gold-immobilized cytochrome P450 2C9. *J. Am. Chem. Soc.*, **128** (26), 8374–8375.

[140] Jesionowski, T., Zdarta, J., Krajewska, B. (2014) Enzyme immobilization by adsorption: a review. *Adsorption*, **20** (5), 801–821.

[141] Nicoli, R., Bartolini, M., Rudaz, S., *et al.* (2008) Development of immobilized enzyme reactors based on human recombinant cytochrome P450 enzymes for phase I drug metabolism studies. *J. Chromatogr. A*, **1206** (1), 2–10.

[142] Kiiski, I.M.A., Pihlaja, T., Urvas, L., *et al.* (2019) Overcoming the pitfalls of cytochrome P450 immobilization through the use of fusogenic liposomes. *Adv. Biosyst.*, **3** (1), 1800245.

[143] Ueda, Y., Morigaki, K., Tatsu, Y., *et al.* (2007) Immobilization and activity assay of cytochrome P450 on patterned lipid membranes. *Biochem. Biophys. Res. Commun.*, **355** (4), 926–931.

[144] Jönsson, A., Svejdal, R.R., Bøgelund, N., *et al.* (2017) Thiol-ene monolithic pepsin microreactor with a 3D-printed interface for efficient UPLC-MS peptide mapping analyses. *Anal. Chem.*, **89** (8), 4573–4580.

[145] Peterson, D.S., Rohr, T., Svec, F., Fréchet, J.M.J. (2002) Enzymatic microreactor-on-a-chip: Protein mapping using trypsin immobilized on porous polymer monoliths molded in channels of microfluidic devices. *Anal. Chem.*, **74** (16), 4081–4088.

[146] Zhu, Y., Chen, Q., Shao, L., *et al.* (2020) Microfluidic immobilized enzyme reactors for continuous biocatalysis. *Reaction Chem. Eng.*, **5**, 9-32.

[147] Oliver-Calixte, N.J., Uba, F.I., Battle, K.N., *et al.* (2014) Immobilization of lambda exonuclease onto polymer micropillar arrays for the solid-phase digestion of dsDNAs. *Anal. Chem.*, **86** (9), 4447–4454.

[148] Losey, M.W., Jackman, R.J., Firebaugh, S.L., *et al.* (2002) Design and fabrication of microfluidic devices for multiphase mixing and reaction. *J. Microelectromech. Syst.*, **11** (6), 709–717.

[149] Slovakova, M., Minc, N., Bilkova, Z., *et al.* (2005) Use of self assembled magnetic beads for on-chip protein digestion. *Lab Chip*, **5** (9), 935–942.

[150] Haapala, M., Saarela, V., Pól, J., *et al.* (2010) Integrated liquid chromatography-heated nebulizer microchip for mass spectrometry. *Anal. Chim. Acta*, **662** (2), 163–169.

[151] Burke, J.M., Smela, E. (2012) A novel surface modification technique for forming porous polymer monoliths in poly(dimethylsiloxane). *Biomicrofluidics*, **6** (1), 016506.

[152] Vlakh, E.G., Tennikova, T.B. (2013) Flow-through immobilized enzyme reactors based on monoliths: II. Kinetics study and application. *J. Sep. Sci*, **36** (6), 1149–1167.

[153] Munirathinam, R., Huskens, J., Verboom, W. (2015) Supported catalysis in continuous-flow microreactors. *Adv. Synthesis Cat.*, **357** (6), 1093–1123.

[154] Sainiemi, L., Nissilä, T., Jokinen, V., *et al.* (2008) Fabrication and fluidic characterization of silicon micropillar array electrospray ionization chip. *Sensors Act. B*, **132** (2), 380–387.

[155] Amato, L., Keller, S.S., Heiskanen, A., *et al.* (2012) Fabrication of high-aspect ratio SU-8 micropillar arrays. *Microelectron. Eng.*, **98**, 483–487.

[156] Hermanson, G. (2013) *Bioconjugate Techniques*, 3rd edn, Academic Press, Boston.

[157] Lee, J.N., Park, C., Whitesides, G.M. (2003) Solvent compatibility of poly(dimethylsiloxane)-based microfluidic devices. *Anal. Chem.*, **75** (23), 6544–6554.

[158] Makamba, H., Kim, J.H., Lim, K., *et al.* (2003) Surface modification of poly(dimethylsiloxane) microchannels. *Electrophoresis*, **24** (21), 3607–3619.

[159] van Midwoud, P.M., Janse, A., Merema, M.T., *et al.* (2012) Comparison of biocompatibility and adsorption properties of different plastics for advanced microfluidic cell and tissue culture models. *Anal. Chem.*, **84** (9), 3938–3944.

[160] Lafleur, J.P., Kwapiszewski, R., Jensen, T.G., Kutter, J.P. (2013) Rapid photochemical surface patterning of proteins in thiol–ene based microfluidic devices. *Analyst*, **138** (3), 845–849.

[161] Holmberg, K., Tiberg, F., Malmsten, M., Brink, C. (1997) Grafting with hydrophilic polymer chains to prepare protein-resistant surfaces. *Colloids Surf. A*, **123–124**, 297–306.

[162] Moskovitz, Y., Srebnik, S. (2005) Mean-field model of immobilized enzymes embedded in a grafted polymer layer. *Biophys J*, **89** (1), 22–31.

[163] Tserepi, A., Gogolides, E., Tsougeni, K., *et al.* (2005) Tailoring the surface topography and wetting properties of oxygen-plasma treated polydimethylsiloxane. *J. Appl. Phys.*, **98** (11), 113502.

[164] Brown, L., Koerner, T., Hugh Horton, J., Oleschuk, R. (2006) Fabrication and characterization of poly(methylmethacrylate) microfluidic devices bonded using surface modifications and solvents. *Lab Chip*, **6** (1), 66–73.

[165] Goddard, J.M., Hotchkiss, J.H. (2007) Polymer surface modification for the attachment of bioactive compounds. *Progress Polym. Sci*, **32** (7), 698–725.

[166] Miralles, V., Huerre, A., Malloggi, F., Jullien, M.-C. (2013) A review of heating and temperature control in microfluidic systems: Techniques and applications. *Diagnostics*, **3** (1), 33–67.

[167] Lee, C.-Y., Chang, C.-L., Wang, Y.-N., Fu, L.-M. (2011) Microfluidic mixing: A review. *Int J Mol Sci*, **12** (5), 3263–3287.

[168] Sajeesh, P., Sen, A.K. (2014) Particle separation and sorting in microfluidic devices: a review. *Microfluid Nanofluid*, **17** (1), 1–52.

[169] Iv, C.W.S., Reyes, C.D., López, G.P. (2015) Microfluidic cell sorting: a review of the advances in the separation of cells from debulking to rare cell isolation. *Lab Chip*, **15** (5), 1230–1249.

[170] Kraly, J.R., Holcomb, R.E., Guan, Q., Henry, C.S. (2009) Review: Microfluidic applications in metabolomics and metabolic profiling. *Anal Chim Acta*, **653** (1), 23–35.

[171] Pedde, R.D., Li, H., Borchers, C.H., Akbari, M. (2017) Microfluidic-mass spectrometry interfaces for translational proteomics. *Trends Biotechnol.*, **35** (10), 954–970.

[172] Fu, L.M., Tsai, C.H. (2008) Electrokinetic sample injection, in *Encyclopedia of Microfluidics and Nanofluidics* (ed. Li, D.), Springer, Boston, MA.

[173] Kim, M.S., Cho, S.I., Lee, K.N., Kim, Y.K. (2005) Fabrication of microchip electrophoresis devices and effects of channel surface properties on separation efficiency. *Sensors Act. B*, **107** (2), 818–824.

[174] Ollikainen, E., Aitta-aho, T., Koburg, M., *et al.* (2019) Rapid analysis of intraperitoneally administered morphine in mouse plasma and brain by microchip electrophoresis-electrochemical detection. *Sci Rep.*, **9** (1), 3311.

[175] Nordman, N., Sikanen, T., Aura, S., *et al.* (2010) Feasibility of SU-8-based capillary electrophoresis-electrospray ionization mass spectrometry microfluidic chips for the analysis of human cell lysates. *Electrophoresis*, **31** (22), 3745–3753.

[176] Fu, L.-M., Yang, R.-J., Lee, G.-B., Liu, H.-H. (2002) Electrokinetic injection techniques in microfluidic chips. *Anal. Chem.*, **74** (19), 5084–5091.

[177] Karlinsey, J.M. (2012) Sample introduction techniques for microchip electrophoresis: A review. *Anal. Chim. Acta*, **725**, 1–13.

[178] Beattie, J.K. (2006) The intrinsic charge on hydrophobic microfluidic substrates. *Lab Chip*, **6** (11), 1409–1411.

[179] Belder, D., Ludwig, M. (2003) Surface modification in microchip electrophoresis. *Electrophoresis*, **24** (21), 3595–3606.

[180] Sikanen, T., Aura, S., Franssila, S., *et al.* (2012) Microchip capillary electrophoresis–electrospray ionization–mass spectrometry of intact proteins using uncoated Ormocomp microchips. *Anal. Chim. Acta*, **711**, 69–76.

[181] Vandaveer, W.R., Pasas-Farmer, S.A., Fischer, D.J., *et al.* (2004) Recent developments in electrochemical detection for microchip capillary electrophoresis. *Electrophoresis*, **25** (21–22), 3528–3549.

[182] Sikanen, T., Franssila, S., Kauppila, T.J., *et al.* (2010) Microchip technology in mass spectrometry. *Mass Spectrom. Rev.*, **29** (3), 351–391.

[183] Wang, X., Yi, L., Mukhitov, N., *et al.* (2015) Microfluidics-to-mass spectrometry: A review of coupling methods and applications. *J. Chromatogr. A*, **1382**, 98–116.

[184] Yuan, X., Oleschuk, R.D. (2018) Advances in microchip liquid chromatography. *Anal. Chem.*, **90** (1), 283–301.

[185] Grinias, J.P., Kennedy, R.T. (2016) Advances in and prospects of microchip liquid chromatography. *TrAC*, **81**, 110–117.

[186] Xie, J., Miao, Y., Shih, J., *et al.* (2005) Microfluidic platform for liquid chromatography–tandem mass spectrometry analyses of complex peptide mixtures. *Anal. Chem.*, **77** (21), 6947–6953.

[187] Zhao, C., Ge, Z., Yang, C. (2017) Microfluidic techniques for analytes concentration. *Micromachines*, **8** (1), 28.

[188] Lichtenberg, J., de Rooij, N.F., Verpoorte, E. (2002) Sample pretreatment on microfabricated devices. *Talanta*, **56** (2), 233–266.

[189] Chen, Q., Wu, J., Zhang, Y., Lin, J.-M. (2012) Qualitative and quantitative analysis of tumor cell metabolism via stable isotope labeling assisted microfluidic chip electrospray ionization mass spectrometry. *Anal. Chem.*, **84** (3), 1695–1701.

[190] Mao, S., Gao, D., Liu, W., *et al.* (2011) Imitation of drug metabolism in human liver and cytotoxicity assay using a microfluidic device coupled to mass spectrometric detection. *Lab Chip*, **12** (1), 219–226.

4

In Vitro–In Vivo Correlation for Pharmaceutical Nano- and Microsystems

Preshita P. Desai[1] and Vandana B. Patravale[2]

[1]Department of Pharmaceutical Sciences, College of Pharmacy, Western University of Health Sciences, Pomona, California, USA
[2]Department of Pharmaceutical Sciences and Technology, Institute of Chemical Technology, N.P. Marg, Mumbai, India

4.1 Introduction

The ultimate objective of any drug product development is to formulate a quality product that is safe and is able to attain the required drug concentration *in vivo* to elicit the desired therapeutic benefit. Hence, quality, safety and efficacy are the three backbones of successful drug development process. For this, from the regulatory perspective, the detailed physicochemical and stability characterization of the drug product, along with the entire *in vivo* performance, is mandatory for market authorization of the product. Further, in the case of generic products it is mandatory to prove the bioequivalence to the respective reference listed drug [1].

From the product properties perspective, it is well understood that the *in vivo* performance of any product depends upon the inherent drug properties, dose, and dosage form. Amongst these, the drug and dose are predetermined and are constant for any product to be developed. Therefore, during product development, the dosage form properties can directly affect the *in vivo* performance and hence the efficacy. In view of this, ideally, multiple

Characterization of Pharmaceutical Nano- and Microsystems, First Edition. Edited by Leena Peltonen.

in vivo performance studies must be conducted during the product development phase to discriminate between various products and to select the best optimum product complying with the target product profile. However, practically, it is a Herculean task to conduct the *in vivo* evaluation for every product developed during the optimization process, and therefore, less cumbersome yet informative *in vitro* methods revealing the important product characteristics are necessary. *In vitro* dissolution study is one such method that reveals valuable information regarding the release of drug from the formulation. Although it is not a direct equivalent of *in vivo* performance, it can be used to discriminate between product properties [2–5].

This leads to the concept that if the *in vitro* release data correlate to the *in vivo* performance data of the respective products by factoring the other parameters influencing the drug's *in vivo* performance, then it can be used as a tool to predict the *in vivo* performance based on *in vitro* tests only. The scientific methodology of generating this correlation is known as *in vitro–in vivo* correlation (IVIVC). It has gained wide attention over the past few decades and the underlying science has evolved over a period. Also, the regulatory agencies worldwide have acknowledged the importance of IVIVC in product development and have released various guidelines from time to time to incorporate its use in various bioequivalence studies, biowaivers, generic product approvals, scale-up and post-approval changes (SUPAC guidelines), and so on. In view of these developments over the past few decades, IVIVC has become well established for conventional and extended release dosage forms (specifically the oral route) [4, 6–9].

With the advent of novel formulation technologies like micro/nanoformulations, the dosage form properties have become more dynamic (small size, high surface area, enhanced bioavailability and permeability, tissue targeting, etc.), and hence development of IVIVC has become more challenging in such cases. To develop a successful IVIVC model for micro/nanoformulations, it is required to understand the unique properties of the micro/nanoformulations under consideration and to analyze their possible impact on various IVIVC development methods. This will enable the developer to identify the critical points, which can then be optimized, leading to a well-correlated IVIVC. In view of this, the chapter gives a detailed overview of IVIVC from conceptual to regulatory perspectives, with special emphasis on its application to micro/nanoformulations. Specifically, the basic principles of IVIVC are discussed along with various models and a step-by-step approach towards IVIVC development. Subsequently, a brief overview of various micro/nanoformulations is given to differentiate them from the conventional dosage forms, and special considerations while developing IVIVC for these formulations are discussed with detailed case studies.

4.2 *In Vitro* Dissolution and *In Vivo* Pharmacokinetics

In vitro drug dissolution studies are performed with the objective of determining the rate and extent of drug release from the dosage form. Typically, a drug product containing a known concentration of drug is introduced into a dissolution apparatus (preferably approved by the pharmacopoeia) containing dissolution media, and the % cumulative drug release

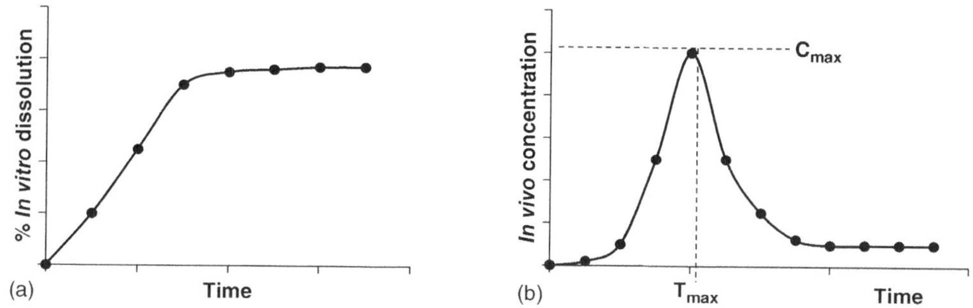

Figure 4.1 *Schematic representation of (a) the* in vitro *dissolution profile and (b) the* in vivo *pharmacokinetic profile for immediate release dosage form.*

vs. time profile is determined by calculating the concentration of drug dissolved in the dissolution medium over a period of time, preferably under sink conditions (for a schematic representation, see Figure 4.1a). The various pharmacopoeial and non-pharmacopoeial dissolution methods reported in the literature are listed in Table 4.1, along with their key features and some applications, especially for micro/nanoformulations. *In vitro* dissolution study is an important quality control and discriminatory test during product development, as it can be used as a tool to compare and differentiate between various products and to select the optimum product. Hence, the selection of the most suitable and discriminatory *in vitro* dissolution method is very important during product development. Further, as *in vitro* dissolution study determines the release of drug from the given dosage form, it is used as an end-product quality control test for batch-to-batch variation.

In vivo pharmacokinetic studies are performed with the objective of analyzing the performance of a drug and/or drug product upon *in vivo* administration. For this, a predetermined dose of drug (alone or in formulation) is administered in an animal model via a predetermined route, and the concentration of drug in body fluids is estimated over a course of time. Specifically, pharmacokinetic studies determine the drug concentration vs. time profile (for a schematic representation, refer to Figure 1b) of the drug from the given dosage form which is used to calculate other pharmacokinetic parameters like maximum plasma concentration (C_{max}), time required to reach maximum plasma concentration (T_{max}), bioavailability (measured as an area under the curve (AUC) of the plasma concentration vs time profile), drug half-life ($t_{1/2}$), and so on.

However, it must be noted that although an *in vitro* dissolution study determines the release of drug from the formulation, it is not representative of the *in vivo* performance. *In vitro* dissolution studies alone cannot be used to determine the *in vivo* pharmacokinetic profile for the product, as release of drug from the dosage is just one amongst the various parameters that influence the *in vivo* drug performance. Briefly, to explain, pharmacokinetic study reveals the entire course of drug fate in the body upon administration, resulting from the combined effects of absorption, distribution, metabolism, elimination, and transformation of drug from its dosage form. Moreover, this absorption, distribution, metabolism, elimination, and transformation depends upon the inherent drug properties,

Table 4.1 *Various pharmacopoeial and non-pharmacopoeial in vitro dissolution methods [10–13]*

Dissolution method	Pharmacopoeial status	Key features	Drawbacks	Applications
Basket type	USP Apparatus I IP Apparatus II BP Apparatus I	Simple, robust, cumulative drug release	Possible basket pore blocking, large volume of dissolution medium (900 ml), pH alteration is possible but tricky by addition or change of dissolution medium	Unit dosage forms, multiple unit dosage forms (e.g., microparticles in capsule), floating dosage forms
Paddle type	USP Apparatus II IP Apparatus I BP Apparatus II	Simple, robust, good mixing dynamics, cumulative drug release	Large volume of dissolution medium (900 ml), filtration system required to separate nanoprecipitates, sinkers are needed for floating formulations. Cone formation.	Unit dosage forms
Reciprocating cylinder	USP Apparatus III	Easy change of dissolution medium over a wide pH range, fractional dissolution results	Error due to dissolution medium evaporation during long-term studies, not suitable for immediate release dosage forms, limited vessel volume	Extended/modified release dosage form
Flow-through cell	USP Apparatus IV BP Apparatus III	Easy change of dissolution medium over a wide pH range, various filters and loading cell designs are available to incorporate modified and nano/micro formulations, can be operated as open or closed system	Possible filter blocking, method development can be tricky, large volumes of dissolution medium, pump accuracy and precision is critical	Extended/modified release dosage forms, micro/nanoformulations, devices, widely used for modified pharmaceutical products

(continued)

Table 4.1 *(continued)*

Dissolution method	Pharmacopoeial status	Key features	Drawbacks	Applications
Paddle over disk	USP Apparatus V	Mimics the transdermal application	Method development can be tricky, patches may not fit in size and hence need a cut that may alter the dissolution	Transdermal products
Cylinder	USP Apparatus VI	Simple apparatus, modified version of USP Apparatus I (steel cylinder instead of basket	Method development can be tricky	Reservoir-type transdermal patches
Reciprocating holder	USP Apparatus VII	Modified dimension version of USP apparatus III with various dimensions	Method development can be tricky, reciprocation distance may not be enough	Osmotic formulations, devices, transdermal formulations
Dialysis membrane dissolution	Nonpharmacopoeial	Ease of sampling and separation from dissolution medium, mixing dynamics and process suitability can be achieved by using with the USP Apparatus II	Meticulous selection of appropriate molecular weight dialysis membrane, membrane pretreatment is necessary, dissolution may vary with the membrane used. Drug–membrane interaction	Suitable for micro/nanoformulation, widely used for modified pharmaceutical products
Microcentrifuge dissolution	Nonpharmacopoeial	Cost-effective, small quantity of product is required	Limited biorelevance, can be used at early development stages	Conventional/modified formulations
Ultracentrifuge dissolution	Nonpharmacopoeial	Suitable separation of drug from excipients especially polymers, small quantity of product is required	Expensive, difficult to develop method	Polymeric formulations
Ultrafiltration dissolution	Nonpharmacopoeial	Suitable separation of drug from excipients, small quantity of product is required	Membrane pretreatment is necessary, dissolution may vary with the membrane used, drug–membrane interaction	Modified release dosage forms

dose, and the dosage form (drug product characteristics). For example, morphine sulfate is available in various dose and dosage forms on the market:

(i) An innovator product of morphine sulfate, KADIAN™ by Allergan Sales LLC, is available in various doses ranging from 10 mg to 200 mg. Although these products have same drug and dosage form, they show different *in vivo* pharmacokinetic profiles as the dose varies [14].
(ii) Morphine sulfate (30 mg dose) is available as an extended release capsule from an innovator company Allergan Sales LLC (KADIAN™) and multiple generic manufacturers like Mpax Laboratories Inc., Par Pharmaceutical Inc., Teva Pharmaceuticals USA, and Upsher-Smith Laboratories LLC. In this case, the drug, dose and the dosage form is similar, resulting in similar *in vitro* drug dissolution profiles, and hence the products are bioequivalent and show similar *in vivo* pharmacokinetic profiles [14].
(iii) Morphine sulfate (30 mg dose) is available in various dosage forms such as an extended release capsule (innovator product KADIAN™ by Allergan Sales LLC), immediate release tablet (West-Ward Pharmaceuticals International Ltd), extended release tablet (MORPHABOND ER by Daiichi Sankyo INC). These products have the same drug and dose but they have different dosage forms. As their dosage form is different, they exhibit different *in vitro* dissolution profiles and hence show different *in vivo* pharmacokinetic profiles [14].

Hence, the *in vivo* pharmacokinetic profile can change from one product to another based on the drug release or drug dissolution from the drug product.

From these examples, it is clear that, although the *in vitro* dissolution profile cannot directly predict the *in vivo* performance, there exists an indirect relationship between the two. Hence, if other factors influencing the *in vivo* performance are factored in, and if a systematic study is conducted to establish this relationship between *in vitro* dissolution profile and *in vivo* pharmacokinetic study, a correlation can be established. Importantly, once this correlation is established it can be used to further predict the *in vivo* performance based on the *in vitro* dissolution studies alone.

This forms the basis for IVIVC. Specifically, the United States Food and Drug Administration (USFDA) defines IVIVC as a predictive mathematical model describing the relationship between the *in vitro* property of an oral dosage form and relevant *in vivo* response [4]. The *United States Pharmacopoeia* (USP) defines IVIVC as the establishment of a relationship between a biological property, or a parameter derived from a biological property produced from a dosage form, and a physicochemical property of the same dosage form [15, 16]. The IVIVC, once established, serves as a low-cost, time-efficient and simple method to be used in replacement of otherwise cumbersome operations from pharmaceutical and regulatory perspectives that are listed below [17, 18]:

- Formulation optimization: As *in vivo* performance can be predicted from *in vitro* dissolution studies, IVIVC can be used as a discriminatory method to optimize and select the most suitable formulation to achieve the *in vivo* target product profile.
- Highly efficient quality control tool to override batch-to-batch variation.
- Convenient replacement for bioequivalence studies: An established IVIVC model can be used instead of cumbersome bioequivalence studies for generic products or in cases of

post-approval changes where the product requires revalidation upon changes in excipient supplies, site of manufacture, small changes in formulation process, and so on (in compliance with SUPAC guidelines).

- Biowaivers: IVIVC is an acceptable alternative to *in vivo* studies and is also recommended by regulatory agencies to get biowaivers for lower strength products and generic products, and hence can reduce the regulatory burden significantly.
- Therapeutic product quality: IVIVC data gives therapeutic understanding to the *in vitro* dissolution data, and hence an *in vitro* release specification can be set and justified based on the therapeutic target profile.

On the other hand, there are a few challenges associated with IVIVC that need to be addressed. These include:

- Extensive *in vitro* and *in vivo* studies to establish IVIVC, which may be difficult at an early stage of product development.
- Critical optimization is required.
- Selection of an appropriate *in vitro* dissolution method can be tricky and requires expertise.
- Stringent regulatory requirements exist for IVIVC.

Furthermore, the extent of IVIVC varies from product to product and depends on how successfully and elaborately *in vitro* and *in vivo* data can be correlated. Based on the extent of correlation, various levels of IVIVC have been identified which are described in the following section.

4.3 Levels of Correlation

In accordance with USFDA Guidelines, five levels of correlation have been identified [4, 15]. Briefly, the level of correlation directly depends on the level of predicting the plasma concentration vs. time profile based on the *in vitro* dissolution for the given dosage form. Higher is the degree of correlation higher is the probability of accurately predicting *in vivo* data from *in vitro* testing. The various levels of correlation are described below

4.3.1 Level A Correlation

This is the highest degree of correlation possible. Level A correlation refers to the point-to-point relationship between the *in vivo* input rate and the *in vitro* dissolution rate of the drug from the given dosage form (Figure 4.2a,b). For this, either the *in vivo* pharmacokinetic data are subjected to mathematical treatment (deconvolution), or *in vitro* data are processed (convolution) to make them comparable to one another by factoring in the other influential parameters. This will be discussed in detail in Section 4.4.

Level A IVIVC is able to predict the complete *in vivo* performance of the dosage form based on the *in vitro* dissolution data. In most cases, Level A correlation exhibits a linear relationship, meaning that the *in vivo* input curve is directly superimposable upon the *in vitro* dissolution curve or can be superimposed by using an appropriate scaling factor. In rare cases, a nonlinear relationship is observed.

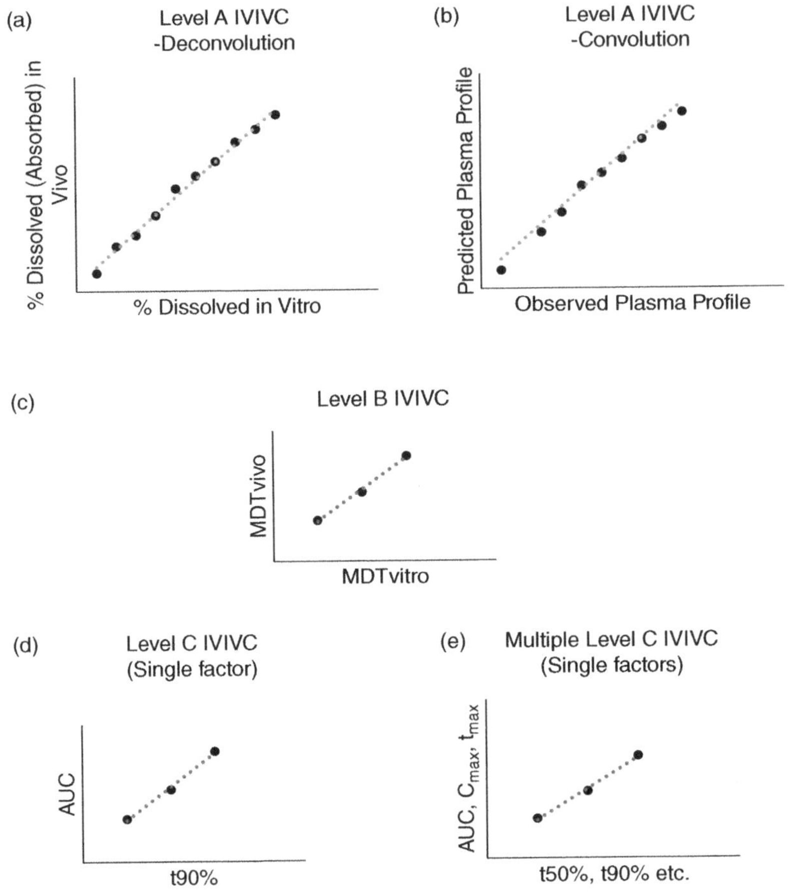

Figure 4.2 *Schematic representations of various levels of IVIVC correlation.*

This type of correlation directly estimates the entire *in vivo* performance and hence is beneficial in the long term because an alteration in manufacturing site, method, or raw material supplier can be justified and *in vivo* performance can be predicted without actual clinical studies. In addition, it is accepted by regulatory bodies worldwide. However, development of Level A correlation demands extensive experimentation, as the point-to-point data from both *in vitro* and *in vivo* performance are compared, analyzed and correlated, which makes it difficult, time-consuming and costly [4, 15, 19, 20].

4.3.2 Level B Correlation

Level B correlation is based on principles of statistical moment analysis. For development of Level B correlation, the entire *in vitro* and *in vivo* dataset is used and the mean *in vitro* dissolution time (MDT_{vitro}) is correlated to the mean *in vivo* residence time (MRT) or the mean *in vivo* dissolution time (MDT_{vivo}) (Figure 2c). To differentiate between Level A and Level B correlation, it must be noted that although both levels require a similar input data (entire *in vitro* dissolution and *in vivo* performance data), the output correlation is different between levels. In addition, Level B correlation is not point-to-point because for the same mean residence time values, multiple *in vivo* curves are possible, and hence Level B correlation cannot determine the actual/real-time *in vivo* plasma concentration vs. time profile [4, 15, 19, 20].

4.3.3 Level C Correlation

Level C correlation is a single point correlation that establishes a relationship between a dissolution profile single point parameter to the *in vivo* plasma performance single point parameter. For example, a specific time point in the *in vitro* dissolution profile, like time required for 50% drug dissolution ($t_{50\%}$) or 90% drug dissolution ($t_{90\%}$), is correlated to AUC, C_{max}, t_{max}, etc. Hence it does not predict the entire *in vivo* input curve, but determines a critical parameter in the *in vivo* performance (Figure 2d). This type of correlation is very useful in the early stages of development, optimization and pilot studies as it is quick, easy, and allows prediction of a critical *in vivo* performance parameter of the formulation. Also, it is useful in cases of modified release formulations like delayed release or extended release targeted formulations. This is the most primary level of correlation and cannot be used to justify biowaiver in case of any variation in product development [4, 15, 19, 20].

4.3.4 Multiple Level C Correlation

A multiple Level C correlation can be described as an extended version of Level C correlation. Specifically, in multiple Level C correlation, one-to-many dissolution profile single point parameters ($t_{50\%}$, $t_{90\%}$, etc.) are correlated to one-to-many *in vivo* plasma performance single point parameters (AUC, C_{max}, t_{max}, etc.). Hence, this allows the opportunity to correlate the entire dissolution profile with *in vivo* performance (multiple level correlation), and if that is achieved a biowaiver can be filed based on this correlation. In addition, it is anticipated that this can be extended to Level A correlation as the entire performance is correlated [4, 15, 19, 20].

4.3.5 Level D Correlation

This is a semi-quantitative or qualitative correlation and is not acceptable for regulatory purposes as it is not a formal correlation. However, it can be useful at a very preliminary level during product and process development [4, 15, 19, 20].

4.4 Models of IVIVC

As described earlier, IVIVC is a mathematical correlation between the *in vitro* dissolution performance and *in vivo* concentration vs. time performance. *In vivo* performance of drugs results from multiple factors that can be broadly classified under two simultaneous procedures, absorption of the drug (directly related to dissolution of the drug) and the overall drug elimination (metabolism, excretion and transformation), whereas *in vitro* dissolution refers only to the release of the drug from the formulation. Hence, to compare *in vitro* data with *in vivo* data, it is mandatory to bring them to same comparable level. Hence, the preliminary data obtained from *in vitro* and *in vivo* studies requires certain mathematical treatment that can enable this comparison.

For example, consider a typical immediate-release oral dosage for which the *in vivo* absorption, elimination and the resultant plasma concentration profile is schematically represented in Figure 4.3. Mathematically, the *in vivo* absorption is an input vector and *in vivo* plasma concentration is an output vector, which can be correlated if the correlation factor or the weighing factor (i.e. the drug elimination) is identified.

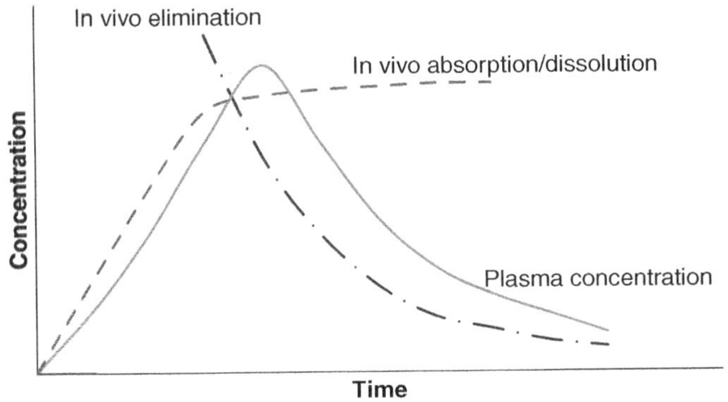

Figure 4.3 Schematic representation of factors influencing in vivo *pharmacokinetic profile for immediate release dosage form.*

In simple mathematical terms, establishing the IVIVC requires determination of this proportionality constant which will allow the direct correlation between the *in vitro* and *in vivo* performance. To obtain this, mathematical correlation can be performed using various models (Table 4.2), the most widely reported being deconvolution and convolution models. Briefly, when an *in vivo* performance is used to deduce the *in vivo* dissolution profile (comparable to *in vitro* dissolution data), it is known as a deconvolution model. In contrast, when an *in vitro* dissolution profile is used to predict the *in vivo* performance (comparable to observed *in vivo* pharmacokinetic data), it is known as a convolution model [2, 3, 18, 21–24].

4.4.1 Deconvolution Model

The deconvolution model is a well-established and most widely reported model for IVIVC. In this model, mathematically treated *in vivo* performance data of the drug are compared with the *in vitro* dissolution profile. Therefore, it is an indirect method wherein firstly the *in vivo* plasma concentration time profile data are subject to mathematical treatment to derive the *in vivo* dissolution data. The *in vivo* dissolution data are then compared against *in vitro* dissolution data by an appropriate model fit and the correlation is established. Based on the aforementioned description, the step-by-step schematic of the deconvolution model for Level A IVIVC is depicted in Figure 4.4.

Table 4.2 *Various mathematical models used for IVIVC [2, 3, 18, 21–25]*

Model and method	Mathematical representation	Key features
Deconvolution		
Model-dependent		
Wagner-Nelson	$F_t = \dfrac{C_t + K_e \int_0^t Cdt}{k_e \int_0^\infty Cdt}$	One-compartment model
Loo-Riegelman	$F_t = \dfrac{C_t + K_{10} \int_0^t Cdt + (X_p)_t/V_c}{k_{10} \int_0^\infty Cdt}$	Multiple compartment model
Mechanistic, physiologically-based	Varies case to case	Suitable for nonlinear kinetics
Differential equation – compartment	Varies case to case	Suitable for nonlinear kinetics
Model-independent		
Numerical	$C_t = \int_0^t C_\delta(t-u) r_{abs}(u)\, du$	No compartment model or absorption kinetics but assumes linear distribution and elimination
Convolution	$C_t = \int_0^t C\delta_t F_t$	
Miscellaneous		
Fractional AUC model	$\text{Fractional AUC}_t = \dfrac{\text{AUC}_{0-t}}{\text{AUC}_{0-t_{last}}}$	Simple, conventional method and can sometimes work as an alternative to Nelson-Wagner deconvolution method
Differential equation method	Varies from case to case	Suitable for nonlinear kinetics

C_t, plasma drug concentration at time t; $C\delta_t$, plasma drug concentrations after an IV dose or oral solution dose at time t; K_e, elimination rate constant; $(X_p)_t$, drug amount in the peripheral compartment as a function of time; V_c, apparent volume of the central compartment; K_{10}, apparent first-order elimination rate constant of the drug from the central compartment (calculated against intravenous dose for same subject); r_{abs}, the absorption rate over the time course; C_δ, concentration time profile; F_t, fraction absorbed at time t. Note: $\int_0^t Cdt = \text{AUC}_{0-t}$ and $\int_0^\infty Cdt = \text{AUC}_{0-\infty}$

To obtain the *in vivo* dissolution profile from the *in vivo* plasma concentration vs. time profile, various methods can be used which are listed in Table 4.2. Amongst these, generally three methods are widely used: model-dependent methods like the Wagner-Nelson method; the Loo-Riegelman method; and model-independent methods such as a numerical method.

The Wagner-Nelson method is used for drugs presenting single compartment model pharmacokinetics and linear elimination. The deconvolution by this method results in a systemic input rate (resulting from the combined effect of dissolution, gastrointestinal (GI) transit

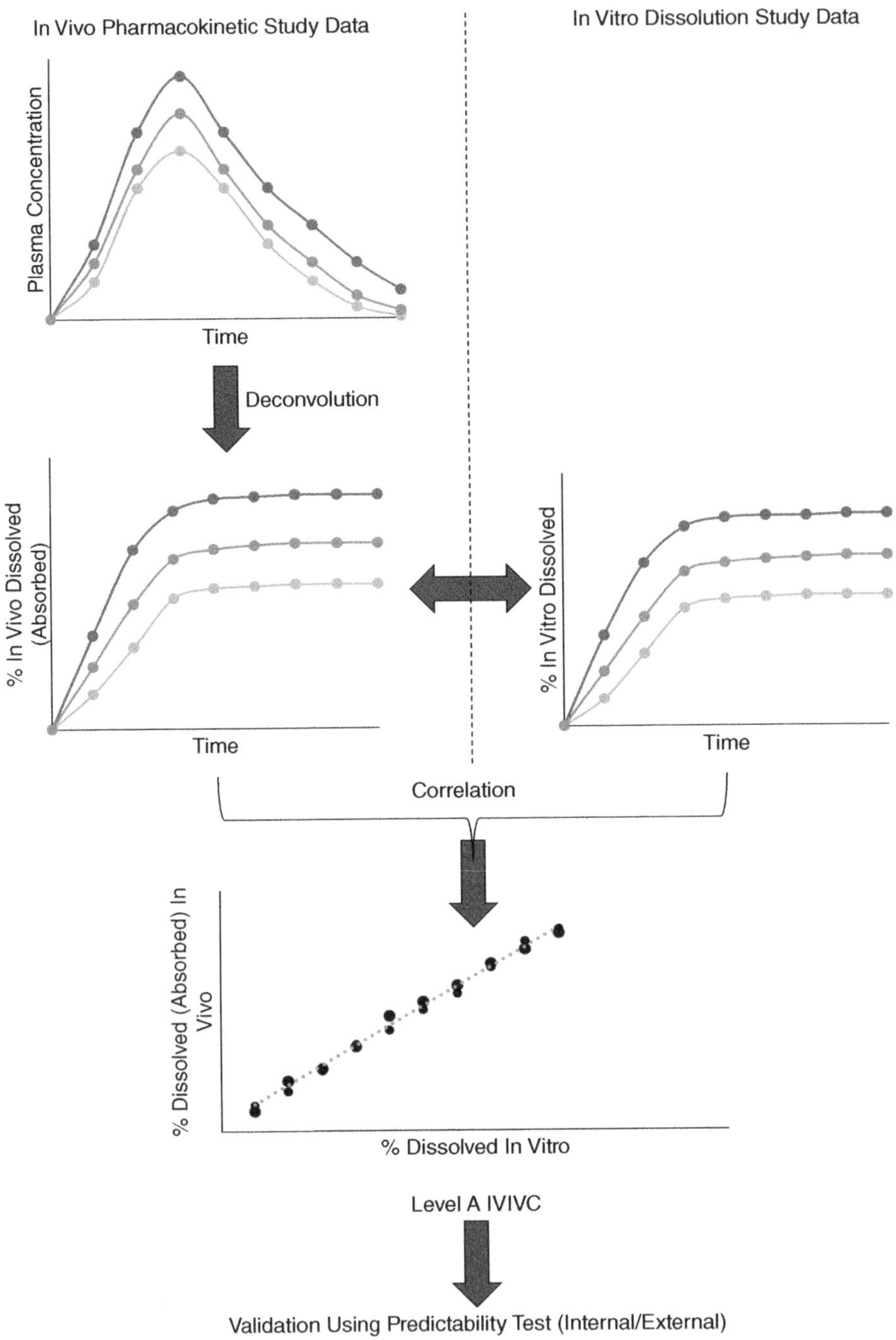

Figure 4.4 *Schematic representation of the step-by-step approach for deconvolution-based Level A IVIVC.*

and permeation, first pass metabolism). As it assumes a one-compartment model, it does not require a parenteral plasma drug concentration profile. The Loo-Riegelman method is used for drugs presenting compartment model pharmacokinetics and linear elimination. The deconvolution by this method results in a systemic input rate (resulting from the combined effect of dissolution, GI transit and permeation, first pass metabolism). As it assumes a multicompartment model, it requires an intravenous plasma drug concentration profile in addition to the data with extravascular route of administration [3, 23, 26].

A numerical deconvolution approach does not require any compartment model or absorption kinetics but assumes linear distribution and elimination. This method is useful when neither of the first two methods are acceptable [3, 23, 26].

Although well-established, there are some fundamental challenges and drawbacks associated with this method, including complex relationships between the overall plasma concentration time profile and the *in vivo* dissolution profile that require expert mathematical and computational treatment. There also exists a risk of bias, as most of the mathematical model-fitting is subjective and lacks statistical confidence levels. To establish this model, multiple products giving different *in vivo* release (slow to fast) are required to develop a suitable dissolution method that can represent the *in vivo* performance, which makes it complicated. Additionally, the *in vivo* data are mandatory in the beginning of this model; this may be difficult in the initial development phase [2, 3, 23, 26]. Despite these challenges, the versatility and confidence of IVIVC development makes it the most acceptable model during process development, optimization and biowaivers.

4.4.2 Convolution Model

In this model, the *in vitro* dissolution data are used as an input to derive the *in vivo* concentration vs. time profile for the drug, by using known drug product pharmacokinetic parameters. Hence, it is a direct and single-step method, unlike the two-step indirect deconvolution model. The mathematical model for this method is specified in Table 4.2. The *in vitro* dissolution profile becomes an input function and the representative/predicted *in vivo* plasma concentration for the given dosage form is an output vector which can be correlated if the correlation factor or the weighing factor (i.e., the known base level plasma concentration time profile) is factored in (e.g., *in vivo* absolute bioavailability, etc.) [2, 3, 18].

It is important to note that these known pharmacokinetic parameters can be obtained either by conducting *in vivo* experiments or by using the literature-reported pharmacokinetic parameter data (volume of distribution, V_d, AUC, etc.) [2, 3, 18]. Although this method is experimentally simple, the model development and data-fitting can be very time-consuming, tricky, and requires a high level of expertise along with very sophisticated software. As a result, while theoretically convolution is a direct single-step approach, practically, deconvolution is very widely used.

4.4.3 Miscellaneous Models

The success of IVIVC using deconvolution and convolution models is sometimes limited when the *in vivo* drug elimination follows nonlinear or saturable kinetics. This is because these mathematical models are based on assumed linear kinetics. Therefore, in some cases, in spite of accurate data, the IVIVC fails due to drug product nonlinear kinetics. As an alternative, in recent years, a differential approach based model has been proposed which can take into account the nonlinear varying factor by time scaling (can be integrated with deconvolution type). This approach needs special consideration, and mathematical treatments can vary from case to case. A software based on NONMEM libraries has been established which simplifies the practical application of this model [3, 18, 27].

Another approach is the fractional AUC approach. It is a simplified method wherein the AUC is calculated using a simple and conventional trapezoidal rule. The fractional AUC is calculated by dividing the cumulative AUC at time t ($AUC_{0\text{-}t}$) by the total cumulative AUC ($AUC_{0\text{-last}}$). The calculated fractional AUC is then compared against the *in vitro* dissolution profile and the IVIVC is generated. This is a useful method in the case of many products, especially those that follow single compartment linear kinetics due to their simplicity. It has been shown that this method is comparable to or even superior to the Nelson-Wagner deconvolution method [3, 22, 28].

4.5 IVIVC Model Validation: Predictability Evaluation

Once the IVIVC model is established, it requires validation to confirm its prediction accuracy. Predictability evaluation (PE) is performed to validate the developed IVIVC method, and is mandatory as far as the regulatory bodies are concerned. This validation process confirms the ability of the IVIVC model to rightly predict the *in vivo* performance based just on the *in vitro* dissolution study over the entire range of dissolution rates, as well as incorporating manufacturing changes (as various batches with different performances are deliberately developed during IVIVC model development). Typically, these studies determine the prediction error, and the model is validated if this prediction error falls within the limits specified by the regulatory bodies. For this, only two *in vivo* pharmacokinetic parameters, the AUC and C_{max}, are predicted for each formulation from the *in vitro* dissolution profile using the established IVIVC model. These predicted values are then compared against the actual values for each formulation, and an average absolute percentage prediction error (% PE) is calculated using the following:

$$\%\ \text{PE} = \frac{(\text{Observed value} - \text{Predicted value})}{\text{Observed value}} \times 100$$

Based on the dataset used for this validation, two types of predictability studies are proposed. When the initial data used for development of the IVIVC model is used for the predictability test, it is known as internal predictability, whereas when additional tests are conducted to generate new data used for the predictability test, it is known as external predictability. Table 4.3 summarizes the key features of both the methods, along with their acceptable limits suggested by the regulatory authorities and recommended use for some specific drugs and drug products [4, 18].

Table 4.3 Regulatory specifications and applications of IVIVC predictability test [4, 18]

Predictability test	Acceptance criteria as per USFDA	Recommended conditions to perform the test by USFDA
Internal predictability	% PE ≤ 10% for C_{max} and AUC and % PE ≤ 15% for each formulation. If not, predictability test is considered inconclusive and external predictability test should be performed.	Internal predictability test is generally performed for all products as first line of validation (general trend).
External predictability	% PE ≤ 10% for C_{max} and AUC. If not, and 10% < % PE < 20%, the predictability test is considered inconclusive and further study is recommended with additional datasets. If not and % PE > 20%, the predictability test is considered inadequate.	• If internal predictability does not meet the acceptance criteria. • If the drug has a very low therapeutic index.

4.6 IVIVC Development Step-by-Step Approach

A successful IVIVC and its acceptance by the regulatory authorities requires a rationalized, well designed and scientifically appropriate study. Various steps involved in IVIVC development are as follows:

Step I: Drug information. The drug to be subjected to IVIVC must be thoroughly studied for its physicochemical properties, available *in vivo* literature (half-life, pharmacokinetic compartment model, etc.) to understand the feasibility and possible approach of IVIVC development. Specifically the focus should be on the biopharmaceutical classification system (BCS) class of the drug. It is recommended that IVIVC can be made successful for the drugs that show dissolution-controlled drug release from the formulation, and hence it is comparatively suitable for BCS class I and BCS class III drugs. Based on the BCS class of drug, absorption rate limiting steps can be identified (as summarized in Table 4.4) and an IVIVC strategy should be planned in order to overcome those limitations [3, 14, 29, 30].

Step II: *In vitro* dissolution studies. *In vitro* dissolution data (at least $n = 3$) for each product is mandatory. The formulations and discriminatory dissolution method are selected in such a way that the formulation should typically show slow, medium and fast release with a minimum variation of 10% in the *in vitro* dissolution profile. If it is a bioequivalence study, innovator/reference product must be included [3, 4, 18].

Step III: *In vivo* performance studies. *In vitro* pharmacokinetic data for each product used for the *in vitro* dissolution study is mandatory. If it is a bioequivalence study, innovator/reference product must be included [3, 4, 18].

Table 4.4 *BCS classification and IVIVC assumptions and expectations [3, 18, 29, 30]*

BCS class	Solubility	Permeability	Rate-limiting factor for absorption	IVIVC assumptions and expectations
I	High	High	Gastric emptying	IVIVC is expected if dissolution rate is slower than gastric emptying time
II	Low	High	Dissolution	IVIVC can be expected if *in vitro* dissolution rate is same as *in vivo* dissolution rate
III	High	Low	Permeability	Permeability is rate-limiting, hence IVIVC with dissolution rate can be tricky
IV	Low	Low	Varies from case to case	Low feasibility of IVIVC

Step IV: IVIVC model development. For regulatory relevance and application, Level A correlation should always be aimed for, as it is a point-to-point correlation. The deconvolution, convolution or other specialized model should be selected appropriately based on the criteria described in above. In many cases, multiple methods can be tried to select the model with the best coefficient of correlation value. In some cases, if the desired correlation is not obtained, suitable optimization of the *in vitro* dissolution method (repeat Step II with modifications) should be done.

Step V: Validation. Validation of the IVIVC model in compliance with regulatory guidelines should be performed.

4.7 Brief Introduction to Micro/Nanosystems and IVIVC Relevance

Nano- and microformulations have gained wide attention over the past few decades due to their unique properties that include, but are not limited to, enhanced absorption, enhanced bioavailability, enhanced intracellular permeability, the ability to target specific organs/tissue or cells, a long circulation time (stealth properties), the ability to control drug release, reduced side-effects with high safety, increased drug stability, reduced dose and dosing frequency, overall enhanced efficacy, biocompatibility, patient compliance, ability to deliver both hydrophilic and hydrophobic actives, and so on [25, 31].

Hence, it should be noted that these advanced micro/nanoformulations exhibit different physicochemical properties from conventional formulations, and hence the result is different for *in vitro* as well as *in vivo* performance. This key difference should always be remembered while developing IVIVC models for these advanced pharmaceuticals. These micro/nanoparticles may comprise polymers, lipid particulate, lipid emulsified inorganic/organic backbones, drug micro/nanosuspensions, or combinations thereof. Various types of micro/nano carriers are depicted in Figure 4.5 (modified from [25]).

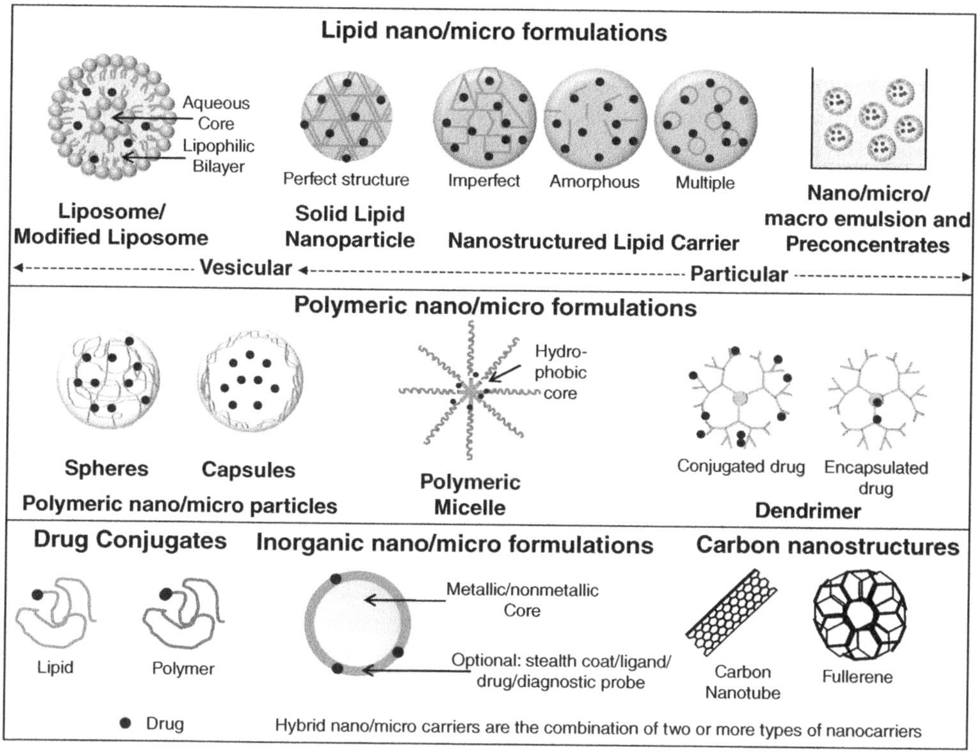

Figure 4.5 *Schematic representations of various advanced micro/nanoformulations (modified from [25]).*

For these specialized micro/nanoformulations, the basic principles of IVIVC remain the same as for any other conventional formulation, but the practical application requires some special consideration and modulation in the *in vivo* and *in vitro* dissolution methods used to develop the IVIVC.

4.7.1 Selection of Appropriate Dissolution Method

A suitable and discriminatory *in vitro* dissolution method is the key for successful IVIVC. The *in vitro* dissolution studies must be able to predict the *in vivo* performance and hence must be biorelevant to the *in vivo* performance. The various *in vitro* dissolution methods are listed in Table 4.1 along with their key features and reported applications.

To achieve the best possible IVIVC, various *in vitro* dissolution methods can be performed and subjected to IVIVC. For example, a research group developed three different micro/nanoformulations for comparison against the pure drug [10]. In the study, nanoamorphous, nanocrystalline, and microamorphous solid dispersions of a model drug, carbamazepine, were developed and compared against the pure crystalline drug. The *in vivo* single dose oral pharmacokinetic studies were performed in mice. The *in vivo* studies showed that nanoformulations showed maximum bioavailability, and the order of *in vivo* performance was nanocrystalline > nanoamorphous > microamorphous > pure crystalline drug.

For the *in vitro* drug dissolution studies, microcentrifuge dissolution, ultrafiltration dissolution, and membrane-permeation dissolution methods were performed. The results of the microcentrifuge dissolution method indicated the rank order of dissolution as nanoamorphous > microamorphous > nanocrystalline > pure crystalline drug. This dissolution rank order was not similar to that obtained with *in vivo* performance, as the nanocrystalline formulation showed the highest *in vivo* absorption. Hence, this method was disregarded. The probable reason for this disparity could be the fact that during dissolution of amorphous dispersions, there is a possibility of nanoprecipitate formation, which could not be separated from the dissolution medium. To address this, the scientists reported that an effective second filtration method must be in place to separate these nanoprecipitates from the dissolved drug, which was not present in the method that was employed here. Hence, the *in vitro* dissolution data showed variation from the *in vivo* performance.

The second method employed for the *in vitro* dissolution study was ultrafiltration dissolution. The rationale behind use of this method was its ability to separate small dissolved molecules from the undissolved polymers (used in preparation of solid dispersion) and precipitates which are more than 3 nm in size. Despite this, the results of the ultracentrifuge dissolution method indicated the rank order of dissolution as nanoamorphous > microamorphous > nanocrystalline > pure crystalline drug, which was similar to results obtained by the previous microcentrifuge method. As the dissolution rank order was not similar to that obtained with *in vivo* performance, this method was also disregarded.

Further, to mimic the *in vitro* method that most resembles and is most biorelevant to the *in vivo* scenario, a membrane-permeation dissolution method was tested. The results revealed that the nanocrystalline system showed maximum dissolution, but the difference in *in vitro* dissolution performance for the remaining three formulations was non-significant. Hence, although the method was suitable and compliant with the *in vivo* data, showing highest performance with the nanocrystalline formulation, fine-tuning of the method was mandatory to make it discriminatory for the remaining three formulations.

In view of this, the dissolution medium was considered as the crucial factor and multiple dissolution mediums were studied. During the first membrane-permeation dissolution method, 0.1 N hydrochloric acid (HCl) and fasted state simulated intestinal fluid (FaSSIF) at pH 6.5 was used as a dissolution media, with 1-decanol as an acceptor solution to enable a drug sink condition. To fine-tune the dissolution medium, multiple factors were assessed, including changing the FaSSIF to canine FaSSIF, which is more similar to rat intestinal fluid. Changing the pH from 6.5 to 5.2 more resembled the pH of mice GI tract, and also minimized the variation due to pH-induced drug release. The acceptor phase sink solution was also altered to a commercial ASB (pION ASB-7.4, Double-Sink™ buffer, P/N 110139) solution that has a physiological pH along with scavenger chemicals with properties similar to those of serum proteins. Additionally, the separation membrane was also changed from Pes polymer membrane to porcine intestine membrane to make the method more biorelevant. The rank order of dissolutions obtained by varying these dissolution medium parameters during the membrane-permeation dissolution method are summarized in Table 4.5.

Among the various dissolution media investigated, three dissolution media showed a rank order similar to that of the *in vivo* studies, and the dissolution studies with canine FaSSIF

Table 4.5 In vitro dissolution rank order for variable dissolution medium studies performed for IVIVC development of carbamazepine formulations [10]

Dissolution medium	Separation membrane	Rank of degree of dissolution	Coefficient of correlation
FaSSIF pH 6.5/decanol	PES polymer membrane	Nanocrystalline > nanoamorphous > pure crystalline drug > microamorphous	0.84
FaSSIF pH 6.5/ASB	PES polymer membrane	Nanocrystalline > nanoamorphous > microamorphous > pure crystalline drug	0.80
FaSSIF pH 5.5/decanol	PES polymer membrane	Nanocrystalline > pure crystalline drug > nanoamorphous > microamorphous	0.81
Canine FaSSIF pH 5.2/decanol	Porcine intestine membrane	Nanocrystalline > nanoamorphous > microamorphous > pure crystalline drug	0.90
Canine FaSSIF pH 5.2/ASB	Porcine intestine membrane	Nanocrystalline > nanoamorphous > microamorphous > pure crystalline drug	0.82

pH 5.2 showed the highest coefficient of correlation (Table 4.5). To summarize, multiple dissolution methods were compared, and the best amongst them were further optimized to achieve more precise and accurate results. This method enabled precise IVIVC with a very high coefficient of determination (R^2) and confirmed the linear relationship [10].

4.7.2 Selection of Appropriate Dissolution Medium

It must be noted that, for successful IVIVC, not only is the selection of an appropriate dissolution method important, but also the selection of an appropriate dissolution medium. This selection becomes more challenging in the case of micro/nanoparticles because of their modified physicochemical properties, such as reduced particle size and increased surface area, which result in an increased rate and extent of solubility. In addition, the selection of a dissolution medium becomes more challenging due to the difficulty in establishing an efficient separation method that is able to distinguish and separate undissolved/dispersed micro/nanoparticles from the dissolved drug.

For example, in one study, hollow-type mesoporous silica nanoparticles (HMSNs) were developed with spherical uniform particles between 50–100 nm, and loaded with silybin meglumine with a loading efficiency of 58.91% ± 0.39% [32]. The single dose oral *in vivo* pharmacokinetic study was conducted in healthy beagle dogs over a period of 72 h and the results showed sustained pharmacokinetic performance complying with a one-compartment model. The deconvolution of the *in vivo* pharmacokinetic data was executed using the Wagner–Nelson method (suitable for a one-compartment model) and the *in vivo* absorption profile was calculated.

In vitro dissolution studies were performed by the dialysis tubing method in paddle dissolution testing apparatus as suggested by the *Chinese Pharmacopoeia 2005*. To select the most optimum dissolution medium to enable Level A correlation, multiple dissolution mediums over a pH range of 1–12 were tested. This included conventional dissolution mediums like artificial gastric juice (pH 1.2) and artificial intestinal juice (pH 6.8), 0.01 M NaOH (pH 12), and a novel dissolution medium comprising Na_2CO_3 solutions at multiple concentrations (0.01 M, 0.06 M, 0.08 M, and 0.1 M).

A significant difference in the *in vitro* dissolution profile was observed with change in the dissolution medium, and the key observations are reported in Table 4.6. Based on the results, the author reported that both pH and ion concentration of the dissolution medium played a critical role in the *in vitro* dissolution studies [32]. The underlying reasons for this difference were listed as follows. First, silica is slightly acidic, which must have shown an acid–base interaction resulting in enhanced dissolution observed with an increase in pH of the medium, and also silybin meglumine is an ammonium salt and hence is stable in alkaline medium. Hence, the extremely slow release in artificial gastric juice and intestinal media and subsequent increase in dissolution with increasing concentration of Na_2CO_3 solutions can be explained. Consistent with this observation, 0.01 M NaOH (pH 12) should have shown a high *in vitro* dissolution but it showed a significantly lower dissolution rate (less than 45% dissolution at end of 72 h). This may have resulted from the fact that 0.01 M NaOH has lower ionic concentration than that of the Na_2CO_3 solutions. Hence, it was reported that ionic concentration is also a controlling factor. Based on the results, the authors confirmed that the *in vitro* release of silybin meglumine from HMSNs can be substantially controlled by varying the concentration of Na_2CO_3 solutions.

Table 4.6 *In vitro dissolution and IVIVC coefficient of correlation results obtained using silybin meglumine HMSNs with multiple dissolution media [32]*

Dissolution medium	pH	Key observations, *in vitro* dissolution profile	IVIVC coefficient of correlation (R^2) value
Gastric juice	1.2	< 20% at 72 h	0.9287
Artificial intestinal juice	6.8	< 20% at 72 h	0.7689
0.01 M Na_2CO_3 solution	11.1	< 15% at 72 h	Not performed
0.06 M Na_2CO_3 solution	11.5	62.8% at 24 h followed by 77.4% at 72 h	0.9741
0.08 M Na_2CO_3 solution	11.6	80.0% at 24 h followed by slow increase up to 92.3% at 72 h	0.9931
0.1 M Na_2CO_3 solution	11.7	94% at 12 h	0.5600
0.01 M NaOH solution	12	< 45% at 72 h	Not performed

Further, to establish IVIVC, the *in vivo* absorption profile was compared with the *in vitro* dissolution profile obtained from the dissolution studies performed using various dissolution media (Table 4.6). The results revealed that 0.08 M Na_2CO_3 solution showed the maximum Level A correlation coefficient (R^2 value of 0.9931), which also showed a highly controlled *in vitro* drug dissolution profile. It must be noted that it was possible to establish

Level A correlation in this study only because multiple dissolution media were screened, wherein only one medium showed a significantly high R^2 value compared with other dissolution media [32].

To summarize, selection of an appropriate dissolution medium for *in vitro* dissolution studies is very critical for successful IVIVC. Some of the key factors that should be considered prior to selection of dissolution media for *in vitro* dissolution studies are listed in Table 4.7.

Table 4.7 *Factors influencing the selection of optimum in vitro dissolution medium for IVIVC*

- Sink or non-sink conditions with proposed dissolution medium (saturation solubility)
- pH of the dissolution medium
- Ionic concentration of the dissolution medium
- Viscosity of the dissolution medium
- Volume of the dissolution medium
- Stability of drug in the dissolution medium
- Dissolution medium–formulation excipient interaction
- Biorelevance of dissolution medium

4.7.3 Selection of Appropriate IVIVC Mathematical Model

As an example, Ali *et al.* [33] developed furosemide solid lipid nanoparticles (FSLNs) exhibiting a size of ~25 nm which was compared against a furosemide suspension. To develop an appropriate IVIVC model, they studied five different deconvolution methods and selected the optimum method that gave the best possible IVIVC correlation. The five different deconvolution methods studied were Wager-Nelson-1-compartment, Loo-Riegelman-2-compartment, Loo-Riegelman-3-compartment, numerical deconvolution single Weibull, and numerical deconvolution double Weibull, using the IVIVCPlus™ module of GastroPlus™ software. The rationale behind studying multiple deconvolution methods was the fact that furosemide is a BCS class IV molecule that shows variable absorption and bioavailability, and the information regarding the exact pharmacokinetic compartment model followed by the drug and other drug parameters was limited.

As supportive data, the pharmacokinetic study data were subject to various non-compartment and compartment analysis using the PKPlus™ module of the GastroPlus™ software. These studies revealed that FSLN exhibited an optimum fit for the two/three-compartment models.

For the deconvolution studies, data were fitted for all the five methods and predictability studies for C_{max} and AUC were conducted to validate the models. For this, the R^2, % PE, standard error of prediction (SEP) and mean absolute error (MAE) were calculated for each method, and the results were compared to the deconvolution method that gave best values for all the above parameters. As representative results, the % PE values for AUC and C_{max} for all models are summarized in Table 4.8. Based on the results, the numerical method was not selected as the correlation predictability value was very off-limit, and Wager-Nelson-1-compartment method was not selected because although it gave an excellent % PE value of AUC, the % PE value of C_{max} was very off-limit. Hence, the Loo-Riegelman method (compartment method allowing incorporation of peripheral compartments) was best suited amongst all the five tested methods. This was also consistent with

Table 4.8 *Predictability test of the % PE values for AUC and C_{max} for FSLNs using various models [33]*

	% PE for C_{max}	% PE for AUC			
	Entire time course	Entire time course	Divided time course		
	0–24 h	0–24 h	0–0.5 h	0.5–3 h	3–24 h
Wager-Nelson-1-compartment	40.17	8.65	49.23	3.71	16.63
Loo-Riegelman-2-compartment	14.87	15.67	54.15	1.04	13.86
Loo-Riegelman-3-compartment	12.62	16.26	51.57	2.66	15.71
Numerical deconvolution single Weibull	82.69	58.78	Not performed		
Numerical deconvolution double Weibull	74.15	50.85	Not performed		

the pharmacokinetic model analysis that showed a best fit with a two/three-compartment model.

Although the R^2 and predicted error values with the Loo-Riegelman method was superior to other methods tested, all the % PE values (data represented in Table 4.8) for the Loo-Riegelman method were not in compliance with regulatory acceptability criteria (refer to Table 4.3), and hence further investigation was necessary.

Interestingly, as stated earlier, the Wager-Nelson-1-compartment method showed very low and desirable % PE values of 8.65%. To understand this effect, the entire plasma concentration time profile was divided into three phases that included 0–0.5 h, 0.5–3 h and 3–24 h (representative of the absorption, distribution, and elimination phases) and again the % PE values for AUC were calculated for the Wager-Nelson-1-compartment, Loo-Riegelman-2-compartment, and Loo-Riegelman-3-compartment models (data shown in Table 4.8). The results showed that the % PE value for AUC for all models showed excellent fit for the 0.5–3 h phase (representative of the distribution phase) followed by the 3–24 h phase (representative of the elimination phase), but failed significantly for the absorption phase. Hence, it was concluded that the IVIVC model was not compliant for the absorption phase, and Ali *et al.* [33] reported that this may have occurred due to a relatively low number of analysis time points during this phase, which could not represent the entire absorption phase. Hence, the time-phase modified Loo-Riegelman compartment models were considered to be most suitable [33].

4.8 Applications of IVIVC for Micro/nanoformulations

IVIVC has wide applications all the way from formulation development to quality control and post-approval changes for the same product; it also plays a significant role in bioequivalency studies for generic formulations. Various applications of IVIVC are summarized in Table 4.9 and some of the key representative applications are described below.

Table 4.9 *IVIVC applications for micro/nanoformulations*

Formulation	Delivery route	*In vitro* method	IVIVC Level	Key features/application	Reference
Lipid-based micro/nanoformulations					
Metronidazole nanoemulsion	Topical	*In vitro* permeation studies	Level A	Targeted topical delivery with enhanced safety	[34]
Simvastatin NLCs	Oral	Dialysis bag method	Level A	Enhanced bioavailability over the solid lipid nanoparticles	[35]
Capsaicin liposomes	Oral	Dialysis bag method	Level A	Model applicable to both free drug and liposomal drug	[36]
Zidovudine solidified reverse micellar microparticles	Intraperitoneal	Dialysis bag method	Level A	Controlled release formulation (using release retardant)	[37]
Polymeric micro/nanoformulations					
Indomethacin gelatin nanoparticles	Oral	Dialysis bag method	Level A	Enhanced anti-inflammatory property	[38]
Trans-resveratrol Eudragit RL 100 nanoparticles	Oral	Dialysis bag method	Level A	Significant enhancement in drug absorption with nanoparticles	[39]
Celecoxib-PVP solid dispersion nanoparticles	Oral	USP rotating paddle apparatus	Multiple Level C (dissolution efficacy correlated with AUC and C_{max})	Significant enhancement in *in vitro* dissolution and bioavailability	[40]
Meloxicam chitosan-coated sodium alginate microparticles	Oral	USP apparatus I	Level A	Extended release of meloxicam	[41]
Clozapine PLGA microparticles	Oral	Not disclosed	Level A	Sustained drug release over a period of 3 weeks	[42]

(*continued*)

Table 4.9 *(continued)*

Formulation	Delivery route	*In vitro* method	IVIVC Level	Key features/application	Reference
Olanzapine microspheres	Subcutaneous	Dialysis bag method	Level A	Sustained release profile	[22]
Dexamethasone PLGA microspheres	Oral	USP apparatus IV		Sustained anti-inflammatory activity	[43]
Risperdal(®) Consta(®) microspheres	Oral	USP apparatus IV		Sustained anti-inflammatory activity	[44]
Huperzine A PLGA microspheres	Intramuscular or subcutaneous	Not disclosed		Sustained release with increase in concentration of PLGA and particle size Correlation for IV route was better than that for SC route	[45]
Nimesulide HPMC microparticles	Oral		Level A	Extended release properties	[46]
Risperidone PLGA microspheres	Intramuscular	Sample-and-separate and USP apparatus IV	Level A	Long-acting injectable atypical antipsychotic drug product	[3]
Risperidone PLGA microspheres	Intramuscular	Sample-and-separate and USP apparatus IV	Level A	Long-acting injectable atypical antipsychotic drug product	[28]
Suspension micro/nanoformulations					
Coenzyme Q10 Lyotropic liquid crystalline nanoparticles	Oral	Dialysis bag method	Level A	Formulation optimization confirmed that glyceryl monooleate-based nanoparticles showed better performance over phytantriol-based nanoparticles	[47]

Puerarin nanocrystals	Oral	*In vitro* dissolution studies in pH 6.8 phosphate buffer		Enhanced safety with oral formulation	[48]
Fenofibrate nanosuspension	Oral	Paddle dissolution apparatus	Level A	Liner correlation between *in vivo* absorption and intestinal absorption fractionabsorption fraction	[49]
Inorganic micro/nanoformulations					
Gold nanoparticles	*In vitro*	*In vitro* cell line studies	Nanoparticle toxicity IVIVC in *in vitro* cell lines vs. animals (zebrafish embryo)	Novel application of IVIVC	[50]
Iron oxide nanoparticles	*In vitro*	*In vitro* cell line studies	Nanoparticle toxicity IVIVC in *in vitro* cell lines vs. animals (zebrafish embryo)		[50]
Hybrid micro/nanoformulations					
Valsartan polymer-based supersaturable SMEDDS	Oral		Multiple level	Enhanced bioavailability	[51]

PVP, polyvinyl pyrrolidone; HPMC, hydroxypropylmethylcellulose; PLGA, polylactic-co-glycolic acid; NLCs, nanostructured lipid carrier; SMEDDS, self-microemulsifying drug delivery system.

4.8.1 Formulation Optimization

Formulation development and optimization is a key step in development of successful products within the specifications of targeted product profiles. Regulatory authorities are now recommending implementation of the Quality by Design (QbD) approach for product development. QbD refers to a systematic and scientific product development wherein the targeted product profile is determined, following which the impact of critical factors related to composition and process on product performance is studied, and the optimum product is selected within the design space with a control strategy [1, 52–56].

IVIVC can be a very valuable tool during this process of product optimization. Using IVIVC, the *in vivo* performance can be predicted based on the *in vitro* dissolution studies and hence the direct impact of compositional and process parameters on real-time *in vivo* performance can be compared. For example, Khaled *et al.* [57] developed encapsulated metoprolol tartrate formulations with drug-to-polymer ratios (w/w) of 1:1, 1:1.5 and 1:2 using a solvent evaporation method and compressed it to form a tablet. The developed formulations were subjected to *in vitro* dissolution studies using the sequential pH change technique, and it was shown to follow a zero-order kinetic model. The IVIVC was generated using the Wagner-Nelson method of deconvolution. The correlation studies were able to establish Level A, Level B and Level C ($t_{50\%}$ vs. AUC) IVIVC with acceptable predictability as per USFDA guidelines. Hence, this method can be used as a discriminatory method to optimize the formulations to match the target profile [57].

In another study, D'Souza *et al.* [22] developed polylactic-co-glycolic acid (PLGA) microparticles of olanzapine using a solvent extraction/evaporation method. For the formulation, they used different compositions of PLGA as described in Table 4.10. The *in vitro* dissolution studies were performed using the modified dialysis bag method.

Table 4.10 *Composition, in vitro dissolution and in vivo pharmacokinetic properties of various PLGA microparticles of olanzapine [22]*

Formulation	PLGA composition	*In vitro* dissolution studies: key observations	*In vivo* pharmacokinetic/ absorption studies: key observations
F1	15 kDa, 75 : 25 PLGA	10% burst release in 24 h, ~21% by day 3, 90% by day 15	10% olanzapine absorption by day 1, almost 35% by day 4
F2	30 kDa, 50 : 50 PLGA		10% olanzapine absorption by day 1, 35% by day 4
F3	82 kDa, 65 : 35 PLGA	Lower initial burst in 24 h, 3–7% release by day 3, 23% by day 8, 74% by day 15	3% olanzapine absorption by day 1, 14% by day 4
F4	131 kDa, 75 : 25 PLGA	Lower initial burst in 24 h, 3–7% release by day 3, 23% by day 8, 44% by day 15	3% olanzapine absorption by day 1, 7% by day 4

The *in vivo* pharmacokinetic studies were performed on Sprague-Dawley rats. The IVIVC model development was conducted using the Nelson-Wagner deconvolution and fractional AUC methods. The results showed that the initial burst release decreases almost linearly with increase in PLGA molecular weight (F1–F3) followed by a plateau, and the total drug release decreased with an increase in PLGA molecular weight. A similar trend was observed with *in vivo* absorption data, wherein the absorption decreased as the PLGA molecular weight increased from F1 to F4. This could be explained as follows. The drug release from PLGA microspheres follows two mechanisms: diffusion and erosion. With low molecular weight PLGA, the hydration was faster, resulting in faster erosion and hence drug release. It was also confirmed that molecular weight was a dominant factor, and varying compositions of PLGA did not have a significant impact. Formulations F3 and F4 with high molecular weight PLGA showed slowed diffusion followed by erosion as hydration happened relatively slowly, and hence overall, a more sustained release profile was observed. However, even if not significant, the lactide:glycolide ratio in PLGA polymer played some role in release kinetics, especially for higher molecular weight PLGA formulations. This could be the possible reason for relatively faster dissolution of the drug from F3, which had a lower concentration of hydrophobic lactic acid compared with F4. The correlation using the Nelson-Wagner deconvolution method showed Level A correlation, which is acceptable to regulatory authorities. In addition, the fractional AUC method showed acceptable correlation and could be used as an alternative. Hence, based on the IVIVC correlation and by deriving valuable information regarding excipient composition and associated drug dissolution properties, successful formulation optimization can be achieved [22].

In another investigation, Shen *et al.* [3] developed PLGA microspheres of risperidone using the same grade PLGA (7525 DLG 6E) as used in the marketed product Risperdal® Consta® (long-acting injectable atypical antipsychotic drug product) using multiple methods of preparation. The objective here was to assess the effect of manufacturing process on formulation properties and performance when the composition is kept constant, and comparing it against the marketed formulation. The formulation process parameters like homogenization, sieving method, and solvents were varied (Table 4.11). The rank order of porosity for these formulations was Risperdal® Consta® ~ F1 < F2 < F3 < F4. The particle size analysis showed almost similar sizes for Risperdal® Consta®, F1, F2 and F3 (~100 μm), whereas F4 showed a relatively smaller particle size (70–80 μm). The *in vitro* drug dissolution studies showed a rank order of Formulation 4 > Formulation 3 > Formulation 1 > Formulation 2. The slow and prolonged dissolution from F1 and F2 must have resulted from their low porosity and slow hydration (very low solubility of water in dichloromethane) leading to slow erosion. The *in vitro* release increased in the case of F3 due to increased porosity and use of ethyl acetate as a solvent. It further increased in F4 due to highest porosity, use of ethyl acetate as a solvent, and smaller particle size (high surface area to volume ratio and small diffusional path). The marketed product showed a different lag time profile to any of the formulations, though the composition was the same. stating that the formulation process had a great impact on the product characteristics. These observations were consistent with the *in vivo* data, and the Level A IVIVC was achieved using the Loo-Riegelman deconvolution method [3].

Table 4.11 *Process variables, in vitro dissolution and in vivo pharmacokinetic properties of various PLGA microspheres of risperidone and its commercially available product [3]*

Sample	Process variables, mixing + sieving + solvent	*In vitro* drug release key observations	*In vivo* drug release key observations (deconvolution)
Risperdal® Consta®	–	Prolonged lag phase than F1–F4, indicating more sustained release in initial phase	Initial lag phase of about 15 days with only 12% drug absorption. Later the fraction of drug absorbed/released increased gradually and a plateau was reached by day 27
F1	Homogenization + dry sieving + dichloromethane	Higher burst release percentages (*ca.* 10–14%) followed by a slightly longer lag phase and reached release plateau around day 40. Dissolution faster than F2 but slower than F3 and F4.	Initial low plasma concentration up to 5 h. Longer lag phase and later the absorption phase peaked at around day 16.
F2	Homogenization + wet sieving + Dichloromethane	Higher burst release percentages (*ca.* 10–14%) followed by a slightly longer lag phase and reached release plateau around day 40. Dissolution slower than F1, F3, F4.	Initial low plasma concentration up to 5 h. Longer lag phase and later the absorption phase peaked at around day 18.
F3	Vortex + wet sieving + ethyl acetate	Lower burst release percentages (<3%) followed by a shorter lag phase and reached release plateau around day 33. Dissolution faster than F1, F2, but slower than F4.	Low initial plasma concentrations due to low burst release *in vitro*. Shorter lag phase and faster absorption profile. The absorption phase peaked at around day 15.
F4	Homogenization + wet sieving + ethyl acetate	Lower burst release percentages (<3%) followed by a shorter lag phase and reached release plateau around day 33. Dissolution faster than F1, F2 and F3.	Low initial plasma concentrations due to low burst release *in vitro*. Shorter lag phase and faster absorption profile. The absorption phase peaked at around day 15.

Therefore, it can be seen that formulation composition and process plays a critical role in product performance, and IVIVC can be used as a tool to discriminate between the formulations and to select the optimum product.

4.8.2 Surrogate for Bioequivalence Studies and Biowaivers

The term biowaiver is used during the regulatory product approval process when the product dossier (product approval application) is granted based on the evidence of equivalence other than *in vivo* equivalence testing [9]. It means that *in vivo* bioavailability or bioequivalence studies are waived and are no longer required for product approval. Biowaiver is a very meaningful process to save time, money and efforts, especially in the case of ANDA (generic products) and scale-up and post-approval change (SUPAC) [34–36].

IVIVC is a very important tool for achieving biowaivers, as it can predict the *in vivo* performance based on *in vitro* dissolution studies. Hence, IVIVC can greatly reduce the burden of *in vivo* studies during product approval if the biowaiver can be granted based on IVIVC. As per USFDA guidelines, an option of biowaiver through IVIVC is possible for drugs that exhibit dissolution as a rate-limiting step (e.g., for BCS Class I and III), and the same is applicable for immediate release and extended release formulations. The biowaiver may be granted under the following scenarios if the drug does not have a narrow therapeutic index:

- Change in manufacturing site.
- Change in excipient (release/non-release controlling).
- Change in manufacturing process.
- Complete removal of non-release controlling excipient.
- Lower strength products if IVIVC is established for highest strength products and *in vitro* dissolution is the same, with the same release mechanism.
- New strength within the dosing range if the new strength is compositionally proportional or qualitatively the same.
- Generic products (if bioequivalence is proved for all strengths of reference listed product, or reference listed product has dose proportionality and they are compositionally proportional or qualitatively the same with similar *in vitro* dissolution profile and release mechanism/bioequivalence is proved for highest and lowest strengths of reference listed product, and all strengths are compositionally proportional or qualitatively the same with similar *in vitro* dissolution profiles and release mechanisms).

If the drug has a narrow therapeutic index it is extremely unlikely to get a biowaiver from regulatory bodies, and if considered for the above cases, an external predictability validation is a must. Detailed guidelines are given by USFDA from time to time in the form of *Guidance for Industry* [4, 9].

4.9 Softwares Used for IVIVC

Mathematical treatment of data is an integral and most tricky part of IVIVC. For this, various softwares are now available with inbuilt model programs that can be used for data comparison and model fitting. While selecting a particular software for IVIVC development, the regulatory compliance of the software must be verified as the data generated using this software should be acceptable to the regulatory agencies. Based on published research in the past decade, the widely used softwares for this purpose are listed in Table 4.12 along with their key features.

Table 4.12 *Softwares used for IVIVC model development*

Software	Provider	Key features	Reference
Phoenix® IVIVC Toolkit™	Certara Inc.	• Compliant with USFDA regulatory guidance. • Suitable for both convolution and deconvolution modeling methods.	[58]
GastraPlus™ (Module: IVIVCPlus™)	Simulations Plus Inc.	• Compliant with USFDA regulatory guidance. Suitable for both conventional convolution and deconvolution methods. • First software to offer mechanistic deconvolution. • Results can be linked to PBPK models.	[59]
IvivcXpress©	Educe Global Inc.	• Compliant with USFDA regulatory guidance. • Suitable for both convolution and deconvolution modeling methods.	[60]

4.10 Conclusion and Future Prospects

With the increase in the number of micro/nanoformulations, a thorough understanding of their unique physicochemical properties and their impact on various steps during IVIVC development is very timely. It must be noted that the basic IVIVC principles remain the same and require some fine-tuning in studies for successful IVIVC. However, no specific guidelines have been released by the regulatory authorities for IVIVC of micro/nanoformulations, which makes it more challenging from a regulatory approval perspective. With the advances seen in this field over a decade, more standardized, sophisticated yet user-friendly interfaces to develop IVIVC for micro/nanoformulations can be expected.

References

[1] Patravale, V., Disouza, I., Rustomjee, M. (2016) *Pharmaceutical Product Development: Insights Into Pharmaceutical Processes, Management and Regulatory Affairs*, 1st edn, CRE Press.

[2] Qureshi, S. (2010) In vitro-in vivo correlation (IVIVC) and determining drug concentrations in blood from dissolution testing – a simple and practical approach. *The Open Drug Delivery Journal*, **4**, 38–47.

[3] Shen, J., Choi, S., Qu, W., *et al.* (2015) In vitro-in vivo correlation of parenteral risperidone polymeric microspheres. *Journal of Controlled Release*, **218**, 2–12.

[4] USFDA (1997) *Guidance for Industry, Extended Release Oral Dosage Forms: Development, Evaluation, and Application of In Vitro/In Vivo Correlations 1997.* https://www.fda.gov/downloads/Drugs/GuidanceComplianceRegulatoryInformation/Guid%20ances/ucm070239.pdf (accessed 18 August 2018).

[5] Murthy, C., Sunkara, G., Young, D.. (2017) *Pharmaceutical Product Development: In Vitro–In Vivo Correlation*, CRC Press.

[6] Kaur, P., Jiang, X., Duan, J., Stier, E. (2015) Applications of in vitro–in vivo correlations in generic drug development: Case studies. *AAPS Journal*, **17** (4), 1035–1039.

[7] Saha, S.K., Chowdhury, A.K.A., Bachar, S.C., *et al.* (2013) Comparative in vitro–in vivo correlation analysis with pioglitazone tablets. *Asian Pacific Journal of Tropical Disease*, **3** (6), 487–491.

[8] Stevens, R.E., Gray, V., Dorantes, A., *et al.* (2015) Scientific and regulatory standards for assessing product performance using the similarity factor, f2. *AAPS Journal*, **17** (2), 301–306.

[9] USFDA (2017) *Waiver of In Vivo Bioavailability and Bioequivalence Studies for Immediate-Release Solid Oral Dosage Forms Based on a Biopharmaceutics Classification System: Guidance for Industry.* https://www.fda.gov/downloads/drugs/guidances/ucm070239.pdf, %20https://www.fda.gov/downloads/Drugs/Guidances/ucm070246.pdf (accessed 18 August 2018).

[10] Warnken, Z., Puppolo, M., Hughey, J., *et al.* (2018) In vitro–in vivo correlations of carbamazepine nanodispersions for application in formulation development. *Journal of Pharmaceutical Sciences*, **107** (1), 453–465.

[11] Boda, K. (2018) *Developing Methods for Apparatus 3 and 7: Agilant Technologies Inc.* https://www.agilent.com/cs/library/eseminars/public/Developing%20Methods%20for%20Apparatus%203%20and%207.pdf (accessed 18 August 2018).

[12] Stippler, E. (2011) *Compendial Dissolution: Theory and Practice 2011.* http://www2.aaps.org/uploadedFiles/Content/Sections_and_Groups/Focus_Groups/In_Vitro_Release_and_Dissolution_Testing/Resources/IVRDTFGStippler2011.pdf (accessed 18 August 2018).

[13] Zhou, M., Shoudt, D., Calderon, G., Feng, M. (2001) Application of USP Apparatus 7 to in vitro drug release in scopolamine transdermal system. *Dissolution Technologies*, 25–29.

[14] USFDA (2018) *Orange Book: Approved Drug Products with Therapeutic Equivalence Evaluations.* https://www.accessdata.fda.gov/scripts/cder/ob/search_product.cfm (accessed 18 August 2018).

[15] United States Pharmacopoeia (2012) In vitro and in vivo evaluation of dosage forms. *United States Pharmacopoeia USP 35-NF30* (Second Supplement), p. 5663–5671.

[16] Vergnaud, J., Rosca, I. (2005) *Assessing Bioavailablility of Drug Delivery Systems: Mathematical Modeling*, CRC Press.

[17] Rettig, H. (2018) *IVIVC – Methods and Applications in MR Product Development.* https://pdfs.semanticscholar.org/presentation/5d8c/26b44b2d6cf1c881c0b480cad2f157ffd699.pdf (accessed 18 August 2018).

[18] Sakore, S., Chakraborty, B. (2011) In vitro–in vivo correlation (IVIVC): A strategic tool in drug development. *Journal of Bioequivalence & Bioavailability*, **2011**, S3.

[19] Emami, J. (2006) In vitro–in vivo correlation: from theory to applications. *Journal of Pharmacy & Pharmaceutical Sciences*, **9** (2), 169–189.

[20] Cardot, J.-M., Beyssac, E, Alric, M. (2007) In vitro–in vivo correlation: Importance of dissolution in IVIVC. *Dissolution Technologies*, **14**, 15–19.

[21] Cardot, J.-M., Davit, B.M. (2012) In vitro–in vivo correlations: tricks and traps. *AAPS Journal*, **14** (3), 491–499.

[22] D'Souza, S., Faraj, J.A., Giovagnoli, S., DeLuca, P.P. (2014) IVIVC from long acting olanzapine microspheres. *International Journal of Biomaterials*, **2014**, 11.

[23] Dunne, A., Gaynor, C., Davis, J. (2005) Deconvolution based approach for Level A in vivo-in vitro correlation modelling: statistical considerations. *Clinical Research and Regulatory Affairs*, **22** (1), 1–14.

[24] Wagner, J. (1974) Application of the Wagner-Nelson absorption method to the two-compartment open model. *Journal of Pharmacokinetics and Biopharmaceutics*, **2** (6), 469–486.

[25] Desai, P., Date, A., Patravale, V. (2012) Overcoming poor oral bioavailability using nanoparticle formulations – opportunities and limitations. *Drug Discovery Today Technologies*, **9** (2), e71–e174.

[26] Margolskee, A., Darwich, A.S., Galetin, A., *et al.* (2016) Deconvolution and IVIVC: Exploring the role of rate-limiting conditions. *AAPS Journal*, **18** (2), 321–332.

[27] Buchwald, P. (2003) Direct, differential-equation-based in-vitro–in-vivo correlation (IVIVC) method. *Journal of Pharmacy and Pharmacology*, **55** (4), 495–504.

[28] D'Souza, S., Faraj, J.A., Giovagnoli, S., DeLuca, P.P. (2014) In vitro–in vivo correlation from lactide-co-glycolide polymeric dosage forms. *Progress in Biomaterials*, **3** (2), 131–142.

[29] Lu, Y., Kim, S., Park, K. (2011) In vitro–in vivo correlation: perspectives on model development. *International Journal of Pharmaceutics*, **418** (1), 142–148.

[30] Tsume, Y., Mudie, D.M., Langguth, P., *et al.* (2014) The Biopharmaceutics Classification System: subclasses for in vivo predictive dissolution (IPD) methodology and IVIVC. *European Journal of Pharmaceutical Sciences*, **57**, 152–163.

[31] Patravale, V., Desai, P., Mapara, S. (2018) Lipid nanocarriers for advanced therapeutic applications, in *Multifunctional Nanocarriers for Contemporary Healthcare Applications* (eds Barkat, A., Harshita, A.B., Beg, S., Ahmad, F.), IGI Global, pp. 85–128.

[32] Cao, X., Deng, W.W., Fu, M., *et al.* (2012) In vitro release and in vitro–in vivo correlation for silybin meglumine incorporated into hollow-type mesoporous silica nanoparticles. *International Journal of Nanomedicine*, **7**, 753–762.

[33] Ali, H., Prasad Verma, P.R., Dubey, S.K., *et al.* (2017) In vitro–in vivo and pharmacokinetic evaluation of solid lipid nanoparticles of furosemide using Gastroplus™. *RSC Advances*, **7** (53), 33314–33326.

[34] Yu, M., Ma, H., Lei, M., *et al.* (2014) In vitro/in vivo characterization of nanoemulsion formulation of metronidazole with improved skin targeting and anti-rosacea properties. *European Journal of Pharmaceutics and Biopharmaceutics*, **88** (1), 92–103.

[35] Tiwari, R., Pathak, K. (2011) Nanostructured lipid carrier versus solid lipid nanoparticles of simvastatin: comparative analysis of characteristics, pharmacokinetics and tissue uptake. *International Journal of Pharmaceutics*, **415** (1–2), 232–243.

[36] Zhu, Y., Wang, M., Zhang, J., *et al.* (2015) Improved oral bioavailability of capsaicin via liposomal nanoformulation: preparation, in vitro drug release and pharmacokinetics in rats. *Archives of Pharmacal Research*, **38** (4), 512–521.

[37] Uronnachi, E., Ogbonna, J., Kenechukwu, F., Attama, A. (2013) Formulation and in vitro/in vivo evaluation of zidovudine contained in solidified reverse micellar delivery system of immune compromised rats. *Journal of Applied Pharmaceutical Science*, **3** (2), 31–35.

[38] Kumar, R., Nagarwal, R.C., Dhanawat, M., Pandit, J.K. (2011) In-vitro and in-vivo study of indomethacin loaded gelatin nanoparticles. *Journal of Biomedical Nanotechnology*, **7** (3), 325–333.

[39] Singh, G., Pai, R.S. (2014) In-vitro/in-vivo characterization of trans-resveratrol-loaded nanoparticulate drug delivery system for oral administration. *Journal of Pharmacy and Pharmacology*, **66** (8), 1062–1076.

[40] Ha, E., Choo, G., Baek, I., Kim, M. (2014) Formulation, characterization, and in vivo evaluation of celecoxib-pvp solid dispersion nanoparticles using supercritical antisolvent process. *Molecules*, **19**, 20325–20339.

[41] Eroglu, H., Burul-Bozkurt, N., Uma, S., Oner, L. (2012) Preparation and in vitro/in vivo evaluation of microparticle formulations containing meloxicam. *AAPS PharmSciTech*, **13** (1), 46–52.
[42] Ishak, R.A.H., Mortada, N.D., Zaki, N.M., *et al.* (2014) Impact of microparticle formulation approaches on drug burst release: a level A IVIVC. *Journal of Microencapsulation*, **31** (7), 674–684.
[43] Zolnik, B.S., Burgess, D.J. (2008) Evaluation of in vivo–in vitro release of dexamethasone from PLGA microspheres. *Journal of Controlled Release*, **127** (2), 137–145.
[44] Rawat, A., Bhardwaj, U., Burgess, D.J. (2012) Comparison of in vitro-in vivo release of Risperdal((R)) Consta((R)) microspheres. *International Journal of Pharmaceutics*, **434** (1–2), 115–121.
[45] Chu, D.F., Fu, X.Q., Liu, W.H., *et al.* (2006) Pharmacokinetics and in vitro and in vivo correlation of huperzine A loaded poly(lactic-co-glycolic acid) microspheres in dogs. *International Journal of Pharmaceutics*, **325** (1–2), 116–123.
[46] Khan, S.A., Ahmad, M., Murtaza, G., *et al.* (2010) In vitro–in vivo correlation study on nimesulide loaded hydroxypropylmethylcellulose microparticles. *Acta Pharmaceutica Sinica*, **45** (6), 772–777.
[47] Swarnakar, N.K., Thanki, K., Jain, S. (2014) Lyotropic liquid crystalline nanoparticles of CoQ10: implication of lipase digestibility on oral bioavailability, in vivo antioxidant activity, and in vitro-in vivo relationships. *Molecular Pharmaceutics*, **11** (5), 1435–1449.
[48] Yi, Y., Tu, L., Hu, K., *et al.* (2015) The construction of puerarin nanocrystals and its pharmacokinetic and in vivo-in vitro correlation (IVIVC) studies on beagle dog. *Colloids and Curfaces B, Biointerfaces*, **133**, 164–170.
[49] Xu, Y., Wang, Y., Li, X.M., *et al.* (2014) Study on the release of fenofibrate nanosuspension in vitro and its correlation with in situ intestinal and in vivo absorption kinetics in rats. *Drug Development and Industrial Pharmacy*, **40** (7), 972–979.
[50] Rizzo, L.Y., Golombek, S.K., Mertens, M.E., *et al.* (2013) In vivo nanotoxicity testing using the zebrafish embryo assay. *Journal of Materials Chemistry B*, **1**.
[51] Yeom, D.W., Chae, B.R., Son, H.Y., *et al.* (2017) Enhanced oral bioavailability of valsartan using a polymer-based supersaturable self-microemulsifying drug delivery system. *International Journal of Nanomedicine*, **12**, 3533–3545.
[52] ICH (2005) *International Conference on Harmonisation of Technical Requirements for Registration of Pharmaceuticals for Human Use, ICH Harmonised Tripartite Guideline, Quality Risk Management Q9 2005*. http://www.ich.org/fileadmin/Public_Web_Site/ICH_Products/Guidelines/Quality/Q9/Step4/Q9_Guideline.pdf (accessed 18 August 2018).
[53] ICH (2008) *International Conference on Harmonisation of Technical Requirements for Registration of Pharmaceuticals for Human Use, ICH Harmonised Tripartite Guideline, Quality Risk Management Q10 2008*. http://www.ich.org/fileadmin/Public_Web_Site/ICH_Products/Guidelines/Quality/Q10/Step4/Q10_Guideline.pdf (accessed 18 August 2018).
[54] ICH (2009) *International Conference on Harmonisation of Technical Requirements for Registration of Pharmaceuticals for Human Use, ICH Harmonised Tripartite Guideline, Pharmaceutical Development Q8(R2) 2009*. http://www.ich.org/fileadmin/Public_Web_Site/ICH_Products/Guidelines/Quality/Q8_R1/Step4/Q8_R2_Guideline.pdf (accessed 18 August 2018).
[55] ICH (2012) *International Conference on Harmonisation of Technical Requirements for Registration of Pharmaceuticals for Human Use, ICH Harmonised Tripartite Guideline, Development and Manufacture of Drug Substances (Chemical Entities and Biotechnological/Biological Entities) Q11, 2012*. http://www.ich.org/fileadmin/Public_Web_Site/ICH_Products/Guidelines/Quality/Q11/Q11_Step_4.pdf (accessed 18 August 2018).
[56] ICH (2014) *International Conference on Harmonisation, Final Concept Paper q12: Technical and Regulatory Considerations for Pharmaceutical Product Lifecycle Management*. http://

www.ich.org/fileadmin/Public_Web_Site/ICH_Products/Guidelines/Quality/Q12/Q12_Final_Concept_Paper_July_2014.pdf (accessed 18 August 2018).

[57] Khaled, A.A., Pervaiz, K., Karim, S., *et al.* (2013) Development of in vitro–in vivo correlation for encapsulated metoprolol tartrate. *Acta Poloniae Pharmaceutica*, **70** (4), 743–747.

[58] Certara (2018) *Phoenix IVIVC Toolkit*, Certara Inc. https://www.certara.com/software/pkpd-modeling-and-simulation-2/phoenix-winnonlin-2/ivivc-toolkit/?ap%5B0%5D=PKPD&ap%5B1%5D=PKPD&ap%5B2%5D=PKPD (accessed 18 August 2018).

[59] SimulationsPlus (2018) *GastroPlus-IVIVCPlus*, Simulations Plus Inc. https://www.simulations-plus.com/software/gastroplus/ivivc/ (accessed 18 August 2018).

[60] EduceGlobal (2018) *IvivcXpress*, Educe Global Inc. http://www.educeglobal.com/ivivc-express.html (accessed 18 August 2018).

5

Characterization of Bioadhesion, Mucin-interactions and Mucosal Permeability of Pharmaceutical Nano- and Microsystems

Ellen Hagesaether[1], Malgorzata Iwona Adamczak[2,3], Marianne Hiorth[2] and Ingunn Tho[2]

[1]*Department of Life Science and Health, Faculty of Health Sciences, Oslo Metropolitan University, Norway*
[2]*Department of Pharmacy, Faculty of Mathematics and Natural Sciences, University of Oslo, Norway*
[3]*GE Healthcare, Pharmaceutical Diagnostics, Oslo, Norway*

5.1 Introduction

Mucosal drug administration has several advantages compared with the more invasive parenteral route. This is particularly the case for local treatment of, for example, infections, pain and dryness, minimizing the systemic side-effects. Also in cases of systemic treatment, several advantages can be identified, such as lower risk, lower cost, usually less pain and less need for qualified healthcare workers to carry out the injections. On the downside, the robust mucosal barrier usually reduces the bioavailability.

Micro- and nanoparticles in the size range of from around 150 nm up to 6 µm can be used to deliver drugs to the mucosal membranes of various organs. The drug may be protected against chemical and enzymatic degradation inside a particle. The effect of

Characterization of Pharmaceutical Nano- and Microsystems, First Edition. Edited by Leena Peltonen.

the drug may be prolonged if drug release from the particle is sustained, particularly if the residence time of the formulation is increased. Increased residence time on the mucosal membrane – mucoadhesion – can also increase the bioavailability of systemically acting drugs, by increasing the time available for drug absorption. This strategy may be even more promising if the formulation is mucopenetrating, thereby bringing the drug in close proximity to the cellular epithelial membrane (Figure 5.1). The particles can even be surface modified to target particular cell types or transporters of the mucosal membrane. Once the absorption site is reached, the particles can be engulfed while still containing the active ingredient, or release the drug to be absorbed separately. If the particles contain excipients affecting the cell membrane, an increased permeability may be achieved. The risk is unwanted toxic effects. Increased absorption is advantageous in the case of systemic treatment, while the opposite is true for local treatment.

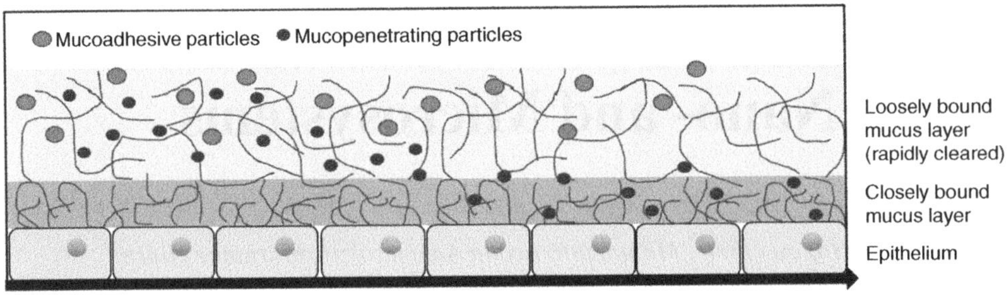

Figure 5.1 *Schematic illustration of mucoadhesive and mucopenetrating particles administered to a mucosal surface.*

The core material, size and coating of micro- and nanoparticles will govern their behaviour. With detailed knowledge of this relationship, particles can be designed to match the desired outcome. This chapter will present the most common and promising *in vitro* and *ex vivo* methods of testing and estimating the *in vivo* behaviour of these small formulations, giving examples of different types of particles and systems along the way. A distinction will be made between mucoadhesion and bioadhesion, where mucoadhesion is preferred when the interaction between particles and the surface is directly related to mucus or mucin.

5.2 Background and Theory

All mucosal membranes are covered by a mucus layer, to which the large glycoprotein mucin gives structure and determines the viscoelastic properties. Generally, mucins can be divided into two categories: cell-associated mucins that range between 100 and 500 nm in length and are anchored to the membrane by a transmembrane domain, and secreted mucins that are up to several microns long and more loosely attached to the membrane [1]. Mucus governs the permeability of particles into the epithelium via two main mechanisms: size filtering and interaction filtering (Figure 5.2). These mechanisms are mainly regulated by the physicochemical properties of the mucus, such as pore size, viscoelasticity, pH, charge and ionic strength [2].

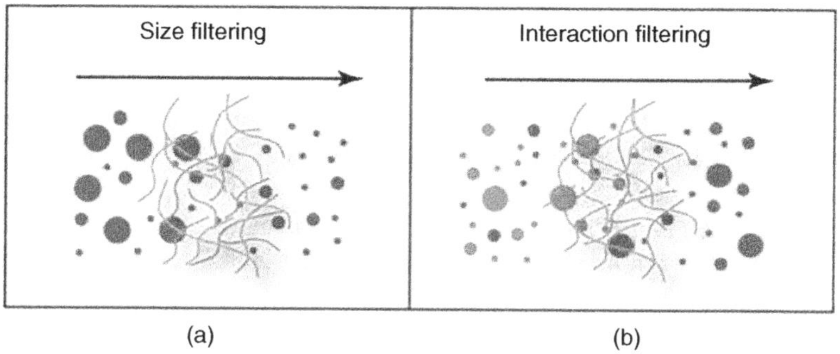

Figure 5.2 *Filtering mechanisms governing permeability of mucus to particles. (a) Size filtering allowing only particles that are smaller than the pore size between mucin fibers to cross. (b) Interaction filtering separates particles based on the surface properties of the particles, e.g., the light grey particles interact strongly with mucus and are trapped, whereas the dark grey particles exhibit only weak interactions and can cross. Reprinted from [3] with permission from Elsevier.*

The mucus layer is crucial for protecting the membrane, and functions as a barrier for micro- and nanoparticles. The trapping of particles in the mucus obstructs their diffusion towards the underlying epithelial cell layer. The mucosal membranes also have rinsing or cleansing mechanisms, reducing the residence time of trapped particles. A particle trapped in mucus (i.e., a mucoadhesive particle) will therefore have an extended, yet limited, residence time, which will vary between the different administration sites.

One mechanism responsible for mucoadhesion is non-covalent interactions, such as hydrogen bonding or ionic interactions, although extensive work has also been put into exploiting the covalent disulfide bonds provided by thiolated polymers [4]. A dry particle can attract water from the underlying mucosa, leading to adhesion by capillary forces. However, this mechanism requires that the formulation is initially dry, which is not always the case. Eventually, the formulation will be fully hydrated and these forces diminish. Entanglements are possible, after the diffusion of hydrated polymer chains on the particle surface into the mucin network. A trapped formulation should release the drug to allow drug diffusion towards the epithelial cell membrane. Another strategy is to exploit mucopenetrating particles.

The inspiration for mucopenetrating particles can be found in nature, more specifically by looking towards mucopenetrating viruses. These viruses are characterized by a hydrophilic surface, yet overall neutral charge, avoiding blockage by hydrophobic and ionic interactions. The size is ~100 nm, avoiding steric blockade. Mucopenetration can thus be achieved by careful surface engineering, alternatively by PEGylating [5]. Another strategy is to associate the particles with mucolytic substances, such as N-acetyl-cysteine or trypsin, yielding active mucopenetrating systems [6].

Epithelial tissues are characterized by abundant cell junctions, such as tight junctions, limiting paracellular transport. Therefore, a particle may not reach further than the epithelial cell membrane, where the drug is subsequently released. Components of the particle can exert an effect on the cell membrane though, increasing the permeability of, among others, the drug. Permeation enhancers like cell-penetrating peptides have been included in particles to exploit this. Cationic substances (like chitosan) and amphiphilic substances also possess this trait, typically weakening tight junctions. This can lead to unintentional toxicity or systemic side-effects if neglected, illustrating the necessity of thorough testing to control the interaction with cell membranes in addition to mucoadhesion and mucopenetration. Additionally, excipients like Poloxamer/Pluronic® possibly inhibit P-glycoprotein (P-gp) [7]. Functionalized nanoparticles targeting a transporter (e.g., PepT1) can overconcentrate a poorly absorbed drug on the membrane surface, increasing the absorption [8].

Even intact particles can be absorbed into cells of the epithelial layer, intentionally or unintentionally from the formulator's point of view. Although nanoparticles will usually be too big for para- or transcellular transport, endocytosis has been demonstrated. M-cells are immune cells of the intestinal mucosa and have a unique ability to perform endocytosis of nanoparticles, particularly after a receptor–ligand interaction. On the basolateral side, the M-cells are co-localized with lymphocytes, thereby delivering their cargo mostly to the lymphatic circulation, before entering the blood circulation via the thoracic duct. Although this capacity is limited, first pass hepatic metabolism is greatly reduced.

5.3 Mucosal Membranes

Every mucosal membrane can be a target for improved administration using micro- and nanoparticles, although for different reasons. The properties of mucus and mucus permeation may vary depending on organ and pathological conditions. In the following sections the most frequently used mucosal membranes for drug delivery are summarized, and Table 5.1 provides an overview of the main features of the mucosal membranes of different organs.

5.3.1 Oral Mucosa

The oral cavity can be used for both local and systemic drug delivery via the buccal route. The oral mucosa is composed of a layer of water, mucin and other proteins with a thickness of 70–100 μm [10]. Saliva is continuously produced and bathes the oral cavity and is also effective in flushing harmful substances away. The buccal and the sublingual route of administration are effective in overcoming first pass metabolism (successfully used for delivery of nitroglycerin) and also when a fast response is an advantage. Local administration of nano- and microparticles could be an advantage for treating diseases in the oral cavity, such as oral mucositis, xerostomia, periodontitis and cancer, but also for preventing dental caries. One of the big challenges when formulating new drug delivery systems for local drug delivery is the flow of saliva, which could be as high as 1–3 ml/min when stimulated [11]. This leads to short residence time for the administration system in the mouth, and hence also for the drug. This leads to reduced efficacy.

One successful formulation for local administration to the oral cavity is based on microspheres and is commercially available under the trade name Arestin®. Arestin® is used for treating periodontitis and is a dry powder of microspheres composed of poly(lactic-co-glycolic) acid (PLGA) and the antibiotic minocycline [25]. The powder is placed in the periodontal pocket with a device. The drug is then slowly released in the periodontal pocket. The success of this formulation is mainly based on the mucoadhesive

Table 5.1 *Main properties of mucus at different mucosal membranes (mainly based on [2, 6, 9])*

Mucosal membrane	Temperature (°C)	pH	Mucus layer thickness (μm)	Mucin concentration (%)	Mucus turnover/ flow rate	Pore size (nm)
Buccal	36.2–36.7 [13]	6.5–7.5	70–100 [10]	0.1–0.5	0.1–1.85 ml/min [16] up to 3 ml/min**	Not found
Gastric	37.0	1–2	30–300	3	4–5 h	500 nm
Small intestine	37.0	5.9–7.5	150–400	1	47–270 min	Not found
Colonic	37.0	6.0–7.0	30–280	< 5	270–300 min	< 200 nm
Pulmonary	32.5–35.5 [18]	7.0 [1]	5–55 [1]	2–4	5–10 cm/min	100 nm – several μm
Nasal	30.2–34.4 [17]	5.5–6.5 [14]	10–15	2–3	5–10 min 6 ml/min*** [12]	150±50 nm [15]
Ocular	34.0 [19]	7.8 [1]	0.02–0.05 [22] 6–7* [23]	0.01	5–10 s	550±50 nm
Vaginal	37.0	3.5–4.5 [24])	20 [20]	1–2	1.5 ml/d**** 6 ml/d*****	340±70 nm (50–1800 nm) [21]

*Including tear film [23].
**Stimulated saliva secretion [11].
***Mucociliary clearance [12].
****Cervical mucus production rate.
*****Vaginal fluid production rate (considerably increased with sexual stimulation).

properties of the microspheres but also the limited flow of saliva in the periodontal pocket.

Oral-lyn™ is a new oral-based systemic delivery of insulin for uptake through the buccal mucosa. In 2005 it was approved for treatment of Type 2 diabetes in Ecuador, and some years later in several other countries. Recombinant insulin is formulated with the patented technology RapidMist® as aerosolized micelles with a size of 6 μm [26]. The drug is administered with an asthma-like inhaler. Nanostructured lipid carriers (NLCs) have also been developed for intra-oral delivery of the anti-emetic agent domperidone [27]. The NLCs showed successful delivery of domperidone through the buccal mucosa in an *ex-vivo* test and were confirmed non-toxic to TR146 buccal cells (Table 5.2).

5.3.2 Gastrointestinal Mucosa

Oral drug administration has some drawbacks despite the huge surface area available for absorption, such as a harsh chemical and enzymatic-degrading environment including first-pass hepatic metabolism, a tight cellular barrier, and fast hydration of formulations. The mucus layer of the gastrointestinal tract is thick, 30–400 μm, causing an unstirred water layer, and the mucus turnover time is 24–48 hours. On a positive note, the constant absorption of water generates a drag towards the cellular epithelial layer for other particles as well.

When lopinavir, used for the treatment of HIV infections, was formulated in glyceryl behenate-based solid lipid nanoparticles (SLNs) of 230 nm, the AUC increased about two-fold in rats compared with a drug solution [40]. The demonstrated reason was targeting and uptake by the lymphatic circulation, reducing first-pass hepatic metabolism. SLNs taken up into the lymphatic circulation also gained direct access to HIV reservoir sites, an additional advantage [40, 41].

When PEGylated neutral PLGA-based nanoparticles of 200 nm displayed RGD molecules (the tripeptide Arg-Gly-Asp) at their surface, the particles concentrated in M-cells due to β_1 integrin targeting. When the particles were loaded with the model antigen ovalbumin and orally administered to mice, an immunoglobulin G (IgG) response was induced, demonstrating the potential for mucosal immunization [42].

5.3.3 Pulmonary Mucosa

The airways are composed of a gradually thinning columnar epithelium populated by ciliated cells with a layer of mucus on top that collectively form the mucociliary escalator [43]. The airways subdivide roughly 16–17 times before the alveoli are reached. The mucociliary apparatus serve as a mechanical barrier by trapping particulates in the surface liquid and clearing them off the bronchial tree by ciliary action [44]. The rheological properties of the mucus may show variations from patient to patient, but also day to day, and appear to be an attribute of both healthy and diseased respiratory tract. If mucus becomes too thick, for example in severe bronchitis or cystic fibrosis where the sputum viscosity can be more than 100 000 times that of water, patients experience great difficulty in mucus clearance, resulting in bacterial overgrowth [1].

Table 5.2 *Overview of cell lines used to study interactions of nano- and microparticles with various mucosal membranes*

Cell line	Description	Comments	References
Gastro-intestinal tract			
Caco-2	The human colon carcinoma Caco-2 line differentiates and polarizes to morphologically and functionally resemble the epithelial cells of human small intestine.	Monolayers of Caco-2 cells are widely used to determine absorption and transport, even recognized by the medicinal regulatory agencies. Drawback is lack of mucus and M-cells. Caco-2 cells might also lack some caveolar structures crucial for caveolae-mediated endocytosis. Available from ATCC.	[28]
HT29-MTX	The HT29 cell-line is, like Caco-2 cells, a human colon adenocarcinoma cell-line that spontaneously differentiates into monolayers of polarized enterocytes connected by tight-junctions. HT29 cells were differentiated into mature goblet cells using methotrexate. Mucous-secreting HT29-MTX subclones were isolated from this cell clone.	MTX treated clones have been shown to postconfluency differentiate into a mixed population of mucus-secreting goblet cells and enterocytes. HT29-MTX provides a model system to study the influence of the mucous layer on nanoparticle diffusion. Available from ECACC	[28]
Raji	B lymphocyte established from Burkitt's lymphoma in an 11 year old male.	Included in co-cultures to model the human Follicle associated epithelium (FAE) expressing M cells. Available from ATCC	[28]
Buccal			
TR 146	The cells originate from a neck node metastasis of a human buccal carcinoma, and has been proposed as a model of human buccal epithelium. The cell line lacks mucus production.	Due to the lack of a mucus producing buccal cell model, an external mucus layer was deposited onto the cell layer. The mucus consisted of porcine gastric mucin dispersed in water with the plasticizer glycerol, which resembled centrifuged natural human mucin more than bovine submaxillary mucin did.	[29]

(continued)

Table 5.2 *(continued)*

Cell line	Description	Comments	References
		The external mucus layer adhered onto the cell layer without compromising the viability. Reliability of the new model was suggested by comparing nanoparticle transport with transport through an excised porcine buccal mucosa using Franz diffusion cell. Available from ECACC	
Nasal			
RPMI 2650	Human nasal epithelial cell line derived from squamous cell carcinoma, capable of growing in monolayer and multilayer and mucus production.	Displayed the same permeability as excised human nasal mucosa for sodium fluorescein (paracellular marker), with analogous trans-epithelial electrical resistance values and production of mucus, when seeded on permeable Thincert® inserts. Available from ATCC	[30]
Airways			
Calu-3	Derived from a bronchial adenocarcinoma. Forms polarized cell layers with tight junctions and express mucus.	Calu-3 provides a model system to study the barrier function of the airway epithelium. Available from ATCC	[31]
16HBE14o-	Developed by transformation of cultured human bronchial-surface epithelial cells. Forms polarized cell layers with tight junctions and apical microvilli and cilia. Produces much smaller amounts of secretory components than Calu-3.	16HBE14o-provides a model system to study the barrier function of the airway epithelium, but lack protective covering of mucus.	[32]
BEAS-2B	Derived from normal human epithelial cells immortalized using an adenovirus. Do not form tight junctions as readily as 16HBE14o- and Calu-3.	Widely used for the study of cell biology, pathologic processes and drug metabolism, but limited potential in modelling drug absorption. Available from ATCC	[33]

CFBE41o-	A cystic fibrosis bronchial epithelial cell line developed by virus transformation. The cell line expresses tight junctions, but does not produce mucus.	Native human mucus, obtained from tracheal tubes of patients undergoing elective surgery, was used as a supplement and the uptake of nanoparticles was dramatically reduced in the presence of mucus. The compatibility of CFBE41o- cells with the human mucus was confirmed.	[34]
MucilAir™	*A* 3D human bronchial epithelial cell model that included a mucociliary apparatus. • Tight junctions • Mucus production • Beating of cilia	Commercial product from *Epithelix* According to epithelix.com, MucilAir™ is available from: • Single donor or **pool of donors** • Different anatomical sites (nasal, tracheal or bronchial) • Patients with **several pathologies**: o Healthy o Chronic Obstructive Pulmonary Disease o Asthma o Cystic Fibrosis o Allergic Rhinitis o Smoker	[33]
Vaginal tract			
HeLa	Human cervical epithelial cells (from Henrietta Lacks)	First human cells successfully cloned. Widely used for different purposes within biomedical sciences, e.g, cytotoxicity studies. Not suitable as *in vitro* permeability model. Available from ATCC	[35]
CaSki	Derived from human cervical carcinoma Carrying human papilloma virus (HPV-16 and -18)	Forms cell monolayers, and expresses phenotypic markers of differentiated endocervical epithelium. Widely used as a model of cervico-vaginal epithelium for studying drug transport mechanisms and permeability assessments. Available from ATCC	[36]

(continued)

Table 5.2 *(continued)*

Cell line	Description	Comments	References
HEC-1A-	Derived from human uterine endometrial adenoma carcinoma	Epithelial layer shows multilayer structure with tight junctions similar to the vaginal epithelium. Used as an *in vitro* permeability model. Available from ATCC	[36]
ME-180	Derived from metastatic human endocervical epithelial cells Carrying human papilloma virus (HPV-39)	Used for development of cytotoxic anticancer drugs but also used in permeability assessment. Available from ATCC	[37]
C-33A	Derived from human cervical epithelial cells Negative for human papilloma virus	Shows epithelial morphology, and are used in permeability assessments Available from ATCC	[38]
EpiVaginal™	*A* 3D human vaginal-ectocervical epithelial cell based tissue modell	Commercial product from *MatTek Corp*. According to mattek.com, EpiVaginal™ is available in: • Partial-Thickness (VEC-100) • Partial-Thickness containing immune-competent dendritic cells (VLC-100) • Full-Thickness (VEC-100-FT) • Full-Thickness containing immune-competent dendritic cells (VLC-100-FT)	[39]

In cystic fibrosis therapy the treatment of bacterial airway infections is essential, but since bacteria are protected from antibiotics inside mucus, the effectiveness of antibacterial treatment is limited. To overcome this, mucus-permeable lipid-core nanocapsules (LNCs) were developed. Ciprofloxacin-loaded LNCs with a mean size of 180 nm showed increased drug permeation by 50% through mucus [45]. Administration of nanoparticles (NPs) with a high antibiotic load capable of penetrating the tight mesh of biofilm/mucus can be an advent to overcome the treatment challenges. Biodegradable and biocompatible polymer PLGA nanoparticles loaded with ciprofloxacin showed promising effects against cystic fibroses *Pseudomonas aeruginosa* lung infection [46].

Optimization of particle characteristics for deposition in the deep regions of the lung after inhalation is the key challenge in delivery to the lungs [43]. Particles in the range of 1–5 μm are ideal for deposition in the alveoli. One successful formulation on the market is the inhalable insulin AFREZZA® from the company MannKind. AFREZZA® is composed of recombinant human insulin mixed with the excipient fumaryl diketopiperazine (FDKP). Insulin powder particles are adsorbed onto uniform-sized (approximately 2 μm) carrier Technosphere particles, which contain mostly crystallized FDKP. Upon inhalation, the Technosphere particles carry the insulin into the alveoli where they dissolve [47]. The Technosphere insulin is a form of short-acting human insulin to be taken by oral inhaler with meals [48].

Pourusina *et al.* prepared porous inhalable parathyroid hormone (1–34)-loaded microparticles suitable for pulmonary delivery by using leucine and hydroxy propyl-β-cyclodextrin (HPβCD) in water to preserve the parathyroid hormone (1–34) stability for spray freeze-drying [49]. siRNA (small interfering RNA) has great potential in the treatment of a variety of lung diseases, and has been formulated for dry powder inhalation by use of TPP-dendrimer nanocarriers [50].

5.3.4 Nasal Mucosa

Particulate drug delivery to the nose can improve local or systemic effects, especially advantageous in relation to mucosal vaccination [51] and as an alternative to parenteral protein and peptide administration [52]. The reason for the latter is mainly accredited to the lower enzymatic activity, avoidance of first-pass hepatic metabolism and less restrictive tight junctions. The mucus gel layer is relatively thin (5–10 μm), but constantly moved due to beating cilia, significantly reducing the residence time to approximately 15–20 minutes [6, 53]. It is important that the formulation is not cilia-toxic nor irritating, posing some restrictions on type of formulations and excipients.

Another intriguing anatomical feature of the nasal mucosa is the olfactory epithelium, with neuronal bundles passing into the central nervous system (CNS) [53]. This direct nose-to-brain delivery of drugs was exploited by Gao *et al.* for delivering vasoactive intestinal peptide across the blood–brain barrier [54]. The brain bioavailability in mice increased when the neuroprotective peptide was incorporated into poly(ethylene glycol)-poly(lactic acid) nanoparticles, 100–120 nm in size, and increased even further when the nanoparticles were modified with the ligand wheatgerm agglutinin. The spatial memory in ethylcholine aziridium-treated rats improved correspondingly.

5.3.5 Ocular Mucosa

Drug delivery to the eye, primarily for local treatment, is a very challenging task due to the anatomical structure of the eye with both hydrophilic and lipophilic features. In addition, the continuous secretion of tear liquid and drainage through the lacrimal tract also lead to rapid dilution and clearance of the drug [55]. This leads to the necessity of administering the formulation several times a day. The mucus layer of the eye is very thin at around 0.02–0.05 μm [22]. When the tear film is included, the thickness is around 6–7 μm [23].

One successful formulation that has been on the market for many years for lowering the pressure of the eye is Blocadren depot®. This formulation can be administered only once a day. The success of this formulation is thought to be the polymer gellan gum being able to react with ions in the tear liquid *in situ*, thereby increasing the viscosity of the formulation and as such the retention time, but also the mucoadhesive properties of the polymer itself [56].

Inulin-based micelles have been investigated for improved delivery of corticosteroids to the posterior eye-segment. The micelles showed improved permeation over corneal epithelial cells due to enhanced permeation activity of the formulation. Also, the micelles showed strong mucoadhesive properties [57].

Nanoemulsions composed of peanut oil, Tween 80 and Cremophor EL and co-surfactants have been loaded with acetazolamide and investigated for use in the treatment of glaucoma. The nanoemulsion was combined with different *in situ* forming gelling systems in which the gellan/xanthan gum was the most stable formulation with sustained release of the drug. *In vivo* trials on albino rabbits showed that the new formulation was superior to products present on the market. This was probably due to both the mucoadhesive properties of the formulation and the enhanced permeation due to the surfactants [58].

5.3.6 Vaginal Mucosa

The non-keratinized stratified squamous epithelium lining the vaginal mucosa is covered by a discontinuous layer of fluid that protects the underlying tissues. Even though it is referred to as mucosal tissue, the vagina does not have secretory glands, and the fluid bathing the vaginal mucosa comprises mainly cervical mucus and a mix of other upper genital tract fluids, vaginal tissue transudates and secretions from vestibular glands [59]. Estrogen stimulates the epithelial cells to secrete glycogen, which is digested by *Lactobacillus vaginalis* inhabiting the vagina to produce lactic acid and maintain a low pH, thereby preventing infections by other microorganisms. The vaginal fluid typically has a pH around 3.5–4.5 in healthy women of reproductive age [24]. Its composition, volume, pH and rheological properties are affected by age, the stage in the menstrual cycle, and sexual arousal, thus influencing the release pattern and efficacy of a drug delivery system administered into the vagina [60]. Due to the self-cleansing action of the vaginal tract, the residence times of dosage forms and delivery systems will be reduced unless they are specially adapted to increase the residence time. Mucopenetrating PEGylated liposomes of 180 nm containing interferon alpha-2b were designed to provide localized therapy of human papilloma virus (HPV) vaginal infections [61].

VivaGel® (SPL7013, astodrimer sodium) from Starpharma is an antimicrobial dendrimer formulated into a mucoadhesive gel that has just recently obtained regulatory approval in the EU for topical treatment of bacterial vaginosis (BV) and to reduce recurrence. Since April 2019 VivaGel® has been available as OTC in Australia, where it is marketed under the brand-name Fleurstat BVgel by Aspen Pharmacare. VivaGel® is further being launched worldwide by Mundipharma under the brand-name Betadine BVgel. VivaGel® has also demonstrated efficacy against human immunodeficiency virus and herpes simplex virus *in vitro* and animal models [62], and is expected to proceed into clinical trials for additional indications. The general concept of one type of nanocarrier in a mucoadhesive gel is frequently seen in development of formulations for vaginal delivery.

Preventing sexual transmission is significant in the fight against HIV, and the use of vaginal or rectal microbicides for topical pre-exposure prophylaxis is being extensively investigated [63]. Polymeric nanoparticles (mean diameter of 170–200 nm) based on poly(ε-caprolactone) (PCL) or PLGA have been designed to deliver dapivirine, one of the most promising microbicide drug candidates [64]. It was shown that cell-associated levels of dapivirine provided by nanocarriers were significantly higher (up to nearly 30-fold) in different immune cells compared with the free drug. In addition it was correlated with moderately enhanced *in vitro* activity against HIV infection in human peripheral blood mononuclear cells (PBMCs) and a co-culture model of DCs and $CD4^+$ T-cells. Another strategy for developing topical anti-HIV microbicides involves combining two or more antiretroviral drugs in one medical product. PLGA nanoparticles were loaded with efavirenz, and the nanoparticles were incorporated into a fast dissolving film together with free tenofavir [65]. A dual system provides flexibility with respect to pharmacokinetics.

5.4 Use of Mucosal Membranes in Studies of Micro- and Nanoparticles

There is a variety of *in vitro* and *ex vivo* methods available for assessment of different aspects of micro- and nanoparticle interaction with mucosal membranes. Most of these methods are used in development and optimization of new mucosal drug delivery systems, and are not standardised or taken into the pharmacopeia. Here we provide an overview of frequently used methods in the scientific literature.

Studies using biological barriers typically involve permeability of drugs or particles through the barrier, either for developing systemic therapy where bioavailability is a key parameter, or for localized therapy where permeability is undesirable [66].

5.4.1 Diffusion Chambers

Permeability experiments are often set up using diffusion chambers, such as Franz diffusion cells (Figure 5.3A), Ussing chambers (Figure 5.3B) or Valia-Chien cells. They have in common that a membrane divides two chambers or compartments, and the membrane can be excised tissue, cell-based models, or other biological models (see below).

The movement of test substance from one side of the membrane, typically the apical side, is monitored by analysing the appearance of substance on the other side, in the basolateral compartment. The movement can be estimated by calculating the flux J (μg/(cm^2 × h)) at steady state:

$$J = \frac{Q}{A \times t}$$

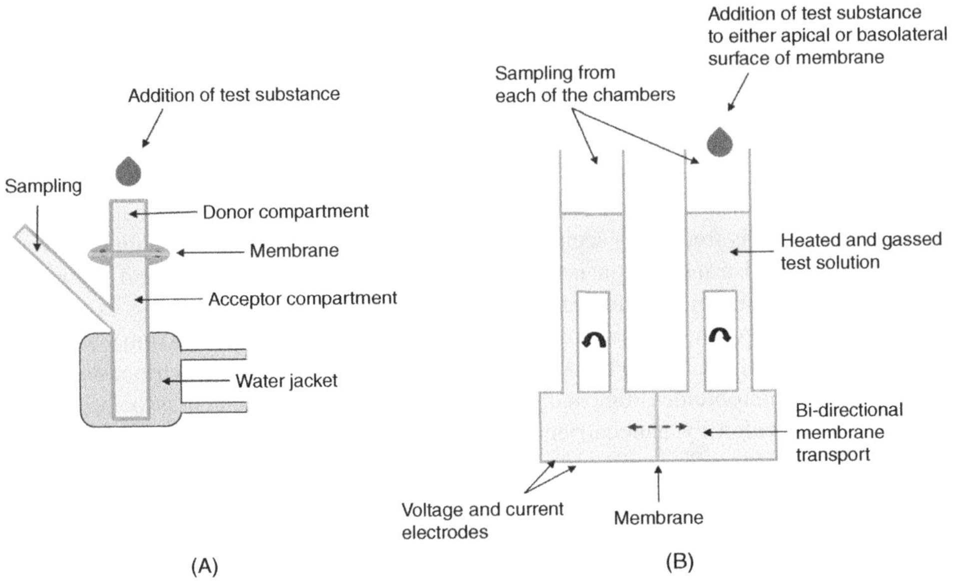

Figure 5.3 *Illustrations of diffusion chambers: (A) Franz diffusion cell and (B) Ussing chamber.*

where Q is the total amount of permeated substance, A is the diffusion area (cm^2) and t is the time of experiment (s).

5.4.2 Permeability Support for Cell-based Systems

Permeable supports like Transwell® from Corning or ThinCert™ from Greiner Bio-One are versatile tools when the diffusion of nanoparticles or encapsulated drug over a cell-monolayer is monitored (Figure 5.4).

The permeable supports can be purchased with filters of different surface areas, pore sizes and coatings to encourage cell adhesion. They can be filled with mucus, or cells can be

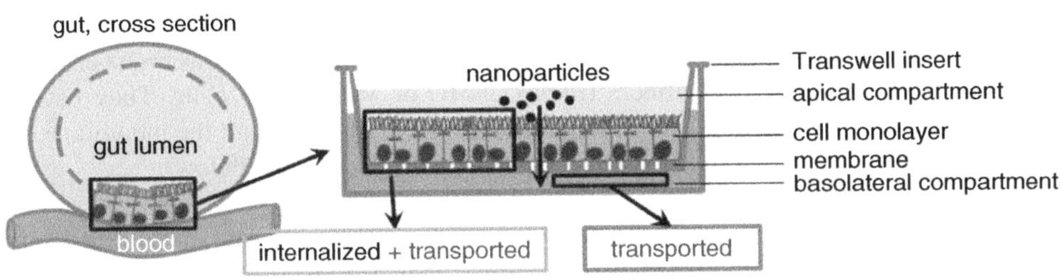

Figure 5.4 *Schematic illustration of permeability setup for study of transport of nanoparticles over a cell-monolayer in a Transwell® system as models of the intestinal barrier. Reprinted from [28] with permission from Elsevier.*

seeded on the filter. A pore size of 3 μm, as opposed to 0.4 μm, is more suitable if transport of particles is expected. Pore sizes of 2–3 μm have been used to avoid retention of the particles by the membrane. However, in general, increasing the membrane pores will increase the likelihood of mucus leaking through as well.

The calculation of movement from the apical compartment to the basolateral compartment is standardized using the apparent permeability coefficient (P_{app}):

$$P_{app}\,(cm/s) = \frac{Q}{A \times C \times t}$$

where Q (total amount of permeated substance), A (diffusion area, cm^2), and t (time, s) are the same as defined above, and C is the concentration (amount/cm^3) in the upper compartment at time zero. Again the coefficient Q/t represents the steady-state flux across the monolayer.

5.5 Selection of Biological Models

In the early development phase, prior to *in vivo* studies, models are required that can, as closely as possible, provide information on the biological performance of the system. These can be animal tissue, cell-based models and artificial models with or without addition of mucus.

5.5.1 Tissue-based Models

Mucosal tissues have been collected at the local slaughterhouse, like cow vaginal tissue, pig gastric or buccal mucosa, sheep nasal mucosa and rabbit small intestine, excised from laboratory rats or from commercially available animal organs. The main challenge using tissue samples is the lack of reproducibility.

5.5.2 Cell-based Models

Choosing a suitable cell-line representing the mucosal membrane of interest for a particular formulation is of importance. Various options are available. In general, the cell model should form an epithelial-like structure or a monolayer with tight junctions, and secrete mucus. Table 5.2 represents examples of different cell-lines that have been used in studies of nano- and microparticles, classified according to which mucosal membranes they are mimicking. Also some commercially available cell-based tissue models (MucilAir™ from Epithelex, and EpiVaginal™ from MatTeck Corp.) are included. The list is not meant to be exhaustive.

In addition to cell-monolayers of mono-cultures, co-cultures of two or more cell-lines have been proposed to more closely resemble tissue. Probably the most sophisticated cellular system simulating the gastrointestinal tract so far consists of a triple co-culture model comprising Caco-2, HT29-MTX and Raji B cells (Figure 5.5), for example [67–69]. Caco-2 monolayers are extensively used as a simple model of the intestinal barrier. The contribution of HT29 clones was demonstrated for dextran-PEGylated microparticles, when the internalization of the encapsulated substance increased due to a favorable effect of

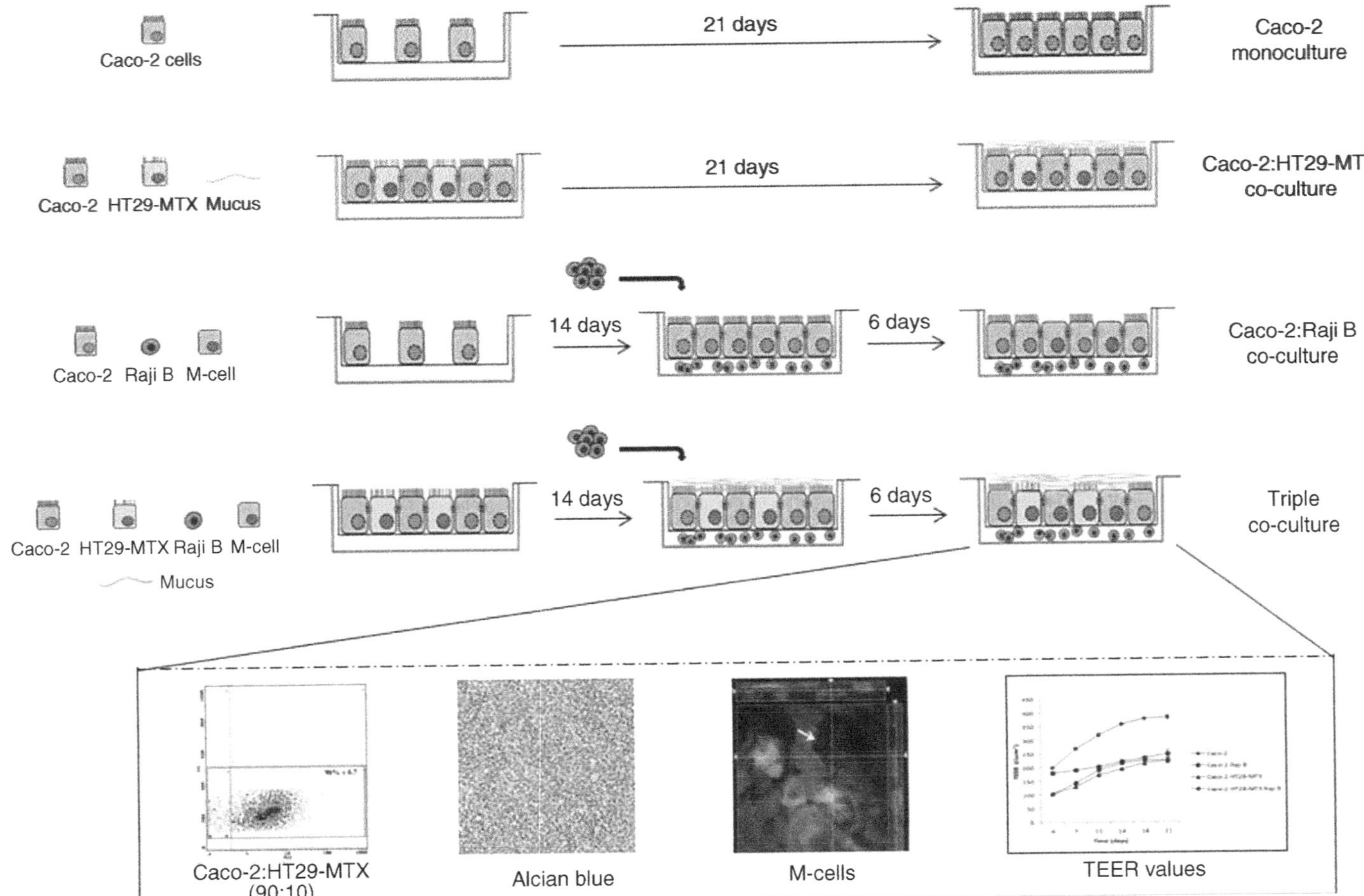

Figure 5.5 *The triple co-culture intestine cell model comprising Caco-2, HT29-MTX and Raji B cells presenting the main features related to the process of drug absorption of human intestinal epithelium: absorptive cells that resemble enterocytes, mucus-producing cells and cells able to induce M-cell phenotype on Caco-2 cells. Reprinted from [68] with permission from Elsevier.*

mucoadhesion. Also, enhanced permeability of insulin encapsulated in multilayered polymeric nanoparticles was more accurately predicted when mucus-producing HT29-MTX cells were included, compared with excised intestinal mucosa of rats. The human Burkitt's lymphoma Raji B line is included in order to model the human follicle-associated epithelium (FAE) expressing M-cells. Including these cells into the co-culture increased the transport of fluorescent PEGylated PLGA-based nanoparticles containing ovalbumin. The transport increased further when the nanoparticles were RGD-labelled, due to interactions between the RGD ligand and the β_1 intregrins detected at the apical surface of the M-like cells in the co-culture. This co-localization was further confirmed *in vivo* in mice after M-cell staining with rhodamine labelled UEA-1 Lectin [42]. The transport of both free and nanoencapsulated insulin increased when the triple co-culture model was used to study the uptake mechanisms and permeability of chitosan/albumin-coated alginate/dextran sulfate nanoparticles for enhanced oral delivery of insulin.

Another interesting cell-based model is the dual-chamber model, which has been developed as an *in vitro* model to study heterosexual transmission of pathogens and the protective effects of microbicides [70]. The model consists of a tight epithelial cell layer of cervical-vaginal cells (e.g., CaSki, HEC-1A, ME-180) grown on a Transwell® insert, and a basal chamber containing sub-epithelial target cells, such as IL-2 stimulated PBMCs which are sub-epithelial target cells for HIV-1 (Figure 5.6). The dual-chamber model is also interesting for studying the protection efficiency of a vaginally applied microbicide formulations against sexually transmitted infections, and may have an unexploited potential as a formulation screening tool.

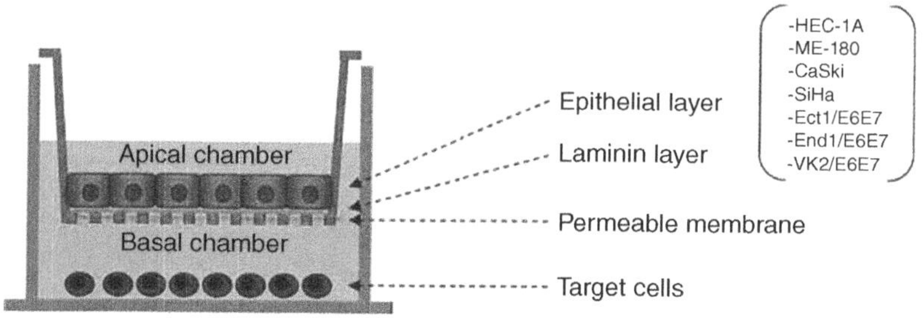

Figure 5.6 *Schematic representation of the dual-chamber model based on the Transwell® setup used to study transmission of pathogens and the protective effects of microbicides. The permeable membrane of the apical chamber is coated with laminin, a component of the basal lamina. Various epithelial cell lines can be cultured in the apical chamber. The basal chamber contains IL-2 stimulated PBMCs, sub-epithelial target cells for HIV-1. Reprinted from [70] with permission from Elsevier.*

5.5.3 Mucus as Models

Simple permeability models have been developed to study the effect of mucus on drug permeability as well as studies of the interaction with nano- and microparticles, where some use native mucus and others use the commercially available mucin in different media [71, 72]. Mucin from porcine stomach and bovine submaxillary glands are commercially available as dry powder which can be reconstituted. However, only the mucin from bovine submaxillary glands forms a solution, while mucin from porcine stomach can be reconstituted as a suspension.

Mucus samples can also be prepared in-house from animals and humans. Mucus has, for example, been scraped off from the small intestinal tissues of freshly slaughtered pigs. Cervico-vaginal mucus has been collected from cow and humans, broncho-alveolar mucus from horse, and tracheal mucus from humans. However, it has been demonstrated that the removal of mucus from its physiological environment can modify its characteristics; it was not possible to obtain a model mucus system with equivalent rheological properties to natural mucus by rehydration of a commercially available dried crude gastric porcine mucin [73].

The type of mucus, any reconstitution or purification steps have been found to influence the results, for example, for mucopenetration of particles. Unfortunately, no validated method for these critical steps exists. Therefore, it becomes challenging to produce a model able to fully mimic physiological mucus, and the differences between native and reconstituted mucus can lead to variations in the resulting drug permeability. Nevertheless, mucus is often used in models to increase the biological relevance of a barrier (Figure 5.7). Used for ranking formulations relative to each other within one laboratory, this can provide quite useful insights.

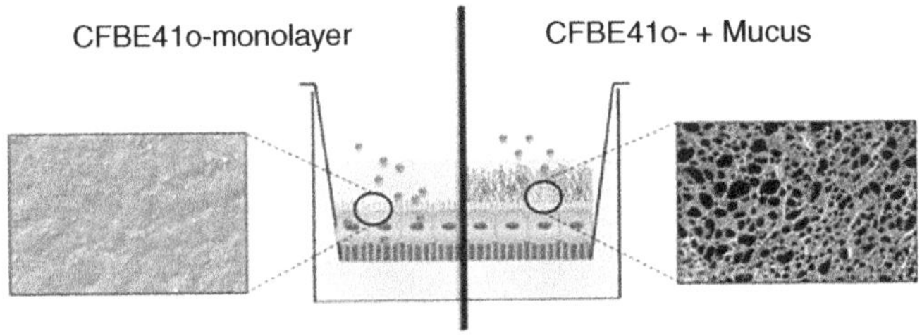

Figure 5.7 *Illustration of a cultivated cell from the airway with mucus (right) and without (left). Reprinted from [34] with permission from Elsevier.*

5.5.4 Artificial Models

Various artificial models have been developed for permeability screening of new drugs and formulations mimicking the environment of mucosal tissue. Compared to cell- and tissue-based models, these are more easily accessible.

The mucus-PVPA (mucus Phospholipid Vesicle-based Permeation Assay) is a further development of the PVPA consisting of a tight barrier of liposomes on a nitrocellulose filter support with a mucin dispersion mimicking the mucus layer [74]. The model has demonstrated the capability to distinguish between the mucoadhesive liposomal formulations (chitosan-coated) and mucopenetrating (PEGylated) liposomes.

The Permeapad® is basically a thin layer of dry phosphatidycholine on a hydrophilic support. The model was recently evaluated for the prediction of buccal absorption of metoprolol by comparing obtained results with permeability across buccal cell monolayers (TR146), porcine buccal mucosa and *in vivo* data from Göttingen mini-pigs [75]. The study did not involve use of mucus, nevertheless, the *in vitro in vivo* correlation (IVIVC) comparing apparent permeability coefficient to the absolute bioavailability of metoprolol administered buccally to mini-pigs was high ($r^2 = 0.98$).

Despite the predictive power of current cell-based and cell-free absorption models for the assessment of drug permeability, high costs and lengthy preparation steps hamper their use [76]. The AMI-system (Artificial Membrane Insert) containing a regenerated cellulose membrane, without any lipids present as substitute of the intestinal wall, has been shown to offer a fast and simple barrier for intestinal drug permeability [76].

None of these simplified permeability models can reflect paracellular or active transport pathways. However, good predictions have been demonstrated for passive permeability compounds [74–76].

5.6 Methods for Testing Biocompatibility

Safety assessments are crucial upon developing new micro- and nanoparticulate formulations. The biocompatibility of a formulation can be assessed on a cellular scale using at least three different concepts: cell viability, cytotoxicity based on plasma membrane integrity, and monolayer integrity.

5.6.1 Viability

Cell viability is based on detecting the activity of intracellular oxidoreductase enzymes, reflecting the number of viable cells. The MTT-based *in vitro* Toxicology Assay Kit produces formazan crystals that can be quantified spectrophotometrically after solubilizing in acidic isopropanol. This method was used for testing the biocompatibility of multilayered polymeric nanoparticles on excised intestinal mucosa of rats [77].

Cell Counting Kit-8 (CCK-8) and CellTiter 96® Aqueous Non-Radioactive Cell Proliferation Assay produce colored formazan products, which are soluble in tissue culture media. The AlamarBlue® assay monitors the reductive environment of cell growth by transforming the dye to a red (reduced) form. The resulting fluorescence can subsequently be read on a plate reader or fluorescence spectrophotometer.

5.6.2 Cytotoxicity

Cytotoxicity is usually assessed by measuring the lactate dehydrogenase (LDH) activity released from the cytosol of damaged cells into the apical medium, which can be quantified with an LDH Cytotoxicity Assay Kit based on formazan formation [33].

5.6.3 Paracellular Permeability

Integrity of the tight junctions is a robust parameter for biocompatibility on confluent cell layers. The integrity can be monitored by measuring changes in the transepithelial electrical resistance (TEER) and by evaluation of cell permeability to paracellular markers like ^{14}C-Mannitol, FITC-dextrans (4, 12 and 40 kDa) and 5(6)-carboxyfluorescein.

Often, more than one method is used to investigate the biocompatibility. For example, when a Caco-2 monolayer was incubated with multilayered polymeric nanoparticles containing insulin, the TEER values dropped to about 50–60% after 4 hours, as opposed to 20% and 27% when Caco-2/HT29 and excised intestinal rat tissue was used, respectively [77]. An MTT test was also conducted.

Mucoadhesive chitosan nanoparticles were improved for wet conditions by conjugating hydrocaffeic acid, after inspiration from mussels, which adhere to almost any surface. Incubation with Caco-2 cells grown on Transwell®s caused a transient opening of monolayer tight junctions, which was monitored using TEER; the transport of 4 kDa FITC-dextran increased three-fold. TEER values returned to normal after removal of the nanoparticles, and the cell viability of non-confluent Caco-2 cells was stable, as monitored with MTT [78].

Further evidence for the influence of chitosan-based nanoparticles on monolayers were provided by fluorescent staining of the ZO-1 protein involved in the regulation of tight junctions [79]. After incubating a Caco-2/HT29-MTX co-culture with the nanoparticles, the ZO-1 protein became segmented and discontinuous, which correlated with higher permeability of the encapsulated FITC insulin across the cell monolayer, as well as across fresh rat ileum mucosal tissue using an Ussing chamber system, and correlated further with blood glucose drops *in vivo* in rats. The MTT assay revealed no effect on cell viability.

Also when trimethyl chitosan encapsulated insulin with a polymeric dissociable coating was incubated with a HT29-MTX monolayer, the TEER values displayed a transient drop correlating with improved insulin transport across the same monolayer. This was reflected in a segmented and discontinuous staining of claudin-4, another protein associated with tight junctions. Again, the MTT assay revealed no effect on cell viability, which was confirmed by no detected weight loss nor any changed organ histology in mice [80].

5.7 Methods for Testing Mucoadhesion

Numerous *in vitro* and *ex vivo* methods for testing the degree of mucoadhesion of micro- and nanoparticles have been published. They differ in *in vivo* relevance, complexity and cost of equipment, source of mucus, and the type and amount of information provided, that is, maximum force of detachment or duration of attachment and the mechanisms involved such as entanglements vs. other physical–chemical interactions and degree of hydrogen bonding.

5.7.1 Atomic Force Microscopy (AFM)

An atomic force microscope is a high-resolution scanning instrument, in which the interactions between the probe and the sample are used to generate the 3D topography of a surface. Local attractive or repulsive forces between the tip and the surface lead to bending (deflection) of a cantilever. This cantilever deflection is converted into an electrical signal to produce the image. Another working mode of AFM is so-called force spectroscopy. In force measurements, the cantilever approaches and retracts the surface in a vertical direction; during this movement, the deflection of the cantilever is measured as a function of distance to the surface. In addition, the tip of the cantilever can be functionalized with molecules of interest in order to measure the interaction with different samples. Both imaging mode and force spectroscopy can provide valuable information in mucoadhesion and bioadhesion measurements.

Patel *et al.* investigated the adsorption of bioadhesive polymers onto human buccal cells using quantitative image analysis [81]. The buccal cells were scanned with AFM before and after incubation with polymer solutions (chitosan, polycarbophil, hydroxypropyl methylcellulose). It was found that cells treated with polymers exhibited higher average surface roughness. Joergensen *et al.* used a similar method to evaluate the mucoadhesive properties of different pectins [82]. Mucin-coated mica was scanned in a liquid cell before and after incubation with pectins, and a roughness analysis was performed. Force spectroscopy measurements were used by Iijima *et al.* in order to investigate the interactions between mucin layers and core–shell nanogel particles over a wide range of pH [83]. Freeze-dried granules of particles were directly adhered to the AFM tip prior to the experiment.

5.7.2 Quartz Crystal Microbalance (QCM)

Quartz crystal microbalance (QCM) is a highly sensitive and versatile instrument for investigation of the adsorption of different compounds at solid–liquid interfaces. The adsorption of material to the sensor surface (piezoelectric quartz crystal with diverse coatings) causes changes in its mass, monitored as changes in resonance frequency of the quartz crystal. In quartz crystal microbalance with dissipation monitoring (QCM-D), simultaneous measurement of dissipation factor provides information about the structural changes of a layer occurring during the adsorption process. From these changes it is possible to learn about kinetics of the adsorption process, calculate mass/thickness of the adjacent layer and its viscoelastic properties. Consequently, the QCM-D technique is a very useful tool for assessment of nano- and microparticle interactions with biological compounds. A mass sensitivity of around 0.5 ng/cm^2 enables accurate quantification of adsorbent.

Mucoadhesion of nano- and microparticles can be evaluated with QCM by using mucin-modified sensors. The QCM sensor with a proper surface chemistry (i.e., gold-coated quartz crystal functionalized with carboxylic acid groups, enabling efficient mucin immobilization) is covered by a layer of mucin (i.e., porcine gastric mucin, bovine submaxillary mucin) until a stable resonance frequency signal is established [84]. Then the mucin-modified sensor is exposed to the particles of interest and changes in frequency/dissipation are recorded. This method has been used to assess the mucoadhesive properties of various polymers (chitosan, polyallylamine hydrochloride, hyaluronic acid), as well as chitosan-coated polycaprolactone nanoparticles loaded with curcumin and other polysaccharide-based nanoparticles for oral drug delivery [85–87]. By additional recording of different overtones of the oscillating system (harmonic frequencies), information from different penetration depths within a deposited film can be collected. A similar overtones response is an indication for penetration of the particles within the mucin film, while different overtones suggest superficial adsorption on the mucin surface [85]. Oh *et al.* investigated the mucoadhesion and mucus permeability of thiolated chitosan particles as a function of pH (4 and 6.8) [88]. Due to the chitosan particle adsorption onto the surface, the mass of a sensor increased, followed by a decrease in frequency and increase in dissipation. Changes in dissipation without initial spreading of overtones indicate chitosan permeation into mucin films.

QCM is also an important tool in bioadhesion measurements. It is often used to study the interactions between the particles for drug delivery and model biological membranes composed of lipid bilayers. Paiva *et al.* studied the interaction of different anaesthetics with liposomal membrane models, showing that the anaesthetics have a fluidizing effect on the model membranes [89].

5.7.3 Rheology

Rheology is a well-known method to determine mucoadhesive properties, especially of polymers. When a polymer and mucin is mixed, the viscosity of the combined sample could be either higher, lower, or the same as the sum of the viscosity of the two components alone. Both higher or lower viscosity could indicate attractive forces either due to enhanced entanglements or hydrogen bonding, or simply due to phase separation [90].

The easiest approach to determine a polymer's mucoadhesive properties is to measure the viscoelastic properties of the polymer in solution, the mucin in solution, and a mixture of the two components by steady shear measurements. The mucoadhesive properties can then be estimated by the equation:

$$\eta_b = \eta_t - \eta_m - \eta_p$$

where η_b is the bioadhesion parameter, η_t is the viscosity of the mixture, η_m is the viscosity of mucin, and η_p is the viscosity of the polymer [91].

Another approach is to do a frequency sweep in the viscoelastic region, measure G' and G'' of each component, and then compare with the measured values of each component of the mixture [4]. Even though very simple, few studies have been conducted using rheology to estimate the mucoadhesive properties of micro- and nanoparticles. In one study the mucoadhesive potential of thiolated particles was estimated by mixing the particles with mucin collected from urinary bladder, and the interactions were measured by investigating how the dynamic viscosity of the samples changed after addition of mucin [4]. In another study, the mucoadhesive potential of nanoparticles of poly-acrylic acid grafted with papain, an enzyme able to cleave linkages of mucin, was studied by investigating the breakdown of mucus [92]. The particles were mixed with pig gastric mucus. If the particles were successfully adhered and entrapped in the mucus layer, the viscosity of the sample decreased due to the breakdown of mucin and hence the mucoadhesive potential was verified.

Rheology could in theory be used to distinguish between mucoadhesion due to entanglements and hydrogen bonding by adding urea. If the interaction is due to hydrogen bonding, the viscosity would decrease when urea is added [93].

5.7.4 Rheology in Combination with Light Scattering (Rheo-SALS)

Rheo-SALS could also have potential in investigating mucoadhesive properties of micro- and nanoparticles. Rheo-SALS measurements have been conducted on pig gastric mucin alone, and both the rheological properties and the scattering intensity varied according to the pH in which the mucin was dissolved [94]. The results showed high scattering intensity at pH 2 and 4 and low scattering intensity at pH 1 and 7 (Figure 5.8). This is due to the different conformation of mucin chains at different pH values, with the mucin chains

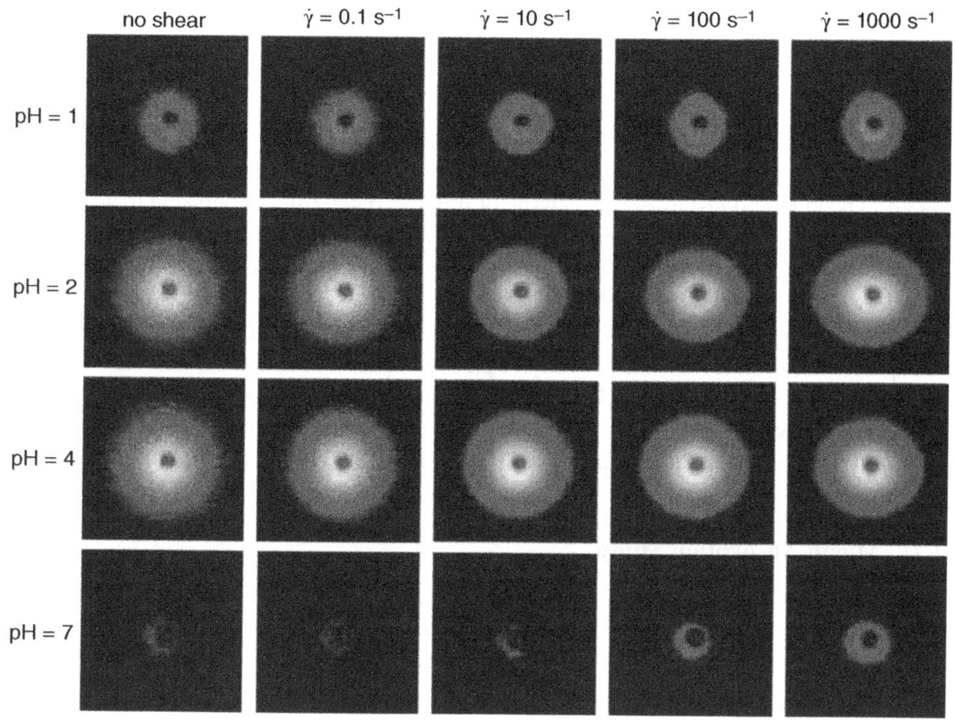

Figure 5.8 *Example of SALS 2D patterns of solutions of mucin (1 wt%) at various pH values and exposed to the stated shear rates. Reprinted from [94] with permission from Elsevier.*

being contracted at pH 1, and also the connectivity within the network. The stronger the associations and higher degree of connectivity, the stronger is the scattering intensity. The samples also showed shear thinning behaviour, with the exception of the sample prepared at pH 7. This technique could probably give an estimate of association and interactions between particles and mucin. A possible drawback could be that the concentration of both the particles and the mucin must be high enough to get reliable rheology measurements.

5.7.5 Dynamic Light Scattering (DLS) and Zeta Potential Measurements

Dynamic light scattering could also be used in combination with zeta potential measurements for estimating the mucoadhesive potential of micro- and nanoparticles. The mucoadhesive potential could be verified by measuring an increase in size and a shift/change in the zeta potential after mixing the components. Liposomes and polymer-coated liposomes have successfully been investigated by this approach. The study indicated that nanoparticles with positive charge (positive liposomes and chitosan-coated liposomes) have the highest mucoadhesive potential [95].

Also the mucoadhesive potential of tetrahydrocannabinol-loaded NLC-formulation was verified by measuring the change in the zeta potential of mucin before and after mixing with the nanoparticles, and also the decrease in the zeta potential of the positive particles after mixing [96].

5.7.6 Mechanical Methods

Using a Texture analyser, various mechanical tests can be used to estimate mucoadhesive capacity. The classical test uses either tissue or mucin-coated filter papers and measures the force and displacement required to detach the formulation from the tissue or mucin after they have been allowed to interact for a predetermined period of time, either conducted with peel, shear or tensile stress [97–99].

The rotating cylinder method is another simple yet useful test for ranking of mucoadhesive capacity between formulations [100]. A piece of hydrated mucosal tissue was mounted on a cylinder. Swollen microparticles were placed gently on the mucosa, before the cylinder with the mucosa was placed in tempered medium and rotated at a predetermined speed for a period of time before the number of particles still attached to the tissue was counted.

5.7.7 Mucin Adsorption Study

The *in vitro* interactions of nanoparticles with mucin can be measured by mixing them with solutions or suspensions of commercially available porcine or bovine mucin. After centrifugation, the amount of free mucin still in the supernatant can be determined using UV measurements (251–261 nm), Bradford protein assay, or the periodic acid schiff colorimetric method. Interactions can thus be quantified at various pH values, concentrations, mixing times, and so on. The results from this simple method were confirmed by *in vivo* nasal clearance and systemic uptake in rabbits and mice, when ranking the mucoadhesion of differently chitosan-coated and uncoated PLGA nanoparticles [101]. A similar approach was used in the development of mucoadhesive liposomes [102, 103].

5.7.8 Wash-off Tests

Incubating formulations with a mucosal surface to follow attachment/detachment upon subsequent rinsing is a versatile and physiologically relevant method to quantify mucoadhesion. The rinsing stage can be performed by immersing the system in water, by washing or by rinsing the mucosal surface placed at an angle with predetermined flow rates. The nanoparticles can be measured in the supernatant or directly on the mucosal surface either by a fluorescent marker like fluorescein or FITC, or by an entrapped drug.

Different types of fluorescently labelled liposomes were introduced to HT29-MTX cell covered wells and blank wells to distinguish mucoadhesion from general adhesion [104]. After incubation, the fluorescence was measured both directly on the surfaces and in the supernatants after first gently washing one time and then again after gently washing a second time, to distinguish high initial adhesion from long-lasting adhesion. Alginate-coated liposomes achieved a high specific (genuine) mucin interaction. The positively charged uncoated liposomes achieved the highest initial mucoadhesion. The chitosan-coated liposomes displayed the highest potential for long-lasting mucoadhesion, but with the drawback of a higher general adhesion (tack) [104].

A similar concept is behind the *ex vivo* retention model used for simulating bioadhesion in the oral cavity under irrigation of saliva [105]. The bioadhesion of spray-dried chitosan microparticles containing metformin was tested in a retention model using porcine buccal mucosa under irrigation at a flow rate of 4 ml/min. Retention of metformin depended on the characteristics of the irrigation media, such as rheology. An irrigation media containing mucin was regarded as beneficial and biomimetic.

5.8 Methods for Testing Mucopenetration

The mucus layer has been demonstrated to act as a barrier for some nanoparticles, for example commercially available Red (580/605) fluorescence-labeled carboxyl polystyrene (CP) particles of 200 nm, peptide ligand modified solid lipid nanoparticles loaded with FITC-labeled salmon calcitonin, and self-assembled polyelectrolyte complex nanoparticles consisting of trimethyl chitosan-coated insulin-loaded dodecylamine-graft-γ-polyglutamic acid micelles. Uptake and transport through the underlying cell layer improved when the mucus layer of Caco-2/HT29-MTX co-cultured cells was removed by N-acetyl-cysteine incubation [106]. Alternatively, putting an external mucin layer on naked oral cell lines (TR 146 and H376) restricted the access, which was confirmed by *ex vivo* transport studies through porcine buccal mucosa [29].

5.8.1 Fluorescent Recovery after Photobleaching (FRAP) and Multiple Image Photography (MIP)

Samples of premixed fresh human cervical mucus with mesh spacing of 20–200 nm, as determined by transmission electron micrographs, and various fluorescently labeled proteins, virus-like particles and polystyrene particles (59–1000 nm) were photobleached (20–50% of the fluorescent molecules). An attenuated laser beam was focused through a modified epifluorescent microscope to a small point in the samples, whereupon the residual fluorescence from the same area in the sample was monitored. Return of the fluorescent signal reveals the rate of diffusion of unbleached fluorescent particles. The polystyrene microspheres showed no recovery after bleaching, as opposed to human papilloma and Norwalk virus-like particles [107].

Diffusion was also observed with multiple image photography (MIP) tracking multiple fluorescent particles individually, using an epifluorescent microscope and a charge-coupled device camera. Successive images were subtracted from one another to reveal particle displacements occurring between images [107].

5.8.2 Permeability Studies

Permeability studies are relatively simple without great requirements for lab equipment nor expertise, perhaps explaining their popularity. For example, Transwell® inserts have been filled with mucus samples. The transport of lipid core nanocapsules of 180 nm and the effect of chitosan and carbopol coating of liposomes was further measured by the amount of either fluorescence or encapsulated drug appearing in the acceptor chamber. Diffusion chambers have been used to study the permeation of various solid lipid nanoparticles.

Critical parameters are type of mucus, any rotation or stirring applied, and the pore size of the mucus-supporting filters. Premixing the particle sample with mucus will reduce viscosity differences and thus avoid bypassing the sample between chamber walls and the mucus barrier [80].

5.8.3 Water-assisted Transport Through Mucus

Water absorption by the intestine can lead to advective transport of particles, which was simulated by connecting the test system to a vessel filled with fluid and continuing to drip fluid into the vessel during the experiment [108]. The test system consisted of a barrel of a syringe which was placed in a vertical position, tip down. Fluorescein diacetate marked nanoparticles of 350–400 nm prepared by ionotropic gelation between different chitosan derivatives, either thiolated or not, and hyaluronan were placed at the bottom of the barrel on top of a 0.2 μm filter. The particles were then covered by mucus scraped off from the small intestinal tissue of a freshly slaughtered pig. Then the tip of the syringe barrel was connected to the vessel by a flexible tubing, causing a bottom-to-surface fluid flow across the mucus layer as fluid was continuously dripped into the vessel at the physiologically relevant speed of 1.24×10^{-4} cm/s, corresponding to a flux of 0.446 ml/(cm^2 h). The liquid level and dripping in the vessel determined whether the fluid hydrated the mucus layer in the barrel, or flowed across it causing a liquid layer to gradually rise over the mucus layer. This *in vivo* relevant method revealed that despite the high mucoadhesion of thiolated nanoparticles, they were able to overcome the mucus barrier with the aid of water movement [108].

5.8.4 Particles with Dynamic Properties

The idea of particles changing character during mucopenetration, to take on traits favorable for interactions with the underlying cell layer, is conceptually intriguing. Various suggestions have been published, such as a zeta potential changing self-emulsifying drug delivery system formulated to release phosphate upon interactions with intestinal alkaline phosphatase [109], or particles with a dissociable "mucus-inert" hydrophilic coating of N-(2-hydroxypropyl) methacrylamide copolymer (pHPMA) derivative, both of which subsequently exposes positive charges [80].

Förster resonance energy transfer (FRET) analysis was used to verify structural changes upon interactions with collected human cervico-vaginal mucus and HT29-MTX cells. When crosslinked trimethyl chitosan nanoparticles containing FITC-insulin were coated with TRITC–pHPMA, the excitation energy of the donor is transferred to the acceptor, whose subsequent emission can be detected in accordance to the donor–acceptor distance. The FRET intensity was shown to decrease rapidly over time in mucus, as opposed to in plain buffer [80]. The pHPMA's favorable impact on mucopenetration was also demonstrated directly by the visualization of fluorescent particles on the HT29-MTX cell layer vs. upper mucus layer by confocal laser scanning microscopy (CLSM), after staining the mucus layer with rhodamine-conjugated ulex europaeus agglutinin I lectin (Rho-UEA-I). *In vivo* pharmacological and pharmacokinetic studies in rats proved the concept.

5.9 Methods for Assessing Cell Interactions

Cell lines are considered to be more reproducible than excised mucosa for the purpose of studies of cell interactions. A robust and reliable cell model is thus a vital part of experiments to assess the effects of micro- and nanoparticles upon adhesion to, uptake into, and further trafficking in cells.

5.9.1 Cell Adhesion

The improved cytoadhesion by functionalizing PLGA microparticles with wheatgerm agglutinin was demonstrated by incubation with a Caco-2 cell layer at 4°C, suppressing both drug release and energy-dependent endocytosis. After removal by washing, the cell-associated fluorescence intensity provided by encapsulated fluorescein was quantified as well as visualized with fluorescence microscopy. At 37°C the encapsulated fluorescein was released from the cell-associated microparticles and accumulated intracellularly, as demonstrated with fluorescence microscopy after staining the nuclei with Hoechst 33342 [110].

The interaction between insulin-loaded alginate/dextran sulfate nanoparticles coated with chitosan and albumin and cellular glycocalyx was proved by pre-treating the cells with 35 mM sodium chlorate for 48 hours. Sodium chlorate inhibited the glycosaminoglycan sulfate ion, subsequently disturbing the electrostatic interactions between the albumin in the coating layer and the glycocalyx on the apical membrane, which is oppositely charged. The transport of insulin across the intestinal monolayers decreased accordingly, demonstrating the importance of the interaction with regard to insulin permeation [69].

5.9.2 Cellular Uptake

Nanoparticles can be taken up into cells by endocytosis, whether or not this is intentional from the formulator's point of view.

The intestinal epithelium is often regarded as impermeable to macromolecules and particles, except in Payer's patches, where the lymphoid follicle-associated epithelium (FAE) contains M-cells that transport antigens and microorganisms (Figure 5.9). The M-cells are specialized cells particularly able to perform endocytosis by phagocytosis, but even enterocytes can perform pinocytosis of particles smaller than ~100 nm. The mechanisms are micropinocytosis, clathrin-dependent endocytosis, caveolae-dependent endocytosis, or clathrin and caveolae-independent endocytosis. Targeting specific lectin receptors expressed on enterocytes and M-cells with, for example, glycoproteins, lectins, mannose, fucose, or galactose, might increase endocytosis. Ligands like folic acid, albumin and RGD peptide might induce caveolin-mediated endocytosis, while ligands like transferrin and mannose might induce clathrin-mediated endocytosis [111].

Size and zeta potential of nanoparticles will also influence which uptake mechanisms will dominate, in addition to the speed of endocytosis and the subsequent fate of the endocytosed particles (i.e., lysis or exocytosis) [53].

Cellular uptake studies are often performed by incubating the tissue model with fluorescent formulations, followed by imaging using CLSM. Cellular structures like nuclei

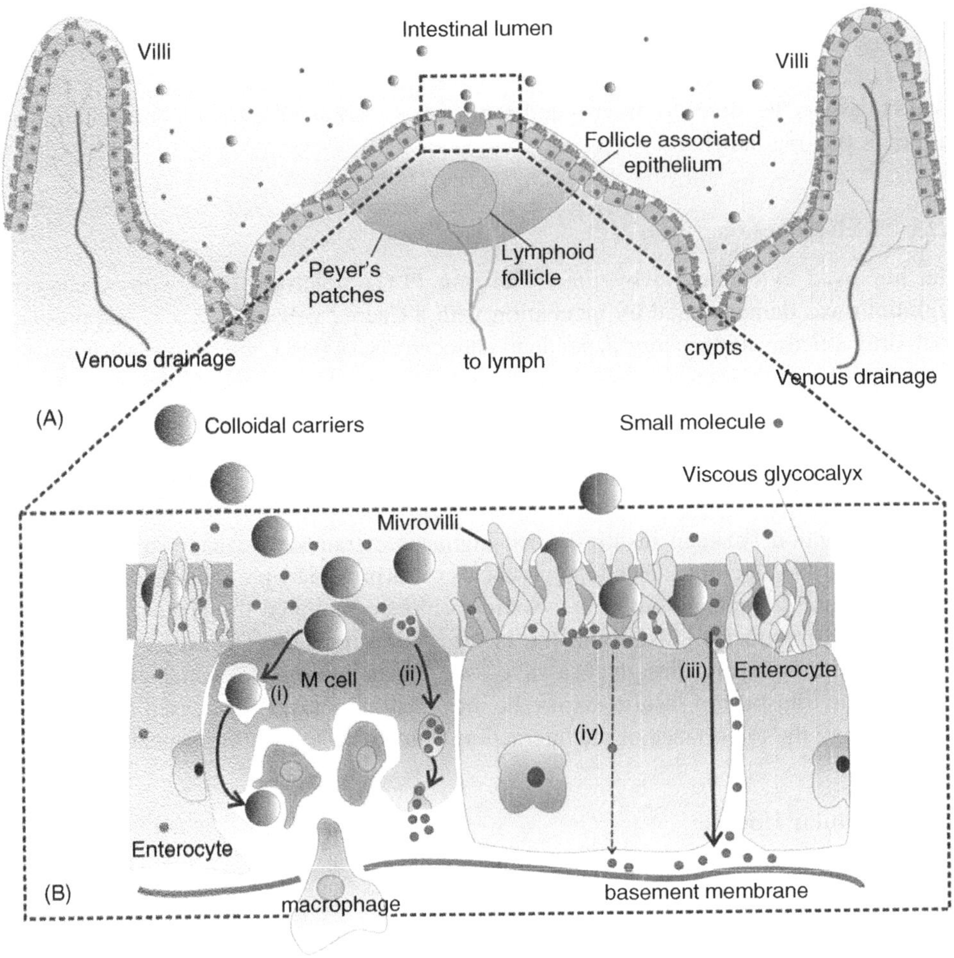

Figure 5.9 *(A) Cross-section of intestinal epithelia showing distribution of M-cells and enterocytes along with Payers's patches and the lymphoid follicle-associated epithelium (FAE). (B) Cells involved in the absorptive processes: (i) phagocytosis, (ii) pinocytosis, (iii) paracellular transport and (iv) transcellular transport, where (i) is the most relevant for colloidal carrier particles. Reprinted from [111] with permission from Elsevier.*

and actin cytoskeleton may be stained using, for example, DAPI or NucBlue® (for live, unfixed cells) or fluorescent antibodies. This technique was used to illustrate the uptake of multilayered polymeric nanoparticles containing insulin into isolated segments of duodenum, jejunum and ileal Peyer's patches of follicle-associated epithelium from rats. Tissue samples were embedded in a cryostat medium and 10 μm cross-sections sliced using a cryostat. Fluorescence from FITC-insulin and RBITC-alginate was co-localized, indicating that insulin remained with the nanostructure [77]. CLSM images were also used to reveal that the cellular uptake of a hydrophobic fluorescent dye, Nile red, was higher in HT29-MTX monolayers than Caco-2 monolayers, when the favourable bioadhesive properties of encapsulation in dextran-pegylated microparticles encapsulation were exploited [112].

For quantification of uptake, cells can be trypsinized after various time periods, washed and recovered by centrifugation before subjected to flow cytometry. Cells that received no particle treatment serve as control samples. This technique was used to assess rat pulmonary arterial endothelial cellular uptake of liposomes containing rhodamine-labeled (red) lipids [113] and FITC-labelled insulin-loaded PLGA nanoparticles (~150 nm) in Caco-2 cells [114]. In the latter study, cell viability was also checked before flow cytometry was performed.

It was shown that the use of liposomes, and especially chitosan-coated liposomes, increased the cellular uptake of alendronate in Caco-2 cells, which correlated with enhanced oral bioavailability of alendronate in rats [115]. In this study, Caco-2 cells were incubated with free, liposomal and chitosan-coated liposomal alendronate, and the drug uptake was quantified by HPLC analysis of the supernatant after cell washing and lysis. The amount of drug uptake was expressed per amount of protein as determined by Bradford protein assay.

Lichtenstein *et al.* found that silver and iron oxide nanoparticles of similar size coated with poly(acrylic acid) mostly localized in the cytosol of cells, with only a minor fraction found in vesicles [28]. The cellular uptake was visualized with transmission electron microscopy and energy-dispersive X-ray spectroscopy of fixed and cut monolayers. Ion beam microscopy of fixed monolayers was also conducted to provide information on intracellular concentration and distribution down to the single cell level. Silver and iron quantification using atomic absorption spectroscopy after cell lysis revealed comparable cellular uptake in Caco-2 cells, and a slight increase in silver uptake when Raji-B lymphocytes were included.

It should be noticed that the endocytic pathways work simultaneously in cells. Therefore, cellular regulatory mechanisms may compensate for deliberate inhibition of certain endocytic pathways by upregulating other endocytic pathways. Inhibition of clathrin and caveolae mediated endocytosis can, for example, lead to increased levels of micropinocytosis. The data interpretation of translocation of particles into cells can therefore be difficult. The biocompatibility of the treatment should always be monitored, for example by TEER, MTT or dehydrogenase activity.

Table 5.3 gives an overview of selected literature studying the most relevant mechanisms for nano- and microparticle uptake into epithelial cell layers; different inhibitors and pathway markers have been used. Further discussions of mechanisms or the intracellular fate of the particles are beyond the scope of this chapter.

5.9.3 Transcellular Transport

Nanoparticles taken up by epithelial cells might translocate to the basolateral side. The transcellular transport of nanoparticles can be assessed *in vitro* by using various permeable supports with a cell monolayer on the membrane. The *ex vivo* alternative is the setup involving excised animal tissue mounted in a diffusion chamber. More commonly, however, only the translocation of the encapsulated drug/marker is monitored.

Lichtenstein and coworkers studied the impact of size, shape and coating on the potential for nanoparticles to penetrate cell membranes and epithelial barriers [28]. If the coating

Table 5.3 *Cellular mechanisms for nano- and microparticle uptake in epithelial cells*

Treatment	Experimental design	Inhibition	Description	References
Low temperature: 4, 10, 16°C vs. 37°C	Experiments are carried out at low temperature, usually 4°C, for a certain time, e.g., 1–3 hours.	Endocytosis	Energy dependent mechanisms	[67, 69, 79, 106, 114, 116]
Methyl-β-CD	Experiments carried out with inhibitor, 10 mM, after 30 min pre-treatment.	Endocytosis	Cholesterol depletion → disrupt clathrin-dependent and clathrin independent, i.e., raft-dependent endocytosis	[114]
Sodium azide or sodium azide + 2-deoxyglucose (50 mM)	Experiments carried out with or without inhibitor, 1 mg/ml, 100 mM, 0.02 % (w/v), with or without a pre-treatment period of 30 min - 1 hour.	Active transport	Energy depletion	[69, 79, 106, 114]
Ethyl-isopropyl-amiloride (EIPA)	Experiments carried out with inhibitor, 50 μM, after a pre-treatment period of 20–60 min.	Macropinocytosis		[67, 79, 106, 114, 116]
Amiloride	0.3 mg/ml, 1 hour pre-treatment			
Dimethyl-amiloride HCl	Experiments carried out with inhibitor, 10 mM, after 30 min pre-treatment			
Nystatin	Experiments carried out with or without inhibitor, 11 μM/25 μg/ml, after a pre-treatment period of 20–30 min.	Macropinocytosis Clathrin-independent endocytosis	Membrane permeable and cholesterol binding → disrupt lipid rafts, caveolae structure and function → inhibitor of caveolae-mediated endocytosis	[67, 79, 114]
Sucrose	Experiments carried out in hypertonic medium of 0.20 and 0.45 M sucrose after a pre-treatment period of 30 min.	Clathrin mediated endocytosis	Lack specificity: Non-specific inhibition of micro-pinocytosis. Inhibition of fluid phase endocytosis has been demonstrated by using the marker horseradish peroxidase.	[116]

(continued)

Table 5.3 *(continued)*

Treatment	Experimental design	Inhibition	Description	References
10 mM acetic acid (pH 5)	Experiments carried out in an acidic cytosol after a pre-treatment period of 10 min.	Clathrin mediated endocytosis		[116]
Potassium depletion ("hypotonic shock")	Cell monolayers were submitted to a 5 min hypotonic shock, and the experiments were subsequently carried out in media devoid of potassium.	Clathrin mediated endocytosis		[67]
Chlorpromazine	Experiments carried out with or without inhibitor, 10 μg/ml – 10 mg/ml, with or without a pre-treatment period of 1 hour.	Clathrin mediated endocytosis	Cationic & amphiphilic, disrupt the assembly of clathrin adaptor protein at the cell surface	[69, 79, 106]
Filipin	Experiments carried out with or without inhibitor, 1 μg/ml-1 mg/ml, after a pre-treatment period of 30 min – 3 hours.	Caveolae mediated endocytosis	Binding of cholesterol → disorganization of caveolin	[69, 79, 106, 116]
Phorbol 12-myristate 13-acetate	Experiments carried out with inhibitor, 10 μM, after 30 min pre-treatment.		Activate protein kinase C → disrupt caveolae and block invaginations	[114]
Protamine sulfate	Experiments carried out with inhibitor, 1 mM, no pre-treatment.	Adsorptive-mediated endocytosis inhibitor		[69]
EGTA	Experiments carried out with inhibitor, 2.5 mM, after a pre-treatment period 2 × 15 min.	Destroy tight junctions		[67]
Co-location using confocal microscopy [117]:				
Cholera toxin subunit B-Alexa Fluor 647 (1 μg/ml)		Pathway marker:	Caveolae-mediated endocytosis	
Transferrin-Texas red (50 μg/ml)		Pathway marker:	Clathrin-mediated endocytosis	
LysoTracker Red		Pathway marker:	Lysosomes	

was lost after cellular uptake, the translocation may be determined by the core material, as demonstrated for silver and iron oxide nanoparticles of similar size, coating and charge. The internalized amount was comparable, but only iron oxide nanoparticles additionally passed the epithelial cell layer.

The transport of FITC salmon calcitonin across excised rat duodenum mucosa without Peyer's patches was studied and found to increase when the drug was encapsulated in SLNs [106]. The permeation increased even further when the nanoparticles were modified with peptide ligands, either for goblet cell targeting or a cell-penetrating peptide. The *in vitro* transcellular transport correlated with the increased drug bioavailability in rats.

Liu and coworkers showed that when trimethyl chitosan particles were coated with a dissociable mucus-inert hydrophilic coating, the transepithelial transport of encapsulated FITC-insulin across a HT29-MTX monolayer increased [60]. The beneficial effect of the coating was confirmed in diabetic rats, where pHPMA-coated NPs generated a hypoglycemic response after oral administration, and increased the relative bioavailability 2.8-fold compared with uncoated TMC-based NPs. The advantageous use of goblet cell-targeting modification was confirmed with nanoparticles of trimethyl chitosan containing micellar insulin, by an increased transport of FITC insulin across a Caco-2/HT29-MTX monolayer, as well as through excised rat ileum mucosal tissue [77]. Removal of mucus by N-acetyl cysteine (NAC) treatment increased the permeation of nanoencapsulated micellar insulin, as opposed to plain micellar insulin. Blood glucose levels in streptozotocin-induced diabetic rats following oral administration decreased in accordance with the result from the transport experiments.

After demonstrating that PEG chain length (2000 vs. 3000 vs. 5000 Da) and core material (PLA vs. PLGA) hardly affected the cellular uptake and intracellular itinerary of nanoparticles in Caco-2 cells, Song and coworkers evaluated the transcellular transport of the incorporated model drug curcumin [118]. Curcumin has low stability, poor water solubility and bioavailability, and they found that the transport of the payload was highest when the PEG chain length was high (5000 Da) and PLA was used as core material, which was attributed to the higher drug loading capacity and slower drug release. Even so, the favorable effect of increasing the PEG chain length was also demonstrated for PEGylated nanostructured lipid carriers (~200 nm) for ocular use [119]. A PEG chain length of 2000–10 000 Da was found to be optimal for the transcorneal permeation of encapsulated ciprofloxacin, while the carbon chain length of phospholipids did not affect penetration. Coating the lipid carriers with chitosan also increased the permeation compared with free drug in solution. However, PEGylated carriers were reported to be more capable of reaching the deeper ocular tissues *in vivo*.

The transcellular permeation of nanoparticles, as opposed to only encapsulated drug, was also demonstrated by Garinot *et al.* [42]. The enhanced transport of PEGylated PLGA-based nanoparticles through an *in vitro* model of a co-culture of Caco-2 with human FAE (follicle-associated epithelium) as compared with monocultures of Caco-2 cells was demonstrated by measuring the number of transported nanoparticles with flow cytometry of basolateral solutions. RGD-labelling of nanoparticles significantly increased their transport by co-cultures, due to interactions between the RGD ligand and the β_1 intregrins detected at the apical surface of co-cultures. When anti-integrin β_1 antibody was added to the apical medium, the transport of the RGD-targeted nanoparticles was inhibited, confirming the RGD advantageous specific targeting of M-cells. *In vivo* studies in mice confirmed the co-localization with M cells and an induced IgG response, reflecting an immune response after oral delivery.

5.10 Concluding Remarks

Micro- and nanoparticles are versatile tools in mucosal drug delivery, whether for local or systemic use. The particles can deliver encapsulated drug to the mucus layer, to the cell membrane or inside cells, or even carry the drug through the mucosal epithelial cells. This can be exploited for improved drug delivery or pose an unwanted risk. In either case, the behaviour should be carefully monitored.

References

[1] Lai, S.K., Wang, Y-Y., Hanes, J. (2009) *Adv. Drug Del. Rev.*, **61**, 158–171.
[2] Leal, J., Smyth, H.D.C., Ghosh, D. (2017) *Int. J. Pharm.*, **532**, 555–572.
[3] Lieleg, O., Ribbeck, K. (2011) *Trend. Cell Biol.*, **21**, 543–551.
[4] Barthelmes, J., Dünnhaupt, S., Unterhofer, S., *et al.* (2013) *Nanomed.*, **8**, 65–75.
[5] Ensign, L.M., Cone, R., Hanes, J. (2012) *Adv. Drug Del. Rev.*, **64**, 557–570.
[6] Netsomboon, K., Bernkop-Schnürch, A. (2016) *Eur. J. Pharm. Biopharm.*, **98**, 76–89.
[7] Bogman, K., Erne-Brand, F., Alsenz, J., Drewe, J. (2003) *J. Pharm. Sci.*, **92**, 1250–1261.
[8] Gourdon, B., Chemin, C., Moreau, A., *et al.* (2017) *Int. J. Pharm.*, **529**, 357–370.
[9] Sosnik, A., das Neves, J., Sarmento, B. (2014) *Progr. Polym. Sci.*, **39**, 2030–2075.
[10] Collins, L.M.C., Dawes, C. (1987) *J. Dent. Res.*, **66**, 1300–1302.
[11] Dawes, C. (1987) *J. Dent. Res.*, **66**, 648–653.
[12] Ugwoke, M.I., Agu, R.U., Verbeke, N., Kinget, R. (2005) *Adv. Drug Del. Rev.*, **57**, 1640–1665.
[13] Sund-Levander, M., Forsberg, C., Wahren, L.K. (2002) *Scand. J. Caring Sci.*, **16**, 122–128.
[14] Beule, A.G. (2010) *GMS Curr. Top. Otorhinolaryngol. Head Neck Surg.*, **9**, Doc 07.
[15] Lai, S.K., Suk, J.S., Pace, A., *et al.* (2011) *Biomaterials*, **32**, 6285–6290.
[16] Dawes, C. (2004) *Caries Res.*, **38**, 236–240.
[17] Lindemann, J., Leiacker, R., Rettinger, G., Keck, T. (2002) *Clin. Otolaryngol.*, **27**, 135–139.
[18] McFadden Jr. E.R., Pichurko, B.M., Bowman, H.F., *et al.* (1985) *J. Appl. Physiol.*, **58**, 564–570.
[19] Fujishima, H., Toda, I., Yamada, M., *et al.* (1996) *Brit. J. Ophthalmol.*, **80**, 29–32.
[20] das Neves, J., Rocha, C.M.R., Goncalves, M.P., *et al.* (2012) *Mol. Pharmaceutics*, **9**, 3347–3356.
[21] Lai, S.K., Wang, Y.-Y., Hida, K., *et al.* (2010) *PNAS*, **107**, 598–603.
[22] King-Smith, P.E., Fink, B.A., Hill, R.M., *et al.* (2004) *Cur. Eye Res.*, **29**, 357–368.
[23] Holly, F.J., Lamp, M.A. (1977) *Surv. Ophthalmol.*, **22**, 69–87.
[24] Owen, D.H., Katz, D.F. (1999) *Contraception*, **59**, 91–95.
[25] Williams, R.C., Paquette, D.W., Offenbacher, S., *et al.* (2001) *J. Periodont.*, **72**, 1535–1544.
[26] Modi, P., Mihic, M., Lewin, A. (2002) *Diabetes Metab. Res. Rev.*, **18**, S38–S42.
[27] Tetyczka, C., Griesbacher, M., Absenger-Novak, M., *et al.* (2017) *Int. J. Pharm.*, **526**, 188–198.
[28] Lichtenstein, D., Ebmeyer, J., Meyer, T., *et al.* (2017) *Eur. J. Pharm. Biopharm.*, **118**, 21–29.
[29] Teubl, B.J., Absenger, M., Fröhlich, E., *et al.* (2013) *Eur. J. Pharm. Biopharm.*, **84**, 386–393.
[30] Goncalves, V.S.S., Mathis, A.A., Poejo, J., *et al.* (2016) *Int. J. Pharm.*, **515**, 1–10.
[31] Mura, S., Hillaireau, H., Nicolas, J., *et al.* (2011) *Int. J. Nanomed.*, **6**, 2591–2605.
[32] Dombu, C.Y., Kroubi, M., Zibouche, R., *et al.* (2010) *Nanotechol.*, **21**, 355102.
[33] Kuper, C.F., Gröllers-Mulderij, M., Maarschalkerweerd, T., *et al.* (2015) *Toxicol. in vitro*, **29**, 389–397.

[34] Murgia, X., Yasar, H., Carvalho-Wodarz, C., *et al.* (2017) *Eur. J. Pharm. Biopharm.*, **118**, 79–88.
[35] Shi, Z., Guo, R., Li, W., *et al.* (2014) *J. Mater. Sci: Mater. Med.*, **25**, 723–731.
[36] das Neves, J., Nunes, R., Machado, A., Sarmento, B. (2015) *Adv. Drug Deliv. Rev.*, **92**, 53–70.
[37] Panwar, R., Sharma, A.K., Kaloti, M., *et al.* (2016) *Appl. Nanosci.*, **6**, 803–813.
[38] Kauschal, G., Trombetta, L., Ochs, R.S., Shao, J. (2006) *Int. J. Pharm.*, **313**, 29–35.
[39] Clark, M.R., McCormick, T.J., Doncel, G.F., Friend, D.R. (2011) *Drug Deliv. and Transl. Res.*, **1**, 175–182.
[40] Aji Alex, M.R., Chacko, A.J., Jose, S., Souto, E.B. (2011) *Eur. J. Pharm. Sci.*, **42**, 11–18.
[41] Ravi, P.R., Vats, R. (2017) *J. Pharm. Pharmacol.*, **69**, 823–833.
[42] Garinot, M., Fievez, B., Pourcelle, V., *et al.* (2007) *Contr. Rel.*, **120**, 195–204.
[43] Patton, J.S., Byron, P.R. (2007) *Nat. Rev. Drug Discov.*, **6**, 67–74.
[44] Wanner, A., Salathé, M., O'Riordan, T.G. (1996) *Am. J. Respir. Crit. Care Med.*, **154**, 1868–1902.
[45] Torge, A., Wagner, S., Chaves, E.G., *et al.* (2017) *Int. J. Pharm.*, **527**, 92–102.
[46] Türeli, N.G., Torge, A., Juntke, J., *et al.* (2017) *Eur. J. Pharm. Biopharm.*, **117**, 363–371.
[47] Klonoff, D.C. (2014) *J. Diabet. Sci. Technol.*, **8**, 1071–1073.
[48] Mikhail, N. (2017) *Cur. Drug Safety*, **12**, 27–31.
[49] Poursina, N., Vatanara, A., Rouini, M.R., *et al.* (2017) *Pharm. Dev. Technol.*, **22**, 733–739.
[50] Bielski, E., Zhong, Q., Mirza, H., *et al.* (2017) *Int. J. Pharm.*, **527**, 171–183.
[51] Khatri, K., Goyal, A.K., Gupta, P.N., *et al.* (2008) *Vaccine*, **26**, 2225–2233.
[52] Jain, A.K., Khar, R.K., Ahmed, F.J., Diwan, P.V. (2008) *Eur. J. Pharm. Biopharm.*, **69**, 426–435.
[53] Mistry, A., Stolnik, S., Illum, L. (2009) *Int. J. Pharm.*, **379**, 146–157.
[54] Gao, X., Wu, B., Zhang, Q., *et al.* (2007) *Contr. Rel.*, **121**, 156–167.
[55] Gaudana, R., Ananthula, H.K., Parenky, A., Mitra, A.K. (2010) *AAPS J.*, **12**, 348–360.
[56] Rozier, A., Mazuel, C., Grove, J., Plazonnet, B. (1989) *Int. J. Pharm.*, **57**, 163–168.
[57] Di Prima, G., Saladino, S., Bongiovì, F., *et al.* (2017) *Eur. J. Pharm. Biopharm.*, **117**, 385–399.
[58] Morsi, N., Ibrahim, M., Refai, H., El Sorogy, H. (2017) *Eur. J. Pharm. Sci.*, **104**, 302–314.
[59] das Neves, J., Sarmento, B. (2015) *Acta Biomat.*, **18**, 77–87.
[60] Vanicé, Z., Skalko-Basnet, N. (2013) *Eur. J. Pharm. Sci.*, **50**, 29–41.
[61] Jøraholmen, M.W., Basnet, P., Acharya, G., Skalko-Basnet, N. (2017) *Eur. J. Pharm. Biopharm.*, **113**, 132–139.
[62] Rupp, R., Rosenthal, S.L., Stanberry, L.R. (2007) *Int. J. Nanomed.*, **2**, 561–566.
[63] Coutinho, C., Sarmento, B., das Neves, J. (2017) *J. Contr. Rel.*, **266**, 119–128.
[64] das Neves, J., Michiels, J., Ariën, K.K., *et al.* (2012) *Pharm. Res.*, **29**, 1468–1484.
[65] Cunha-Reis, C., Machado, A., Barreiros, L., *et al.* (2016) *J. Contr. Rel.*, **243**, 43–53.
[66] Jøraholmen, M.W., Vanić, Ž., Tho, I., Škalko-Basnet, N. (2014) *Int. J. Pharm.*, **472**, 94–101.
[67] des Rieux, A., Fievez, V., Théate, I., *et al.* (2007) *Eur. J. Pharm. Sci.*, **30**, 380–391.
[68] Araújo, B., Sarmento, B. (2013) *Int. J. Pharm.*, **458**, 128–134.
[69] Lopes, M., Shrestha, N., Correia, A., *et al.* (2016) *J. Contr. Rel.*, **232**, 29–41.
[70] Gali, Y., Arien, K.K., Praet, M., *et al.* (2010) *J. Virol. Meth.*, **165**, 186–197.
[71] Legen, I., Kristl, A. (2001) *Drug Dev. Ind. Pharm.*, **27**, 669–674.
[72] Khanvilkar, K., Donovan, M.D., Flanagan, D.R. (2001) *Adv. Drug Deliv. Rev.*, **48**, 173–193.
[73] Kocevar-Nared, J., Kristl, J., Smid-Korbar, J. (1997) *Biomaterials*, **18**, 677–681.
[74] Falavigna, M., Klitgaard, M., Brase, C., *et al.* (2018) *Int. J. Pharm.*, **537**, 213–222.
[75] Bibi, H.A., Holm, R., Bauer-Brandl, A. (2016) *Eur. J. Pharm. Sci.*, **93**, 399–404.
[76] Berben, P., Brouwers, J., Augustijns, P. (2018) *J. Pharm. Sci.*, **107**, 250–256.
[77] Woitski, C.B., Sarmento, B., Carvalho, R.A., *et al.* (2011) *Int. J. Pharm.*, **412**, 123–131.
[78] Soliman, G.M., Zhang, Y.L., Merle, G., *et al.* (2014) *Eur. J. Pharm. Biopharm.*, **88**, 1026–1037.

[79] Zhang, P., Xu, Y., Zhu, X., Huang, Y. (2015) *Int. J. Pharm.*, **496**, 993–1005.
[80] Liu, M., Zhang, J., Zhu, X., *et al.* (2016) *J. Contr. Rel.*, **222**, 67–77.
[81] Patel, D., Smith, J.R., Smith, A.W., *et al.* (2000) *Int. J. Pharm.*, **200**, 271–277.
[82] Joergensen, L., Klösgen, B., Simonsen, A.C., *et al.* (2011) *Int. J. Pharm.*, **411**, 162–168.
[83] Iijima, M., Yoshimura, M., Tsuchiya, T., *et al.* (2008) *Langmuir*, **24**, 3987–3992.
[84] das Neves, J., Bahia, M.F., Amiji, M.M., Sarmento, B. (2011) *Expert Opin. Drug Del.*, **8**, 1085–1104.
[85] Oh, S., Borrós, S. (2016) *Colloid Surface B*, **147**, 434–441.
[86] Mazzarino, L., Coche-Guérente, L., Labbé, P., *et al.* (2014) *J. Biomed. Nanotechnol.*, **10**, 787–794.
[87] Chayed, S., Winnik, F.M. (2007) *Eur. J. Pharm. Biopharm.*, **65**, 363–370.
[88] Oh, S., Wilcox, M., Pearson, J.P., Borrós, S. (2015) *Eur. J. Pharm. Biopharm.*, **96**, 477–483.
[89] Paiva, J.G., Paradiso, P., Serro, A.P., *et al.* (2012) *Colloids. Surf. B. Biointerfaces*, **95**, 65–74.
[90] Mackie, A.R., Goycoolea, F.M., Menchicchi, B., *et al.* (2017) *Macromol. Biosci.*, **17**, 1600534.
[91] Hassan, E.E., Gallo, J.M. (1991) *Pharm. Res.*, **7**, 491–495.
[92] Müller, C., Perera, G., König, V., Bernkop-Schnürch, A. (2014) *Eur. J. Pharm. Biopharm.*, **87**, 125–131.
[93] Kjøniksen, A-L., Hiorth, M., Roots, J., Nyström, B. (2003) *J. Phys. Chem. B*, **107**, 6324–6328.
[94] Maleki, A., Lafitte, G., Kjøniksen, A-L., *et al.* (2008) *Carbohydr. Res.*, **343**, 328–340.
[95] Klemetsrud, T., Kjøniksen, A-L., Hiorth, M., *et al.* (2016) *J. Liposome Res.*, Early online, DOI: 10.1080/08982104.2016.1255640
[96] Hommoss, G., Min Pyo, S., Müller, R.H. (2017) *Eur. J. Pharm. Biopharm.*, **117**, 408–417.
[97] Edsman, K., Hägerström, H. (2005) *J. Pharm. Pharmacol.*, **57**, 3–22.
[98] Hagesaether, E., Hiorth, M., Sande, S.A. (2009) *Eur. J. Pharm. Biopharm.*, **71**, 325–331.
[99] Vanić, Ž., Planinsek, O., Škalko-Basnet, N., Tho, I. (2014) *Eur. J. Pharm. Biopharm.*, **88**, 443–454.
[100] Hagesaether, E., Bye, R., Sande, S.A. (2008) *Int. J. Pharm.*, **347**, 9–15.
[101] Pawar, D., Mangal, S., Goswami, R., Jaganathan, K.S. (2013) *Eur. J. Pharm. Biopharm.*, **85**, 550–559.
[102] Naderkhani, E., Erber, A., Skalko-Basnet, N., Flaten, G.E. (2014) *J. Pharm. Sci.*, **103**, 661–668.
[103] Andersen, T., Bleher, S., Flaten, G.E., *et al.* (2015) *Mar. Drugs*, **13**, 222–236.
[104] Adamczak, M.I., Hagesaether, E., Smistad, G., Hiorth, M. (2016) *Int. J. Pharm.*, **498**, 225–233.
[105] Madsen, K.D., Sander, C., Baldursdottir, S., *et al.* (2013) *Int. J. Pharm.*, **448**, 373–381.
[106] Fan, T., Chen, C., Guo, H., *et al.* (2014) *Eur. J. Pharm. Biopharm.*, **88**, 518–528.
[107] Olmsted, S.S., Padgett, J.L., Yudin, A.I., *et al.* (2001) *Biophysical J.*, **81**, 1930–1937.
[108] Fabiano, A., Zambito, Z., Bernkop-Schnürch, A. (2017) *Int. J. Pharm.*, **517**, 279–285.
[109] Suchaoin, W., de Sousa, I.P., Netsomboon, K., *et al.* (2016) *Int. J. Pharm.*, **510**, 255–262.
[110] Wang, X-Y., Koller, R., Wirth, M., Gabor, F. (2012) *Int. J. Pharm.*, **436**, 738–743.
[111] Pawar, V.K., Meher, J.G., Sing, Y., *et al.* (2014) *J. Contr. Rel.*, **196**, 168–186.
[112] Cassano, R., Ferrarelli, T., Schätzlein, A.G., *et al.* (2013) *Eur. J. Pharm. Biopharm.*, **84**, 540–548.
[113] Gupta, N., Al-Saikhan, F.I., Patel, B., *et al.* (2015) *Int. J. Pharm.*, **488**, 33–43.
[114] Reix, N., Parat, A., Seyfritz, E., *et al.* (2012) *Int. J. Pharm.*, **437**, 213–220.
[115] Han, H-K., Shin, H-J., Ha, D.H. (2012) *Eur. J. Pharm. Sci.*, **46**, 500–507.
[116] O'Neill, M.J., Guo, J., Byrne, C., *et al.* (2011) *Int. J. Pharm.*, **413**, 174–183.
[117] Song, Q., Yao, L., Huang, M., *et al.* (2012) *Biomaterials*, **33**, 6769–6782.
[118] Song, Q., Wang, X., Hu, Q., *et al.* (2013) *Int. J. Pharm.*, **445**, 58–68.
[119] Balguri, S.P., Adelli, G.R., Janga, K.Y., *et al.* (2017) *Int. J. Pharm.*, **529**, 32–43.

6

Cell–Nanoparticle Interactions: Toxicity and Safety Issues

Flavia Fontana[1], Nazanin Zanjanizadeh Ezazi, Nayab Tahir[1,2,3] and Helder A. Santos[1,4]

[1]*Drug Research Program, Division of Pharmaceutical Chemistry and Technology, Faculty of Pharmacy, University of Helsinki, Finland*
[2]*College of Pharmacy, University of Sargodha, Pakistan*
[3]*Faculty of Pharmacy and Alternative Medicine, The Islamia University of Bahawalpur, Pakistan*
[4]*Helsinki Institute of Life Science, HiLIFE, University of Helsinki, Finland*

6.1 Introduction

6.1.1 Role of Nanoparticles in Modern Medicine and Applications

In the past 20 years, the application of nanotechnology to medicine has brought advantages in the medical field, both as treatment and as diagnostic agents [1–3]. The term nanotechnology refers to the manipulation of materials at atomic and molecular scales (ranging from 1–100 nm) [4–6]. However, a wider definition of nanotechnology and nanoparticles concerns elements in the size range between 1 and 1000 nm [7]. Nanoparticles (NPs) display advantageous intrinsic properties different from the bulk material, due to their size (e.g., increased surface-to-volume ratio, increased surface energy, and quantum effects) [8–10]. As a result of these properties, nanomaterials are being intensively investigated for drug delivery and diagnostics, particularly for the treatment of cancer, and cardiovascular and autoimmune diseases [2, 6, 11, 12].

The administration of drug-loaded NPs results in multiple advantages, compared with a systemic administration: (i) the formulation of poorly water-soluble drugs in particles leads

Characterization of Pharmaceutical Nano- and Microsystems, First Edition. Edited by Leena Peltonen.

to an increase in their dissolution rate; (ii) the encapsulation of drugs within nanomaterials reduces the systemic diffusion of the drug, leading to lower systemic side-effects; and (iii) the particles protect the cargo from degradation. Moreover, by fine-tuning the properties of the nanocarrier, drugs can be passively targeted to the organ of interest, while the modification of the nanosystems with targeting moieties (e.g., antibodies, aptamers, and small molecules) enables the active targeting of tissues and cells of interest [9].

Several polymeric, lipidic, and inorganic (silicon-, silica-, carbon-, and metal-based) materials have been proposed as drug delivery systems (DDSs). Some nano-DDS have been used in the clinic for more than 20 years (e.g., Doxil, liposomal formulation of doxorubicin was approved by the FDA in 1995), with many more carriers evaluated in different phases of clinical trials [9, 10, 13]. Recently, the efficacy of the nanosystems has been questioned based on meta-analysis, showing that only 0.7% of the injected dose is effectively accumulating on the disease site (tumor), together with concerns about the industrial scale-up of new nanosystems [14, 15]. Moreover, the study of the characteristics of a nanosystem *per se* is not sufficient to predict its behavior in living organisms. The interactions between cells and nanomaterials need to be investigated and taken into account to gain the best advantage from the favorable properties of nanomaterials [16, 17].

6.1.2 Cell–NP Interactions

There are several properties of NPs that influence their interactions with cells. These properties are presented below and summarized in Figure 6.1.

6.1.2.1 Size

The interaction and internalization of nanoparticles within the cells is dependent on the particles' size, according to four different mechanisms: pinocytosis (for particles smaller than 10 nm), micropinocytosis (for particles smaller than 10 nm), clathrin/caveolar-mediated endocytosis (usually receptor-dependent pathway; used for uptake of NPs opsonized by the proteins of the blood, up to 200 nm, with an optimal size of 50 nm), and phagocytosis (for the internalization of particles bigger than 500 nm) [18]. In particular, it is possible to calculate the theoretical efficiency of NP uptake by receptor-mediated endocytosis, based on the spatial distribution of the receptor on the membrane and the enthalpic and entropic characteristics of the interaction between the nanoparticle and the cell membrane, leading to identification of the most suitable nanoparticle size for each receptor [19]. However, it is worth noting that the factors influencing the interaction of the particles with the cells most often cannot be reduced to a single factor, for example, the optimal size for an efficient uptake is different between particles made from different materials, and differs also according to the cell type investigated in the study [18].

6.1.2.2 Shape

Nanoparticles can present different shapes, from spherical, to quasi-hemispherical, to rod-like, to filamentous [20]. The different shapes lead to different uptake in cells

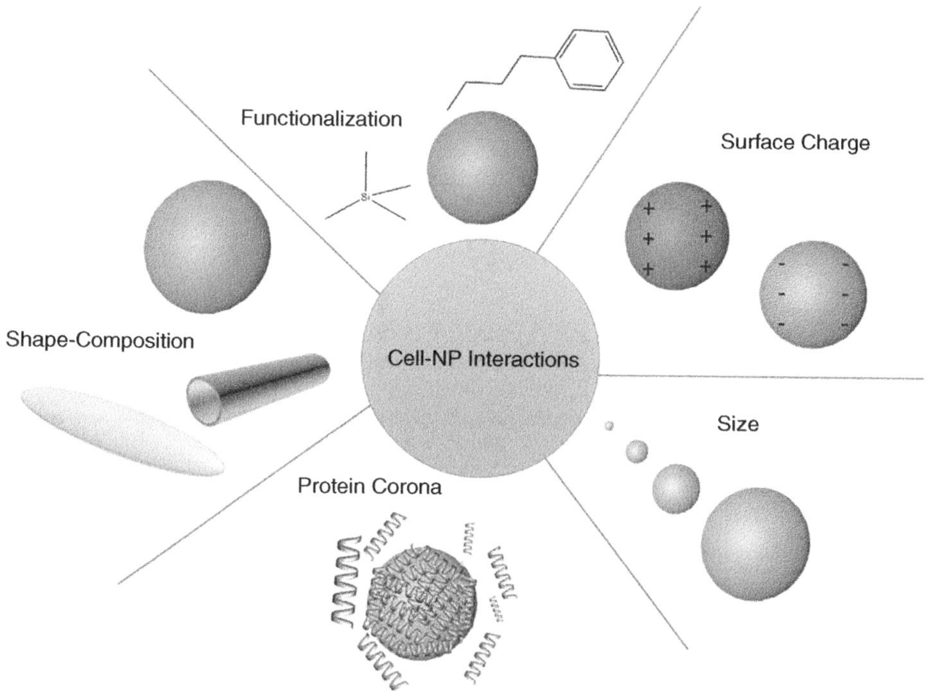

Figure 6.1 Factors influencing the interactions between NPs and cells. The main factors reported are: size of the nanosystem, surface charge, presence of functional groups on the surface, shape of the material, composition of the material, and composition of the protein corona formed upon incubation of the particles with biological media.

[21, 22]. Spherical particles are taken up more efficiently than rod-like or filamentous particles because they require less effort from the cells to wrap membrane around the NPs (Figure 6.2) [23]. In particular, in the case of elongated particles, like carbon nanotubes, failed uptake may result in frustrated endocytosis, especially by cells of the reticular endothelial system (RES), increasing the circulation time of the particles *in vivo* [24, 25].

6.1.2.3 Surface Charge

NPs can present negative, neutral or positive surface charge, depending on the characteristics of the material. Positively charged particles present a strong cell uptake, due to the formation of electrostatic interactions between the positively charged particles and the negatively charged cell membrane, which facilitates interactions with receptors and endocytosis [23, 26]. This unspecific interaction, however, leads also to increased toxicity of positively charged nanosystems compared with negatively charged ones [27]. Neutral and anionic particles, in contrast, exhibit prolonged circulation *in vivo*, with lower non-specific uptake and lower uptake from the immune cells of the RES [28].

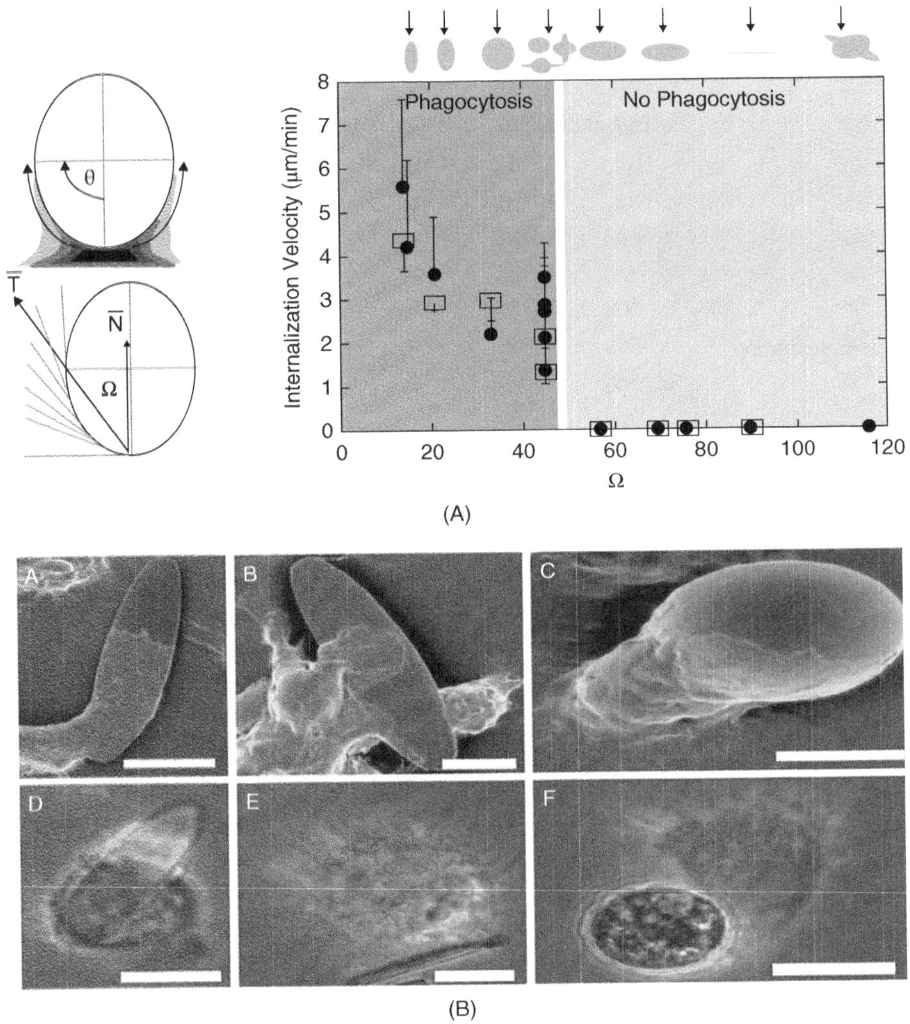

Figure 6.2 Influence of NP shape on the phagocytosis. (A) Internalization velocity of the particles related to the geometric properties of the particles (Ω). The more elongated particles exhibit low to no phagocytosis. (B) A–C, pictures of the interaction between cells and particles presenting different shapes taken with a scanning electron microscope (SEM); D–E, images of cells incubated with NPs: the cells have been stained for actin, to highlight the polymerization of actin, and the formation of actin rings. Reproduced with permission from [20]; copyright © 2011, Elsevier B.V.

6.1.2.4 *Surface Functionalization and Hydrophobicity*

The surface of nanomaterials is routinely functionalized with fluorescent or radioactive probes, or with targeting molecules. Often, in order to covalently bind these moieties, chemical modifications of the surface are necessary. The composition of the surface, its charge and hydrophobicity are all factors having an impact on the interactions between the particles and cells. In particular, Manshian *et al.* demonstrated a strong impact of NP hydrophobicity on cellular viability, as measured by damage to the plasma membrane and induction of autophagy [29].

6.1.2.5 *Protein Corona*

The interaction between a NP and a biological fluid (e.g., blood) will inevitably result in the formation of a protein corona adsorbed around the NP [30]. This corona influences the efficiency of the active targeting of NPs to the tissue of interest by masking the targeting moiety [31]. The characteristics of each NP (i.e., surface functionalization, charge, and composition) will result in the adsorption of different proteins, and in a NP-specific corona; the presence of distinct proteins and the glycosylation status of those proteins influences the entity of the interaction between the nanosystems and the cells [32]. In particular, as demonstrated by Ritz *et al.*, the presence of Apo H in the corona increases the cellular uptake, while the presence of Apo A4 or C3 results in a decrease in the uptake [33].

6.1.3 NP Toxicity

All the abovementioned factors influence the interaction between cells and NPs, potentially leading to cellular toxicity. In this chapter, we will first introduce the common mechanisms of cellular toxicity induced by NPs, followed by the assays employed to evaluate this toxicity. Then we will illustrate case-specific interactions between cells and several types of NPs, including metal oxides (zinc oxide, cerium oxide, and iron oxide), liposomes, polymeric, and silica-/silicon-based NPs. Table 6.1 summarizes the studies presented throughout the text.

6.2 Mechanisms of NP-Induced Cellular Toxicity

The interaction between cells and NPs influences the genomics, proteomics, and metabolomics of the cells, and often results in toxicity, leading to apoptosis, necrosis or autophagy. The main mechanisms of toxicity are given below.

6.2.1 Damage to the Plasma Membrane

The interaction of cationic NPs with the membrane induces the rapid formation of holes within the membrane that require a longer time to reseal, leading to influx of Ca^{2+} ions within the cells [76, 77]. NPs may also interact with the ion channels on the membrane, modifying their function [78]. Moreover, the shape of the nanosystem may lead directly to disruption in the cellular membrane, as in the case of fibrillar (needle-like) NPs that puncture the membrane [79].

6.2.2 Alterations or Disruptions in the Cytoskeleton

The uptake of NPs can directly induce modifications in the cytoskeleton, leading to reduced cellular proliferation (due to reduced cellular division) and motility [80, 81]. Furthermore, NPs like iron oxide can interact directly with the proteins of the cytoskeleton (e.g., actin), disturbing their action [78].

Table 6.1 *Overview of studies assessing the toxicity of different nanosystems, including the outcome, the toxic concentrations, and the cell type in which the systems were tested*

Particles	Application	System	Interaction outcome and toxic concentration	Cells	Ref.
ZnO	Direct seeding; secondary organs uptake	NPs	Toxic in 25 μg/ml and higher; 24 h treatment	HEK 293 cell	[34]
ZnO	Direct seeding; *in vitro*	NPs	Toxic higher than 8 μg/ml; 24 h treatment	TM-4, GC2-spd	[35]
ZnO	Direct seeding; neurobiology	NPs	Toxic decrease in mitochondrial and lysosomal activity in 20 μg/ml and higher; 24 h treatment	C6 Glial astrocytes	[36]
ZnO + UV	Direct seeding; *in vivo*; skin care products	NPs	Toxic higher than 1 μg/ml; toxic in combination with ZnO + UV in 1 μg/ml	Primary mouse keratinocytes (PMKs)	[37]
ZnO	Direct seeding; *in vitro*	Nanorods	Safe to 10 μg/ml	HeLa (CCL-2)	[38]
ZnO	Direct seeding; cancer	NPs	Toxic at 10 μg/cm^2	Human colon carcinoma cells (LoVo)	[39]
CeNP	Direct seeding; stem cell treatment in tissue engineering and regenerative medicine	NPs	Safe up to 320 μg/ml; 24 h treatment	Human dental pulp stem cells, hDPSCs	[40]
Biocomposite TMC-CeO_2	*In vitro* studies	Nanocomposite	Safe; only 14% reduction in 200 μg/ml	BRL-3A immortalized rat liver cells	[41]
Iron oxide	Direct seeding; bone tissue engineering	NPs	Promote differentiation of bone; safe up to 300 μg/ml	Human bone-derived mesenchymal stem cells (hBMSC)	[42]
Iron oxide; CEIO–PEG–FA	Cancer therapy	NPs	Safe up to 500 μg/ml	Normal human fibroblast	[43]

Iron oxide-Doxorubicin	Cancer therapy	NPs	Toxic	Human colon (HT29) cell	[44]
Dimercaptosuccinate (DMSA)-coated iron oxide	Neurobiology	Coated NPs	Concentration dependent; safe up to 450 μM	Microglial cells	[45]
DMSA-coated iron oxide	Neurobiology	Coated NPs	Safe	Cerebellar granule neuron	[46]
Cationic liposomes	*In vitro* studies; delivery of plasmids to fish	NP-DOPE/DOTAP	Toxic after 20 μg/ml	Macrophage-like cells	[47]
Cationic liposome	Nanomedicine	NP-DOTAP/CHOL	Toxic at 100 μg/ml and higher	Liver (HepG2) and lung (A549) cells	[48]
Temperature-responsive liposomes	siRNA delivery	NP-[P(NIPAAm-co-DMAPAAm)]	Safe; low toxicity	HeLa cells	[49]
PLGA	Anti-inflammatory and hepatoprotective	NPs	Toxicity from 100 μg/ml	HepG2 and A549	[50]
PLGA	Cancer therapy	NPs	High endocytic uptake; safe up to 500 μg/ml	Caco-2 and HT-29 cells	[51]
Carboxy and amine functionalized polystyrene	Cancer cell targeting and biomedical applications	NPs	Safe and high phagocytic uptake in the cells up to 100 μg/ml	Human macrophages and THP-1 cells	[52]
PLGA coated with PVA and vitamin E TPGS	Chemotherapy	NPs	Uptake and intracellular localization; safe at 100–500 μg/ml	Caco-2	[53]
Polystyrene	Oral chemotherapy and brain tumor imaging	NPs	Size-dependent uptake and surface binding	Caco-2 and MDCK cells	[54]
Chitosan grafted and chitosan hydrochloride	Gene delivery	NPs	Toxic, high cellular uptake and bio-distribution	Murine peritoneal macrophage, SMMC-7721, HEK 293, 786-O, HFL-I, and A549	[55]
Albumin	Cancer therapy	Quantum dots	Safe and highly internalized	HeLa cells	[56]

(*continued*)

Table 6.1 *(continued)*

Particles	Application	System	Interaction outcome and toxic concentration	Cells	Ref.
Polystyrene	Vaccine and DNA delivery	NPs	Adsorption on protein corona	Monkey kidney epithelial cells (BS-C-1)	[57]
PLGA and chitosan	Anti-inflammatory	NPs	Toxic for concentrations higher than 100 μg/ml	THP-1 macrophages	[58]
Gold coated with amphiphilic block polymer	Cancer therapy	NPs	High cytotoxicity due to intracellular translocation of the nanoparticles, toxic for concentrations higher than 20 nM	HUVECs, 3T3 fibroblast and C17.2 cells	[59]
2-dimethylamino ethyl methacrylic acid (DEAEMA)	Immunotherapy	Core–shell NPs	Toxicity and high cell internalization for concentrations higher than 25 μg/ml	Dendritic cells	[60]
Poly(methacrylic acid) (PMAA)	Cancer therapy	Hydrogel	Rapid and higher internalization up to concentrations of 200 μg/ml	HeLa cells	[61]
Poly(lactic acid)-poly(ethylene glycol)-poly(l-lysine) (PLA-PEG-PLL), PLA and PLA-PEG-PLA block polymers	siRNA delivery	NPs and NPs-hydrogel	Higher uptake due to interaction with cytoskeleton; safe up to 500 μg/ml	HepG2, HeLa cells and monocytic cells	[62]
PLGA	Cancer therapy	NPs	Enhanced cytotoxicity due to cell adhesion and interact with cell migration	Caco-2 cells	[63]
Polyamidiamine (PAMAM)	Protein and DNA delivery	Dendrimers	Adsorption at cell membrane	–	[64]

Polyamidiamine (PAMAM)	Cancer therapy	Dendrimers	Galectin-3 induced cellular aggregation	A549, DU-145, and HT-1080	[65]
Polyphenylene	Cancer therapy	Dendrimers	Cell surface interaction with different ligands	Human lung carcinoma cells	[66]
Poly amindoamine	Cancer therapy	Dendrimers	Cell membrane adsorption through lactate dehydrogenase and luciferase enzymatic activity	KB and Rat2 cells	[67]
Silica particles	Cancer therapy	NPs	Size-dependent increase in cytotoxicity and cell uptake; safe up to concentrations of 200 μg/ml	HepG2, mouse embryonic fibroblast (NIH/3T3) and A549 cells	[68]
Silica	Cancer therapy	NPs	Toxicity above concentrations of 741 μg/ml	A549, HT-29, and MKN-28 cells	[69]
Silica	Neuro-regenerative therapy	NPs	Neural toxicity above concentrations of 292 μg/ml	Neuronal cells (GT1-7)	[70]
Mesoporous silica	Blood disorders	NPs	Membrane deformities and hemolysis	RBCs	[71]
Silica	Cancer therapy	NPs	Enhanced uptake	Vascular endothelial (HUVEC) and cervix carcinoma (HeLa) cells	[72]
Mesoporous silica	—	NPs	Particle adhesion and cell internalization; safe at a concentration of 100 μg/ml	A375 cells	[73]
Porous silicon	Chemotherapy and immunotherapy	NPs	Immunotoxicity starting from concentrations higher than 200 μg/ml	B-cells, T-cells, U937 monocytes and RAW264.7 macrophages	[74]
Mesoporous silicon	Chemotherapy	NPs and nanocomposites	Safe and high cellular uptake for concentration of 100 μg/ml	MDA-MB-231 cells	[75]

6.2.3 Mitochondrial Toxicity

The direct interaction of NPs with the mitochondria may result in increased permeability of the mitochondrial membrane, with the generation of oxidative stress and indirect toxicity towards the whole cell [82].

6.2.4 Nuclear Damage

Only small NPs (smaller than 5 nm) can freely accumulate within the nucleus. Depending on the characteristics of the NPs (mainly surface charge), the particles can exhibit a high affinity for chromatin, interfering with cellular division [83].

6.2.5 Reactive Oxygen Species (ROS)

The interaction of most NPs with cells results in the formation of ROS due to the surface characteristics of the NPs (e.g., electronically active surfaces). The ROS generated induce oxidative stress which leads to damage within the mitochondria, and for longer incubation times, to cell death by necrosis [82, 84].

6.2.6 Interference in the Signaling Pathways

NPs have been reported to interact with different signaling pathways, such as mitogen-activated protein kinase, nuclear factor κB, bone morphogenic protein, and transforming growth factor β; these interactions may lead to toxic effects [78].

6.3 *In Vitro* Assays to Evaluate Cell–NP Interactions

The methods employed to evaluate preclinically the interactions between cells and NPs are detailed below and presented in Figure 6.3.

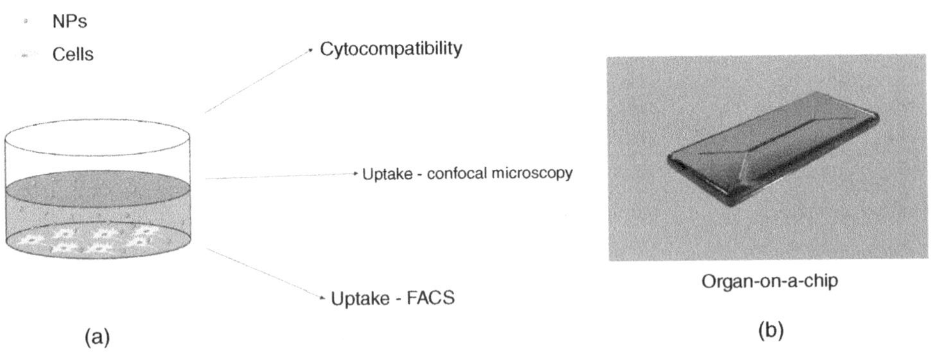

Figure 6.3 *In vitro assays for the evaluation of cell–NP interactions. (a) Traditional assays based on 2D cell culture. (b) Innovative assay based on organ-on-a-chip platforms.*

6.3.1 Traditional Assays

Usually, the first preclinical screening of newly developed nanomaterials is performed in 2D cell cultures. Cellular viability and cell proliferation are evaluated through multiple assays [85]: methods based on mitochondrial activity (like tetrazolium reduction and resazurin reduction); methods based on protease activity; methods based on assessment of the ATP concentration; and methods based on membrane integrity. The modifications in the cellular morphology are usually evaluated by SEM, while the cellular uptake of NPs is evaluated qualitatively by confocal microscopy and quantitatively by fluorescence-activated cell sorting (FACS) [73, 86].

6.3.2 Innovative Assays

Recently, 3D cell models (spheroids) have been introduced in the evaluation of the cellular uptake of NPs, particularly in cancer research [87, 88]. Moreover, organs-on-a-chip have been developed to provide a suitable intermediate platform between the *in vitro* and the *in vivo*. These platforms partly reproduce on a chip the physiology of a selected organ (e.g., lung), allowing a more precise evaluation of the interactions between the tissue and NPs [89]. Finally, *in silico* models for cell–NP interactions and for the behavior of the NPs within the human body are being developed [90, 91].

6.4 Metal Oxide Nanoparticles

Several different metal oxide particles have been studied in recent years as novel engineered platforms for biomedical applications, such as drug delivery and cancer therapy, due to their unique physicochemical properties [92, 93]. They have shown to have antimicrobial behavior [92, 94], nontoxicity and high stability [94], but also to induce ROS production and oxidative stress [95]. Although there are numerous studies describing the interactions between each metal oxide NP and different cells, usually there is no indication about whether the interaction assessed is safe or toxic for the cells. In cell–NP interactions there are many different factors that could be involved in toxicity or compatibility of each metal oxide, such as shape, size, dose of particles, temperature, cell type, organ function, and application [96]. In the next sections, we will review the most common metal oxide particles, together with their interactions with different cell types.

6.4.1 Zinc Oxide

Maybe the most studied metal oxide is zinc oxide (ZnO) due to its strong oxidation behavior and high chemical activity [34]. In addition, this semiconductor has exceptional electronic and optoelectronic properties at nanoscale, which makes it useful in a variety of applications, such as biosensors, sunscreens, makeup, nail products [95], dental filling, or even wound healing [38] and therapeutic intervention [95]. One of the most important properties of ZnO is its antibacterial nature [92, 94, 97]. ZnO NPs can diffuse inside the bacteria and induce toxicity by releasing ROS and Zn^{2+} ions. The released Zn^{2+} ions in growth media can be absorbed on the cell membrane, resulting in mechanical damage to the wall [97]. In general, in the nanoscale, particles can be endocytosed in the cells, damage organelles, such as mitochondria, and release ROS, causing inflammation and cell death. Figure 6.4 illustrates a schematic of immunotoxicity displayed by ZnO NPs [95].

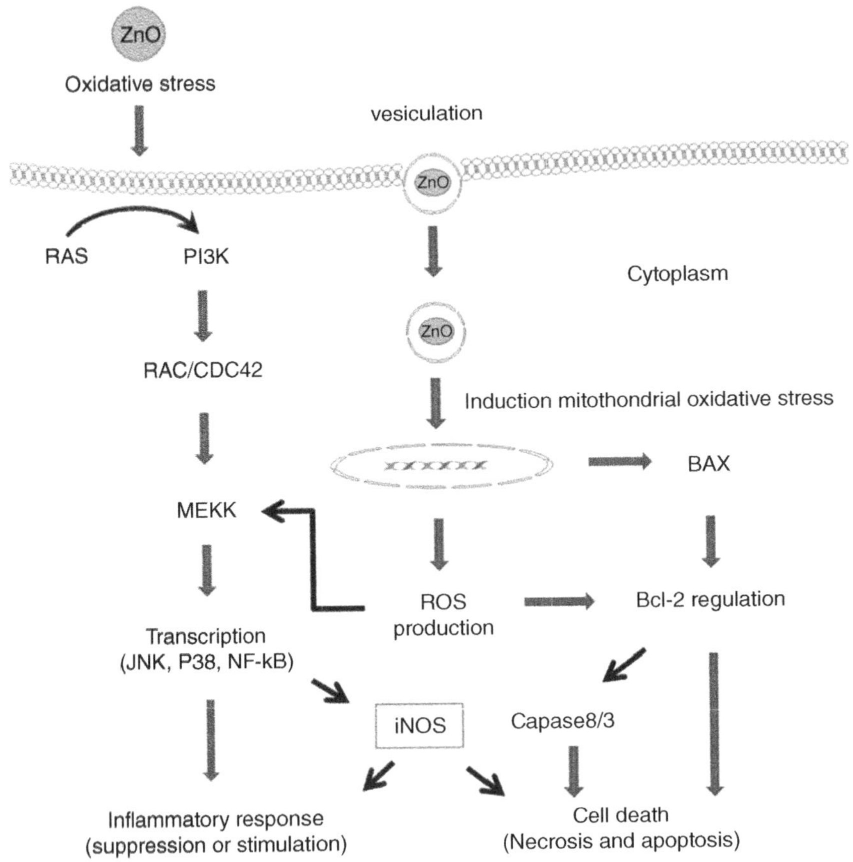

Figure 6.4 Mechanism of ZnO NP immunotoxicity. The release of Zn ions induces the production of ROS, resulting in cell death and inflammation. ZnO NPs enter in the cells and in the mitochondria according to two separate endocytosing mechanisms, produces ROS and cytochrome c (which binds to apoptotic protease-activating factor 1 impact caspase 3 and causes cell death). ROS release also starts the activation of the mitogen-activated protein kinase (MAPK) pathway that activate the transcription factors Jun N-terminal kinase (JNK), p38 MAPK and nuclear factor kappa-light-chain-enhancer of activated B cells (NF-κβ), causing inflammation. Reproduced with permission from [95]; copyright © 2014, the Korean Society of Toxicogenomics and Toxicoproteomics, and Springer.

Condello *et al.* have investigate the ZnO NP toxicological pathway within human colon carcinoma cells (LoVo) by detecting the release of Zn^{2+} ions using Zinquin fluorescent dye. They have considered two possible mechanisms for entering the cells, passive diffusion or

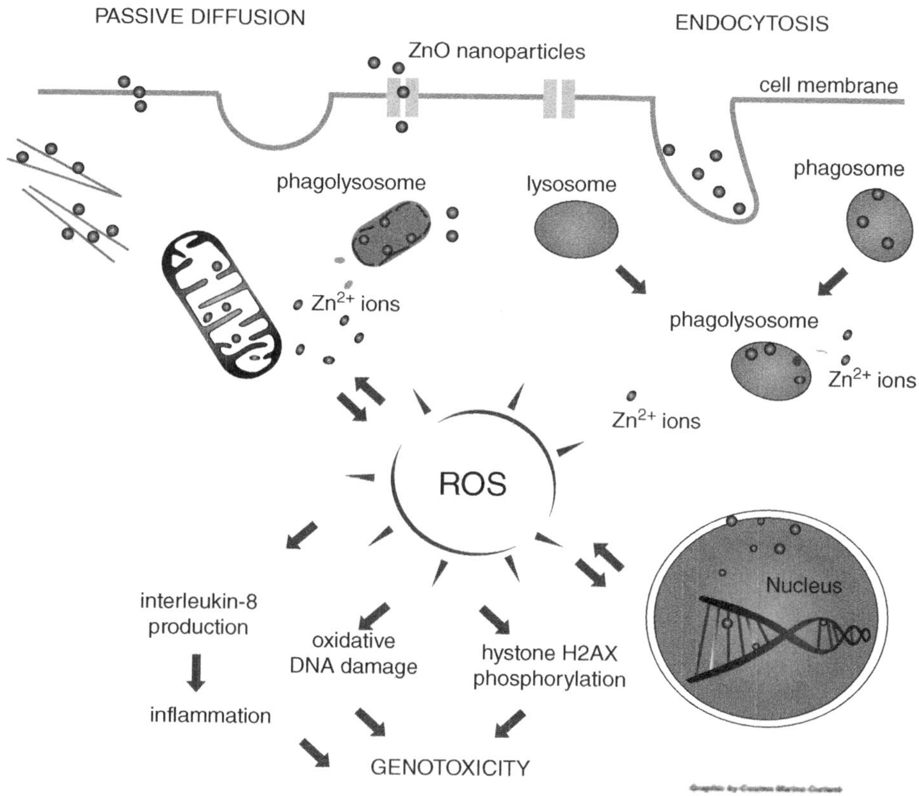

Figure 6.5 Schematic of the different toxicity mechanisms of ZnO NPs, according to the different uptake pathways (passive diffusion or endocytosis). Reproduced with permission from [39]; copyright © 2016, Elsevier B.V.

endocytosis (Figure 6.5), which cause oxidative stress and DNA damage within 6 h. In the endocytosis process, Zn NPs bind to the membrane and are swallowed into the cells with the creation of a phagosome compartment. The phagosome then merges with lysosomes, creates the phagolysosome characterized by lower pH, which induces the release of Zn^{2+} ions, causing ROS production, apoptosis, and DNA fragmentation. In passive diffusion, the particles cross the cell membrane (within 30 min of the treatment) in an energy-free fashion, reach the mitochondria and nucleus, and induce toxicity in LoVo cells [39, 98].

Studying the interactions between kidney cells and metal oxide nanosystems is essential, in cases where kidneys represent the targeted or secondary targeted organ. The secondary organs are still able to take up the particles: kidney is one of the most important of them, since it represents one of the organs involved in NP excretion from the body [34]. Reshma and Mohanan investigated the viability of HEK-293 cells in contact with ZnO NPs by 3-(4,5-dimethylthiazol-2-yl)-2,5-diphenyltetrazolium bromide) tetrazolium (MTT) assay, neutral red assays (for lysosomal activity) and live/dead assay, and they studied the cell morphology changes and ROS production. The morphology showed dramatic changes to a round shape, starting from interactions with particles at a concentration of 75 μg/ml after

24 h of exposure. Furthermore, they found a toxic concentration of NPs responsible for kidney cell death starting from 25 μg/ml. The mechanism of toxicity implies diffusion to the nucleus, causing condensation of nuclease and production of ROS. This influences the mitochondrial and lysosomal activity due to Zn^{2+} ion release [34].

The toxicity of ZnO NPs was also reported by Liu *et al.* where they demonstrated the toxicity of particles *in vitro* from a concentration of 8 μg/ml for mouse Sertoli cell line (TM-4) and spermatocyte cell line (GC2-spd, male reproductive cells) [35]. They reported ROS production causing the downregulation of the expression of BTB proteins, weakening the integrity of the blood–testis barrier and initiating DNA damage in germ cells.

The bio-interaction of ZnO NPs with rat C6 glial astrocytes in brain showed uptake of the particles by these cells. After 3 to 6 h, NPs produce a strong oxidative stress in cells and induced apoptosis, damaging the astrocytes [36]. After 24 h of incubation, 40 and 80 μg/ml of particles decreased the cell viability, while 80 μg/ml of the ZnO NPs increased the ROS to 53% after 3 h of incubation. However, increasing the concentration of the particles led to a decrease in the ROS production due to a scavenging effect on free radicals in glial cells. Moreover, under UV light exposure, the ZnO NPs, which are routinely used in sunscreen, induce higher ROS generation, with skin cell death.

As Pala *et al.* reported, the viability of primary mouse keratinocytes (PMKs) decreased after incubation with ZnO NPs. *In vivo* studies on the skin of SKH-1 hairless mice showed DNA damage and cell death after exposure to ZnO NPs and UV, evaluated by immunostaining with 8-hydroxy-2'-deoxyguanosine and terminal deoxynucleotidyl transferase dUTP nick end labeling (TUNEL) [37]. However, Gopikrishnan *et al.* showed a good biocompatibility of ZnO nanorods, synthesized from zinc acetate and hexamethylenetetramine, when in contact with HeLa cells (CCL-2), even at high concentrations (10 μg/ml), showing a low oxidative stress [38].

6.4.2 Cerium Oxide

Nano cerium oxide (CeO_2) particles (nanoceria) are used in applications such as bone, cardiac, and neural tissue engineering [40], anticancer and anti-inflammation in biomedicine [99]. This material is used as an antioxidant to decrease cellular aging, due to its redox properties based on oxidation states of Ce^{3+} and Ce^{4+}. The surface activity of this metal oxide enables functionalization with different molecules [41, 99].

Mahapatra *et al.* have investigated the compatibility of different type of morphologies of CeO_2 to stem cells from human dental pulp, reporting the importance of CeO_2 NPs in regenerative medicine. CeO_2 NPs are supportive for the cell migration together with high compatibility. The intracellular uptake lowered the ROS and protected the DNA and other components [40]. They have shown the internalization of cubic NPs (CeN), small-sized nanorods (CeS), medium-sized nanorods (CeM), and highly elongated nanowires (CeW) into the stem cells by fluorescein isothiocyanate (FITC)-fluorescence images, as presented in Figure 6.6. CeW particles had lower uptake and time-dependence compared with the rest of the samples due to their morphology and the oxidation status of Ce^{3+} [40].

CeO_2 NPs showed dose-dependent toxicity for concentrations higher than 50 μg/ml towards BRL-3A immortalized rat liver cells. However, these particles can be combined with other biomaterials to improve the cell interaction. For example, Mohammad *et al.*

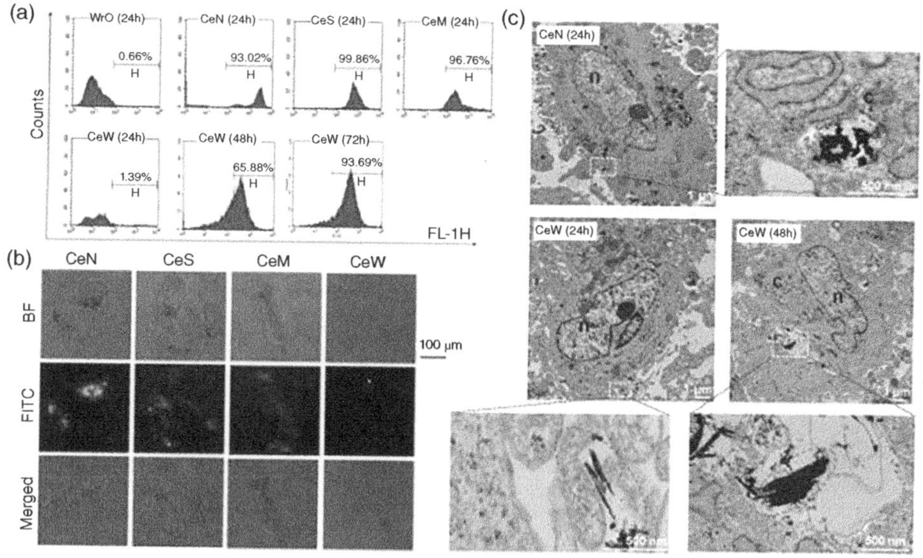

Figure 6.6 Cellular uptake of CeO_2 nanomaterials. (a) Flow cytometric analysis of the uptaken fluorescent particles; CeW samples have been analyzed for 48 and 72 h, indicating higher internalization with culture time. (b) Confocal microscopic images of cells at 24 h, displaying the internalization of the particles (light grey signal). (c) TEM images confirming CeN and CeW internalization at 24 and 48 h. Reproduced with permission from [40], copyright © 2017, Elsevier B.V.

demonstrated that the biocomposite N,N,N-trimethyl chitosan/CeO_2 (TMC-CeO_2) can improve the viability even at concentrations of 250 μg/ml, while protecting the cells from oxidation [41].

6.4.3 Iron Oxide

Iron oxides, such as magnetite (Fe_3O_4), maghemite (γ-Fe_2O_3), and hematite (α-Fe_2O_3), are biocompatible ferromagnetic materials with superparamagnetism properties, which can be controlled by an external magnetic field, and are being used in imaging, biosensors, cancer therapy, drug and gene delivery [100], and neurobiology [45]. Recently, iron oxide have been employed in bone tissue engineering studies where it has been proved that it can promote the osteogenic differentiation of human bone-derived mesenchymal stem cells (hBMSC) after intercellular uptake and activation of the mitogen activated protein kinase (MAPK) signal pathway [42]. As the concentration increased up to 300 μg/ml, the alkaline phosphatase (ALP) activity increased, and the quantitative and qualitative alizarin red assay also showed higher mineralization [42].

Iron oxide NPs can be combined with other materials in engineered systems for biomedical applications. Carbon encapsulated iron oxide NPs (CEIO-NPs) have been produced by the carbon arc method and further modified with poly(ethylene glycol)-folic acid (PEG-FA) for detecting cancer cells without displaying any toxic effect to healthy fibroblast cells

[43]. Moreover, these particles have been conjugated with the anticancer drug doxorubicin to evaluate the antiproliferative effect in human colon HT29 cancer cells, and showed an improved cellular uptake, resulting in toxicity to the cancer cells [44]. These particles have largely been studied for brain applications and also neurobiology drug delivery [45]. Luther *et al.* have investigated the endocytic uptake pathway of iron oxide nanoparticles (IONPs) by microglial cells, the immune cells of the brain. Dimercaptosuccinate-coated IONPs did not show toxicity up to 450 μM. As shown in Figure 6.7, iron oxide NPs (stained by Perls'

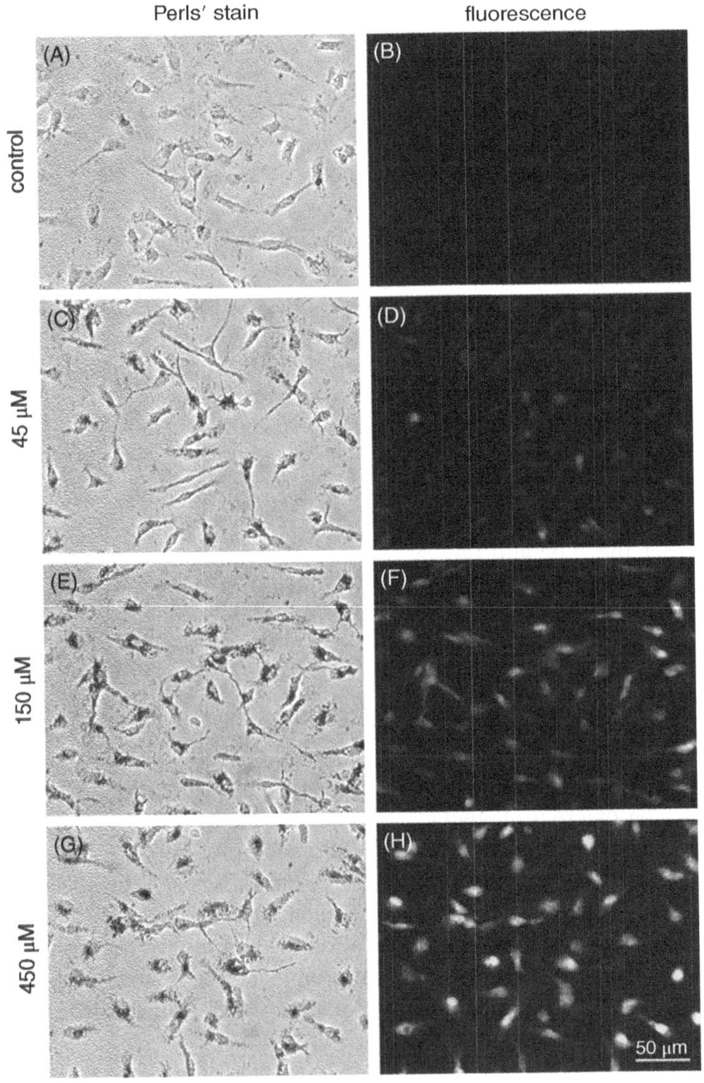

Figure 6.7 *Concentration-dependent uptake of IONPs by microglial cells shown using Perls' staining and fluorescence. Reproduced with permission from [45]; copyright © 2013, Elsevier B.V.*

and fluorescence) prove the higher uptake of NPs at 450 μM compared with lower concentrations and control samples, indicating the concentration dependency in the uptake process for IONPs. In addition, they have shown that 500 nmol iron/mg protein represent the maximum cytocompatible concentration after accumulation. The effects of this accumulation have also been tested on cerebellar granule neuron for 6 h and no sign of toxicity was detected, probably due to the antioxidative behavior of the particles in cerebellar granule neurons. Furthermore, the release of iron took place slowly compared with microglial cells, where the particles migrated to lysosomes and started to degrade due to the low pH [45, 46].

6.5 Non-metallic Nanoparticles

6.5.1 Liposomes

Liposomes are self-assembled colloidal phospholipidic systems, usually made of lecithin, cholesterol, and phosphatidylserine [101]. Liposomes have a dimension range between 20 nm and 2 μm, and present hydrophilic and hydrophobic properties (amphipathic), which make them useful in targeted and stimuli-responsive drug delivery vehicles [102]. They can be artificially produced as unilamellar or multilamellar lipid bilayers, surrounding aqueous compartments, as shown in Figure 6.8 [102–104], which help protect the drug stability and enhance its solubility [102]. Their unique properties, such as biodegradability and biocompatibility [103], render them useful in different applications in drug/gene delivery, biology or food industry, and antibacterial applications, such as ofloxacin antibiotic loaded liposomes [101]. Liposomes are able to enhance the drug concentration at the site of interest, increasing the accumulation in McCoy fibroblast cells compared with free drugs, while decreasing the toxicity of antimicrobial agents [101].

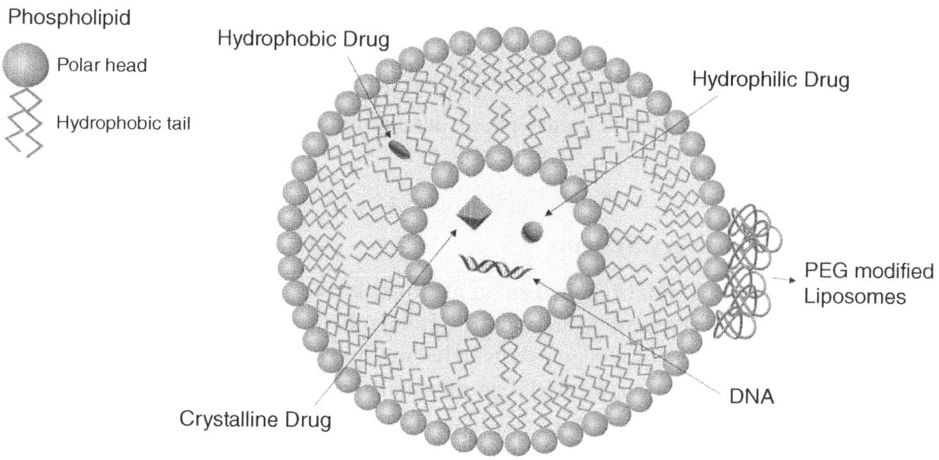

Figure 6.8 *Schematic structure of a liposome. Hydrophilic drugs, DNA or drug crystals can be loaded in the aqueous compartment, while hydrophobic drugs sit in the phospholipidic double layer that constitutes the membrane. Moreover, liposomes can be modified with PEG to prolong the circulation time.*

In addition, surfaces of liposomes can easily be modified with different molecules to prolong the blood circulation time or to avoid recognition by immune cells, such as pegylated liposome systems [48, 102]. These liposomes are commonly prepared by drying lipids from organic solvents, then dispersing them in the aqueous phase, followed by further purifications [103]. Roursgaard *et al.* compared poly(methylmethacrylate)-b-poly (aminoethyl methacrylate)-b-poly(ethyleneglycol) (PMMA-b-PAEMA-b-PEG) micelles with non-PEGylated liposomes made of N-[1-(2,3-dioleyl)propyl]-N,N,Ntrimethylammonium-chloride (DOTAP) and cholesterol [48]. They reported that the non-PEGylated liposomes are more toxic than the PEGylated micelles to liver (HepG2) and lung (A549) cells, starting from concentrations of 100 μg/ml and higher. The mechanism of toxicity was due to impaired mitochondrial function, producing oxidative stress caused by lipid oxidation.

Liposomes presenting different sizes ranging from 50 to 800 nm in diameter and different charges (positive, negative and neutral structures) have been studied to investigate their toxicity on different cell lines *in vitro* [105–107]. Epstein-Barash *et al.* showed that in the best mononuclearphagocyte system (MPS) cell internalization takes place with small and negatively charged systems, such as 85 nm negatively charged liposomes composed of 1,2-distearoyl-sn-glycero-3-phosphocholine (DSPC), distearoylphosphatidylglycerol (DSPG), and cholesterol (DSPC/DSPG/CHOL, molar ratio 3:1:2) [105]. Phosphatidylserine and phosphatidylglycerol, given their negative charge, also showed good internalization.

Cationic liposomes are used in different applications, such as gene delivery, due to their low immunogenicity [107]. In addition, the electrostatic interactions between positively charged liposomes and the negatively charged cell membranes can help the cell to take up the drug delivery system [105, 108]. However, the large and positively charged systems showed toxicity and apoptosis [105, 107]. Romøren *et al.* proved that macrophage-like cells had dose-dependent toxicity due to membrane activity of cationic lipids, resulting in malfunction and destruction of the integrity of cells [47]. Wang *et al.* improved the liposome nanocarriers described previously to enhance short interfering RNA (siRNA) delivery by developing [P(NIPAAm-co-DMAPAAm)] temperature-responsive liposomes. The modified system showed low cytotoxicity at 37°C to HeLa cells, while inducing a high gene silencing activity due to higher cellular uptake, depending on microtubule-dependent pathway [49].

6.5.2 Polymeric Delivery Systems

Polymers are long, straight or branched chain macromolecules formed by the covalent linkage of repeating monomer units from natural, semisynthetic, and synthetic sources [109]. These polymeric materials have been extensively employed as drug delivery carriers, either individually or in combination, to modulate the final characteristics of the desired carrier system. These nanomaterials are also getting attention in the food industry, agriculture and in biomedicine. However, the extensive application elicited the need to understand the unidentified and unpredicted effects on the living organism [110]. Molecular and cellular interactions of these nanomaterials and nanocarriers with the biological systems under normal physiology and in different pathological circumstances determined these unwanted effects.

The cellular interaction of these carriers might be responsible for a variety of biological responses. These responses include the production of abnormal inflammatory markers, abnormal cell migration, cell division, and cell differentiation [111, 112]. Co-enzyme Q10-loaded PLGA NPs have been prepared using an emulsion–diffusion–evaporation technique by Swarnakar *et al.* [50]. These particles were tested for stability, cellular uptake, subcellular localization, anti-inflammatory and hepatoprotective activity. It was found that the optimized NPs were stable for almost 6 months, and were taken up by the most reactive species generative organs of the body. However, the co-enzyme-loaded NPs were found to be 10 times more toxic than the free co-enzyme, due to the deleterious response produced by the free reactive moieties. Similarly, the non-invasive imaging of the hepatocytes indicate a 10-fold higher effect owing to higher accumulation of the NPs in the hepatocytes, and intracellular localization studies indicated that more than 30% NPs were present in the mitochondria.

The cell–NP interactions of PLGA NPs capped with didodecyl dimethyl ammonium bromide (DMAB) in epithelial colorectal adenocarcinoma (Caco-2) and human colorectal adenocarcinoma cells (HT-29) was studied [51]. The Deserno's model indicated size-dependent cell internalization in both cell lines. Below 100 nm the cell uptake followed the receptor mediated endocytic mechanism, whereas larger particles might involve the surface potential induced by various electrostatic forces. Win and Feng [53] demonstrated the cellular uptake and localization of PLGA NPs modified by polyvinyl alcohol (PVA) and d-α-tocopheryl polyethylene glycol 1000 succinate (TPGS) coating, using the polystyrene NPs as standard. The TPGS coated PLGA-NPs showed 1.4 times higher cell uptake compared with the PVA modified PLGA-NPs, whereas it was 4–6 fold higher compared with the standard polystyrene NPs. In addition to increasing the absorption, vitamin E TPGS inhibited the P-gp mediated efflux of the NPs, which resulted in high particle uptake. Moreover, the particle size also influenced the uptake mechanism. It was found that the small-sized (100 nm) NPs exhibited 1.3- and 1.8-fold higher uptake, compared with NPs having sizes of 500 and 1000 nm, respectively. These small-sized NPs had better interfacial interaction with cell membranes, whereas the larger particles were taken up by a fluid phase pericytosis process [53].

Cell–NP interactions using the same-sized carboxy (-COOH) and amine (-NH_2) functionalized polystyrene (PS) NPs were described by Lunov *et al.* [52] in human macrophages, and in undifferentiated and poly (methacrylic acid) (PMA) differentiated monocytic THP-1 cells. All the prepared formulations were considered safe up to concentrations of 100 μg/ml. Despite the changes in the mechanism, the composition of the medium affected the degree of internalization. Macrophages showed four times more cell uptake of negatively charged NPs in a serum-containing medium. NP interactions with the CD64 receptor in the serum were responsible for the phagocytic uptake of the NPs in macrophages, whereas the internalization of amine-modified PS NPs followed the dynamin II-dependent endocytosis process THP-1 cell (Figure 6.9). *In vivo* administration of these particles strengthen the *in vitro* results, showing that the carboxy-functionalized PS NPs were primarily located in the liver, the major site for the macrophages, and the amine-modified NPs target the tumor cells. Similar results were demonstrated by Kulkarni and Feng for the efficiency of cellular uptake of fluorescent PS NPs of different sizes (20, 50, 100 and 200 nm) in Caco-2 and MDCK cells [54].

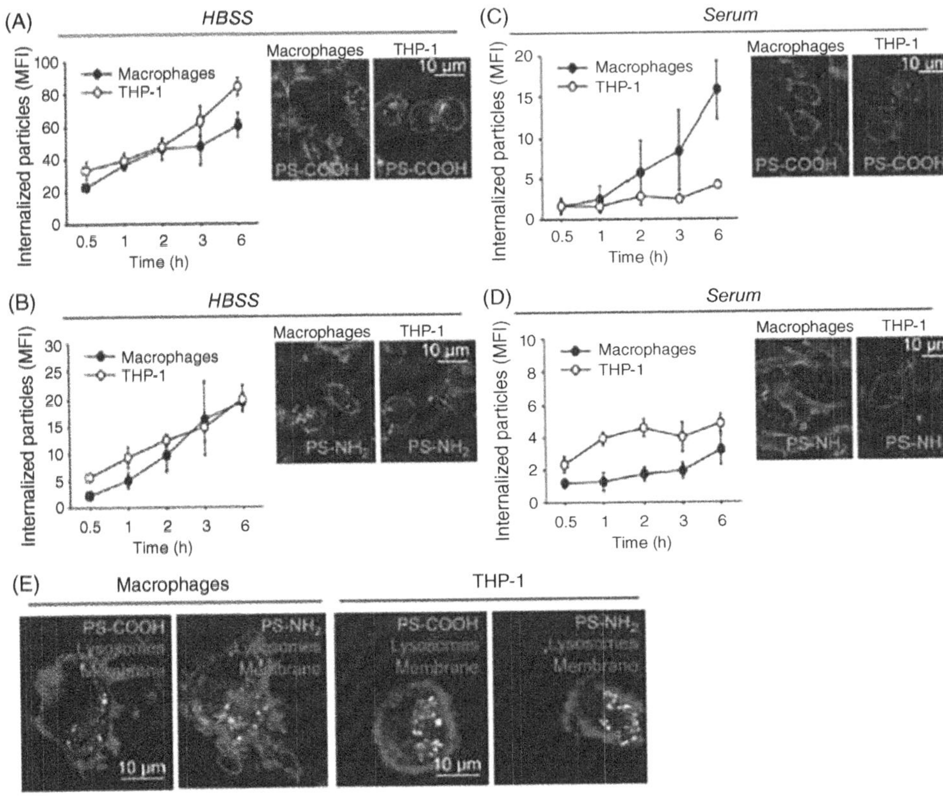

Figure 6.9 Cell uptake mechanism of carboxy- and amine-modified PS-NPs. Carboxy-modified PS-NPs and amine-modified PS-NPs in HBSS (A and B), and in serum (C and D). Accumulation in macrophage lysosomes (E) and THP-1 lysosome (F). Reproduced with permission from [52]; copyright © 2011, American Chemical Society.

He *et al.* prepared carboxymethyl grafted chitosan and chitosan hydrochloride grafted NPs with positive and negative charges to visualize the biodistribution and cellular internalization [55]. Results indicated a rapid clearance of larger NPs (>300 nm), but higher accumulation, biodistribution and enhanced cytotoxic effects of smaller sizes (<150 nm) and negatively charged NPs.

Concerning the effect of the protein corona on NPs, Treuel *et al.* formulated human serum albumin-modified quantum dots coated with dihydrolipoic acid [56]. The modification of the particles was responsible for the addition of some carboxyl groups, which resulted in a three-fold decrease in the binding affinity of the NPs in HeLa cells. The mechanistic study indicated that the uptake predominantly followed an active, pinocytic mechanism and the quantum dots were localized in the endosome and lysosomes.

Fleischer and Payne investigated the effects of protein–NP interactions at the cellular level by using amine- and carboxyl-modified PS NP with surface albumin as model carriers [57]. The binding assay showed greater binding of cationic NPs compared with the anionic counterpart. A competitive assay elucidated that cationic NP complexes bound to the scavenger receptor, while anionic complexes bound to the albumin receptor. Fluorescence microscopy and circular dichroism spectroscopy identified that the difference in binding properties was as a result of the changes in the structure of serum albumin after complexation with differently charged PS NPs.

Similarly, Grabowski *et al.* prepared the positively, negatively and neutral NPs with PLGA, chitosan, and Poloxamer 188 [58]. These particles were investigated for toxicity in human-like THP-1 macrophage cells in term of production of the ROS, mitochondrial activity, initiation and promotion of the apoptosis, and other anti-inflammatory pathways. The particles without stabilizer and at lower concentration (up to 0.1 mg/ml) showed no sign of toxicity, but a significant increase in cytotoxicity was observed at concentration $\geq$1 mg/ml. The extent of cell uptake was highest among the PLGA-poloxamer, but fastest in stabilizer-free PLGA NPs.

Hühn *et al.* studied the effect of charge on cellular uptake and cytotoxicity in HUVECs, 3T3 fibroblast and C17.2 cells, by cell viability, change in morphology, and oxidative stress generation studies [59]. The NPs were coated with an amphiphilic block polymer with negatively and positively charged hydrophobic side-chains. The results indicated that cationic NPs were more cytotoxic with compared anionic NPs, because of the lower colloidal stability of anionic NPs.

Hu *et al.* developed core–shell NPs, with a central core of 2-dimethylamino ethyl methacrylic acid (DEAEMA) linked with polyethylene glycol dimethylacrylic acid and 2-aminoethyl methacrylic acid shell, that showed a 2.8 times increase in particle size by changing the pH from 7.4 to 5.5 [60]. The endosomal escape and disruption was studied in the dendritic cells (CD2.4) by confocal fluorescence imaging with transmission electron microscopy (TEM) analysis. The pH-sensitive fluorescent indicator LysoTracker Red DND-99 was added to visualize the endo/lyososomal components of the CD2.4 cells (Figure 6.10). The DEAEMA particles showed punctate fluorescence, while the PDEAEMA NPs showed diffuse fluorescence inside the cell.

Xue *et al.* fabricated intracellulary degradable PMAA hydrogel NPs to evaluate the effect of the nanoparticle shape on cellular uptake, cytotoxicity, and elimination of the metabolites from the body [61]. The spherical particles had 12% higher uptake compared with cubic hydrogel particles after 10 h incubation with HeLa cells. The membrane adhesion process significantly influenced the uptake of the particles. However, the rapid degradation of the particles in high intracellular concentration of glutathion (GSH) (5 mM) resulted in the formation of low molecular weight fragments, facilitating rapid clearance from the body, which made the particle noncytotoxic.

Liu *et al.* investigated the effect of stiffness of the fabricated NPs and NP-hydrogel on the cell functions and cell uptake [62]. Hydrogel NPs composed of poly(2-hydroxyethyl methacrylate) (HEMA), using N,N-methylene-bis-acrylamide (BIS) as crosslinker in four different ratios, were prepared by an emulsion-polymerization technique. The softer NPs were taken up by HepG2 cells at a greater rate and extent than the stiffer NPs. Cellular uptake studies also indicated that softer particles interact with the cytoskeleton, while the stiffer NPs affected the adhesive ability of the cells.

Chernenko *et al.* investigated the degradation products and the accumulation of polymeric NPs in the cell using Raman microscopic analysis [113]. Ester linkage hydrolysis resulted in the degradation of these polymers. It was also evident from the study by Katsikari *et al.* that the degradation product of PLGA contributed to the cytotoxicity effect

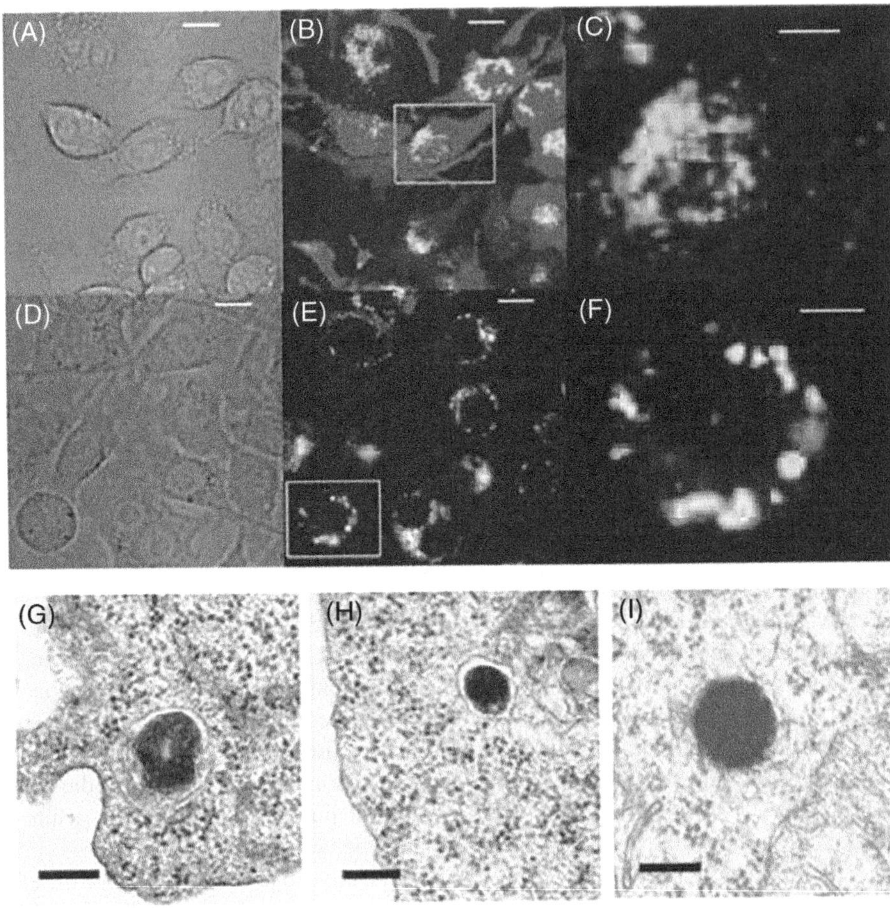

Figure 6.10 *(A–F) Endolysosomal escape of the core–shell NPs in the dendritic cells. (G–I) Cell cross-sectional images showing the internalized NP in the membrane bound lysosomal chamber and cytosolic compartments. Reproduced with permission from [60]; copyright © 2007, American Chemical Society.*

in the Caco-2 cells by influencing the cell adhesion and migration process [63, 114]. Dimchevska *et al.* fabricated biodegradable poly(D,L-lactide-co-glycolide)-poly(ethylene oxide)-poly(D,L-lactide-co-glycolide) (PLGA-PEO-PLGA) NPs and evaluated the cell internalization, gene expression and nano-biointerface interactions in human colon adenocarcinoma SW-480 cells [115]. The optimized formulations showed no significant influence on cell viability. However, the concentration of the structural components affected the interaction of these NPs with the cell membranes and degree of internalization. Gene expression analysis revealed that SN-38-loaded NPs increased the expression of the ubiquitin D gene by 18 times (Figure 6.11), which activated the caspase mediated apoptosis process.

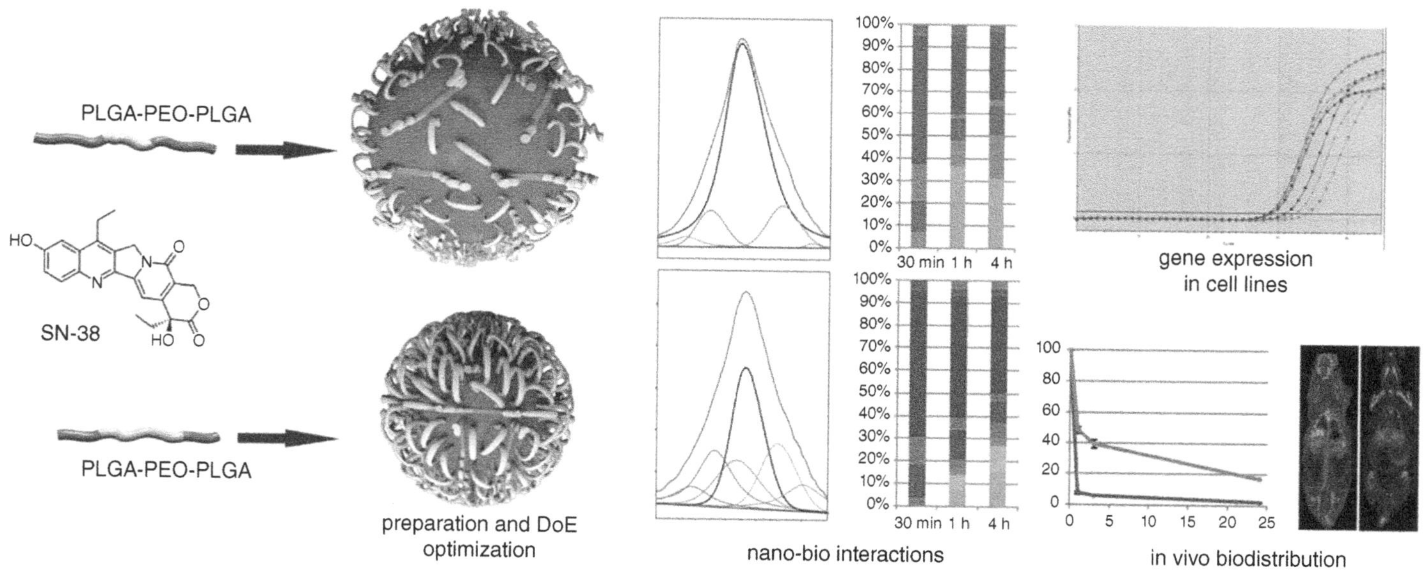

Figure 6.11 The preparation, optimization and nano-biointeractions of the NPs, showing the gene expression and in vivo biodistribution. Reproduced with permission from [115]; copyright © 2017, Elsevier B.V.

6.5.3 Dendrimers

Dendrimers are polymeric globular three-dimensional topologies with extensively branched and well organized nanometric macromolecular architecture with very low polydispersity index. The central core with tree-like branched arrangements derived its name from the Greek word "dendron", meaning tree. These radial branches, or generations, act as a source of modification for different functional moieties [116]. These multiple branches and large number of generations offer multiple interactional sites, and host guest complexations utilized for a wide range of applications. Various polymers, such as polyamidoamine (PAMAM), poly(propylene imine) (PPI), poly-L-lysine, melamine, poly(etherhydroxylamine) (PEHAM), poly(esteramine) (PEA) and polyglycerol have been extensively employed in the development of the dendritic platforms [117]. Dendrimers are so far not yet employed in biomedical applications; however, the possibility of high drug entrapment in the intramolecular central core position, and the ability to interact with the cells via the endoreceptor and exoreceptor interactions with the dendritic branches, provide an excellent alternative to polymeric delivery systems [118].

The most fascinating aspect of dendrimers, their dendrites, correlates with cellular internalization activity and effectivity in biological systems. These interactions are based on the size of the dendrimers, number and density of the generations present, and the terminal functional groups [119]. Otremba and Ravoo prepared polyamidiamine (PAMAM) dendrimers modified with boronic acid terminal groups and investigated the interaction with the biological membranes by isothermal titration calorimetry and optical density measurements, using thecatechole-adamantane decorated cyclodextrin vesicles as a model [64]. It was suggested that the vesicle aggregation followed the host (vesicles)/guest (dendrimers) complexation process, depending on the concentration of functional group and number of generations.

The oppositely charged dendrimers interact with biological membranes, cell organelles, and proteins. The mechanism of these interactions involves an increase in permeability, and release of catalytic enzymes, such as luciferase and lactate dehydrogenase [120]. Michel *et al.* prepared lactose-modified multivalent PAMAM dendrimers and studied their interaction with three different cell lines, including A549, DU-145, and HT-1080 cells [65]. It was shown that smaller dendrimers with small numbers of lactose functional groups showed less galectin-3-induced cellular aggregation compared with larger dendrimers having more than 100 lactose units. The homotypic cell aggregation analysis revealed that smaller particles inhibited these aggregations by providing greater numbers of binding sites, which diverted the different binding ligands, whereas the larger dendrimers enabled the presence of an excess of ligands that easily found the receptor and resulted in cell aggregation (Figure 6.12).

Hammer *et al.* demonstrate the effect of different functional groups and ratios of polar and nonpolar species on the cell surface interaction and cellular uptake in biological systems by using polyphenylene dendrimers (PPDs) [66]. The cellular uptake efficiencies and the cytotoxicity were significantly reduced by changing the polar to nonpolar group ratios from 1:1 to 2:1. These changes contributed to the instauration of repulsive forces between the PPDs and the cellular components, and changes in the balance required for the formation of optimal liphophilic sheets for the ideal uptake of these dendrimers.

Zhong *et al.* studied the effect of the route of administration and the degree of PEGylation of generation 3 polyamidoamine dendrimers on cellular uptake in the

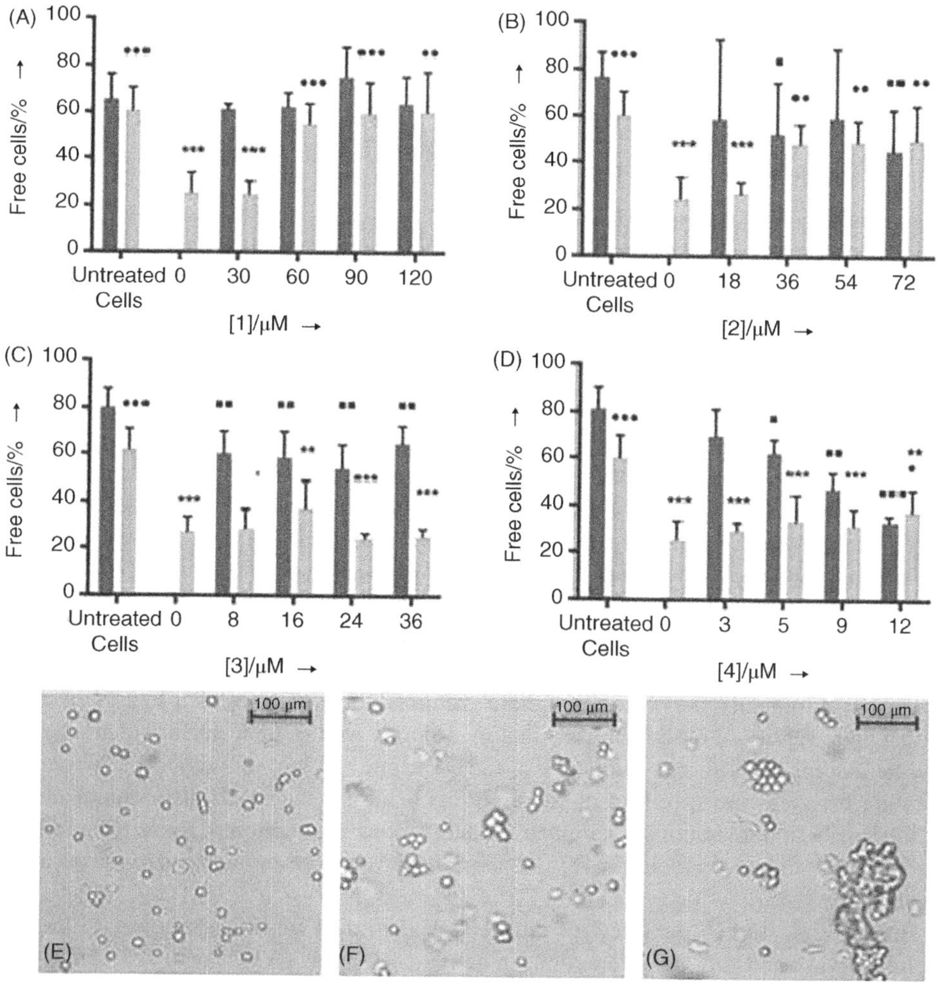

Figure 6.12 *Influence of lactose-functionalized dendrimers on normal and galectin-3-induced homotypic A549 cells aggregates. (A–D) Aggregation results normal (dark grey) and galectin treated (light grey), and micrographs of these cell aggregates. Reproduced with permission from [65]; copyright © 2014, Wiley-VCH Verlag GmbH & Co. KGaA, Weinheim.*

lung cells and distribution within biological systems [121]. Interactions with the lung cells showed that the PEGylated dendrimers were internalized 20% more compared with non-PEGylated dendrimers. The PEGylated dendrimers conjugated with the cationic polymers resulted in a reduction of the affinity for different cellular components. The interaction of the polyamidoamine dendrimers with the KB and Rat2 cell membranes was investigated by Hong *et al.*, using enzymatic assays and fluorescence microscopy [67]. It was demonstrated that these dendrimers create holes in the cells membrane, consistently with their uptake, evaluating the release of lactate dehydrogenase and luciferase enzymes. Further evaluation by incubating the PAMAM dendrimers with Caco-2 cells indicated that the interaction between the cationic dendrimers and anionic cell membrane contributed to the internalization and endocytosis process [122].

6.5.4 Silicon/Silica-based Drug Delivery Systems

Silica- and silicon-based materials and nanocarriers have been extensively studied as drug delivery systems. Various types of silica particles have been fabricated, including fume silica, porous silica and mesoporous silica particles. The solid silica NPs have been extensively employed in optical imaging processes. Mesoporous nanoparticles (MSNs) have been prepared by the sol-gel method and extensively employed for drug delivery, because of their high surface area, uniform porosity, better stability in aqueous media, higher pore volume, better *in vivo* compatibility, and degradibility [73]. The therapeutic or diagnostic moieties can be loaded inside the large number of nanosized pores (porous silica) or physically attached onto the surface of these particles (non-porous silica NPs) [123]. The presence of weak Si–Si and Si–O linkages in these materials contribute to better biodegradibility compared with polymers and other carbon-based nanomaterials [124]. Due to the presence of the silanol group backbone having negative charge at normal pH (7.4), silica NPs have better interaction with the membrane of phospholipids via electrostatic forces that contribute to excellent cell–NP interactions and facilitate the cell internalization process. The nucleophilic oxygen atoms present in the silanol group interact with the carbonyl group of cellular proteins. Additionally, the presence of these groups further enhances their functionalization with a wide variety of other materials that incorporate well-defined surface characteristics for the loading of different compounds [125]. Despite the fact that ease of surface modification broadens the spectrum of silica NP applications, the mechanistic understanding of biodistribution, metabolism, elimination and cellular interaction of these NPs and their degradation products in biological systems are decisive for the application of these carriers in the clinic [126].

Huang *et al.* studied the effect of different aspect ratios for uniformly sized MSNs on cellular uptake and cell–NP interactions in human melanoma (A375) cells [73]. TEM analysis revealed that different shaped MSNs were nonspecifically taken up by the A375 cells and then conjugated with the endosome for intracellular transportation. Cell adhesiveness was determined by analyzing the intercellular cell adhesion molecule-1 (ICAM-1) and melanoma cell adhesion molecule (MCAM). Western blot analysis indicated that MSN shape did not affect the expression of ICAM-1, but significantly decreased the expression of MCAM with an increase in the aspect ratio of MSNs [73].

Kim *et al.* fabricated monodisperse silica NPs to demonstrate the effect of NP size and concentration in human hepatoma cells (HepG2), human alveolar carcinoma cells (A549), and mouse embryonic fibroblast (NIH/3T3) cells [68]. The interaction of these NPs was evaluated by determining the extent of cell internalization and cytotoxicity through the production of ROS, morphological changes, membrane peroxidation, and mitochondrial damage. The results showed a dose- and size-dependent cytotoxicity in NIH/3T3 cells and A549 cells, while no significent reduction in the viability was observed in HepG2. The enhanced toxicity in A549 cells might be due to an excessive generation of ROS that contributed to the higher extent of cell death. However, HepG2 cells were least sensitive to the cytotoxic effects of these NPs. TEM images of the A549 cells after 24 h exposure to the 200 μg/ml of NPs of 20, 60 and 200 nm revealed that the NPs with diameter of 60 nm were excessively distributed in membrane-bound vesicles, as well as in cytoplasm (Figure 6.13). Such a distribution of the silica NPs might be due to endocytic and non-endocytic transport mechanisms, suggesting this size range as optimum for maximum cell uptake.

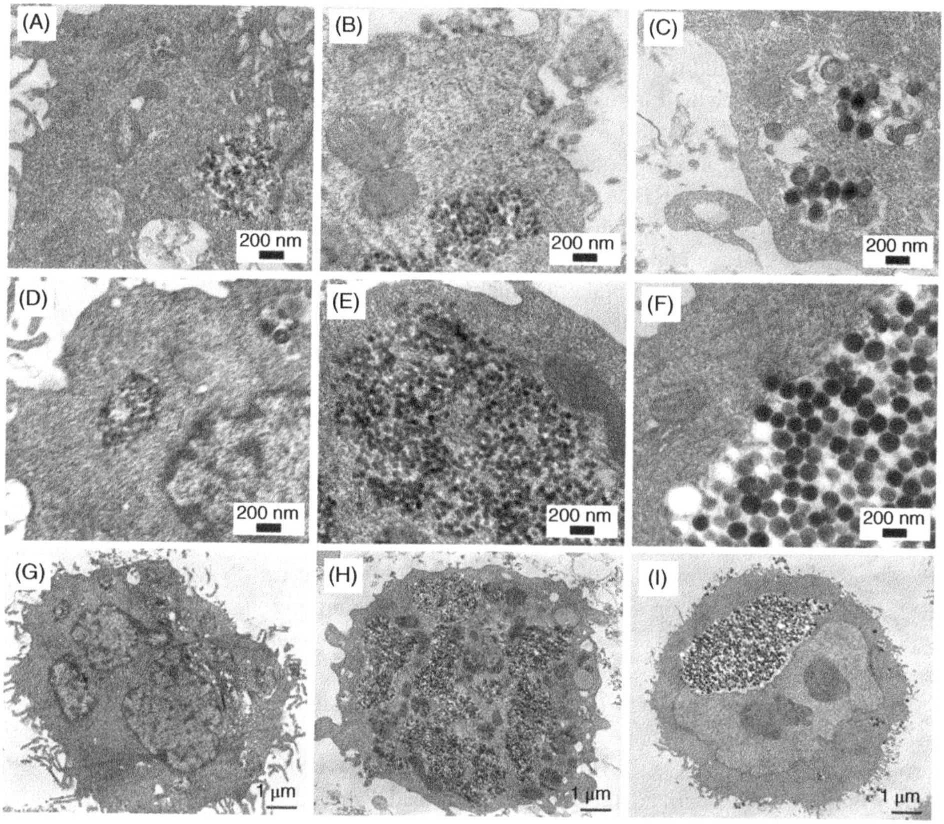

Figure 6.13 *TEM images of the A549 cells after exposure to NPs of 20 nm (A and D), 60 nm (B and E), and 200 nm (C and F), presenting the cell morphology and extent of cellular uptake in the A549 cells. Reproduced with permission from [68]; copyright © 2015, Elsevier B.V.*

Zhang *et al.* demonstrated that the production of ROS species involved the degradation of the silaxan ring in the silica NP structure [127]. The NP-induced ROS generation process in turn affected the intrinsic defence system of the cell, modifying the structure of lipid and nucleic material, damaging the cell machinery and ultimately initiating the cell apoptotic pathway. The development of cytotoxicity involved a set of different processes, including a change in the permeability of the membrane by cellular uptake of these porous and non-porous NPs. Similarly, Chang *et al.* studied the cytotoxic effects of solid silica NPs in different cell lines under different growth and exposure conditions [69]. It was concluded that the cytotoxicity of these NPs in lung and colon carcinoma cells, and in human fibroblast cells, was dependent on the type of cells, growth media, and the growth rate of that particular cell line. Ariano *et al.* also investigated and compared the influence of silica NPs on the biological responses of the matured neuronal cells (GT1-7) with a change in dose and particle size [70]. At a lower concentration of 100 μg/ml both NP formulations (50 and 200 nm particle size) had insignificant cytotoxic effects. However, at higher concentration (292 μg/ml), smaller silica particles affected the neural cell viability by initiating the apoptotic pathway, identified by the amount of fragmented nuclei. The most prominent mechanism for this neural toxicity involved the alteration in the calcium homeostatis: smaller particles affected the membrane permeability and promoted inward movement of extracellular calcium, which in turn activated the different toxicity mechanisms.

The compounds present on the surface of NPs also elicit various cellular responses. For example, the hemotoxicity of mesoporous NPs was evaluated by hemolytic activity on human red blood cells (RBCs) by Lin *et al.* [71]. The smaller particles (100 nm) were found to be more toxic then the larger (300–500 nm) NPs. These NPs adsorbed easily onto the surface of the RBCs and internalized, causing significant membrane deformities that resulted in increased cellular uptake and hemolysis of the RBCs. Different *in vitro* studies evaluated the toxic potential of porous and non-porous silica particles in various cell lines. Blechinger *et al.* investigated the toxicity mechanism and the uptake kinetics of these NPs, labeled with perylene dye, in human vascular endothelial (HUVEC) and cervix carcinoma (HeLa) cells [72]. In early interaction with the cells the cellular uptake was 10 times greater in the HUVEC cells compared with the HeLa cells. Investigation with different inhibitors concluded that the clathrin-induced endocytosis pathway was most prominent in the uptake process. However, the opposite ratio of cell uptake was observed after 24 h of cell–NP interactions. The perylene functionalization did not affect the toxicity of these NPs, suggesting that it imparted no significant effects on the cell–NP interactions. These differences in the interaction and intercellular trafficking among the different cells provide an excellent opportunity for cell specific threapy in different carcinomas and autoimmune diseases [128].

Shahbazi *et al.* demonstrated the influence of surface chemistry of porous silicon NPs (PSi NPs) on cell interaction and immunotoxicity by using different types of PSi NPs (e.g., thermally oxidized PSi (TOPSi), thermally carbonized PSi (TCPSi), (3-aminopropyl) triethoxysilane functionalized thermally carbonized PSi (APSTCPSi), thermally hydrocarbonized PSi (THCPSi) and undecylenic acid functionalized THPSi (UnTHCPSi)) [74]. The effects of these PSi NPs were evaluated in B-cells, T-cells, U937 monocytes and RAW264.7 macrophages. It was found that the cytotoxicity of all these PSi NPs was charge-, concentration- and surface-dependent. The mechanism of toxicity was also found to be cell-specific. Generation of ROS and tumor necrosis factor (TNF-α) was prominent in RAW264.7 macrophages. However, the toxicity on T-cells was promoted by the formation of reactive nitrogen species (RNOS). The evaluation of the hemocompatibility of these PSi NPs at different concentrations indicated time-dependent hemolytic process, where APSTCPSi and UnTHCPSi NPs were more toxic compared with TOPSi, TCPSi and THCPSi NPs. The difference in cytotoxicity was based on the highly hydrophilic nature and strong negative charge of TOPSi and TCPSi NPs that limited the interaction and binding of these NPs with the cell membranes.

In another study, Shahbazi *et al.* investigated the immunostimulant effects of PSi NPs on human monocyte-derived dendritic cells (MDDCs) and lymphocytes [129]. Results demonstrated that all PSi NPs were safe and biocompatible; however, TOPSi and THCPSi NPs activated the immune system more effectively by stimulating the antigen-presenting cells.

Almeida *et al.* prepared a multifunctional nanocomposite containing the PSi NPs with gold NPs on a polymeric matrix to potentiate the tumor cell–NP interactions and enhance the endosomal escape of the PSi-NPs [75]. *In vitro* biocompatibility, cell interaction, cytotoxicity, and cell uptake analyses were performed. *In vitro* cytotoxicity assay in MDA-MB-231 cells showed no sign of toxicity in the concentration range from 12.5 to 200 g/ml. The cell–PSi NP association and internalization was evaluated by incubating the different PSi NPs with MDA-MB-231 cells. The blank UnTHCPSi nanoparticles showed the least cellular uptake, due to the negatively charged surface, poor adhesiveness to biological membranes, and instability in aqueous solutions due to the hydrophobic surface. However, undecylenic acid-functionalized THCPSi@cysteine-polyethyleniminepoly (methyl vinyl ether-alt-maleic acid) (UnCPP) and undecylenic acid-functionalized Au NPs@cysteine-polyethyleniminepoly(methyl vinyl ether-alt-maleic acid) (UnΛuCPP) NPs exhibited a high degree of cell membrane association and higher cellular uptake [75].

Shahbazi *et al.* prepared the thermally carbonized porous silicon nanoparticle (APSTCPSi-NPs) surface conjugated with poly(methyl vinyl ether-co-maleic acid) copolymer and studied the effect of surface modification on system stability, cytotoxicity and cellular uptake in MDA-MB-132 and MCF-7 breast cancer cell lines [130]. NP protein binding analysis showed minimum protein binding. The intracellular transport and location of these PSi NPs analysed by confocal imaging demonstrated that most of the APSTCPSi NPs were present near the cell membrane, whereas enhanced internalization of PMVE-MA conjugated PSi NPs was attributed to the bioadhesive properties of the PMVE-MA copolymer. This bioadhesion facilitated the fusion of these particles with the biological membranes.

6.6 Conclusions and Future Perspectives

The study of the interactions between cells and NPs is of fundamental importance in the preclinical assessment of the properties of nanomaterials. Several general indications concerning size, surface charge, hydrophobicity and the effect of the protein corona help with a rational design of future nanoparticulate systems suitable for the desired application. Moreover, the application of *in silico* modeling to evaluate such interactions will avoid the optimization of the NPs on a trial-and-error basis. Here, we reviewed the interactions between cells and different types of NPs, polymeric, liposomes, metal oxides, and silica/silicon-based delivery systems. We highlighted the outcome of the interactions, particularly regarding the cytotoxicity. Further studies will allow a more precise understanding of the effects of the interactions, both short- and long-term, on the human body – a goal for the applicability of any NPs in pharmaceutics and biomedicine.

Acknowledgements

Nayab Tahir acknowledges the International Research Support Initiative Program, Higher Education Commission of Pakistan, for providing the funds for the research visit to

Prof. Santos' lab at the Faculty of Pharmacy, University of Helsinki. Prof. Santos also acknowledges financial support from the Academy of Finland (Decision No. 304844), the University of Helsinki Research Funds, HiLIFE Research Funds, Sigrid Jusélius Foundation (Decision No. 4704580), and the European Research Council under the European Union's Seventh Framework Programme (FP/2007–2013, Grant No. 310892).

References

[1] Roco, M.C. (2003) Nanotechnology: convergence with modern biology and medicine. *Curr Opin Biotechnol*, **14**, 337–346.
[2] Shi, J., Votruba, A.R., Farokhzad, O.C., Langer, R. (2010) Nanotechnology in drug delivery and tissue engineering: from discovery to applications. *Nano Lett*, **10**, 3223–3230.
[3] Doane, T.L., Burda, C. (2012) The unique role of nanoparticles in nanomedicine: imaging, drug delivery and therapy. *Chem Soc Rev*, **41**, 2885–2911.
[4] Roco, M., Williams, R., Alivisatos, P. (2000) *Nanotechnology Research Directions: Vision for the Next Decade*. National Science and Technology Council, Washington, DC.
[5] I.I.S. Organization (2010) Nanotechnologies, in: TC 229. https://www.iso.org/committee/381983/x/catalogue/
[6] Farokhzad, O.C., Langer, R. (2009) Impact of nanotechnology on drug delivery. *ACS Nano*, **3**, 16–20.
[7] Schroeder, A., Heller, D.A., Winslow, M.M., *et al.* (2011) Treating metastatic cancer with nanotechnology. *Nat Rev Cancer*, **12**, 39–50.
[8] Rossi, M., Cubadda, F., Dini, L., *et al.* (2014) Scientific basis of nanotechnology, implications for the food sector and future trends. *Trends in Food Science & Technology*, **40**, 127–148.
[9] Chen, G., Roy, I., Yang, C., Prasad, P.N. (2016) Nanochemistry and nanomedicine for nanoparticle-based diagnostics and therapy. *Chem Rev*, **116**, 2826–2885.
[10] Kim, B.Y., Rutka, J.T., Chan, W.C. (2010) Nanomedicine. *N Engl J Med*, **363**, 2434–2443.
[11] Veiseh, O., Tang, B.C., Whitehead, K.A., *et al.* (2015) Managing diabetes with nanomedicine: challenges and opportunities. *Nat Rev Drug Discov*, **14**, 45–57.
[12] Ryu, J.H., Lee, S., Son, S., *et al.* (2014) Theranostic nanoparticles for future personalized medicine. *J Control Release*, **190**, 477–484.
[13] Bobo, D., Robinson, K.J., Islam, J., *et al.* (2016) Nanoparticle-based medicines: A review of FDA-approved materials and clinical trials to date. *Pharm Res*, **33**, 2373–2387.
[14] Chan, W.C.W. (2017) Nanomedicine 2.0. *Acc Chem Res*, **50**, 627–632.
[15] Bastogne, T. (2017) Quality-by-design of nanopharmaceuticals – a state of the art. *Nanomedicine*, **13**, 2151–2157.
[16] Doherty, G.J., McMahon, H.T. (2009) Mechanisms of endocytosis. *Annu Rev Biochem*, **78**, 857–902.
[17] Banerjee, A., Berzhkovskii, A., Nossal, R. (2014) Efficiency of cellular uptake of nanoparticles via receptor-mediated endocytosis. arXiv preprint, arXiv:1411.7348.
[18] Oh, N., Park, J.H. (2014) Endocytosis and exocytosis of nanoparticles in mammalian cells. *Int J Nanomedicine*, **9** (Suppl. 1), 51–63.
[19] Zhang, S., Li, J., Lykotrafitis, G., *et al.* (2009) Size-dependent endocytosis of nanoparticles. *Adv Mater*, **21**, 419–424.
[20] Venkataraman, S., Hedrick, J.L., Ong, Z.Y., *et al.* (2011) The effects of polymeric nanostructure shape on drug delivery. *Adv Drug Deliv Rev*, **63**, 1228–1246.
[21] Dasgupta, S., Auth, T., Gompper, G. (2014) Shape and orientation matter for the cellular uptake of nonspherical particles. *Nano Lett*, **14**, 687–693.

[22] Albanese, A., Tang, P.S., Chan, W.C. (2012) The effect of nanoparticle size, shape, and surface chemistry on biological systems. *Annu Rev Biomed Eng*, **14**, 1–16.
[23] Verma, A., Stellacci, F. (2010) Effect of surface properties on nanoparticle–cell interactions. *Small*, **6**, 12–21.
[24] Geng, Y., Dalhaimer, P., Cai, S., *et al.* (2007) Shape effects of filaments versus spherical particles in flow and drug delivery. *Nat Nanotechnol*, **2**, 249–255.
[25] Clift, M.J., Endes, C., Vanhecke, D., *et al.* (2014) A comparative study of different in vitro lung cell culture systems to assess the most beneficial tool for screening the potential adverse effects of carbon nanotubes. *Toxicol Sci*, **137**, 55–64.
[26] Thurn, K.T., Brown, E., Wu, A., *et al.* (2007) Nanoparticles for applications in cellular imaging. *Nanoscale Res Lett*, **2**, 430–441.
[27] Frohlich, E. (2012) The role of surface charge in cellular uptake and cytotoxicity of medical nanoparticles. *Int J Nanomedicine*, **7**, 5577–5591.
[28] Alexis, F., Pridgen, E., Molnar, L.K., Farokhzad, O.C. (2008) Factors affecting the clearance and biodistribution of polymeric nanoparticles. *Mol Pharm*, **5**, 505–515.
[29] Manshian, B.B., Moyano, D.F., Corthout, N., *et al.* (2014) High-content imaging and gene expression analysis to study cell-nanomaterial interactions: the effect of surface hydrophobicity. *Biomaterials*, **35**, 9941–9950.
[30] Lundqvist, M., Stigler, J., Elia, G., *et al.* (2008) Nanoparticle size and surface properties determine the protein corona with possible implications for biological impacts. *Proc Natl Acad Sci USA*, **105**, 14265–14270.
[31] Mirshafiee, V., Mahmoudi, M., Lou, K., *et al.* (2013) Protein corona significantly reduces active targeting yield. *Chemical Communications*, **49**, 2557–2559.
[32] Wan, S., Kelly, P.M., Mahon, E., *et al.* (2015) The "sweet" side of the protein corona: effects of glycosylation on nanoparticle-cell interactions. *ACS Nano*, **9**, 2157–2166.
[33] Ritz, S., Schottler, S., Kotman, N., *et al.* (2015) Protein corona of nanoparticles: distinct proteins regulate the cellular uptake. *Biomacromolecules*, **16**, 1311–1321.
[34] Reshma, V.G., Mohanan, P.V. (2017) Cellular interactions of zinc oxide nanoparticles with human embryonic kidney (HEK 293) cells. *Colloids and Surfaces B-Biointerfaces*, **157**, 182–190.
[35] Liu, Q., Xu, C., Ji, G., *et al.* (2016) Sublethal effects of zinc oxide nanoparticles on male reproductive cells. *Toxicol In Vitro*, **35**, 131–138.
[36] Sruthi, S., Mohanan, P.V. (2015) Investigation on cellular interactions of astrocytes with zinc oxide nanoparticles using rat C6 cell lines. *Colloids Surf B Biointerfaces*, **133**, 1–11.
[37] Pal, A., Alam, S., Mittal, S., *et al.* (2016) UVB irradiation-enhanced zinc oxide nanoparticles-induced DNA damage and cell death in mouse skin. *Mutat Res Genet Toxicol Environ Mutagen*, **807**, 15–24.
[38] Gopikrishnan, R., Zhang, K., Ravichandran, P., *et al.* (2010) Synthesis, characterization and biocompatibility studies of zinc oxide (ZnO) nanorods for biomedical application. *Nano-Micro Letters*, **2**, 31–36.
[39] Condello, M., De Berardis, B., Ammendolia, M.G., *et al.* (2016) ZnO nanoparticle tracking from uptake to genotoxic damage in human colon carcinoma cells. *Toxicol In Vitro*, **35**, 169–179.
[40] Mahapatra, C., Singh, R.K., Lee, J.H., *et al.* (2017) Nano-shape varied cerium oxide nanomaterials rescue human dental stem cells from oxidative insult through intracellular or extracellular actions. *Acta Biomater*, **50**, 142–153.
[41] Mohammada, F., Arfinb, T., Al-Lohedana, H.A. (2017) Enhanced biological activity and biosorption performance of trimethyl chitosan-loaded cerium oxide particles. *Journal of Industrial and Engineering Chemistry*, **45**, 33–43.

[42] Wang, Q., Chen, B., Cao, M., *et al.* (2016) Response of MAPK pathway to iron oxide nanoparticles in vitro treatment promotes osteogenic differentiation of hBMSCs. *Biomaterials*, **86**, 11–20.

[43] Sadhasivam, S., Savitha, S., Wu, C.J., *et al.* (2015) Carbon encapsulated iron oxide nanoparticles surface engineered with polyethylene glycol-folic acid to induce selective hyperthermia in folate over expressed cancer cells. *Int J Pharm*, **480**, 8–14.

[44] Augustin, E., Czubek, B., Nowicka, A.M., *et al.* (2016) Improved cytotoxicity and preserved level of cell death induced in colon cancer cells by doxorubicin after its conjugation with iron-oxide magnetic nanoparticles. *Toxicol In Vitro*, **33**, 45–53.

[45] Luther, E.M., Petters, C., Bulcke, F., *et al.* (2013) Endocytotic uptake of iron oxide nanoparticles by cultured brain microglial cells. *Acta Biomater*, **9**, 8454–8465.

[46] Petters, C., Dringen, R. (2015) Accumulation of iron oxide nanoparticles by cultured primary neurons. *Neurochem Int*, **81**, 1–9.

[47] Romorene, K., Thu, B.J., Bols, N.C., Evensen, O. (2004) Transfection efficiency and cytotoxicity of cationic liposomes in salmonid cell lines of hepatocyte and macrophage origin. *Biochim Biophys Acta*, **1663**, 127–134.

[48] Roursgaard, M., Knudsen, K.B., Northeved, H., *et al.* (2016) In vitro toxicity of cationic micelles and liposomes in cultured human hepatocyte (HepG2) and lung epithelial (A549) cell lines. *Toxicol In Vitro*, **36**, 164–171.

[49] Wang, J., Ayano, E., Maitani, Y., Kanazawa, H. (2017) Enhanced cellular uptake and gene silencing activity of siRNA using temperature-responsive polymer-modified liposome. *International Journal of Pharmaceutics*, **523**, 217–228.

[50] Swarnakar, N.K., Jain, A.K., Singh, R.P., *et al.* (2011) Oral bioavailability, therapeutic efficacy and reactive oxygen species scavenging properties of coenzyme Q10-loaded polymeric nanoparticles. *Biomaterials*, **32**, 6860–6874.

[51] Xu, A., Yao, M., Xu, G., *et al.* (2012) A physical model for the size-dependent cellular uptake of nanoparticles modified with cationic surfactants. *International Journal of Nanomedicine*, **7**, 3547.

[52] Lunov, O., Syrovets, T., Loos, C., *et al.* (2011) Differential uptake of functionalized polystyrene nanoparticles by human macrophages and a monocytic cell line. *ACS Nano*, **5**, 1657–1669.

[53] Win, K.Y., Feng, S.-S. (2005) Effects of particle size and surface coating on cellular uptake of polymeric nanoparticles for oral delivery of anticancer drugs. *Biomaterials*, **26**, 2713–2722.

[54] Kulkarni, S.A., Feng, S.-S. (2013) Effects of particle size and surface modification on cellular uptake and biodistribution of polymeric nanoparticles for drug delivery. *Pharmaceutical Research*, **30**, 2512–2522.

[55] He, C., Hu, Y., Yin, L., *et al.* (2010) Effects of particle size and surface charge on cellular uptake and biodistribution of polymeric nanoparticles. *Biomaterials*, **31**, 3657–3666.

[56] Treuel, L., Brandholt, S., Maffre, P., *et al.* (2014) Impact of protein modification on the protein corona on nanoparticles and nanoparticle–cell interactions. *ACS Nano*, **8**, 503–513.

[57] Fleischer, C.C., Payne, C.K. (2014) Nanoparticle–cell interactions: molecular structure of the protein corona and cellular outcomes. *Acc. Chem. Res*, **47**, 2651–2659.

[58] Grabowski, N., Hillaireau, H., Vergnaud, J., *et al.* (2015) Surface coating mediates the toxicity of polymeric nanoparticles towards human-like macrophages. *International Journal of Pharmaceutics*, **482**, 75–83.

[59] Hühn, D., Kantner, K., Geidel, C., *et al.* (2013) Polymer-coated nanoparticles interacting with proteins and cells: focusing on the sign of the net charge. *ACS Nano*, **7**, 3253–3263.

[60] Hu, Y., Litwin, T., Nagaraja, A.R., *et al.* (2007) Cytosolic delivery of membrane-impermeable molecules in dendritic cells using pH-responsive core-shell nanoparticles. *Nano Letters*, **7**, 3056–3064.

[61] Xue, B., Kozlovskaya, V., Liu, F., *et al.* (2015) Intracellular degradable hydrogel cubes and spheres for anti-cancer drug delivery. *ACS Applied Materials & Interfaces*, **7**, 13633–13644.

[62] Liu, W., Zhou, X., Mao, Z., *et al.* (2012) Uptake of hydrogel particles with different stiffness and its influence on HepG2 cell functions. *Soft Matter*, **8**, 9235–9245.

[63] Katsikari, A., Patronidou, C., Kiparissides, C., Arsenakis, M. (2009) Uptake and cytotoxicity of poly (d, l-lactide-co-glycolide) nanoparticles in human colon adenocarcinoma cells. *Materials Science and Engineering: B*, **165**, 160–164.

[64] Otremba, T., Ravoo, B.J. (2017) Dynamic multivalent interaction of phenylboronic acid functionalized dendrimers with vesicles. *Tetrahedron*, **73**, 4972–4978.

[65] Michel, A.K., Nangia-Makker, P., Raz, A., Cloninger, M.J. (2014) Lactose-functionalized dendrimers arbitrate the interaction of galectin-3/MUC1 mediated cancer cellular aggregation. *ChemBioChem*, **15**, 2106–2112.

[66] Hammer, B.A., Wu, Y., Fischer, S., *et al.* (2017) Controlling cellular uptake and toxicity of polyphenylene dendrimers by chemical functionalization. *ChemBioChem*, **18** (10), 960–964.

[67] Hong, S., Bielinska, A.U., Mecke, A., *et al.* (2004) Interaction of poly (amidoamine) dendrimers with supported lipid bilayers and cells: hole formation and the relation to transport. *Bioconjugate Chemistry*, **15**, 774–782.

[68] Kim, I.-Y., Joachim, E., Choi, H., Kim, K. (2015) Toxicity of silica nanoparticles depends on size, dose, and cell type. *Nanomedicine: Nanotechnology, Biology and Medicine*, **11**, 1407–1416.

[69] Chang, J.-S., Chang, K.L.B., Hwang, D.-F., Kong, Z.-L. (2007) In vitro cytotoxicity of silica nanoparticles at high concentrations strongly depends on the metabolic activity type of the cell line. *Environmental Science & Technology*, **41**, 2064–2068.

[70] Ariano, P., Zamburlin, P., Gilardino, A., *et al.* (2011) Interaction of spherical silica nanoparticles with neuronal cells: Size-dependent toxicity and perturbation of calcium homeostasis. *Small*, **7**, 766–774.

[71] Lin, Y.-S., Haynes, C.L. (2010) Impacts of mesoporous silica nanoparticle size, pore ordering, and pore integrity on hemolytic activity. *Journal of the American Chemical Society*, **132**, 4834–4842.

[72] Blechinger, J., Bauer, A.T., Torrano, A.A., *et al.* (2013) Uptake kinetics and nanotoxicity of silica nanoparticles are cell type dependent. *Small*, **9**, 3970–3980.

[73] Huang, X., Teng, X., Chen, D., *et al.* (2010) The effect of the shape of mesoporous silica nanoparticles on cellular uptake and cell function. *Biomaterials*, **31**, 438–448.

[74] Shahbazi, M.-A., Hamidi, M., Mäkilä, E.M., *et al.* (2013) The mechanisms of surface chemistry effects of mesoporous silicon nanoparticles on immunotoxicity and biocompatibility. *Biomaterials*, **34**, 7776–7789.

[75] Almeida, P.V., Shahbazi, M.-A., Correia, A., *et al.* (2017) A multifunctional nanocomplex for enhanced cell uptake, endosomal escape and improved cancer therapeutic effect. *Nanomedicine*, **12** (12).

[76] Chen, J., Hessler, J.A., Putchakayala, K., *et al.* (2009) Cationic nanoparticles induce nanoscale disruption in living cell plasma membranes. *J Phys Chem B*, **113**, 11179–11185.

[77] Arvizo, R.R., Miranda, O.R., Thompson, M.A., *et al.* (2010) Effect of nanoparticle surface charge at the plasma membrane and beyond. *Nano Letters*, **10**, 2543–2548.

[78] Mu, Q., Jiang, G., Chen, L., *et al.* (2014) Chemical basis of interactions between engineered nanoparticles and biological systems. *Chem. Rev*, **114**, 7740–7781.

[79] Doshi, N., Mitragotri, S. (2010) Needle-shaped polymeric particles induce transient disruption of cell membranes. *J R Soc Interface*, **7** (Suppl 4), S403–410.

[80] Holt, B.D., Short, P.A., Rape, A.D., *et al.* (2010) Carbon nanotubes reorganize actin structures in cells and ex vivo. *ACS Nano*, **4**, 4872–4878.

[81] Tarantola, M., Schneider, D., Sunnick, E., *et al.* (2009) Cytotoxicity of metal and semiconductor nanoparticles indicated by cellular micromotility. *ACS Nano*, **3**, 213–222.

[82] Pan, Y., Leifert, A., Ruau, D., *et al.* (2009) Gold nanoparticles of diameter 1.4 nm trigger necrosis by oxidative stress and mitochondrial damage. *Small*, **5**, 2067–2076.

[83] Zakhidov, S., Marshak, T., Malolina, E., *et al.* (2010) Gold nanoparticles disturb nuclear chromatin decondensation in mouse sperm in vitro. *Biochemistry (Moscow) Supplemental Series A: Membrane and Cell Biology*, **4**, 293–296.

[84] George, S., Pokhrel, S., Xia, T., *et al.* (2009) Use of a rapid cytotoxicity screening approach to engineer a safer zinc oxide nanoparticle through iron doping. *ACS Nano*, **4**, 15–29.

[85] Riss, T.L., Moravec, R.A., Niles, A.L., *et al.* (2016) Cell viability assays, in *Assay Guidance Manual* (eds Sittampalam, G.S., Grossman, A., Brimacombe, K., *et al.*), Eli Lilly & Company and the National Center for Advancing Translational Sciences, Bethesda, MD. https://www.ncbi.nlm.nih.gov/books/NBK144065/

[86] Suzuki, H., Toyooka, T., Ibuki, Y. (2007) Simple and easy method to evaluate uptake potential of nanoparticles in mammalian cells using a flow cytometric light scatter analysis. *Environ Sci Technol*, **41**, 3018–3024.

[87] Huang, K., Ma, H., Liu, J., *et al.* (2012) Size-dependent localization and penetration of ultrasmall gold nanoparticles in cancer cells, multicellular spheroids, and tumors in vivo, *ACS Nano*, **6**, 4483–4493.

[88] Mehta, G., Hsiao, A.Y., Ingram, M., *et al.* (2012) Opportunities and challenges for use of tumor spheroids as models to test drug delivery and efficacy. *Journal of Controlled Release*, **164**, 192–204.

[89] Bhise, N.S., Ribas, J., Manoharan, V., *et al.* (2014) Organ-on-a-chip platforms for studying drug delivery systems. *J Control Release*, **190**, 82–93.

[90] Wong-Ekkabut, J., Baoukina, S., Triampo, W., *et al.* (2008) Computer simulation study of fullerene translocation through lipid membranes. *Nat Nanotechnol*, **3**, 363–368.

[91] Koay, E.J., Ferrari, M. (2014) Transport oncophysics in silico, in vitro, and in vivo. *Preface. Phys Biol*, **11**, 060201.

[92] Rasouli, N., Movahedi, M., Doudi, M. (2017) Synthesis and characterization of inorganic mixed metal oxide nanoparticles derived from Zn-Al layered double hydroxide and their antibacterial activity. *Surfaces and Interfaces*, **6**, 110–115.

[93] Jha, R.K., Jha, P.K., Chaudhury, K., *et al.* (2014) An emerging interface between life science and nanotechnology: present status and prospects of reproductive healthcare aided by nano-biotechnology. *Nano Rev*, 5.

[94] Krol, A., Pomastowski, P., Rafinska, K., *et al.* (2017) Zinc oxide nanoparticles: Synthesis, antiseptic activity and toxicity mechanism. *Adv Colloid Interface Sci*, **249**, 37–52.

[95] Ignacio, R.M.C., Kim, C-S., Kim, S.-K. (2014) Immunotoxicity of metal oxide nanoparticle: zinc oxide. *Mol Cell Toxicol*, **10**, 237–244.

[96] Milani, Z.M., Charbgoo, F., Darroudi, M. (2017) Impact of physicochemical properties of cerium oxide nanoparticles on their toxicity effects. *Ceramics International*, **43**, 14572–14581.

[97] Sirelkhatim, A., Mahmud, S., Seeni, A., *et al.* (2015) Review on zinc oxide nanoparticles: Antibacterial activity and toxicity mechanism. *Nano-Micro Letters*, **7**, 219–242.

[98] De Berardis, B., Civitelli, G., Condello, M., *et al.* (2010) Exposure to ZnO nanoparticles induces oxidative stress and cytotoxicity in human colon carcinoma cells. *Toxicol Appl Pharmacol*, **246**, 116–127.

[99] Pulido-Reyes, G., Martin, E., Gu Coronado, J.L., *et al.* (2017) Physicochemical and biological interactions between cerium oxide nanoparticles and a 1,8-naphthalimide derivative. *J Photochem Photobiol B*, **172**, 61–69.

[100] Valdiglesias, V., Fernandez-Bertolez, N., Kilic, G., *et al.* (2016) Are iron oxide nanoparticles safe? Current knowledge and future perspectives. *J Trace Elem Med Biol*, **38**, 53–63.

[101] Fresta, M., Spadaro, A., Cerniglia, G., *et al.* (1995) Intracellular accumulation of ofloxacin-loaded liposomes in human synovial fibroblasts. *Antimicrob Agents Chemother*, **39**, 1372–1375.

[102] Bimbo, L.M., Peltonen, L., Hirvonen, J., Santos, H.A. (2012) Toxicological profile of therapeutic nanodelivery systems. *Curr Drug Metab*, **13**, 1068–1086.

[103] Akbarzadeh, A., Rezaei-Sadabady, R., Davaran, S., *et al.* (2013) Liposome: classification, preparation, and applications. *Nanoscale Res Lett*, **8**, 102.

[104] Lasic, D.D. (1998) Novel applications of liposomes. *Trends Biotechnol*, **16**, 307–321.

[105] Kelly, C., Jefferies, C., Cryan, S.A. (2011) Targeted liposomal drug delivery to monocytes and macrophages. *J Drug Deliv*, 27241.

[106] Epstein-Barash, H., Gutman, D., Markovsky, E., *et al.* (2010) Physicochemical parameters affecting liposomal bisphosphonates bioactivity for restenosis therapy: internalization, cell inhibition, activation of cytokines and complement, and mechanism of cell death. *J Control Release*, **146**, 182–195.

[107] Lv, H., Zhang, S., Wang, B., *et al.* (2006) Toxicity of cationic lipids and cationic polymers in gene delivery. *J Control Release*, **114**, 100–109.

[108] Wiethoff, C.M., Smith, J.G., Koe, G.S., Middaugh, C.R. (2001) The potential role of proteoglycans in cationic lipid-mediated gene delivery. Studies of the interaction of cationic lipid-DNA complexes with model glycosaminoglycans. *J Biol Chem*, **276**, 32806–32813.

[109] Araújo, F., Shrestha, N., Granja, P.L., *et al.* (2015) Safety and toxicity concerns of orally delivered nanoparticles as drug carriers. *Expert Opinion on Drug Metabolism & Toxicology*, **11**, 381–393.

[110] Ranjan, S., Dasgupta, N., Chakraborty, A.R., *et al.* (2014) Nanoscience and nanotechnologies in food industries: opportunities and research trends. *Journal of Nanoparticle Research*, **16**, 2464.

[111] Jiang, P., Yu, D., Zhang, W., *et al.* (2015) Influence of bovine serum albumin coated poly (lactic-co-glycolic acid) particles on differentiation of mesenchymal stem cells. *RSC Advances*, **5**, 40924–40931.

[112] Mahmoudi, M., Sant, S., Wang, B., *et al.* (2011) Superparamagnetic iron oxide nanoparticles (SPIONs): development, surface modification and applications in chemotherapy. *Advanced Drug Delivery Reviews*, **63**, 24–46.

[113] Chernenko, T., Matthäus, C., Milane, L., *et al.* (2009) Label-free Raman spectral imaging of intracellular delivery and degradation of polymeric nanoparticle systems. *ACS Nano*, **3**, 3552–3559.

[114] Deng, J., Gao, C. (2016) Recent advances in interactions of designed nanoparticles and cells with respect to cellular uptake, intracellular fate, degradation and cytotoxicity. *Nanotechnology*, **27**, 412002.

[115] Dimchevska, S., Geskovski, N., Koliqi, R., *et al.* (2017) Efficacy assessment of self-assembled PLGA-PEG-PLGA nanoparticles: Correlation of nano-bio interface interactions, biodistribution, internalization and gene expression studies. *International Journal of Pharmaceutics*, **533**, 389–401.

[116] Madaan, K., Kumar, S., Poonia, N., *et al.* (2014) Dendrimers in drug delivery and targeting: Drug–dendrimer interactions and toxicity issues. *Journal of Pharmacy & Bioallied Sciences*, **6**, 139–150.

[117] Dufes, C., Uchegbu, I.F., Schätzlein, A.G. (2005) Dendrimers in gene delivery. *Advanced Drug Delivery Reviews*, **57**, 2177–2202.

[118] Nanjwade, B.K., Bechra, H.M., Derkar, G.K., *et al.* (2009) Dendrimers: emerging polymers for drug-delivery systems. *European Journal of Pharmaceutical Sciences*, **38**, 185–196.

[119] Caminade, A.-M., Yan, D., Smith, D.K. (2015) Dendrimers and hyperbranched polymers. *Chemical Society Reviews*, **44**, 3870–3873.

[120] Chen, H.-T., Neerman, M.F., Parrish, A.R., Simanek, E.E. (2004) Cytotoxicity, hemolysis, and acute in vivo toxicity of dendrimers based on melamine, candidate vehicles for drug delivery. *Journal of the American Chemical Society*, **126**, 10044–10048.
[121] Zhong, Q., Merkel, O.M., Reineke, J.J., da Rocha, S.R. (2016) Effect of the route of administration and PEGylation of poly (amidoamine) dendrimers on their systemic and lung cellular biodistribution. *Molecular Pharmaceutics*, **13**, 1866–1878.
[122] Kitchens, K.M., Foraker, A.B., Kolhatkar, R.B., *et al.* (2007) Endocytosis and interaction of poly (amidoamine) dendrimers with Caco-2 cells. *Pharmaceutical Research*, **24**, 2138–2145.
[123] Ambrogio, M.W., Thomas, C.R., Zhao, Y.-L., *et al.* (2011) Mechanized silica nanoparticles: a new frontier in theranostic nanomedicine. *Accounts of Chemical Research*, **44**, 903–913.
[124] Caltagirone, C., Bettoschi, A., Garau, A., Montis, R. (2015) Silica-based nanoparticles: a versatile tool for the development of efficient imaging agents. *Chemical Society Reviews*, **44**, 4645–4671.
[125] Kwon, S., Singh, R.K., Perez, R.A., *et al.* (2013) Silica-based mesoporous nanoparticles for controlled drug delivery. *Journal of Tissue Engineering*, **4**, 2041731413503357.
[126] Beck, M., Mandal, T., Buske, C., Lindén, M. (2017) Serum protein adsorption enhances active leukemia stem cell targeting of mesoporous silica nanoparticles. *ACS Applied Materials & Interfaces*, **9** (22), 18566–18574.
[127] Zhang, H., Dunphy, D.R., Jiang, X., *et al.* (2012) Processing pathway dependence of amorphous silica nanoparticle toxicity: colloidal vs pyrolytic. *Journal of the American Chemical Society*, **134**, 15790–15804.
[128] Zhao, F., Zhao, Y., Liu, Y., *et al.* (2011) Cellular uptake, intracellular trafficking, and cytotoxicity of nanomaterials. *Small*, **7**, 1322–1337.
[129] Shahbazi, M.-A., Fernández, T.D., Mäkilä, E.M., *et al.* (2014) Surface chemistry dependent immunostimulative potential of porous silicon nanoplatforms. *Biomaterials*, **35**, 9224–9235.
[130] Shahbazi, M.-A., Almeida, P.V., Mäkilä, E., *et al.* (2014) Poly (methyl vinyl ether-alt-maleic acid)-functionalized porous silicon nanoparticles for enhanced stability and cellular internalization. *Macromolecular Rapid Communications*, **35**, 624–629.

7

Intestinal Mucosal Models to Validate Functionalized Nanosystems

Cláudia Azevedo[1,2,3], Inês Pereira[1,2] and Bruno Sarmento[1,2,4]

[1] *INEB – Instituto de Engenharia Biomédica, Universidade do Porto, Portugal*
[2] *i3S – Instituto de Investigação e Inovação em Saúde, Universidade do Porto, Portugal*
[3] *Instituto de Ciências Biomédicas Abel Salazar, University of Porto, Portugal*
[4] *CESPU, Instituto de Investigação e Formação Avançada em Ciências e Tecnologias da Saúde, Gandra, Portugal*

7.1 Introduction

Nanoformulations have been widely studied in the research field, but unfortunately they have not been broadly translated to the clinics/market [1]. The reason behind this is the difficulty in predicting the behavior of these formulations in the human body [2]. Every substance that is designed to be administered in the human body, including nanosystems, has to have several parameters assessed. This includes not only efficacy, but also safety, stability, and method of internalization [3]. To do so, conventional cell models are used, whether in a 2D or 3D configuration, and then animal models, before passing to clinical trials. Cell models are the standard beginning and have been used for almost 100 years. 3D models present themselves as more similar in terms of architecture to what cells experience *in vivo*, but still they are not the most used. The lack of accuracy when physiological processes are being mimicked make these models insufficient when researchers are trying to test a new formulation [4]. *In vivo* models are then necessary to complement these experiments. The problems with these models are the high cost and time needed, as well as all associated ethical problems [5]. Even though there is a major gap between cell models and

Characterization of Pharmaceutical Nano- and Microsystems, First Edition. Edited by Leena Peltonen.

animal models where most formulations fail, another gap exists between animal models and what really happens in humans [6].

When designing nanosystems and then testing them on these models, it is crucial to know the intestinal morphology and the main transport mechanisms for internalization of nanosystems, as discussed in the following sections.

7.2 Intestinal Mucosal Characteristics

The intestinal epithelium is responsible for the control of the passage of drugs, through digestion, absorption of nutrients and water homeostasis. The intestine is thus important for the permeability of nanosystems, which depends on intestinal morphology, biochemical factors, and dosage [7].

7.2.1 Intestinal Morphology

Morphologically, the small intestine is divided into mucosa, submucosa, muscularis propria, and serosa [8]. The mucosa is 90% covered by enterocytes and consequently by microvilli, having an important impact on absorption (Figure 7.1) [9]. This area is composed of epithelium, lamina propria and muscularis mucosae. In more detail, the epithelium, the layer in contact with lumen, is where cell proliferation, differentiation and apoptosis may occur, due to the presence of proliferative crypts. These differentiated cells can be absorptive enterocytes or mucus secretory cells (goblet cells, enteroendocrine cells and Paneth cells). In terms of function, *enterocytes* perform the most important one – the absorption of nutrients, since they are highly polarized cells and can transport them from the apical, through the epithelium, to the basolateral side. Other than those, *goblet cells* are responsible for lubricating and protecting the gut by secreting mucins, which produce mucus; *enteroendocrine cells* can secrete hormones and coordinate intestinal functioning; finally, *Paneth cells* secrete proteins responsible for innate immunity [10]. Away from intestinal crypts there are isolated lymphoid follicles, Peyer's patches, which contain the specialized microfold cells (M cells). M cells are essential for the transepithelial transport of macromolecules or particles and microorganisms. In addition, M cells have the characteristic of possessing sparse microvilli on their apical side, not possessing a glycocalix layer, and forming a pocket with lymphocytes or macrophages at the basolateral side. These specific characteristics make these cells an ideal place for antigen sampling, which facilitates contact with the immune system [11, 12]. The second layer, lamina propria, works as a connective tissue presenting a set of cells (macrophages, dendritic cells, B and T cells, and blood capillaries), while the third layer is composed of smooth muscle [10].

The submucosa is formed by a connective tissue with lymphatic and blood vessels, ganglion and nerve cells. The muscularis propria layer is made of smooth muscles, and finally, the serosa layer is composed of mesothelium – a continuous connective tissue with the abdominal peritoneum [8].

However, as is already known, nanosystems cannot be well absorbed when orally administrated, because of its rapid enzymatic degradation in the gastrointestinal tract (GIT). In fact, several macromolecular proteins are degraded before crossing the

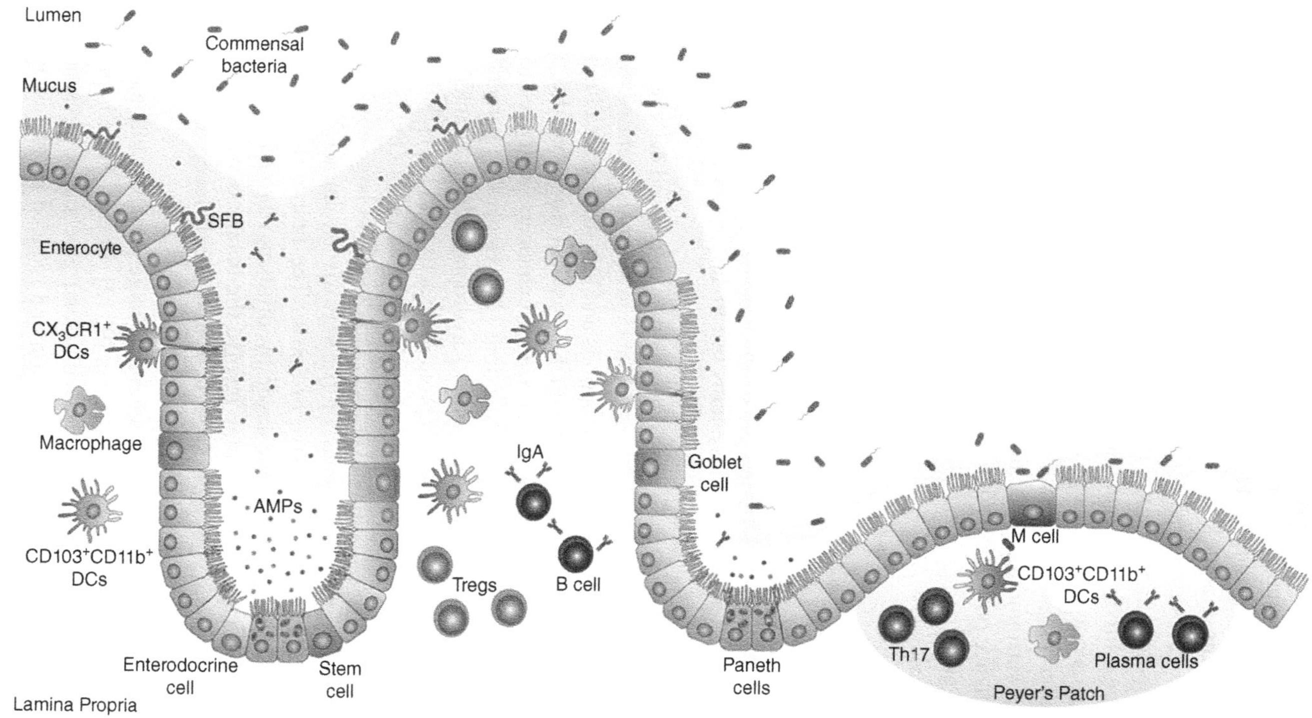

Figure 7.1 *Intestinal mucosal surface at the steady state. The intestinal epithelial barrier is a highly organized mucosal surface that prevents the entry of microbes into the lamina propria. The epithelium comprises a single layer of intestinal epithelial cells (IECs) covered by a stratified mucus layer. Unlike the outer mucus layer that is colonized by commensal species, the inner mucus layer is mostly devoid of bacteria; it contains immunoglobulin A (IgA) and antimicrobial peptides (AMPs) that avert the commensal species from interacting with the surface of the IECs. Yet, few opportunistic bacteria such as segmented filamentous bacteria (SFB) can breach the mucus barrier and enter into contact with the IECs. There are five IEC lineages derived from epithelial stem cells that proliferate and give rise to daughter cells. These cells include enterocytes, mucus-producing goblet cells, hormone-producing enteroendocrine cells, AMP-producing Paneth cells at the base of the crypts, and finally, M cells that sample antigens from the intestinal lumen in order to present them to nearby immune cells. A high number of T cells, macrophages, IgA secreting B and plasma cells are present in the lamina propria and the Peyer's patches. In addition, CD103+CD11b- and CD103+CD11b+ dendritic cells (DCs) promote the development of T helper (Th17) and T regulatory (Treg) cells respectively, while CX3CR1+ DCs sample the lumen antigenic content. Reprinted with permission from [13].*

intestinal epithelium. There are three important points that must be considered: enzymatic degradation, the physiological barrier, and nanosystem structure. The harsh conditions of the GIT particularly contribute to nanosystem degradation. This effect can occur due to enzymes such as proteolytic enzymes present in lysosomes, brush-border peptidases in the villi, and pancreatic proteases in the duodenal region, as well as bacterial flora from the complex hydrogel-containing mucin, known as mucus of intestine. All these factors can modulate the solubility at gastrointestinal pH, lipophilicity, pKa and molecular weight of nanosystems, and consequently their permeability [14]. Also epithelial cells, bound by tight junctions, plus the transit time, fluid dynamics, physiological response, gut mucosa, transporters and receptors, may contribute as physiological and anatomical barriers [15]. Nanosystems can also suffer physicochemical modifications caused by low pH, high temperature, reactions with surrounding molecules, as well as nanosystem weight, shape and charge, which also influence poor absorption and degradation [16, 17]. Finally, factors regarding luminal complexation, drug metabolism, liver uptake and bile excretion are related to permeability efficiency [18].

7.2.2 Transport Mechanisms

Once ingested, small molecules can cross the epithelial layer by passive diffusion, while the larger molecules must cross paracellularly or even by the transcellular route. Figure 7.2 shows different transport mechanisms that nanosystems use to surpass intestinal epithelium [19].

Paracellular is the term for transport that occurs between two juxtaposed cells, where a channel connects the luminal space to the basolateral membrane by tight junctions, which maintain structural integrity. The presence of tight junctions and low surface area limit the transport of nanosystems, since they cannot pass through the intestinal barrier via this route [20]. Thus nanosystems can then be associated with permeation enhancers, such as chitosan, polyacrylate, and thiolated to reversibly open tight junctions [16]. Hence, the paracellular pathway occurs to one side of the cell and mainly transports hydrophilic drugs, thanks to the diffusion through these channels, and it is deficient in proteolytic activity [11, 21]. However, tight junctions represent less than 1% of intestinal mucosa [22], which is not the best way to transport nanosystems if the purpose is to deliver them rapidly and in large amounts.

On the other hand, the *transcellular pathway* or *transcytosis* is characterized by a movement throughout the cell and can be passive or active. The first is used by large and hydrophobic particles, which can cross the layer by diffusing through the lipidic bilayers [11]. The transcellular active pathway may involve the uptake of vesicles at the apical membrane by M cells or endocytosis, via specific or nonspecific receptor-mediated transcytosis.

The transcellular pathway used by M cells can internalize large particles, like nanosystems, while smaller particles are captured by enterocytes [9, 12]. In fact, M cells present a lack of mucus secretion, an underdeveloped microvillus and glycocalyx, presence of apical microfolds, an increased intracellular vacuolization and reduced proteases, which are properties that favor the transport of nanosystems [22, 23]. After contact, nanosystems may interact with M cells via specific or nonspecific receptor-mediated mechanisms: clathrin-mediated endocytosis, macropinocytosis, phagocytosis or actin-dependent uptake

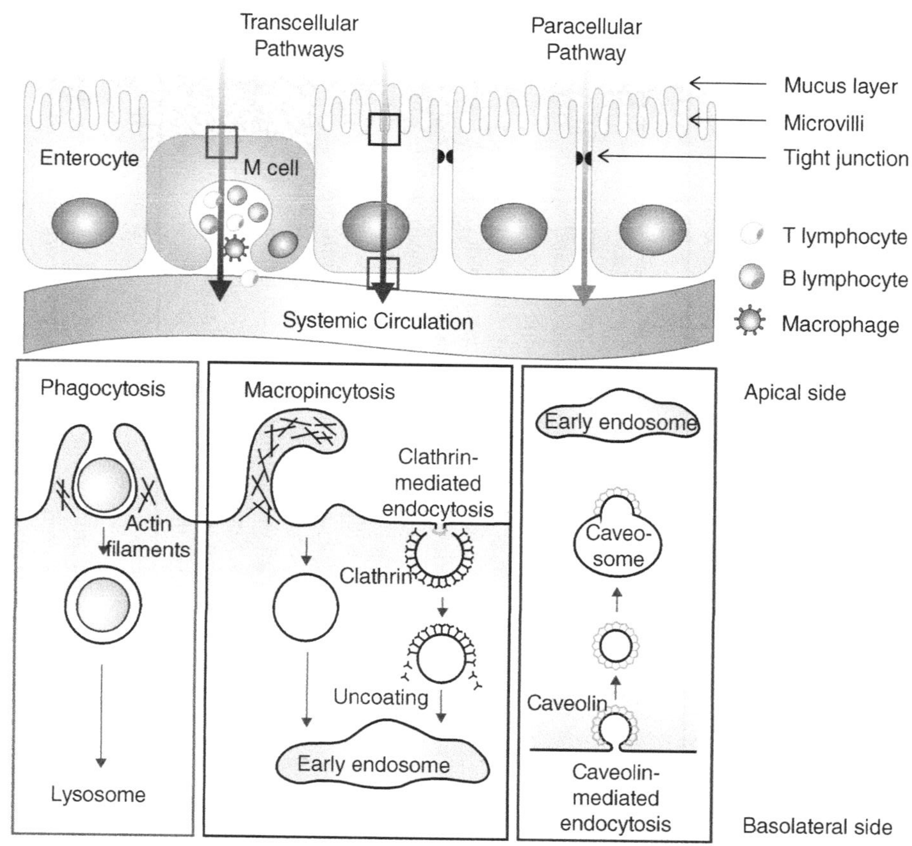

Figure 7.2 *Schematic illustration of the different delivery pathways. Reprinted with permission from [19].*

processes [22]. Since M cells are specialized for antigen uptake, they are a potential portal for oral delivery of drugs [21, 24]. However, similar to tight junctions, M cells also represent around 1% of intestinal epithelium, which may not be the best option to increase drug absorption [11]. The interactions of nanosystems with the intestine and how to target M-cells is reviewed elsewhere [25].

In cells, *endocytosis* (phagocytosis – uptake of large particles; pinocytosis – uptake of fluids and solutes) represents the classic mechanism of nanosystem uptake, which involves engulfment in membrane invaginations and formation of intracellular vesicles [26, 27]. This pathway can be divided into five types: clathrin-mediated uptake, caveolae-mediated uptake, clathrin- and caveolae-independent uptake, macropinocytosis, and phagocytosis. Clathrin- and caveolae-independent uptakes have not yet been observed in epithelial cells (Table 7.1). Also, enterocytes normally take up nanosystems by clathrin-mediated uptake or phagocytosis [28].

Table 7.1 *Summary of endocytic mechanisms present in epithelial cells*

Mechanism	Description	Reference
Phagocytosis	Large particles are engulfed by the cellular membrane through a receptor-mediated process. It is limited to macrophages, dendritic cells and M cells.	[23, 29]
Macropinocytosis	A large volume of fluid, containing nanosystems, is engulfed. It is an active and actin-dependent process, but is non-specific receptor mediated, and so not restricted to a cellular type.	[22, 23]
Clathrin-mediated endocytosis	It is a receptor-mediated uptake mechanism, where small vesicles are formed coated by clathrin, and then fused with early endosomes.	[23]
Caveolae-mediated endocytosis	Small caveolae-coated vesicles (50–80 nm) are formed in a non-specific uptake process. They can escape from the endolysosomes and lead to direct exocytosis. Represents the mechanism with least involvement in drug delivery and uptake, due to the small size of the vesicles.	[22, 23]

A strategy to mimic the intestinal morphology and transport mechanisms, and to study the permeability of nanosystems, is used for *in vitro, ex vivo, in situ* or even *in vivo* models, as discussed in the following sections.

7.3 *In Vitro* Models

Despite the fact that *in vitro* models do not mimic real *in vivo* situations, they can easily simulate a multiple transport system and be reproduced in the laboratory. These types of studies are becoming more important to the pharmaceutical industry, since the obtained results can influence the initial steps of drug development. This method is characterized as being less labor-intensive and cost-intensive, and raising fewer ethical considerations compared with *in vivo* studies. Additionally, the use of *in vitro* models allows for an early evaluation of drugs, without using *in vivo* models and large amounts of drug to analyze its administration, distribution, metabolism, excretion and toxicity (ADMET) [10]. In this sense, it is possible to have *in vitro* models where gastric and intestinal fluid can be simulated, and dialysis membrane systems and Transwell® can test the permeability of compounds.

The *simulated gastric and intestinal fluid model* can be used as a preliminary study to analyze the reaction of the nanosystem to the GIT environment. The used medium has to be representative of the stomach and intestinal fluid, presenting characteristics such as enzymes and ionic strength, which influence the interaction and stability of nanosystems. According to the *United States Pharmacopeia* (USP) the simulated gastric fluid (SGF) medium should be at pH 1.2 and the simulated intestinal fluid (SIF) medium at pH 6.8. These mediums are easily prepared and normally experiments are performed at 37 °C with agitation (100 rpm). At different time points the supernatant is removed by centrifugation and quantified. It is known that these conditions reduce the number of nanosystems, besides changing their size and morphology and consequently bioavailability [30].

On the other hand, the *dialysis membrane system* tests the transport of nanosystems across the barriers of the gut. In this method, nanosystems are added to a dialysis membrane with the same conditions as the last technique, in order to represent the GIT characteristics and movements. These two models can be done without using human cell lines, which can be a drawback since they will not represent the epithelial behavior [31].

Lastly, the *Transwell® system* is the most commonly used model to evaluate the permeability of drugs from lumen to blood circulation, since it realistically mimics the *in vivo* intestine. Briefly, this system is composed of a membrane with pores that separate the apical from the basolateral sides. The epithelial cells are seeded at the apical side and when confluent with an ideal transepithelial electrical resistance (TEER), the experiment can be performed. Nowadays, this system can be performed with only one cell line (monoculture) or combined with other cell lines (co-culture), like HT29-MTX or Raji [10].

7.3.1 Monoculture Models

In this type of model a single cell line is seeded, forming a monolayer. There are several cell lines being used in *in vitro* models of the intestine, but the most widely used to recreate the human environment is the Caco-2 cell line. According to the USFDA, the Caco-2 cell line integrates the biopharmaceutical classification system (BCS), since it has the advantageous ability to screen and predict nanosystems' solubility and bioavailability, as well as their interaction with the gut [32]. Caco-2 is a high-resistance cell line from a human colorectal carcinoma and is considered the gold-standard permeability screen for compounds in drug discovery, due to its morphological and functional characteristics as well as the presence of cytochrome P450 3A4 (CYP3A4) enzymes, sucrase-isomaltase, mature enterocyte alkaline phosphatase, drug transporters, and efflux proteins. Initially (3–4 days) the cells are undifferentiated and do not present microvilli or their indicator (sucrase immunoreactivity). When confluent, they form a polarized monolayer with microvilli, connected by tight junctions (Figure 7.3A). The strength and tightness of the monolayer, meaning confluence and viability, can be identified measuring TEER. Higher TEER values indicate tighter membranes and consequently the modulation of paracellular transport. Also, an indicator of

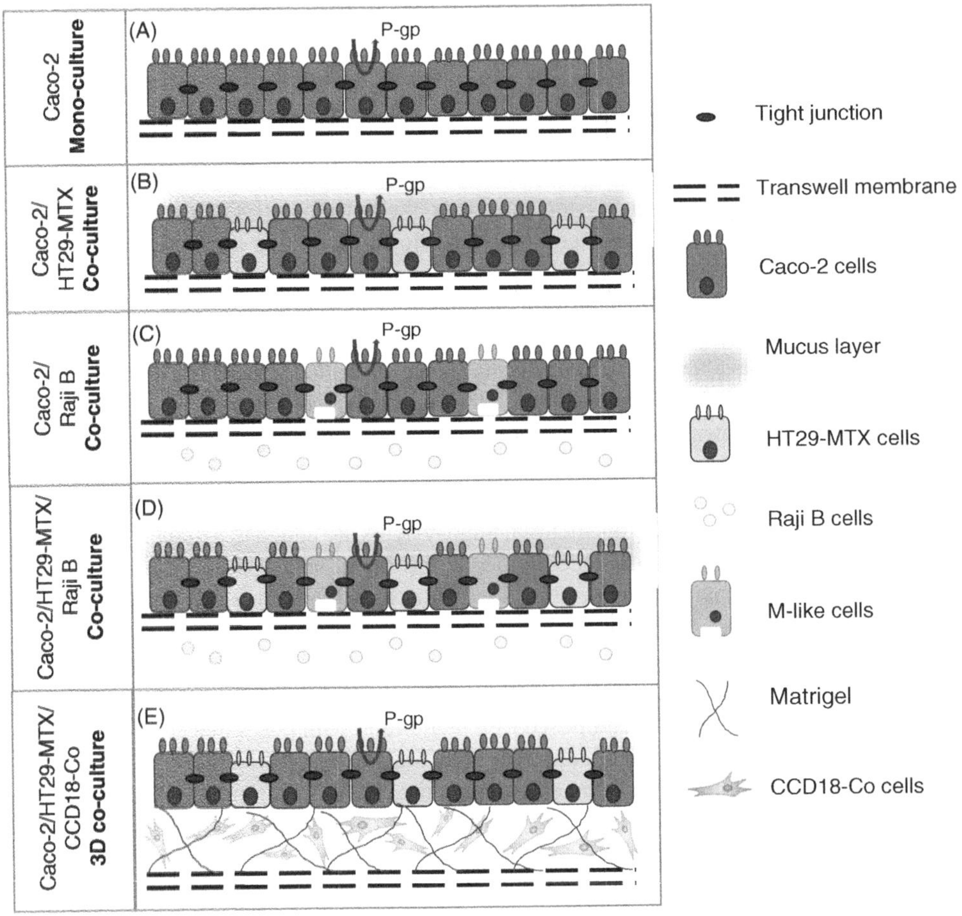

Figure 7.3 *Illustration of (A) Caco-2 monoculture model; (B) Caco-2/HT29-MTX co-culture model; (C) Caco-2/Raji B co-culture model; (D) Caco-2/HT29-MTX/Raji B co-culture model; and (E) 3D co-culture model developed by [70].*

the membrane integrity is the paracellular transport of small molecules like lucifer yellow, FITC-dextran or mannitol [10]. Caco-2 cells do present some drawbacks, namely the low or absent expression of the CYP3A4 metabolizing enzyme, high and variable expression of P-glycoprotein (P-gp) and low paracellular transport of hydrophilic compounds, because of the tighter junctions [33]. Also, the permeability of drugs with slow absorption is poor in this model, compared with rates in humans, and may be due to the lack of a 3D architecture, as explained in Section 7.3.3. Plus, this model lacks mucus production, which excludes the main barrier to nanosystem absorption.

Furthermore, there have been some suggestions of alternative cell lines to find a model that better mimics the human system, some of them being used to analyze nanosystem permeability (Table 7.2). In this line of thought, it is known that TC-7, Caco-2/15 and BBE are some examples of Caco-2 subclones, and consequently are valid alternatives *a priori*.

For instance, TC-7 cells are the second most used model, and compared with Caco-2 cells, they are more homogeneous, present higher enzymatic (sucrase isomaltase) and taurocholic acid active transport levels; however they are not a good predictor for lipophilic nanosystems and only predict the pass diffusion of compounds [34]. Also, authors verified that culturing these cells according to fast protocols (3–7 days) is not advantageous, since cells are not able to induce mature P-gp efflux [35]. In addition, human colonic adenocarcinoma cell line (T84) is a possibility for permeability studies, particularly regarding the transport of IgG and albumin in a FcRn-dependent manner [36]. Studies show that the T84 cell line produces mucin in culture, and releases it in response to various secretagogues [37]. It is known that the absorption mainly occurs at the small intestine, so cells from colon have, *per se,* the drawback of differing in their paracellular permeability – since they present a tighter conformation, the paracellular absorption of compounds is lower.

Also, cells that are not of human origin were developed in order to overcome some drawbacks of the Caco-2 cells. IPEC-J2 is originally from porcine small intestinal enterocytes and is a good alternative to predict the behavior of drugs due to the similarities between pigs and humans [38]. The 2/4/A1 cell line is from rat intestinal epithelium and is able to form a monolayer with differentiated cells, with membrane enzymes, transporter proteins, and tight junctions, which like human intestine lead to paracellular transport. Actually, tight junctions from 2/4/A1 cells are not so tight, mimicking the human intestine transport better than Caco-2 cells, being a good predictor for poorly absorbed nanosystems. Despite similarities with the human cell line, 2/4/A1 cells do not have active transport of drugs, which limits the model [39]. Also, IEC-18 is a low-resistance small intestine cell line from rats, but in this case from native ileal crypts, with fetal alkaline phosphatase. These cells can also be used to study the effects of enzymes and receptors during drug permeability, showing the transcellular transport of lipophilic compounds, which is concordant with Caco-2 cells [40, 41]. However, IEC-18 are less differentiated than Caco-2 cells, which means that some of the carrier-mediated transport is absent. Nevertheless, IEC-18 is an useful model to study paracellular transport and, in addition, a good molecular size selector [42].

Despite having a non-intestinal origin, the LLC-PK1 cell line derived from pig kidney epithelial cells are polarized cells with low expression of transporters and easier to transfect than Caco-2. Also, LLC-PK1 is a good tool to study the bidirectional transport of compounds and their toxicity [43]. Studies have demonstrated that this cell line presents P-gp, responsible for polarised transport of drugs [44]. Additionally, the Madin-Darby canine kidney (MDCK), from the distal tubule and collecting duct, is highly correlated with Caco-2 cells, working as a strong alternative for studying drugs that are absorbed passively [45]. Authors showed the utility of MDCK cells in probing interactions between P-gp and their substrates. Comparing the permeability of anti-malaria compounds in Caco-2 and MDCK cells, they verified the overexpression of P-gp in MDCK-MDR1 cells, which allows high permeability [46]. A recent study inclusively analyzed the mechanisms of transport of Deoxynivalenol (DON), using inhibitors, verifying the presence of organic anion and cation transporters. P-gp was also identified as the major efflux protein, due to the higher basolateral-to-apical observed transport [47]. Regarding nanosystem permeability, the transport of nanoemulsions of self-nano-emulsifying drug delivery systems (SNEDDS) was also proved through MDCK cells [48], besides other examples mentioned in Table 7.2. Also, these cells only need 3–6 days to become confluent and form tight junctions [45].

Conversely, the fact that they are neither intestinal nor human confers different levels of transporter expression as well as different metabolic activity [49].

Moreover, it is also possible to do permeability studies using non-oncogenic and non-transformed cells, such as primary cells and stem cells. For instance, primary human intestinal epithelial cells (hInEpCs) and induced pluripotent stem cells (iPSCs)-derived intestinal cells, besides having neotanal Fc receptor, also ensure marker expression and a similar or better monolayer formation [50]. In fact, it is much easier to evaluate drug absorption and metabolism in these cells, due to the presence of enzymes, receptors and expression gene levels, which do not differ from human enterocytes. Researchers developed enterocyte-like cells from human iPSCs and verified an increased efficiency, higher expression level of CYP3A4 and peptide transporter 1, than Caco-2 cells. Plus, the generated cells are able to form a barrier and have induction potency of CYP3A4, proved by using 1, 25-dihydroxyvitamin D3 or rifampicin [51]. Other possible candidates include adult intestinal stem cells (ISCs) that are characterized as having proliferation and differentiation ability, which leads to the maturation of enterocytes. Human small intestinal epithelial cell (HIEC) lines are differentiated from ISCs and are morphologically composed of polarized columnar epithelia, presenting dense microvilli and tight junctions. Comparing HIEC with Caco-2 cells, studies verified no significant differences in drug-metabolizing enzymes and transporters. Regarding permeability, it was higher in HIEC cells when using compounds that are paracellularly transported and nucleoside transporter substrates, while there were no differences in transcellular transport [52]. A recent study showed the advantages of this cell line, describing their accuracy to predict the absorption of several drugs with different properties, such as the presence/absence of efflux transporters [53]. Despite having a lot of potential, stem and primary cells are difficult to obtain and to maintain, and are limited by their viability and extensive differentiation time.

7.3.2 Co-culture Models

7.3.2.1 The Caco-2/HT29-MTX Model

HT29 cell line is originally from human colorectal adenocarcinoma and can mimic goblet cells. The HT29 cells need a different medium compared with Caco-2 cells, so they can differentiate and acquire a heterogeneous phenotype. To surpass this barrier, authors verified that by adding 10^{-6} M methotrexate (MTX), the HT29-MTX cell line is able to form a polarized monolayer with goblet cells and differentiate into mature goblet cells, able to produce mucus, after 21 days in culture [54]. The differentiation character can be confirmed by MUC cell markers. Mucus is a visoelastic gel produced by mucin glycoproteins and is responsible for preventing pathogen entrance, also acting as a barrier to nanosystem absorption, which affects the nanosystem retention time [8]. In fact, authors have previously demonstrated that mucus clearly decreases the uptake and transport of nanosystems [55]. It is known that nanosystems (with positive charges) can interact with mucus (negatively charged) through electrostatic interactions, as well as hydrophobic or van der Waals interactions [56]. The role of mucus in cell-based models is already well described elsewhere [57]. So, this parameter must be taken into consideration when evaluating nanosystem permeability. When cultured with Caco-2 cells, there is the necessity to study different ratios to prevent alterations in barrier properties caused by the overgrowth of HT29-MTX cells [58].

When Caco-2 cells at 10^5 cells/cm^2 are combined with HT29-MTX in a 9:1 proportion, the model (Figure 7.3B) gets closer to human intestine characteristics, generating more predictable results [59]. Furthermore, HT29-MTX do not express P-gp and present less tight junctions than Caco-2 (decreasing TEER values), which facilitate the passive permeability of nanosystems and represent a more physiological model [60].

7.3.2.2 *The Caco-2/Raji B Model*

It is also described that Raji B cells, from a human Burkitt's lymphoma, when combined in a 1:2 proportion with Caco-2 cells at 10^5 cells/cm^2 induce M cell phenotype (sparse microvilli and high transcytosis activity) (Figure 7.3C). This happens due to the interaction between Caco-2 cells and lymphocytes and the presence of galectin-9 or SLAA, a signal of differentiation. In this model, Raji B lymphocyte cells are only added at the basolateral side after 14 days of seeding Caco-2 cells at the apical side. A protocol was recently published suggesting an inverted *in vitro* M-cell model, where it is possible to do transport experiments every 10 h, after 3 weeks [61]. Despite not being in contact with Caco-2 cells, it is possible to observe cells with M-cell characteristics. As mentioned before, M cells serve as antigen-sampling cells and, unlike enterocytes, they can translocate particles without digesting them. Indeed, studies confirm that M cells play a major role in translocation and permeability of nanosystems, showing that these parameters are higher in this type of co-culture compared with the Caco-2 monoculture [62]. All in all, as a drawback, this model does not take into account the mucus, which is an important intestinal characteristic.

7.3.2.3 *The Caco-2/HT29-MTX/Raji B Model*

Taking all of the above information into consideration, Sarmento's group developed a triple co-culture model, with the three cell lines: Caco-2, HT29-MTX and Raji B cells (Figure 7.3D). This group tested the model with normal and inverted orientations, verifying higher permeation with a normal orientation [63]. This is actually a more accurate and realistic model due to its similarity to human intestine characteristics: enterocytes (Caco-2 cells), mucus production by goblet cells (HT29-MTX cells) and M cells (Raji B cells). The proportions mentioned before were based on physiological ratios and previous works [64].

In addition, it has already been shown that Caco-2/HT29-MTX co-culture and the triple Caco-2/HT29-MTX/Raji B models are useful for predicting the intestinal transport of free drugs and nanosystems and to classify them, acording to BCS, with low or high permeability. Furthermore, they have a good correlation with *ex vivo* models, which will be explained later [65].

7.3.3 3D Co-culture Models

All the models mentioned before are based on a 2D surface which, once again, does not represent *in vivo* cells with a heterogeneous environment containing several cellular events [10]. So, there was a necessity to create a 3D model with physiological structures to reproduce more accurately the intestinal functions, namely growth, differentiation, interactions between cells themselves and with the matrix, and also mechanical forces. In addition, cells in a 3D architecture are easily differentiated and polarized when on top, compared with those that are near the villus base, which may be an advantage for nanosystems that remain longer in the lumen (slow absorption) [66]. Like atenolol, a hydrophilic and slowly absorbed drug, certain nanosystems can migrate to the bottom of cells and be transported through cell–cell tight junctions [67]. Furthermore, 3D models have less tight junctions, compared with a Caco-2 monolayer, resulting in a better permeability which leads to less expression of P-gp and is consistent with *in vivo* observations [68].

Some authors have verified that MDCK cells in Matrigel can form a 3D architecture when the viscosity is low and the cellular contractile forces are strong. The 3D structure can then be changed in a viscosity-dependent manner. One way to use it is to do a treatment with genipin (GP), since it allows the crosslink between amino groups of the matrigel peptide chains [69]. The 3D model can also include macrophages, increasing the uptake of particles, as well as other type of cells or even polymeric scaffolds to comprise the intestinal microvilli architecture. For instance, a recent model was proposed, where 10^4 cells/cm^2 of intestinal myofibroblasts (CCD18-Co cells) are entrapped in a thick layer of Matrigel and then 10^5 cells/cm^2 of Caco-2 cells are added [70]. The differentiation of CCD18-Co cells can be confirmed by the observation of α-SMA or CD90 through immunocytochemistry. It was verified that the matrix is important to maintain the architecture and support the model. On the other hand, fibroblasts contribute to epithelial cell growth (Figure 7.3E) [70]. There is also a model using Caco-2, HT-29 and T84 cell lines as well as proinflammatory stimuli, to mimic an inflamed intestinal mucosa [71]. Also, a model with the traditional Caco-2 cell line and human immune cell lines (human macrophages (THP-1) and human dendritic cells (MUTZ-3)) was developed to evaluate the toxicity/safety of nanosystems [72]. The 3D co-culture model presents, in fact, additional complexity that may contribute to development of nanosystems to treat inflammatory diseases or even to study the interaction of the nanosystems with an inflamed intestine. Some authors have already investigated the genetic and physiological properties of the 3D model, verifying that by knocking down MUC17 it is possible to reduce the protein expression of tight junctions (occludin) and increase the permeability. These authors created a 3D collagen scaffold enabling modeling of the exact geometry and density of the intestine in an *in vivo* situation, and more recently a synthetic and biodegradable hydrogel [71, 73].

7.3.4 Gut-on-a-Chip

These microfluidic devices are characterized by their ability to have cultured cells in constant contact with fluid flow, with the main goal of mimicking physiological functions of the desired tissue or organ [74]. Even though reconstructing a whole organ on a chip is not possible, it is also not the goal. Having small units that can recreate to some extent what happens at tissue or organ level is the aim [75]. A chip is produced similarly to computer chips, using lithography techniques, and can be engineered in order to be simple, with only a monolayer of cell culture, or, becoming more complex, using more than one cell type, usually separated by a porous membrane. Besides the fluid flow stress, other physical forces can be introduced into these chips, such as mechanical compression or cyclic stress. All these parameters can be regulated and adjusted as desired in order to better mimic the microenvironment where the cells would be. In the specific case of the intestine, Kim *et al.* [76] recreated a gut-on-a-chip, where Caco-2 cells were cultured on top of a PDMS porous membrane. The cells were exposed to continuous flow from the channel on top and on the bottom of the membrane. A type of bacteria was also cultured with Caco-2 in order to better mimic the influence of the microflora in the barrier functions. Latest publications describe a gut-on-a-chip as constituted with four chambers, two where the culture medium circulates and two vacuum chambers that, when active, stretch the membrane, simulating the peristaltic motion that cells feel in the normal intestine (Figure 7.4). The results showed a promising model, even though the complexity of the entire intestine was not simulated [76].

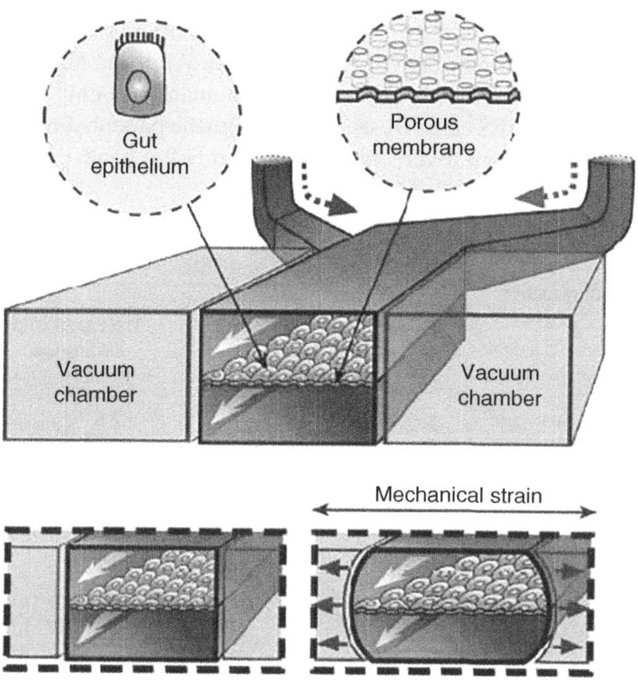

Figure 7.4 *Representation of a gut-on-a-chip. Reprinted with permission from [76].*

Recently, a 3D structure of iPSCs was incorporated in a small microengineered chip, making an amenable model for study. This model is characterized for its polarization and having all the intestinal epithelial subtypes, as well as its response to exogenous stimuli [77]. Plus, an organ-on-a-chip platform has lately been developed, compatible with standard equipment and easy to use. Using this, the intestinal barrier integrity through matrix-polarized epithelial gut tubes was recreated in a microfluidic environment, expressing polarization and transporters. The pharmacological response can be documented by imaging techniques. According to the authors, this system stands out because it is sensible, compatible, biomimetic, allows multitesting and reduction of reagents, cells and time.

Lastly, this system is already patented and commercialized for science and pharma, and can be applied to different organs or diseases with complex compositions [78]. Chips like this are being proposed as methodologies to test the permeability of nanosystems, but there are still not many published papers about it, since this technique is only in its infancy [79, 80]. Several improvements have to be made, but the potential of this technology cannot be ignored [81].

The possibility of connecting chips with different organs allows the evaluation of the effects of the drug or nanoparticle tested in organs where its effect was not predicted. For example, several drugs have been removed from clinical trials or the market because of cardiac effects that were not predicted. In this light, testing the effect of the drugs only on the target tissue is not logical. A "circuit" composed of chips from organs like gut, liver, lung, heart, kidney and bone, as shown in Figure 7.5, would give insights into the effects of the tested formulations in non-target organs [83, 84]. After the human-on-a-chip goal, the perfect and most ambitious goal is to create the human-on-a-chip with cells from the patient themselves, namely iPS cells. By using cells from the patient, the difference between patients can be erased and each treatment can be tested beforehand, guaranteeing the best possible care [75, 85].

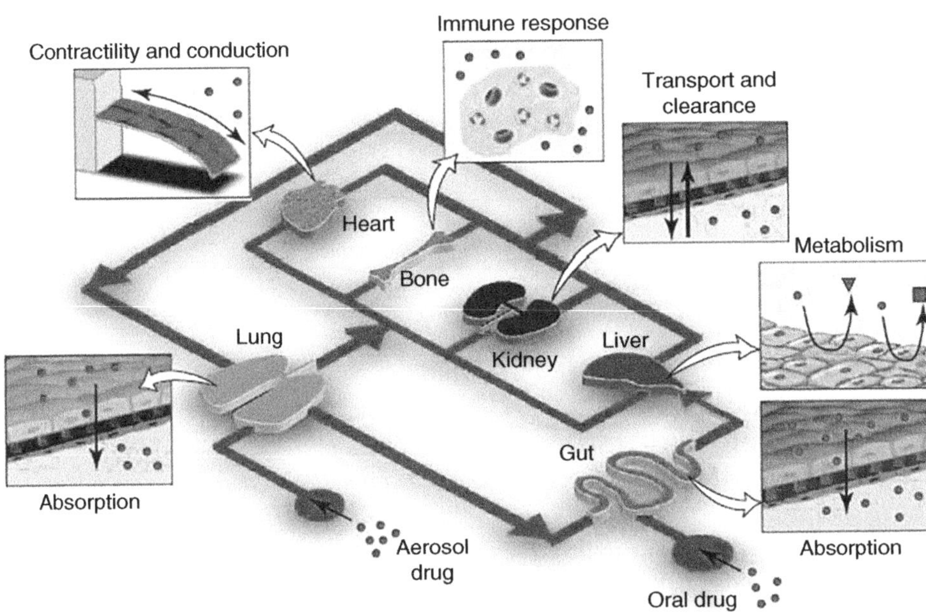

Figure 7.5 *Illustration of a human-on-a-chip. Reprinted with permission from [82].*

To sum up, Table 7.2 shows examples of different nanosystems and the *in vitro* model used in each of them, to screen the intestinal permeability.

Table 7.2 *Examples of in vitro models that can be used by nanosystems in permeability studies*

Model	Cell line	Nanosystem	Observations	Ref
Monoculture	Caco-2 (human)	Ag or Au nanoparticles	After apical treatment, both internalization and transfer were greater for smaller Ag nanoparticles.	[18]
		Fe_3O_4 nanoparticles	Higher uptake under inflammatory conditions.Upregulation of clathrin in IL-1β-stimulated Caco-2 cells.Internalization via clathrin.	[86]
		Carboxylate-modified polystyrene nanoparticles	The extracellular matrix proteins may impact on transmucosal delivery.	[87]
Monoculture – Alternatives to Caco-2	T84 (human)	Ag nanoparticles	Exposure to AgNP compromises the integrity of the intestinal epithelium.	[88]
	TC-7 (human)	Poly(isobutylcyanoacrylate) NPs coated with chitosan and thiolated chitosan	Lower nanoparticle toxicity in TC-7 and HT29-MTX but strongly toxic in HeLa cells.	[89]
	IPEC-J2 (pig)	PAMAM dendrimers	Dendrimer-polyamine conjugates as carriers for antigen/drug delivery.	[90]
	LLC-PK1 (pig)	Different types of carbon black and titanium dioxide nanoparticles	A type of carbon black enhances the ROS production.	[91]
	MDCK (dog)	PEI/PIP3 polyplexes	PIP3 improves the internalization of basolateral receptors. Polyplexes with an efficient endosomal escape.	[92]
		Polystyrene nanoparticles (PNP)	Evaluation of PNP flux. Translocation via clathrin-mediated endocytosis.	[93]
Co-culture	Caco-2, HT29-MTX	Liposomes carrying hydrophilic or hydrophobic bioactive molecules	Study of mucus-liposomes interactions.	[94]
		Ag nanoparticles	Mucus layer protects against ROS and decrease in IL-8 release.	[95]
	Caco-2, Raji B	TiO_2 NPs	Shows the relevant role for M-cells in nanoparticle absorption.	[62]
	Caco-2, HT29-MTX, Raji B	Dual chitosan/albumin-coated alginate/dextran sulfate NPs	Reduced permeability after decreased temperature. Transport by clathrin-mediated endocytosis. The interaction between glycocalix and the NPs is critical for insulin permeation.	[96]
		Polystyrene NPs	NPs were capable of penetrating the intestinal barrier mainly via specialized M cells.	[97]
		Nanostructured polyelectrolyte complexes	Study of the mucoadhesivity.	[98]
3D co-culture	Caco-2, HT29-MTX, C CD18-Co	PAMAM dendrimer and gold nanoparticles	To analyze the nephrotoxicity.	[99]
	Caco-2, HT-29, T84	Polystyrene nanoparticles	Interaction of the nanosystems with an inflammated intestine	[71]

7.4 *Ex Vivo* Intestinal Models for *In Vitro/In Vivo* Correlation of Functionalized Nanosystems

In vitro tissue-based models, also called *ex vivo*, are used in intestinal permeability studies in a way to move from *in vitro* to *in vivo* models. In this type of model, living functional tissues or organs are cultivated in a controlled environment after being isolated from organisms. These models present features that are not normally present in cell models, such as the adequate presence of mucus, paracellular permeability, or the expression of transport proteins and drug metabolism. The main drawbacks of these and all *in vitro* models is the impossibility to recreate the gastric empty rate, GI pH or transit rate. Despite this, *ex vivo* models have been used to analyze nanosystem permeability (Table 7.3).

7.4.1 Diffusion Chambers

The most performed method using intestinal tissues is diffusion chambers, namely the Ussing chamber or Franz cells. The main difference between these two is the direction of the flux. In the first it is horizontal, or side-by-side, and in the second it is vertical [100].

7.4.1.1 *Ussing Chamber*

The Ussing chamber (Figure 7.6A) was invented in 1951, and used with intestinal tissue from 1988 [101, 102]. The size, volume and area of exposed tissue are parameters that can vary. In this methodology, a section of intestinal tissue is excised, separated into small fragments and opened to form a flat surface. Then the tissue is placed in the chamber between two halves. The chamber is filled with physiological buffer and gassed with carbogen as well as warmed at 37 °C. Two chambers, the donor and receiver, are divided by the tissue, one of them representing the serosal side and the other the mucosal side. The drug or nanosystem to be tested in a permeability test can be placed in one of the chambers and samples should be removed from the other chamber, replacing the volume taken with fresh buffer [103, 104]. The TEER is again one of the parameters that can be measured by placing the electrodes on each side of the membrane [105]. The main advantage of this model is the maintenance of the gut architecture, membrane transporters and enzymes when compared with cell-based models. It is also possible to study regional differences in the GIT by using different sections of it. This model has been used inclusively to test nanoparticle permeability capacity [106].

The Ussing chamber can also be used with human tissues instead of animal tissues. Since tissues can easily be obtained from tumor resections or gastric bypass surgeries, more recent studies have been using this type of tissue [107]. One disadvantage, though, is the fact that the experiments should be performed near a surgery clinic since the tissues can quickly deteriorate outside the body [108]. Nanosystems can then be tested using this technique in a more sensitive and complex manner, compared with *in vitro* cell models.

7.4.1.2 *Franz Cell*

On the other hand, the Franz cell (Figure 7.6B) is mostly used for skin permeation studies rather than intestinal permeability, but is used nevertheless. This model is not much different

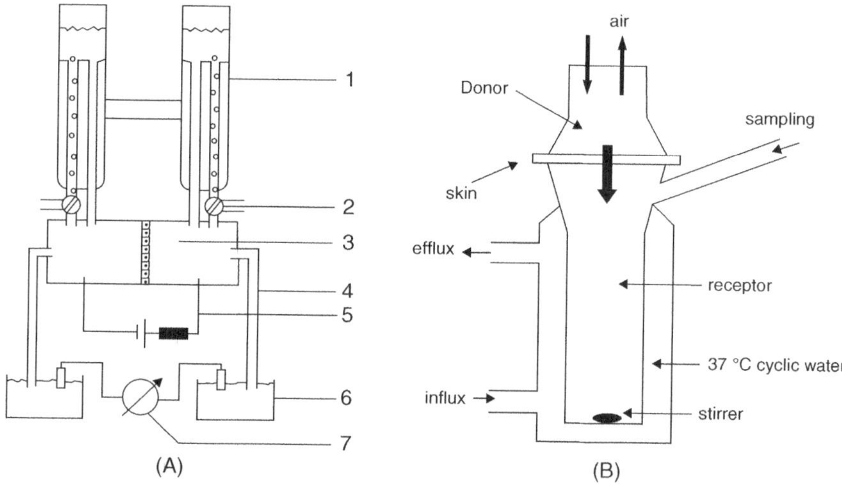

Figure 7.6 *Representation of diffusion chambers: (A) Ussing chamber with (1) U-shaped glass tube, (2) three-way valve to adjust flow rate of gassing, (3) half-chamber with tissue sample, (4) agar bridge, (5) current injector, (6) KCl-filled vessel, (7) voltmeter; (B) a Franz diffusion cell. Reprinted with permission from [109, 110].*

from the Ussing chamber, since the tissue membrane is still placed between the donor and receptor chambers. The main significant difference, besides the direction of the flow, is the size of the chambers, since in the Ussing chamber the sizes are similar and in the Franz cell the donor compartment is smaller than the receptor [111]. This difference in volumes can induce discrepancies in the permeability compared with the Ussing chamber, but can be resolved by using a magnetic stirrer in the receptor chamber in order to reduce the unstirred water layer. It is important to use the tissue as soon as harvested and ensure that sink conditions are maintained. A few studies have been done using intestinal tissue in Franz cells to evaluate nanoparticle permeability, for example chitosan nanoparticles, as mentioned in Table 7.3 [112].

7.4.2 Everted Intestinal Sac Model

Another method used in *ex vivo* intestinal permeability studies is the everted intestinal sac. This model can be applied in the study of mechanisms and kinetics of drug absorption and metabolism, including pro-drug conversion in the GIT, efflux transport, drug resistance and interactions [113]. A section of the intestine of the animal is removed, washed, inverted with a rod, and one of the ends is tied. The inside of the sac is filled with oxygenated buffer and the other end closed after. Then it is placed in the container. This container holds the sample to be tested, and its presence within the sac is then evaluated, within the determined time-points. There are two different techniques that can be applied in this method: one where both sides are closed, creating a closed sac; and the other where the end of the tissue is cannulated in the tubes. The main advantages of this model is the low cost, speed, the possibility of testing several drugs, the presence of the mucus layer, the large surface area and the possibility of studying both passive and active transport. The main disadvantages are the risks of damaging the tissue, the release of proteins and/or enzymes as well as the muscularis mucosa which might give misleading results, and the time for which the tissue remains viable, at just two hours [114, 115]. The everted sac is the main method used to test nanosystem permeability. Chitosan or PLGA nanoparticles are two examples of the particles tested [116–119].

7.4.3 Non-everted Intestinal Sac Model

In order to avoid tissue damage, this model does not evert the gut sac after excision. The intestinal sac is removed from the animal, washed thoroughly, and one of the ends of the sac is tied. The sample solution is placed inside the sac and the other end is closed. Then the sac is placed in the release medium. This release medium should be able to mimic the conditions existing in the GIT. This method is more limited than the everted sac in terms of information that can be accessed, but it is simple and rapid to perform, while still capable of testing drug absorption. It is not only simpler but also requires smaller amounts of the drug to be tested, and samples are easier to quantify and can be collected at various time points [111, 120]. In terms of disadvantages, this method is quite similar to the everted sac, where the viability of the tissue is the main issue during handling. This method is used frequently to test nanoparticle permeability, particularly lipid nanoparticles [121–123].

7.4.4 Everted Intestinal Ring

Similar to the everted intestinal sac, the everted intestinal ring is also used in *ex vivo* intestinal permeability studies. Most often used to assess the uptake and metabolism of the tested compounds, in this model the intestinal tissue is harvested, everted and cut into small rings. Then it is put into an incubator with proper incubation media and agitation. Simplicity is the main advantage of this method, allied with the ability to test several compounds [124]. However, this technique also presents disadvantages. Since both connective tissue and muscular layers are exposed in the media, the diffusion of the tested compounds could not occur on the mucosal side. Also, the time of incubation is quite reduced, at around 10 min [104, 124]. Some examples of nanosystems that have used this technique are presented in Table 7.3.

7.5 *In Situ* Models

For *in situ* models, the animal is anaesthetized and the experience is performed on a segment of the intestine. Since the blood supply and nerves are intact, this model is a good simulator of an *in vivo* environment. Thus, between the *in situ* models, nanosystem permeability can be analyzed through intestinal perfusion, intestinal loop or intestinal vascular cannulation to reflect drug absorption, predicting more accurately the *in vivo* permeability (Table 7.4).

Table 7.3 *Examples of ex vivo models that can be used by nanosystems in permeability studies*

Model	Tissue	Nanosystem	Observations	Ref.
Diffusion chambers: Ussing chamber	Mice jejunum	Chitosan nanoparticles (CS NP)	Catechin and epigallocatechin gallate in CS NP enhances *in vitro* intestinal absorption.	[125]
	Rat jejunum	Paclitaxel-loaded Cyclodextrin-poly(anhydride) nanoparticles	NP increased the permeability of paclitaxel.	[126]
Diffusion chambers: Franz cell	Rat duodenum, jejunum and ileum	Insulin-loaded TMC nanoparticles	TMC NP had higher permeability values in the jejunum compared with CS NPs.	[112]
Everted intestine sac	Rat intestine	Gemcitabine-loaded chitosan nanoparticles	The NP can increase the transport of the drug, even though it might affect the integrity of the intestinal epithelium.	[127]
		Alginate-coated chitosan nanoparticle	Increase of paracellular uptake of enoxaparin.	[128]
		Thiolated chitosan-modified PLA-PCL-TPGS nanoparticles (TNP)	TNP and CNP increased significantly the amount of paclitaxel transported.	[129]
Non-everted intestine sac	Rat intestine	Lipid nanoparticles	NP improved the drug permeation.	[122]
Everted intestine ring	Rat intestine	Solid–liquid nanoparticles (SLN)	Uptake was reduced in the presence of chlorpromazine or nystatin since these inhibit the endocytosis pathways, used in the uptake of SLN.	[130]
	Rat jejunum	Oleoyl alginate ester (OAE) nanoparticles	Uptake pathway is dependent on the particle size.	[131]

Usually, these models provide insights about intestinal physiology, allowing the study of barriers, homeostasis, immune responses and transport [132].

7.5.1 Intestinal Perfusion

In this method, a perfusion tube is inserted into the intestine and the fluid containing the nanosystem is poured into the intestinal cavity with a peristaltic pump. Depending on whether the system is continuous or not, the perfusion can be circular or single-pass.

In circular perfusion, the intestinal segment is exposed and connected to two tubes in each end: the upper side is connected to the peristaltic pump to introduce the fluid, and the lower side is connected to the container, where the fluid containing nanosystems is collected. Normally, phenol red is also added as a reference for the viability of the intestine. The concentrations and the effective permeability (P_{eff}) are measured before and after the perfusion, while the release kinetics and the absorption rate constant (k_a) are analyzed through the excision of the perfused intestinal segment. According to the literature this method is mainly used in the early stages of drug development, and absorption is normally higher due to the longer retention time in the intestine [133]. Although involving a simple operation, samples are complex to test and during the experience animal failure might occur. Also, the phenol red may be absorbed, which may affect the intestinal transport and the results.

On contrast, in single-pass perfusion the fluid containing the nanosystems does not return to the initial container. In this case, the method is used for determining the concentration decrease and consequently the pharmacokinetic parameters. The permeability and kinetics can be calculated taking into consideration the bottle's weight (supply and collecting bottles) and content. Single-pass perfusion is a simple operation and enables the study of drugs that are rapidly absorbed. Thus, it is a good method to categorize drugs into a BCS permeability class, as determined *in vivo* [32]. However, as for the last method, samples are complex to test and animal failure may occur. Plus, the measurements by weight are not that accurate [132]. There are already strategies to modulate the P-gp mediated efflux. For instance, combining a P-gp inhibitor (verapamil) with cytostatic drugs increases their permeability instead of an inducer (rifampicin) [134].

This approach is also valid with nanosystems. In fact, authors have used D-alpha-tocopheryl poly(ethylene glycol) succinate 1000 (TPGS) and alkyl-PEO (Brij78) as excipients capable of inhibiting efflux pumps. These excipients with P-gp modulator activity were shown to be a promising option to improve the oral bioavailability, intestinal absorption and water solubility of hydrophobic drugs [135].

For curiosity, a very recent study shows an engineered small intestine system composed of a perfusion chamber (with a segment of intestine), peristaltic pump, pH meter, donor, receiver, and buffer circulation compartments. To overcome passive diffusion due to the absence of energy transporters, a proportionality constant was added, making this system a reliable alternative to the *in situ* single-pass intestinal perfusion technique, especially for compounds with passive diffusion properties [136].

Table 7.4 *Examples of in situ models that can be used by nanosystems in permeability studies*

Model	Segment	Nanosystem	Observations	Ref
Intestinal perfusion: Circular perfusion	Jejunum	lumefantrine loaded binary solid lipid nanoparticles (LF-SLNs)	Enhancement in the permeability of LF from LF-SLN.	[137]
	Duodenum, jejunum, ileum, and colon	Polybutylcyanoacrylate nanoparticles (PBCNs) with curcumin	The absorption was mainly in ileum and colon, through passive diffusion mechanism.	[138]
Intestinal perfusion: Single-pass perfusion	Jejunum (proximal to the duodenum)	Solid lipid nanoparticles (SLN) and mixed micelles (MM) with γ-Tocotrienol	SLN has tenfold higher permeability than MM, enhancing the oral bioavailability by threefold.	[139]
	Duodenum, jejunum and ileum	Doxorubicin hydrochloride loaded noxaparin sodium-PLGA hybrid nanoparticles	Improved membrane permeability and intestinal wall. Caveolin- and clathrin-mediated endocytosis pathways as the main mechanisms.	[140]
Intestinal loop	Ileum	Titanium dioxide nanoparticles	NPs present in the apical regions and in deeper parts of the cytoplasm. Accumulation in the liver.	[141]
	Jejunum	PEGylated solid lipid nanoparticle	The PEGylation promotes cellular uptake, mucus penetration and stability. Internalized by active transport processes.	[142]
Intestinal vascular cannulation	Duodenum	Nanoemulsified vitamin E and green tea microstructures	These nanosystems were shown to be safe and to increase absorption, according to physiological parameters.	[143]

7.5.2 Intestinal Loop

As suggested by the name, this is an intestinal loop with clamps, where the fluid containing nanosystems is injected. After 24 h the animal is euthanized and the segment is removed and kept on ice for analysis. The morphology can be determined by microscopy. Dissolving the tissue into acetonitrile, it is also possible to quantify the remaining nanosystems and calculate the drug content and its absorption at different sites (intestine, jejunum, ileum, and colon). It is easy to perform, compared with intestinal perfusion methods, but difficult to analyze the samples due to the digestive fluids that might be present [115].

7.5.3 Intestinal Vascular Cannulation

This method involves the extraction of an artery or vein blood to simulate direct intestinal absorption of the fluid, containing nanosystems administered intragastrically from the lumen into the circulation. In this case, a gentle surgery is used to overcome the direct vein cannulation to prevent vein engorgement or the acute loss of blood. The absorption is calculated using the amount of the nanosystem absorbed, independent of the animal size and the used blood volume, being a good simulator of intestinal absorption. However, it is difficult to handle, and the sample can be susceptible to catabolism [132].

7.6 *In Vivo* Models

Even though *in vitro, ex vivo* and *in situ* models keep improving, *in vivo* models are always necessary in drug delivery. The release kinetics and biodistribution are two main features tested through these techniques since they can contribute to the failure of therapeutic effects of nanosystems. Since oral delivery of nanosystems mostly occurs in the GIT, most organ analysis and imaging techniques are performed in the stomach, small intestine and colon. For imaging, radioactive labeling and IVIS are preferentially chosen. Radioactive labeling is a technique able to monitor in real time the circulation of nanoparticles in gamma scintigraphy, through Technetium labelling [144]. Using this method, it is possible to visualize and quantify the particles moving through the GIT after oral administration. This technique is normally non-invasive, and the GIT can be removed and analysed after, for better accuracy. The nanosystems used with this method have to be labeled with a gamma-emitting radioisotope chosen according to its half-life and radiation. Technetium (^{99}mTc) is one of the most used because it has a low radiation dose and a half-life of 6 h. The radiation exposure is a major drawback of this method [145]. Single-photon emission computed tomography/computed tomography (SPECT-CT) and positron emission tomography (PET) are two alternative techniques that also use radiolabeling. These two techniques allow a 3D result with higher resolution and correction attenuation that is not present in gamma scintigraphy [146]. The difference between the two is that SPECT-CT measures the single photon release with radioisotopes and PET measures positron release [147]. On the other hand, IVIS is an optical imaging technique also used for *in vivo* analysis of oral delivery systems. Using this technique, it is possible to obtain *in vivo* imaging but using fluorescence instead of radiation. Since mammalian tissue has low bioluminescence, the results present a high signal-to-noise ratio. Some of the fluorescent markers used are sulforhodamine B or Cy5.5 [148, 149]. As mentioned before, the organs analysis can precede imaging techniques. The organs analysis is normally performed after oral feeding, when the animals are sacrificed. The organs are removed and fixed with paraformaldehyde. Histological analysis allows the semi-quantification done by gel permeation chromatography (GPC). Through the dissolution of the tissues, the distribution of nanosystem size can be evaluated. However, for particles smaller than 0.5 μm, histological analysis is not adequate. After organ removal, particles can also be observed by light, fluorescent or confocal microscopy [150].

Also the immune response, or anaphylactic reaction, which is the most common response to allergens, is an important parameter when it comes to *in vivo* analysis. Since any part of a delivery system is a potential allergen, anaphylactic reaction is a parameter to be monitored. The tolerance of a delivery system can be assessed through the occurrence or not of anaphylaxis. Several symptoms appear in animal models that can indicate an immune response, such as slowing of movement, or respiratory arrest of anaphylaxis [151]. "Symptom reversal", known as phenotypic correction, is a technique to assess whether the system is effective by monitoring the changes in enzyme or protein levels or even changes in reaction or behavior to a stimulus [152]. Oral delivery of insulin or haemophilia treatment uses this type of technique to evaluate the efficacy of the systems [153]. Locomotive activity, exploratory patterns or anxiety can also be parameters to observe in order to evaluate whether neural symptoms appear [154].

When dealing with gene delivery studies, one of the most sensitive methods of biodistribution analysis is genomic DNA evaluation. PCR is performed with DNA extracted from tissue samples and later purified. For proteins extracted from tissues, Western blot is the chosen technique. These tissues are removed from the sacrificed animal and homogenized. Nevertheless, the blood and the lymphatic system are also important to analyse in order to correctly evaluate the therapeutic and immune effect of nanosystems [33]. Blood is one of the most used fluid to assess the presence of delivery systems. Several assays can be performed with blood, such as the Fluorogenic Thrombin Generation Test [150].

7.7 Conclusion

When nanosystems are designed, it is important that these are absorbed and transported across the intestinal membrane. In this light, this chapter approaches different models to evaluate the permeability and absorption of oral nanosystems (Figure 7.7). The most widely used is the Caco-2 monoculture model. However, there are other cell lines as alternatives to this model, with non-human and/or non-intestinal origin. Over recent years, co-culture models have been developed in order to use different cell lines on the same model and to mimic and approach an *in vivo* situation, with production of mucus and the presence of M cells. On the other hand, the recent gut-on-chip models seem to be very promising, since parameters such as physical forces, mechanical compression and cyclic stress can be modulated and improve the cellular microenvironment.

Furthermore, *ex vivo* models are also advantageous due to the use of living functional tissue or organs. In this way, the gut architecture is maintained as well as membrane transporters and enzymes, compared with the previous models. The most accepted model is the diffusion chamber (Ussing chamber or Franz cell). Generally, all of these models

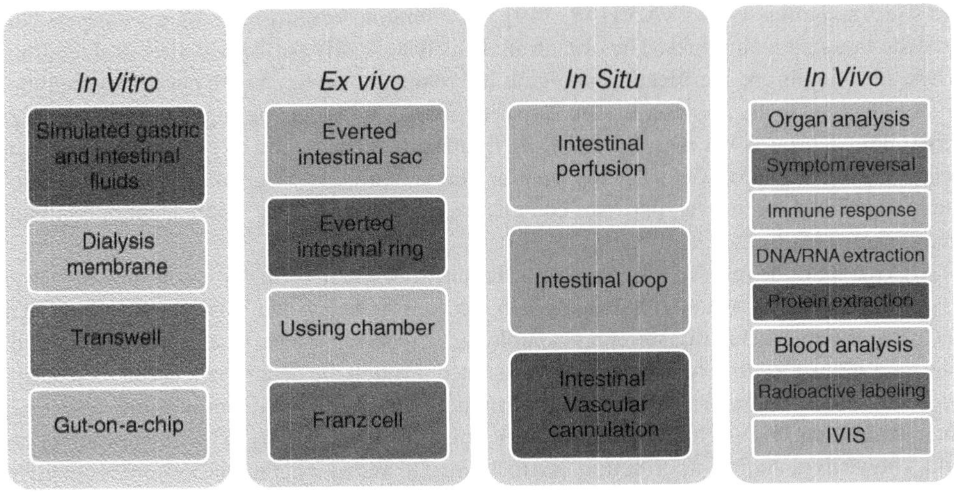

Figure 7.7 *Schematic of different intestinal mucosal models (examples, advantages and drawbacks) to study and validate nanosystem permeability.*

present drawbacks, namely the impossibility of recreating the gastric empty rate, GI pH or transit rate.

Regarding *in situ* models, the study can be in a specific site with the physiological factors and metabolism, the single-pass perfusion being the most preferred among researchers. However, anaesthesia and hydrostatic pressure can interfere with the intestinal absorption. Despite these methods being more convenient, they obviously lack in accuracy. This can be overcome with the true reflection of *in vivo* methods in a dynamic system. Still, in this case it is not possible to study the variables involved in the absorption separately. In short, an innovative and more accurate method to test nanosystems and beyond is needed in order to obtain a better correlation with *in vivo* data. Meanwhile, the conjugation of different models must be considered for a more accurate study of intestinal nanosystem absorption.

Acknowledgements

This chapter is a result of the project NORTE-01-0145-FEDER-000012, supported by Norte Portugal Regional Operational Programme (NORTE 2020), under the PORTUGAL 2020 Partnership Agreement, through the European Regional Development Fund (ERDF). This work was also financed by FEDER – Fundo Europeu de Desenvolvimento Regional funds through the COMPETE 2020 Operacional Programme for Competitiveness and Internationalisation (POCI), Portugal 2020, and by Portuguese funds through FCT – Fundação para a Ciência e a Tecnologia/Ministério da Ciência, Tecnologia e Ensino Superior under the framework of the project "Institute for Research and Innovation in Health Sciences" (POCI-01-0145-FEDER-007274). Cláudia Azevedo would like to thank to Fundação para a Ciência e a Tecnologia (FCT), Portugal for financial support (SFRH/BD/117598/2016), and Maria Helena Macedo for the attentive revision and pertinent comments. This chapter was also partially supported by CESPU/IINFACTS under the PI4AC 2018 project NanoCEA-CESPU-2018.

References

[1] Venditto, V.J., Szoka, F.C. (2013) Cancer nanomedicines: so many papers and so few drugs! *Advanced Drug Delivery Reviews*, **65** (1), 80–88.

[2] Bhise, N.S., *et al.* (2014) Organ-on-a-chip platforms for studying drug delivery systems. *Journal of Controlled Release*, **190**, 82–93.

[3] Desai, N. (2012) Challenges in development of nanoparticle-based therapeutics. *The AAPS Journal*, **14**, 282–295.

[4] Folch, A., Toner, M. (2000) Microengineering of cellular interactions. *Annual Review of Biomedical Engineering*, **2** (1), 227–256.

[5] van der Worp, H.B., *et al.* (2010) Can animal models of disease reliably inform human studies? *PLOS Medicine*, **7** (3), e1000245.

[6] Stirland, D.L., *et al.* (2013) Mind the gap: A survey of how cancer drug carriers are susceptible to the gap between research and practice. *Journal of Controlled Release*, **172** (3), 1045–1064.

[7] Nunes, R., Silva, C., Chaves, L. (2016) 4.2 – Tissue-based in vitro and ex vivo models for intestinal permeability studies A2 - Sarmento, B., in *Concepts and Models for Drug Permeability Studies*, Woodhead Publishing, 203–236.

[8] Braakhuis, H.M., *et al.* (2015) Progress and future of in vitro models to study translocation of nanoparticles. *Arch Toxicol*, **89** (9), 1469–1495.

[9] Pawar, V.K., *et al.* (2014) Targeting of gastrointestinal tract for amended delivery of 3Q4 protein/peptide therapeutics: Strategies and industrial perspectives. *Journal of Controlled Release*, **196**, 168–183.

[10] Pereira, C., *et al.* (2016) 3.3 – Cell-based in vitro models for intestinal permeability studies, in *Concepts and Models for Drug Permeability Studies*, Woodhead Publishing, 57–81.

[11] Pereira, C., *et al.* (2014) Targeting membrane transporters and receptors as a mean to optimize orally delivered biotechnological based drugs through nanoparticle delivery systems. *Current Pharmaceutical Biotechnology*, **15**, 650–658.

[12] Lopes, M.A., *et al.* (2014) Intestinal absorption of insulin nanoparticles: Contribution of M cells. *Nanomedicine: Nanotechnology, Biology, and Medicine*, **10**, 1139–1151.

[13] Muniz, L., Knosp, C., Yeretssian, G. (2012) Intestinal antimicrobial peptides during homeostasis, infection, and disease. *Frontiers in Immunology*, **3** (310).

[14] Hu, M., Li, X. (2011) *Oral Bioavailability: Basic Principles, Advanced Concepts, and Applications*, Wiley.

[15] El-Kattan, A., Varma, M. (2012) Oral absorption, intestinal metabolism and human oral bioavailability, in *Topics on Drug Metabolism* (ed. J. Paxton), InTech, Rijeka.

[16] Alai, M.S., Lin, W.J., Pingale, S.S. (2015) Application of polymeric nanoparticles and micelles in insulin oral delivery. *Journal of Food and Drug Analysis*, **23**, 351–358.

[17] Banerjee, A., *et al.* (2016) Role of nanoparticle size, shape and surface chemistry in oral drug delivery. *Journal of Controlled Release*, **238** (Suppl. C), 176–185.

[18] Imai, S., *et al.* (2017) Cellular internalization, transcellular transport, and cellular effects of silver nanoparticles in polarized Caco-2 cells following apical or basolateral exposure. *Biochemical and Biophysical Research Communications*, **484** (3), 543–549.

[19] Chen, M.-C., *et al.* (2011) A review of the prospects for polymeric nanoparticle platforms in oral insulin delivery. *Biomaterials*, **32** (36), 9826–9838.

[20] Azevedo, C., Macedo, M.H., Sarmento, B. (2017) Strategies for the enhanced intracellular delivery of nanomaterials. *Drug Discovery Today*, **23** (5), 944–959.

[21] Sarmento, B., Ferreira, D., Vasconcelos, T. (2009) Polymer-based delivery systems for oral delivery of peptides and proteins, in *Delivery Technologies for Biopharmaceuticals: Peptides, Proteins, Nucleic Acids and Vaccines* (eds Jorgensen L., Nielsen, H.M.), John Wiley & Sons, Ltd., pp. 207–227.

[22] Woitiski, C.B., *et al.* (2008) Strategies toward the improved oral delivery of insulin nanoparticles via gastrointestinal uptake and translocation. *Biodrugs*, **22** (4), 223–237.

[23] Fonte, P., *et al.* (2015) Polymer-based nanoparticles for oral insulin delivery: Revisited approaches. *Biotechnology Advances*, **33**, 1342–1354.

[24] Araújo, F., *et al.* (2014) The impact of nanoparticles on the mucosal translocation and transport of GLP-1 across the intestinal epithelium. *Biomaterials*, **35**, 9199–9207.

[25] Hussain, N. (2016) Regulatory aspects in the pharmaceutical development of nanoparticle drug delivery systems designed to cross the intestinal epithelium and M-cells. *International Journal of Pharmaceutics*, **514** (1), 15–23.

[26] Blanco, E., Shen, H., Ferrari, M. (2015) Principles of nanoparticle design for overcoming biological barriers to drug delivery. *Nature Biotechnology*, **9**, 941–951.

[27] Jhaveri, A., Torchilin, V. (2015) Intracellular delivery of nanocarriers and targeting to subcellular organelles. *Expert Opinion on Drug Delivery*, **13** (1).

[28] Michael Danielsen, E., Hansen, G.H. (2016) Small molecule pinocytosis and clathrin-dependent endocytosis at the intestinal brush border: Two separate pathways into the enterocyte. *Biochimica et Biophysica Acta (BBA) - Biomembranes*, **1858** (2), 233–243.

[29] Agarwal, R., Roy, K. (2013) Intracellular delivery of polymeric nanocarriers: a matter of size, shape, charge, elasticity and surface composition. *Therapeutic Delivery*, **4** (6), 705–723.

[30] Pindáková, L., *et al.* (2017) Behaviour of silver nanoparticles in simulated saliva and gastrointestinal fluids. *International Journal of Pharmaceutics*, **527** (1), 12–20.

[31] Modi, S., Anderson, B.D. (2013) Determination of drug release kinetics from nanoparticles: overcoming pitfalls of the dynamic dialysis method. *Mol Pharm*, **10** (8), 3076–3089.

[32] Kim, J.S., *et al.* (2006) The suitability of an in situ perfusion model for permeability determinations: utility for BCS class I biowaiver requests. *Mol Pharm*, **3** (6), 686–694.

[33] Gamboa, J.M., Leong, K.W. (2013) In vitro and in vivo models for the study of oral delivery of nanoparticles. *Adv Drug Deliv Rev*, **65** (6), 800–810.

[34] Turco, L., *et al.* (2011) Caco-2/TC7 cell line characterization for intestinal absorption: how reliable is this in vitro model for the prediction of the oral dose fraction absorbed in human? *Toxicol In Vitro*, **25** (1), 13–20.

[35] Zeller, P., *et al.* (2015) Multiparametric temporal analysis of the Caco-2/TC7 demonstrated functional and differentiated monolayers as early as 14 days of culture. *Eur J Pharm Sci*, **72**, 1–11.

[36] Foss, S., *et al.* (2016) Enhanced FcRn-dependent transepithelial delivery of IgG by Fc-engineering and polymerization. *Journal of Controlled Release*, **223**, 42–52.

[37] McCool, D.J., *et al.* (1990) The T84 human colonic adenocarcinoma cell line produces mucin in culture and releases it in response to various secretagogues. *Biochem J*, **267** (2), 491–500.

[38] Vergauwen, H. (2015) The IPEC–J2 cell line, in *The Impact of Food Bioactives on Health*, (ed. Verhoeckx, K.), Springer, pp. 125–134.

[39] Tavelin, S., *et al.* (2003) An improved cell culture model based on 2/4/A1 cell monolayers for studies of intestinal drug transport: characterization of transport routes. *Pharm Res*, **20** (3), 373–381.

[40] Ma, T.Y., *et al.* (1992) IEC-18, a nontransformed small intestinal cell line for studying epithelial permeability. *J Lab Clin Med*, **120**, 329–341.

[41] Versantvoort, C.H.M., *et al.* (2002) Monolayers of IEC-18 cells as an in vitro model for screening the passive transcellular and paracellular transport across the intestinal barrier: comparison

of active and passive transport with the human colon carcinoma Caco-2 cell line. *Environmental Toxicology and Pharmacology*, **11** (3), 335–344.

[42] Duizer, E., *et al.* (1997) Comparison of permeability characteristics of the human colonic Caco-2 and rat small intestinal IEC-18 cell lines. *Journal of Controlled Release*, **49** (1), 39–49.

[43] Steinmassl, D., *et al.* (1995) LLC-PK1 epithelia as a model for in vitro assessment of proximal tubular nephrotoxicity. *In Vitro Cellular & Developmental Biology – Animal*, **31** (2), 94–106.

[44] van der Sandt, I.C.J., *et al.* (2000) Specificity of doxorubicin versus rhodamine-123 in assessing P-glycoprotein functionality in the LLC-PK1, LLC-PK1:MDR1 and Caco-2 cell lines. *European Journal of Pharmaceutical Sciences*, **11** (3), 207–214.

[45] Irvine, J.D., *et al.* (1999) MDCK (Madin-Darby canine kidney) cells: A tool for membrane permeability screening. *J Pharm Sci*, **88** (1), 28–33.

[46] Jin, X., *et al.* (2014) Comparison of MDCK-MDR1 and Caco-2 cell based permeability assays for anti-malarial drug screening and drug investigations. *Journal of Pharmacological and Toxicological Methods*, **70** (2), 188–194.

[47] Li, X., *et al.* (2017) Carrier-mediated and energy-dependent uptake and efflux of deoxynivalenol in mammalian cells. *Scientific Reports*, **7** (1), 5889.

[48] Khatri, P., Shao, J. (2017) Transport of lipid nano-droplets through MDCK epithelial cell monolayer. *Colloids and Surfaces B: Biointerfaces*, **153** (Suppl. C), 237–243.

[49] Braun, A., *et al.* (2000) Cell cultures as tools in biopharmacy. *European Journal of Pharmaceutical Sciences*, **11**, S51–S60.

[50] Kauffman, A., *et al.* (2013) Alternative functional in vitro models of human intestinal epithelia. *Frontiers in Pharmacology*, **4** (79).

[51] Ozawa, T., *et al.* (2015) Generation of enterocyte-like cells from human induced pluripotent stem cells for drug absorption and metabolism studies in human small intestine. *Scientific Reports*, **5**, 16479.

[52] Takenaka, T., *et al.* (2014) Human small intestinal epithelial cells differentiated from adult intestinal stem cells as a novel system for predicting oral drug absorption in humans. *Drug Metab Dispos*, **42** (11), 1947–1954.

[53] Takenaka, T., *et al.* (2016) Application of a human intestinal epithelial cell monolayer to the prediction of oral drug absorption in humans as a superior alternative to the Caco-2 cell monolayer. *Journal of Pharmaceutical Sciences*, **105** (2), 915–924.

[54] Lesuffleur, T., *et al.* (1990) Growth adaptation to methotrexate of HT-29 human colon carcinoma cells is associated with their ability to differentiate into columnar absorptive and mucus-secreting cells. *Cancer Res*, **50**.

[55] Akbari, A., Lavasanifar, A., Wu, J. (2017) Interaction of cruciferin-based nanoparticles with Caco-2 cells and Caco-2/HT29-MTX co-cultures. *Acta Biomaterialia*, **64**, 249–258.

[56] Araujo, F., *et al.* (2017) Chemical modification of drug molecules as strategy to reduce interactions with mucus. *Adv Drug Deliv Rev*, **124**, 98–106.

[57] Lechanteur, A., das Neves, J., Sarmento, B. (2017) The role of mucus in cell-based models used to screen mucosal drug delivery. *Advanced Drug Delivery Reviews*, **124**, 50–63.

[58] Mahler, G.J., *et al.* (2009) Characterization of a gastrointestinal tract microscale cell culture analog used to predict drug toxicity. *Biotechnol Bioeng*, **104** (1), 193–205.

[59] Chen, X.-M., Elisia, I., Kitts, D.D. (2010) Defining conditions for the co-culture of Caco-2 and HT29-MTX cells using Taguchi design. *Journal of Pharmacological and Toxicological Methods*, **61** (3), 334–342.

[60] Pontier, C., *et al.* (2001) HT29-MTX and Caco-2/TC7 monolayers as predictive models for human intestinal absorption: Role of the mucus layer. *Journal of Pharmaceutical Sciences*, **90** (10), 1608–1619.

[61] Beloqui, A., *et al.* (2017) A human intestinal M-cell-like model for investigating particle, antigen and microorganism translocation. *Nat. Protocols*, **12** (7), 1387–1399.

[62] Cabellos, J., *et al.* (2017) Contribution of M-cells and other experimental variables in the translocation of TiO_2 nanoparticles across in vitro intestinal models. *NanoImpact*, **5** (Suppl. C), 51–60.

[63] Antunes, F., *et al.* (2013) Establishment of a triple co-culture in vitro cell models to study intestinal absorption of peptide drugs. *European Journal of Pharmaceutics and Biopharmaceutics*, **83** (3), 427–435.

[64] Araújo, F., Sarmento, B. (2013) Towards the characterization of an in vitro triple co-culture intestine cell model for permeability studies. *International Journal of Pharmaceutics*, **458** (1), 128–134.

[65] Lozoya-Agullo, I., *et al.* (2017) Usefulness of Caco-2/HT29-MTX and Caco-2/HT29-MTX/Raji B coculture models to predict intestinal and colonic permeability compared to Caco-2 monoculture. *Mol Pharm*, **14** (4), 1264–1270.

[66] Yu, J., *et al.* (2012) In vitro 3D human small intestinal villous model for drug permeability determination. *Biotechnol Bioeng*, **109** (9), 2173–2178.

[67] Artursson, P. (1991) Cell cultures as models for drug absorption across the intestinal mucosa. *Crit Rev Ther Drug Carrier Syst*, **8** (4), 305–330.

[68] Hidalgo, I.J. (2001) Assessing the absorption of new pharmaceuticals. *Curr Top Med Chem*, **1** (5), 385–401.

[69] Imai, M., *et al.* (2015) Three-dimensional morphogenesis of MDCK cells induced by cellular contractile forces on a viscous substrate. *Scientific Reports*, **5**, 14208.

[70] Pereira, C., *et al.* (2015) Dissecting stromal-epithelial interactions in a 3D in vitro cellularized intestinal model for permeability studies. *Biomaterials*, **56**, 36–45.

[71] Leonard, F., Collnot, E.-M., Lehr, C.-M. (2010) A three-dimensional coculture of enterocytes, monocytes and dendritic cells to model inflamed intestinal mucosa in vitro. *Molecular Pharmaceutics*, **7** (6), 2103–2119.

[72] Susewind, J., *et al.* (2015) A 3D co-culture of three human cell lines to model the inflamed intestinal mucosa for safety testing of nanomaterials. *Nanotoxicology*, **10**, 53–62.

[73] Sung, J.H., *et al.* (2011) Microscale 3-D hydrogel scaffold for biomimetic gastrointestinal (GI) tract model. *Lab Chip*, **11** (3), 389–392.

[74] Huh, D., *et al.* (2012) Microengineered physiological biomimicry: organs-on-chips. *Lab Chip*, **12** (12), 2156–2164.

[75] Bhatia, S.N., Ingber, D.E. (2014) Microfluidic organs-on-chips. *Nature Biotechnology*, **32**, 760.

[76] Kim, H.J., *et al.* (2012) Human gut-on-a-chip inhabited by microbial flora that experiences intestinal peristalsis-like motions and flow. *Lab Chip*, **12** (12), 2165–2174.

[77] Workman, M.J., *et al.* (2017) Enhanced utilization of induced pluripotent stem cell–derived human intestinal organoids using microengineered chips. *Cellular and Molecular Gastroenterology and Hepatology*, **5** (4), 669–677.

[78] Trietsch, S.J., *et al.* (2017) Membrane-free culture and real-time barrier integrity assessment of perfused intestinal epithelium tubes. *Nature Communications*, **8** (1), 262.

[79] Esch, E.W., Bahinski, A., Huh, D. (2015) Organs-on-chips at the frontiers of drug discovery. *Nat Rev Drug Discov*, **14** (4), 248–260.

[80] Valencia, P.M., *et al.* (2012) Microfluidic technologies for accelerating the clinical translation of nanoparticles. *Nat Nanotechnol*, **7** (10), 623–629.

[81] Balijepalli, A., Sivaramakrishan, V. (2017) Organs-on-chips: research and commercial perspectives. *Drug Discov Today*, **22** (2), 397–403.

[82] Huh, D., Hamilton, G.A., Ingber, D.E. (2011) From 3D cell culture to organs-on-chips. *Trends Cell Biol*, **21** (12), 745–754.

[83] Esch, M.B., *et al.* (2014) Body-on-a-chip simulation with gastrointestinal tract and liver tissues suggests that ingested nanoparticles have the potential to cause liver injury. *Lab on a Chip*, **14** (16), 3081–3092.

[84] Tottey, W., *et al.* (2013) The human gut chip "HuGChip", an explorative phylogenetic microarray for determining gut microbiome diversity at family level. *PLoS One*, **8** (5), e62544.
[85] Ingber, D.E. (2016) Reverse engineering human pathophysiology with organs-on-chips. *Cell*, **64** (6), 1105–1109.
[86] Zhou, G., *et al.* (2017) Enhanced uptake of Fe_3O_4 nanoparticles by intestinal epithelial cells in a state of inflammation. *Molecules*, **22** (8), 1240.
[87] Vllasaliu, D., *et al.* (2014) Basement membrane influences intestinal epithelial cell growth and presents a barrier to the movement of macromolecules. *Experimental Cell Research*, **323** (1), 218–231.
[88] Williams, K.M., *et al.* (2016) Size and dose dependent effects of silver nanoparticle exposure on intestinal permeability in an in vitro model of the human gut epithelium. *J Nanobiotechnology*, **14** (1), 62.
[89] Pradines, B., *et al.* (2015) Cell line-dependent cytotoxicity of poly(isobutylcyanoacrylate) nanoparticles coated with chitosan and thiolated chitosan: Insights from cultured human epithelial HeLa, Caco2/TC7 and HT-29/MTX cells. *Int J Pharm*, **491** (1–2), 17–20.
[90] Pisal, D.S., *et al.* (2008) Transport of surface engineered polyamidoamine (PAMAM) dendrimers across IPEC-J2 cell monolayers. *Drug Delivery*, **15** (8), 515–522.
[91] L'Azou, B., *et al.* (2008) In vitro effects of nanoparticles on renal cells. *Part Fibre Toxicol*, **5**, 22.
[92] Wang, C., *et al.* (2016) Entry of PIP3-containing polyplexes into MDCK epithelial cells by local apical-basal polarity reversal. *Sci Rep*, **6**, 21436.
[93] Fazlollahi, F., *et al.* (2011) Polystyrene nanoparticle trafficking across MDCK-II. *Nanomedicine*, **7** (5), 588–594.
[94] Li, Y., *et al.* (2017) Mucus interactions with liposomes encapsulating bioactives: Interfacial tensiometry and cellular uptake on Caco-2 and cocultures of Caco-2/HT29-MTX. *Food Research International*, **92** (Suppl. C), 128–137.
[95] Georgantzopoulou, A., *et al.* (2016) Effects of silver nanoparticles and ions on a co-culture model for the gastrointestinal epithelium. *Part Fibre Toxicol*, **13**, 9.
[96] Lopes, M., *et al.* (2016) Dual chitosan/albumin-coated alginate/dextran sulfate nanoparticles for enhanced oral delivery of insulin. *Journal of Controlled Release*, **232** (Suppl. C), 29–41.
[97] Schimpel, C., *et al.* (2014) Development of an Advanced Intestinal in Vitro Triple Culture Permeability Model To Study Transport of Nanoparticles. *Molecular Pharmaceutics*, **11** (3), 808–818.
[98] Boni, F.I., *et al.* (2018) Mucoadhesive nanostructured polyelectrolytes complexes modulate the intestinal permeability of methotrexate. *European Journal of Pharmaceutical Sciences*, **111** (Suppl. C), 73–82.
[99] Astashkina, A.I., *et al.* (2014) Nanoparticle toxicity assessment using an in vitro 3-D kidney organoid culture model. *Biomaterials*, **35** (24), 6323–6331.
[100] Fröhlich, E., Roblegg, E. (2016) Oral uptake of nanoparticles: human relevance and the role of in vitro systems. *Archives of Toxicology*, **90** (10), 2297–2314.
[101] Ussing, H.H., Zerahn, K. (1951) Active transport of sodium as the source of electric current in the short-circuited isolated frog skin. *Acta Physiologica*, **23** (2–3), 110–127.
[102] Grass, G.M., Sweetana, S.A. (1988) In vitro measurement of gastrointestinal tissue permeability using a new diffusion cell. *Pharmaceutical Research*, **5** (6), 372–376.
[103] Barthe, L., Woodley, J., Houin, G. (1999) Gastrointestinal absorption of drugs: methods and studies. *Fundamental & Clinical Pharmacology*, **13** (2), 154–168.
[104] Nunes, R., Silva, C., Chaves, L. (2015) Tissue-based in vitro and ex vivo models for intestinal permeability studies, in *Concepts and Models for Drug Permeability Studies: Cell and Tissue Based In Vitro Culture Models* (ed. Sarmento, B.), Woodhead Publishing, pp. 203–236.

[105] Gitter, A.H., Fromm, M., Schulzke, J.-D. (1998) Impedance analysis for the determination of epithelial and subepithelial resistance in intestinal tissues. *Journal of Biochemical and Biophysical Methods*, **37** (1), 35–46.

[106] Brun, E., *et al.* (2014) Titanium dioxide nanoparticle impact and translocation through ex vivo, in vivo and in vitro gut epithelia. *Particle and Fibre Toxicology*, **11** (1), 13.

[107] Lundquist, P., Artursson, P. (2016) Oral absorption of peptides and nanoparticles across the human intestine: Opportunities, limitations and studies in human tissues. *Advanced Drug Delivery Reviews*, **106**, 256–276.

[108] Sjöberg, Å., *et al.* (2013) Comprehensive study on regional human intestinal permeability and prediction of fraction absorbed of drugs using the Ussing chamber technique. *European Journal of Pharmaceutical Sciences*, **48** (1), 166–180.

[109] Li, H., Sheppard, D.N., Hug, M.J. (2004) Transepithelial electrical measurements with the Ussing chamber. *Journal of Cystic Fibrosis*, **3**, 123–126.

[110] Cui, L., *et al.* (2001) Enhancing effect of electret on transdermal drug delivery. *Journal of Electrostatics*, **51–52**, 153–158.

[111] Jain, A.S., *et al.* (2016) In vitro and ex vivo evaluations of lipid anti-cancer nanoformulations: Insights and assessment of bioavailability enhancement. *AAPS PharmSciTech*, **17** (3), 553–571.

[112] Sandri, G., *et al.* (2010) Insulin-loaded nanoparticles based on N-trimethyl chitosan: in vitro (Caco-2 model) and ex vivo (excised rat jejunum, duodenum, and ileum) evaluation of penetration enhancement properties. *Aaps Pharmscitech*, **11** (1), 362–371.

[113] Alam, M.A., Al-Jenoobi, F.I., Al-Mohizea, A.M. (2012) Everted gut sac model as a tool in pharmaceutical research: limitations and applications. *Journal of Pharmacy and Pharmacology*, **64** (3), 326–336.

[114] Barthe, L., *et al.* (1998) An improved everted gut sac as a simple and accurate technique to measure paracellular transport across the small intestine. *European Journal of Drug Metabolism and Pharmacokinetics*, **23** (2), 313–323.

[115] Luo, Z., *et al.* (2013) Ex vivo and in situ approaches used to study intestinal absorption. *Journal of Pharmacological and Toxicological Methods*, **68** (2), 208–216.

[116] Saremi, S., *et al.* (2011) Thiolated chitosan nanoparticles for enhancing oral absorption of docetaxel: preparation, in vitro and ex vivo evaluation. *International Journal of Nanomedicine*, **6**, 119.

[117] Bayat, A., *et al.* (2008) Nanoparticles of quaternized chitosan derivatives as a carrier for colon delivery of insulin: ex vivo and in vivo studies. *International Journal of Pharmaceutics*, **356** (1), 259–266.

[118] Mainardes, R.M., *et al.* (2006) Development of praziquantel-loaded PLGA nanoparticles and evaluation of intestinal permeation by the everted gut sac model. *Journal of Nanoscience and Nanotechnology*, **6** (9-1), 3057–3061.

[119] Bothiraja, C., Pawar, A., Deshpande, G. (2016) Ex-vivo absorption study of a nanoparticle based novel drug delivery system of vitamin D3 (Arachitol Nano™) using everted intestinal sac technique. *Journal of Pharmaceutical Investigation*, **46** (5), 425–432.

[120] Ruan, L.P., *et al.* (2006) Prediction of human absorption of natural compounds by the non-everted rat intestinal sac model. *European Journal of Medicinal Chemistry*, **41** (5), 605–610.

[121] Soni, K., Mujtaba, A., Kohli, K. (2017) Lipid drug conjugate nanoparticle as a potential nanocarrier for the oral delivery of pemetrexed diacid: Formulation design, characterization, ex vivo, and in vivo assessment. *International Journal of Biological Macromolecules*, **103** (Suppl. C), 139–151.

[122] Neupane, Y.R., *et al.* (2013) Lipid drug conjugate nanoparticle as a novel lipid nanocarrier for the oral delivery of decitabine: ex vivo gut permeation studies. *Nanotechnology*, **24** (41), 415102.

[123] Jain, A.S., *et al.* (2016) Docetaxel in cationic lipid nanocapsules for enhanced in vivo activity. *Pharmaceutical Development and Technology*, **21** (1), 76–85.

[124] Leppert, P.S., Fix, J.A. (1994) Use of everted intestinal rings for in vitro examination of oral absorption potential. *Journal of Pharmaceutical Sciences*, **83** (7), 976–981.

[125] Dube, A., Nicolazzo, J.A., Larson, I. (2010) Chitosan nanoparticles enhance the intestinal absorption of the green tea catechins (+)-catechin and (−)-epigallocatechin gallate. *European Journal of Pharmaceutical Sciences*, **41** (2), 219–225.

[126] Agüeros, M., *et al.* (2009) Combined hydroxypropyl-β-cyclodextrin and poly(anhydride) nanoparticles improve the oral permeability of paclitaxel. *European Journal of Pharmaceutical Sciences*, **38** (4), 405–413.

[127] Derakhshandeh, K., Fathi, S. (2012) Role of chitosan nanoparticles in the oral absorption of Gemcitabine. *International Journal of Pharmaceutics*, **437** (1), 172–177.

[128] Bagre, A.P., Jain, K., Jain, N.K. (2013) Alginate coated chitosan core shell nanoparticles for oral delivery of enoxaparin: In vitro and in vivo assessment. *International Journal of Pharmaceutics*, **456** (1), 31–40.

[129] Jiang, L., *et al.* (2013) Thiolated chitosan-modified PLA-PCL-TPGS nanoparticles for oral chemotherapy of lung cancer. *Nanoscale Research Letters*, **8** (1), 66.

[130] Zhang, Z., *et al.* (2010) The characteristics and mechanism of simvastatin loaded lipid nanoparticles to increase oral bioavailability in rats. *International Journal of Pharmaceutics*, **394** (1), 147–153.

[131] Li, Q., Liu, C.-G., Yu, Y. (2015) Separation of monodisperse alginate nanoparticles and effect of particle size on transport of vitamin E. *Carbohydrate Polymers*, **124** (Suppl. C), 274–279.

[132] Liu, W., *et al.* (2016) Developments in methods for measuring the intestinal absorption of nanoparticle-bound drugs. *Int J Mol Sci*, **17** (7).

[133] Liu, Y., *et al.* (2009) Optimization and in situ intestinal absorption of self-microemulsifying drug delivery system of oridonin. *Int J Pharm*, **365** (1–2), 136–142.

[134] Prasad, N., Bhasker, S. (2012) Characterization of intestinal transport of vincristine in rats applying in situ single pass intestinal perfusion. *Pharmacologia*, **3** (11), 617–621.

[135] Ji, H., *et al.* (2016) Curcumin-loaded solid lipid nanoparticles with Brij78 and TPGS improved in vivo oral bioavailability and in situ intestinal absorption of curcumin. *Drug Delivery*, **23** (2), 459–470.

[136] Parthasarathi, S., Bhushani, J.A., Anandharamakrishnan, C. (2018) Engineered small intestinal system as an alternative to in-situ intestinal permeability model. *Journal of Food Engineering*, **222** (Suppl. C), 110–114.

[137] Garg, A., *et al.* (2017) In-situ single pass intestinal permeability and pharmacokinetic study of developed Lumefantrine loaded solid lipid nanoparticles. *International Journal of Pharmaceutics*, **516** (1), 120–130.

[138] Sun, M., *et al.* (2012) Evaluation of an oral carrier system in rats: bioavailability and gastrointestinal absorption properties of curcumin encapsulated PBCA nanoparticles. *Journal of Nanoparticle Research*, **14** (2), 705.

[139] Abuasal, B.S., *et al.* (2012) Enhancement of intestinal permeability utilizing solid lipid nanoparticles increases γ-tocotrienol oral bioavailability. *Lipids*, **47** (5), 461–469.

[140] Wang, J., *et al.* (2017) Development of novel self-assembled ES-PLGA hybrid nanoparticles for improving oral absorption of doxorubicin hydrochloride by P-gp inhibition: In vitro and in vivo evaluation. *European Journal of Pharmaceutical Sciences*, **99**, 185–192.

[141] Onishchenko, G.E., *et al.* (2012) Effects of titanium dioxide nanoparticles on small intestinal mucosa in rats. *Bulletin of Experimental Biology and Medicine*, **154** (2), 265–270.

[142] Yuan, H., *et al.* (2013) Improved transport and absorption through gastrointestinal tract by PEGylated solid lipid nanoparticles. *Mol Pharm*, **10** (5), 1865–1873.

[143] Saratale, R.G., *et al.* (2018) Absorption kinetics of vitamin E nanoemulsion and green tea microstructures by intestinal in situ single perfusion technique in rats. *Food Research International*, **106**, 149–155.

[144] McDowell, A., *et al.* (2005) Gastrointestinal transit in the common brushtail possum measured by gamma scintigraphy. *International Journal of Pharmaceutics*, **302** (1–2), 125–132.

[145] Wilding, I., Coupe, A., Davis, S. (2001) The role of γ-scintigraphy in oral drug delivery. *Advanced Drug Delivery Reviews*, **46** (1–3), 103–124.

[146] Muehllehner, G., Karp, J.S. (2006) Positron emission tomography. *Physics in Medicine & Biology*, **51** (13), R117.

[147] Cheng, Y., *et al.* (2008) Highly efficient drug delivery with gold nanoparticle vectors for in vivo photodynamic therapy of cancer. *Journal of the American Chemical Society*, **130** (32), 10643–10647.

[148] Lee, D., *et al.* (2007) In vivo imaging of hydrogen peroxide with chemiluminescent nanoparticles. *Nature Materials*, **6**, 765.

[149] Hu, X., *et al.* (2016) Evidence does not support absorption of intact solid lipid nanoparticles via oral delivery. *Nanoscale*, **8** (13), 7024–7035.

[150] Chen, Z., *et al.* (2006) Acute toxicological effects of copper nanoparticles in vivo. *Toxicology Letters*, **163** (2), 109–120.

[151] Roy, K., *et al.* (1999) Oral gene delivery with chitosan–DNA nanoparticles generates immunologic protection in a murine model of peanut allergy. *Nature Medicine*, **5**, 387.

[152] Cui, F., *et al.* (2006) Biodegradable nanoparticles loaded with insulin–phospholipid complex for oral delivery: preparation, in vitro characterization and in vivo evaluation. *Journal of Controlled Release*, **114** (2), 242–250.

[153] Bowman, K., *et al.* (2008) Gene transfer to hemophilia A mice via oral delivery of FVIII–chitosan nanoparticles. *Journal of Controlled Release*, **132** (3), 252–259.

[154] Mittal, G., *et al.* (2011) Development and evaluation of polymer nanoparticles for oral delivery of estradiol to rat brain in a model of Alzheimer's pathology. *Journal of Controlled Release*, **150** (2), 220–228.

8

Biodistribution of Polymeric, Polysaccharide and Metallic Nanoparticles

Nazlı Erdoğar[1], Gamze Varan[2], Cem Varan[2] and Erem Bilensoy[1]

[1]Department of Pharmaceutical Technology, Faculty of Pharmacy, Hacettepe University, Turkey
[2]Department of Nanotechnology and Nanomedicine, Institute of Graduate Studies in Science, Hacettepe University, Turkey

8.1 Introduction

The specific properties of nanoparticles (NPs) make them potential candidates for a wide range of applications in nanomedicine, including diagnostics and therapeutics [1, 2]. NPs can be applied to increase bioavailability, improve selectivity, enhance drug efficacy, or protect a drug from enzymatic degradation. Nevertheless, NPs can cause toxicity resulting from increasing drug delivery to tissues. The pharmacokinetic profile and the amounts of encapsulated drug in NPs are often different. Therefore, it is very important to observe the biological distribution of NPs in order to understand the efficacy and toxicity of NPs, which are determinants of whether NP-based technology is clinically successful or not [3]. For example, understanding the ratio of localized NPs in the area of interest is important to assess how NPs effectively target either passively or actively. It is also necessary to monitor the unintentional accumulation in other tissues, and characterize the long-term fate of the NPs, to understand their biosafety profile [4].

Biodistribution of NPs depends on properties such as size, shape, charge, and surface modification, which can affect particle interaction with cells and serum proteins. In addition

Characterization of Pharmaceutical Nano- and Microsystems, First Edition. Edited by Leena Peltonen.

to these properties, surface modification is used to change the biodistribution behavior of NPs. Other factors such as administration route, dosing, biodegradation, metabolism and excretion are important in term of biological behaviors of NPs *in vivo*.

This chapter highlights the impact of the mechanisms affecting biodistribution, such as physicochemical properties, administration route, dosing and coating, on behaviors of NPs in systemic circulation, effects of mononuclear phagocytotic systems and cellular interactions on the targeted tissue.

8.2 Biodistribution and Pharmacokinetics

Nano-sized drugs have begun to be used clinically for the treatment of several diseases, especially cancer, and interest in this area continues to increase. The identification of pharmacokinetic parameters, such as distribution and excretion of the NPs in the body, is very important for predicting the efficacy and toxicity of the formulation. Pharmacokinetics refers to the examination of the profiles formed by biological fluids such as plasma, urine, saliva, lymph and spinal fluid after administration of the active substances to the organism in pure or dosage form. In other words, pharmacokinetics is defined by mathematical equations of events such as absorption, distribution, metabolism, and excretion that arise as the active substances enter the organism through various pathways and dosage forms. Important parameters of pharmacokinetics are maximum concentration (C_{max}), half-life ($t_{1/2}$), clearance (Cl), area under the curve (AUC), and mean residence time (MRT). These parameters give very important information about the duration and fate of the drug in the body. Using these parameters and mathematical equations, pharmacokinetic profiles are determined. A formulation designed for prolonged blood circulation is expected to increase in AUC and MRT but decrease in Cl value. In contrast, a decreased $t_{1/2}$, AUC, and MRT with increased Cl are usually expected for a formulation that is quickly eliminated from the organism [5].

The pharmacokinetic properties and bioavailability of the drug relates to toxicity and efficacy. The accumulation of free/nanoparticulate drug in the targeted tissue directly, by escaping from the opsonization of the drug, increases the efficiency of the drug. In contrast, systemic toxicity is inevitable if the drug is distributed throughout the body instead of the targeted organ. For this reason, the unique properties of nanosized drug carrier systems are particularly important in drug development studies. Especially in conventional chemotherapy, the drug is removed from the body a very short time after intravenous (i.v.) application, and the biodistribution of the remaining drug after clearance is nearly equal in healthy and tumorous tissues. The amount of drug reaching the tumor tissue remains lower than the therapeutic dose and is not sufficient for tumor therapy. The drug dosing frequency is increased to provide the necessary drug accumulation in the tumor area. As an effect of the drug dosage regimen, toxicity increases especially in fast-growing healthy cells such as hair follicles, and side-effects of the drug are observed. Nanoparticulated drug delivery systems such as polymeric NPs, metallic NPs, micelles, liposomes, dendrimers and drug–polymer conjugates change the pharmacokinetic profiles and biodistribution of active molecules, as they provide improved drug solubility as well as particle size, surface charge and surface modification. In the following sections, each parameter is explained in detail with literature examples.

8.3 Mechanisms Affecting Biodistribution

Many factors such as size, shape and surface properties of NPs, and the physiological environment, influence the biodistribution. In particular, particle size, shape, surface charge, and surface coating of NPs has a critical effect on biodistribution and clearance. In this section, the effect of these critical parameters on bioavailability and clearance of polymeric, polysaccharide and metallic NPs will be presented in the light of literature examples.

8.3.1 Nanoparticle Properties

8.3.1.1 Effect of Particle Size

Particle size is a key parameter in the biodistribution of NPs, which can be designed to precise dimensions and high monodispersity to attain therapeutic efficacy *in vivo* [6, 7]. This parameter affects many factors, including attachment, adhesion, phagocytosis, biological half-life, distribution and accumulation of NPs. It also affects the endocytosis, cellular uptake, and efficiency of particle processing in the endocytic pathway as it influences their internalization mechanism [8–12]. For instance, NPs with diameters 10–20 nm spread out in various tissues by passing endothelial tight-junctions and are rapidly excreted from kidneys upon i.v. administration [13]. In addition, small NPs have the ability to re-enter the bloodstream and reduce the amount retained within the tumor. NPs with diameters 100–200 nm have been shown to utilize enhanced permeability and retention (EPR) effects and escape filtration by the liver and spleen. NPs above 150 nm accumulate more in the liver and spleen than smaller NPs [14]. Due to the size range of endothelia with vascular fenestrations, larger NPs accumulate in the liver [14]. Particles in the micrometer range (2–5 μm) show a tendency to aggregate under physiological conditions and accumulate in capillaries of the lungs, which provides a distinct advantage when targeting one of the predominant sites of metastatic disease. For these particles, the clearance rate is also high. Particles are rapidly taken up by the mononuclear phagocytic system (MPS) and have high accumulation in the liver, spleen and bone marrow (Figure 8.1) [15].

It has been found that the circulation and biodistribution of NPs, such as metallic NPs (gold, silver, iron oxide, zinc oxide, titanium dioxide, etc.), polymeric NPs, silica NPs, and polysaccharide NPs, are size-dependent after being systemically administered [5].

A particular understanding of the pharmacokinetics of metallic NPs is important for favorable risk assessment and safe biomedical applications. Gold (Au) NPs can be used in cancer treatment and may be applied as drug carriers, gene delivery and contrast agents. Takeuchi *et al.* evaluated the particle size effect in biodistribution and excretion of Au NPs (20 nm, 50 nm, and 100 nm) after i.v. injection. Au concentrations were measured in different organs using inductively coupled plasma atomic emission spectrometry. 20 nm NPs were accumulated especially in the lungs and brain, but low accumulation was observed in the heart, stomach and pancreas [16]. In another study, the tissue distribution of 7 nm Au NPs is predominant in the lungs, followed by brain distribution which was higher than that of 20 nm nanoparticles, indicating disintegration of NPs after inhalation, and showed the impact of primary size of NPs on subsequent biodistribution [17]. Following i.v. injection of three different sizes of Au NPs (1.4 nm, 18 nm, 80 nm), researchers evaluated whether NPs cross the placental barrier during pregnancy or accumulate in the fetal organism by different

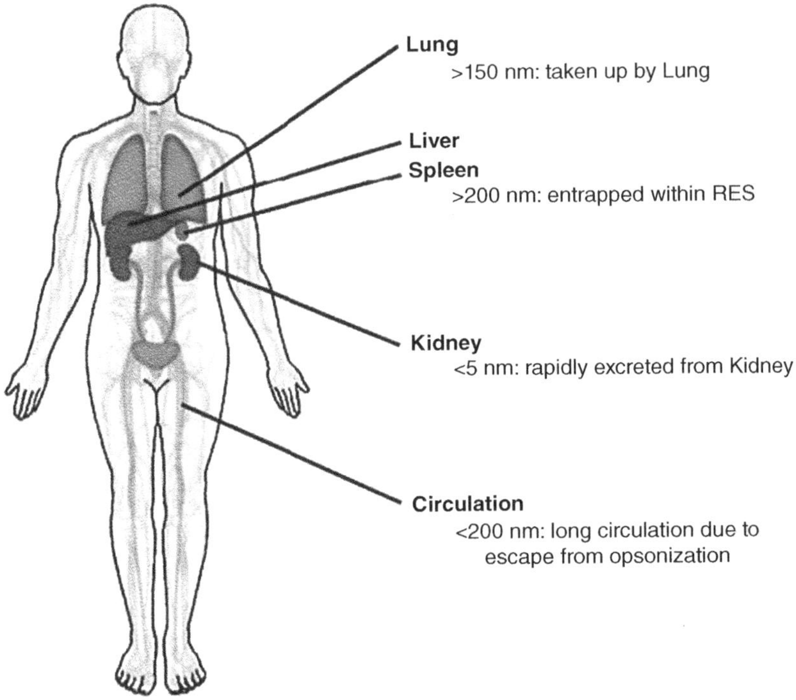

Figure 8.1 *Effects of particle size on biodistribution.*

transport mechanisms in a size-dependent manner. The study showed that no Au NPs had passed from the amniotic fluid to the fetus within 24 hours. However, results showed that Au NPs pass through the placenta using transtrophoblastic channels and/or transcellular pathways [18]. In another study, higher uptake and deeper penetration was observed for smaller NPs (3–4 nm) than for larger (13–14 nm) or PEGylated Au NPs in placenta, which barely passed the trophoblast barrier layer. These results indicate that size is a precondition for the development of targeted therapies in pregnancy [19]. These studies used Au NPs in different targeted therapies with modulation of particle size over long periods.

Pharmacokinetic studies have shown that the liver and spleen are the main organs of accumulation of silver (Ag) NPs, followed by the kidneys, when given via oral, i.v., subcutaneous or inhalation routes. Ag deposit was determined in several cells, including macrophages, hepatocytes and sinusoid endothelium cells in the liver [20]. In the kidney, Ag deposition was observed especially in the renal glomerular basement membrane and mesangium, and a gender-related difference in the accumulation of Ag was reported in the kidney of rats following repeated oral administration, with a two-fold increase in females compared with males [21]. Besides kidneys, a high distribution to the spleen has been reported after repeated (5 days) i.v. administration in a size-dependent manner, with 20 nm NPs distributed mainly to the liver, followed by kidneys and spleen, whereas 80 or 110 nm Ag NPs were distributed primarily to the spleen, followed by the liver and lung [22]. The mechanism was shown by Li and Huang who found that NPs smaller than the liver fenestrae pores (approximately 100 nm) are better taken up by the liver, while larger NPs are absorbed with greater affinity in the spleen [3]. In another study, after inhalation of 15 and 410 nm Ag NPs, Ag concentrations in vesicles and nuclei of lung cells were significantly higher in 15 nm than in 410 nm groups 7 days after exposure, and NPs with different sizes have different release rates of Ag, resulting in pulmonary toxicity [23]. These results proved that tissue distribution of Ag NPs was size-dependent and also suggested the potential for size-dependent toxicity and health risks.

Polysaccharide NPs (chitosan, hyaluronic acid, dextran) are also frequently used as a carrier system for drug delivery and imaging studies in the literature [24].

8.3.1.2 Effect of Surface Charge

The surface charge and hydrophobicity can dramatically affect opsonization, phagocytosis, blood circulation and biodistribution of NPs. Surface properties also influence stability and cellular interaction of NPs [25, 26].

The surface charge is determined as the zeta potential (ZP). ZP has a key role in the pharmacokinetic and biodistribution of NPs. Negatively charged NPs (ZP $\leq$ 10 mV) usually exhibit strong MPS uptake. On the other hand, positively charged NPs (ZP > 10 mV) will induce serum protein aggregation. Neutral NPs (within $\pm$10 mV) also exhibit the least MPS interaction and the longest circulation [3]. After i.v. administration, it is also known that positively charged NPs form aggregations that result in electrostatic interaction with negatively charged serum proteins. The dissociated NPs subsequently redistribute to the liver. Thus, positive NPs show a rapid blood clearance phase with a high dose accumulating in the lung and liver. Also, positively charged NPs have a high non-specific internalization rate and short blood circulation half-life. In general, macrophage uptake is directly proportional to the absolute value of surface charge. NPs with ZP values close to 0 can effectively avoid the uptake by phagocytic cells, thus exhibiting a significantly delayed blood clearance [27].

It has been reported that coating by zwitterionic polymers is helpful in evading uptake by MPS and achieving long blood circulation time. It is also shown that this coating process increases the biocompatibility and stability of NPs [28].

Xiao *et al.* found that NPs with high positive (>10 mV) or negative surface charge ($\leq$10 mV) were efficiently opsonized and cleared by macrophage cells from the blood circulation [29]. Similar results were found by Levchenko *et al.* They showed that NPs exhibiting ZP $\leq$ 40 mV exhibited 90% clearance in 10 min, compared with 10% clearance for the neutral NPs ($\pm$10 mV), and increased liver uptake was implicated in the accelerated clearance [30]. Increase in plasma protein adsorption with increasing absolute values of surface charge for negatively charged NPs was also shown by Zhang *et al.*, who achieved similar results to these studies. They showed that positively charged lipoplex formed aggregates in serum [31]. They also showed that these NPs caused transient embolism in the lungs and they were cleaned from the body in the liver. In the literature, PEGylation served to shield the charge effect, suggesting that PEG may minimize opsonization not only through steric hindrance but also charge shielding [30, 31].

8.3.1.3 *Effect of Particle Shape*

The shape of NPs also affects their biodistribution and clearance [32, 33]. NPs are generally prepared as spheres, but they can be prepared as ellipsoids, discs, cubes, cylinders, rods or cones (Figure 8.2). They can also be prepared in other complex shapes such as red blood cell-like biconcave discoids. This complex shape is shown to have effects on transport through the vasculature, circulation half-life, targeting efficacy, phagocytosis, and endocytosis, which directly influences the targeting ability of NPs [34–38].

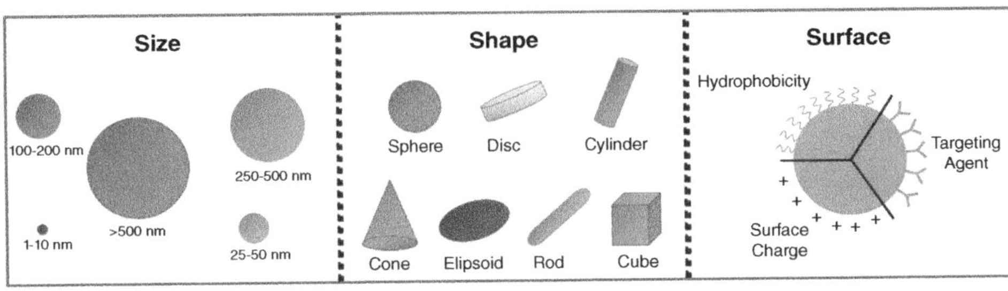

Figure 8.2 *Illustration of the different sizes, shapes, and surface properties of NPs.*

Studies have shown that optimization of the shape of NPs can improve targeting effectiveness. Particle shape is also an important property for the performance of NPs as a drug carrier, for cellular uptake, transport and degradation [39]. For instance, human cervical cancer cells (HeLa) easily take up rod-shaped NPs with diameter higher than 100 nm and aspect ratio (AR) higher than 3 [11]. In contrast, rod-shaped Au NPs are taken up less well by macrophages than spherical NPs [40].

The shape of NPs is a significant parameter for biodistribution. The effect of the shape on biodistribution can also directly affect their therapeutic behavior. Interestingly, non-spherical NPs tend to show deviating hydrodynamic behavior in blood vessels with shear flow [41]. Furthermore, shape determines the adhesion pattern of NPs, resulting in improved efficacy of drug carriers with non-spherical shapes [42]. For example, discoidal NPs exhibit specific tumbling and margination dynamics through interacting with the vessel wall substantially more than spherical NPs, resulting in particle binding and adhesion to the endothelium. The circulation half-life of a particle is also affected by shape. For example, Discher *et al.* have established that filamentous polymer micelles have long circulation times (>1 week) compared to spherical NPs (2–3 days), due to the tendency of these NPs to align with blood flow [43].

In vivo biodistribution of NPs of different geometries has been studied using healthy laboratory animals and various administration routes such as i.v. and oral. Studies have shown that NPs with different shapes accumulated at different rates in different organs. In a comprehensive study, liver concentration of discoidal NPs was found to be minimal among all the shapes. The concentration of these discoidal NPs in the other organs was also more than any of the other shapes. This phenomenon may result from these NPs escaping from opsonization and accumulation in other organs because of the ability of discoidal NPs to drift towards the vessel wall. It has also been shown that cylinder-shaped silica NPs mainly accumulated in the liver, while spherical NPs exhibited high MPS uptake [44]. Anselmo *et al.* showed that both rod-shaped and spherical NPs improved lung targeting in mice when coated with ICAM antibody. NPs coated with IgG antibody also showed enhanced liver and spleen uptake [45]. In another study, biodistributions of spherical, cubic and rod-shaped Au NPs were determined. Results showed that these three shapes of NPs exhibited enhanced liver and spleen concentrations [46]. Anti-intracellular adhesion molecule-1 coated rod-shaped and spherical NPs were compared with each other in another study. Rod-shaped NPs showed 2-fold higher concentration in lungs than spherical NPs. The rod-shaped and spherical NPs, when coated with IgG antibodies, showed almost similar liver concentrations. However, spleen concentration was found to be higher for rod-shaped NPs due to size-based physical filtration [47].

In another study, the effect of aspect ratio (AR) and administration route on *in vivo* distribution of mesoporous silica NPs was investigated. After oral administration, spherical NPs exhibited high spleen uptake, while rod-shaped NPs mainly concentrated in the lungs and liver. Furthermore, liver concentration of these NPs decreased when the AR of NPs was increased from 1 to 5. These NPs, when administered intravenously, revealed a different biodistribution profile. Short rod-shaped NPs were found to accumulate in the liver, while long rod-shaped NPs were trapped in the spleen [48].

The number of studies showing the effects of biodegradation for non-spherical NPs is very limited. There is a need for comprehensive studies in which the effect of the biodegradation by particle shape is examined in detail, using different administration routes such as i.v. and oral.

8.3.2 Dosing and Toxicity

It is very important to elucidate *in vivo* toxicity by elucidating the long-term accumulation and excretion mechanisms of NPs, especially metallic NPs.

Monge-Fuentes *et al.* conducted different toxicity studies for dimercaptosuccinic acid-coated iron oxide NPs (IONPs) in monkeys [49]. Studies with these metallic NPs have shown that they accumulate in the lung and brain, but accumulation mechanisms have not been elucidated [50]. No significant toxicity was detected during an *in vivo* study for a period of 4 months with experimental animals [50]. Di Bona *et al.* found that IONPs pass the placental barrier in mice when injected intraperitoneally. They also showed that these NPs accumulate in the fetus and increase the risk of fetal death due to greater accumulation of NPs in the fetal liver [51].

Hirst *et al.* examined the biodistribution and potential applications of cerium oxide NPs [52]. Tang *et al.* investigated biodistribution of Ag NP in rats. According to their results, Ag NPs have been found to be transferred to the blood and distributed through the brain after subcutaneous injection. Otherwise, blood–brain barrier destruction and neuronal degeneration have been observed in the same rats [53]. Ma *et al.* determined brain damage due to accumulation of NPs after high-dose administration of titanium dioxide. The concentration of NPs in the brain was highly correlated with the dosage [54].

Chen *et al.* determined toxicity of Au NPs for an *in vivo* mice model. Au NPs with diameters between 3–100 nm were intraperitoneally injected into mice at a dose of 8 mg/kg/week. Injected NPs with diameters between 8–37 nm exhibited side-effects such as fatigue and weight loss, and half of the mice died by day 21 [55]. These results may have resulted from the high dose used [56]. In another study by Cho *et al.*, the toxicity of PEGylated Au NPs was determined. *In vivo* effects of PEGylated Au NPs at three different doses (0.17, 0.85, and 4.26 mg/kg) were examined, and found that apoptosis in liver tissue increased with increasing dose [57]. The same research group established that PEG-coated Au NPs with diameters between 4–13 nm upregulated the CYP1A1 metabolic enzyme in the liver, but 100 nm NPs did not. These results were attributed to a higher surface area dose from the smaller NPs [58]. However, another study showed that PEG-coated Au NPs with diameters 50 nm injected in rats at a concentration of 57 μg/ml caused changes to the spleen, liver, lungs, and kidneys [59].

Studies with different animal models and different NP formulations are rather difficult to compare to each other. However, it can be said that size, shape and surface coatings of NPs influence the biodistribution, excretion and toxicity of NPs. NPs of all sizes accumulate quickly in the liver and spleen and remain sequestered for extended periods. Appropriate dosing also needs to be determined, as short- and long-term toxicity may depend upon it.

8.3.3 Effect of Coating

One of the factors affecting the biodistribution of polymeric, polysaccharide and metallic NPs is surface coating. In the literature, the surfaces of these NPs are modified with poly(ethylene glycol) (PEG), polymers such as poly(ethylene oxide) (PEO), poly(propylene glycol) (PPG), poly(propylene oxide) (PPO), poloxamer derivatives, chitosan, and various targeting agents, which in turn affect NP biodistribution.

Opsonization of NPs has one of the most important effects on biodistribution and organ accumulation. Opsonization is a defense mechanism that the body shows against foreign material. It is the process by which foreign material, including NPs, is covered by opsonin proteins. These opsonin-coated foreign materials (or NPs) are recognized and phagocytosed by the mononuclear phagocytic system (MPS) cells. The vast majority of these cells comprise tissue macrophages and blood monocytes. These cells accumulate primarily in organs such as the spleen, bone marrow, and liver [60, 61]. For this reason, NPs unmodified by any polymer such as PEG are opsonized and phagocytosed by macrophages in a few seconds, and then they are accumulated especially in the liver.

The surfaces of NPs can be modified with hydrophilic polymers such as PEG and PPG to prevent opsonization and accumulation of NPs in MPS organs. This process is called PEGylation. "Stealth" NPs, which cannot be recognized by the MPS, are prepared by this process. PEGylation increases the circulation time of NPs in the bloodstream, as well as decreasing the immunogenicity of NPs. PEG is approved for therapeutic use in humans and classified as "Generally Recognized As Safe" (GRAS) by the USFDA. Despite the many PEGylated nanoparticulate systems, the most well-known and FDA-approved PEGylated formulation is DOXIL®. The bioavailability of doxorubicin increases 87-fold with this PEGylated formulation. The half-life of doxorubicin also increases to 72 hours, and the circulation half-life to 36 hours [62–64].

NPs can be PEGylated covalently, or by grafting, entrapping or adsorbing PEG chains. Type of PEGylation affects the duration of the NPs' blood circulation. Studies have shown that PEG chains covalently bound to the surface of NPs remain longer in the bloodstream than those adsorbed onto them [8, 65–67]. Therefore, copolymers of PEG and other polymers with a long circulation time in the bloodstream are frequently used in the preparation of NPs. In one study, Oliveira *et al.* prepared plitidepsin-loaded polymeric NPs by using PEG-block-poly(γ-benzyl-L-glutamate) (PEG-b-PBLG) and poly(trimethylene carbonate)-block-poly(glutamic acid). They investigated the pharmacokinetics, bioavailability and therapeutic effects of these NPs on the xenograft renal cancer mouse model. *In vivo* experiments with mice have shown that accumulations of all formulations in the tumor region were similar. However, PEG-b-PBLG NPs showed less accumulation in the kidney and liver compared with PEG-free NPs [68]. The poloxomer is also a triblock copolymer containing PEG. One of the most widely used derivatives is Poloxamer 407, which consists of two units of PEG forming the hydrophilic moiety, and polypropylene glycol forming the hydrophobic center. It is often used in the preparation of NPs. Redhead *et al.* [69] coated the surface of PLGA NPs with two different poloxomeric derivatives (Poloxamer 407 and Poloxamer 907). In this way, they aimed to prevent the capture of NPs by liver macrophages. They used Rose Bengal as the model drug, and studied the circulation time of this drug in the bloodstream. The remaining amount of drug given as a solution was only 8% after 5 minutes. However, when the drug was administered in poloxomer-coated NPs, this increased to 30% after 1 hour. It has been emphasized that this increase in blood concentration may be related to the prolongation of the half-life of the drug resulting from drug-loaded NPs escaping from the MPS. This study has shown that poloxomer-coated polymeric NPs are important drug delivery systems for drug targeting [69].

The molecular weight of PEG affects the circulation time of NPs. In particular, PEG chains with molecular weights of 2 kDa and above are important in inhibiting the opsonization of NPs [67]. In one study, Sadzuka *et al.* [70] have shown that liposome formulations coated with PEG, having a molecular weight of 2 kDa, increase the accumulation of irinotecan hydrochloride and the active metabolite of irinotecan in tumor tissue. In addition, it has been observed that PEG-uncoated liposome formulations accumulate in the liver after i.v. injection, and that blood concentrations of the PEG-coated liposomes are found to be higher 8 hours after injection compared with PEG-uncoated liposomes in male CDF_1 mice [70].

It is known that PEG with high molecular weight increases the circulation time of NPs in the bloodstream. In a study of the bioavailability of micelle formulations prepared with PEG at different molecular weights, an increase in molecular weight of PEG increases the circulation half-life. With 5 kDa PEG containing micelle, this value was found to be 4.6 min, and with 20 kDa PEG it was about a 4-fold increase to 17.8 min [71]. Similarly, the circulation half-life was determined after systemic application of Au NPs, with PEG chains of molecular weight 2, 5 and 10 kDa. When the PEG molecular weight is increased from 2 kDa to 10 kDa, the circulation half-life of Au NPs increased by about 10 times [72]. In another study with PEG–PLA co-polymers with different PEG molecular weights, the particle size and circulation half-life of prepared NPs were investigated. The increase in molecular weight of PEG was shown to increase the particle size and circulation half-life of NPs [66]. Lipka *et al.* [73] investigated the bioavailability and circulation times in the bloodstream of Au NPs after i.v. and intratracheal injection. The particle size and zeta potential of the prepared Au NPs are influenced by the molecular weight of the PEG chains in the structure. The average particle size of the Au NPs containing 10 kDa PEG was increased by 10 nm and zeta potentials increased by 18 mV compared with non-PEGylated NPs. When non-PEGylated and PEGylated (containing 750 Da or 10 kDa PEG chains) Au NPs were applied by i.v. injection, non-PEGylated or PEGylated NPs with 750 Da PEG mainly accumulated in the liver and spleen. The accumulation amount of NPs containing 10 kDa PEG chains in the liver was half that compared with non-PEGylated NPs, 24 hours after injection. In intratracheal administration, all NPs were found to accumulate in the lung. At the end of 24 hours, it was also found that NPs containing 10 kDa PEG were absorbed better than other NPs, with higher and more rapid plasma uptake. It was also observed that the duration of systemic circulation of Au NPs containing 10 kDa PEG chains is longer for i.v. injection. This study confirmed that the effect of PEG chains under 10 kDa in Au NPs on the biodegradation and circulation times of NPs did not appear to be evident. The study showed that these effects were only possible with PEG chains above 10 kDa. PEGylation has also been shown to be effective in reducing the accumulation-dependent toxicity of Au NPs [73].

The density of PEG chains on the surface of NPs affects their bioavailability. NPs without PEG chains on the surface accumulate generally in the liver and spleen; an increase in PEG chains on the surface of NPs causes this accumulation to shift to the spleen. In a study of liposomes containing PEG at different ratios, it has been shown that increasing the PEG ratio from 5% to 10% reduces the risk of aggregation of liposomes and they are stable for longer [74]. In another study, docetaxel-loaded PEG-coated lipid nanocapsules were prepared. When pharmacokinetics and biodistribution of lipid nanocapsules containing PEG at different ratios (6, 10, 15% mol) was examined, the increase in PEG density was found to increase the blood concentration time [75]. The effect of PEG density and molecular weight on the pharmacokinetic properties of PEG-coated PLA nanocapsules has been investigated in a biodistribution study. In this study, PEG chains of 5 kDa and 20 kDa and 10–30% w/w PEG ratios were used to prepared PLA nanocapsules, and the *in vitro–in vivo* stabilities of nanocapsules were determined. The length and density of the PEG chains directly increased the half-life of nanocapsules [76]. In a study by Yang *et al.* [77], the effect of molecular weight and concentration of PEG chains on the interaction of NPs with the immune system cells and on the circulation time was investigated. In this study, it was observed that the increase in PEG concentration and molecular weight significantly inhibited the uptake of NPs by the immune system cells both *in vitro* and *in vivo* [77]. Walkey *et al.* prepared Au NPs containing different PEG ratios. In this study, the absorption of prepared NPs with serum proteins and the macrophage uptake were examined. It was seen that the increase in PEG density and the adsorption of NPs of serum proteins were inversely proportional. Accordingly, PEG chain densities were also observed to control the uptake of NPs by macrophages [78].

In spite of all these advantages of PEGylation, there are studies showing that PEGylation in some cases affects the immune system and causes hypersensitivity reactions. Studies on Doxil®, a PEGylated liposome formulation, have shown that it has caused hypersensitivity reactions explained as Type-1 allergic reactions. In a study of patients treated with Doxil, it showed that activation of the complement (C) system was effective in this case. In this study of 29 patients with solid tumors, Doxil-induced hypersensitivity reactions developed in patients predisposed to C activation [79]. In a study with PEGylated liposomes carrying plasmid DNA, the efficacy and safety of the formulation in repeated administrations *in vivo* was examined. Repeated administrations of the formulations showed hypersensitivity reactions in mice. It has been suggested that this reaction may be an improved antibody response to PEG due to the immunostimulatory effect of plasmid DNA. It has also been suggested that modification of the alkyl chains of the PEG-lipid conjugate may reduce this effect [80].

Active targeting agents play important roles in the biodistribution of NPs and pharmacokinetic profiles of active molecules. Inorganic or polymeric NPs have been modified for specific receptors overexpressed on tumor nodules. The structure and concentrations of receptors on the tumor surface depend on the cancer type. Coating the surface of the NPs with specific molecules gives a high affinity to receptors on the cancer cell membrane, enhancing tumoral uptake and providing tumor-specific targeting. Nascimento *et al.* developed EGFR-modified chitosan NPs containing siRNA/cisplatin for lung cancer therapy, and investigated the effect of active targeting on biodistribution and pharmacokinetic parameters. EGFR-targeted chitosan NPs were reported to reach the tumor tissue immediately from plasma, and they remain in the tumor for 96 hours. In addition, active targeted chitosan NPs exhibited tumor targeting activity six times higher when compared to naked chitosan NPs [81].

One study examined the effect of surface chemistry on serum protein adsorbed mesoporous silica NPs targeted with a selective antibody against B220-positive leukemic stem cells. The uptake kinetics of NPs by positive cells was found to be higher than the uptake of NPs by B220-negative leukemia stem cells. It was emphasized that adsorption of NPs' surface charge and serum proteins are important factors in increasing drug selectivity for targeting drug delivery by using NPs [82].

8.4 Conclusion

Illumination of the biodistribution of NPs is essential for effective and safe drug treatment. By illuminating the factors affecting the biodistribution, it is possible to form appropriately targeted NPs without side-effects. Of course, the most important factors affecting biodistribution are particle diameter, surface charge, particle shape, and surface properties.

By optimizing these factors (size less than 100 nm, ZP less than 10 mV, and coating with PEG), nanoparticles can escape from MPS uptake. On the other hand, blood circulation can be prolonged, allowing increased biodistribution of NPs in target tissue, by using these factors. NPs larger than 50 nm show less penetration in tissues. In contrast, NPs smaller than 50 nm can be taken up by the liver and may cause liver toxicity. Neutral NPs have the potential to deliver drugs into tumors, thanks to long circulation times. PEG-coated NPs have hydrophilic surfaces, and thus have a prolonged circulation time. PEGylation also suppresses NP interactions with cells.

Shape plays an important role in biodistribution and the *in vivo* properties of NPs. In addition, the surface charge of NPs affects cellular interactions due to negatively charged membrane components. Non-spherical NPs can escape from phagocytosis, thus remaining in circulation as a result.

Another important subject of research has been control of the release profile of NPs. If controlled drug release is possible in the target tissue specifically, the problem of tissue penetration can be solved especially for small molecule drugs. Although different release strategies have been developed for NPs, success in clinical practice is still limited. Today, there are several NP formulations in clinical trials, which reduce or eliminate the side-effects of conventional therapy and therefore improved efficacy.

Thanks to the new nanoparticles, more efficient delivery of drugs to the targeted tissue is anticipated, and the development of new carrier systems for drug release will follow. Inexpensive and non-toxic materials must be developed to reduce the costs of nano-sized drug delivery systems based therapy.

References

[1] Agasti, S.S. Rana, S., Park, M.H., *et al.* (2010) Nanoparticles for detection and diagnosis. *Adv Drug Deliv Rev*, **62**, 316–328.
[2] Kennedy, L.C., Bickford, L.R., Lewinski, N.A., *et al.* (2011) A new era for cancer treatment: gold-nanoparticle-mediated thermal therapies. *Small*, **7**, 169–183.
[3] Li, S.D., Huang, L. (2008) Pharmacokinetics and biodistribution of nanoparticles. *Mol Pharm*, **5**, 496–504.
[4] Li, M., Al-Jamal, K.T., Kostarelos, K., Reineke, J. (2010) Physiologically based pharmacokinetic modeling of nanoparticles. *ACS Nano*, **4**, 6303–6317.
[5] Lin, Z., Monteiro-Riviere, N.A., Riviere, J.E. (2015) Pharmacokinetics of metallic nanoparticles. *Wiley Interdiscip Rev Nanomed Nanobiotechnol*, **7**, 189–217.
[6] Alexis, F., Pridgen, E., Molnar, L.K., Farokhzad, O.C. (2008) Factors affecting the clearance and biodistribution of polymeric nanoparticles. *Molecular Pharmaceutics*, **5**, 505–515.
[7] Powers, K.W., Palazuelos, M., Moudgil, B.M., Roberts, S.M. (2007) Characterization of the size, shape, and state of dispersion of nanoparticles for toxicological studies. *Nanotoxicology*, **1**, 42–51.
[8] Ernsting, M.J., Murakami, M., Roy, A., Li, S.D. (2013) Factors controlling the pharmacokinetics, biodistribution and intratumoral penetration of nanoparticles. *J Control Release*, **172**, 782–794.
[9] Shinde Patil, V.R., Campbell, C.J., Yun, Y.H., *et al.* (2001) Particle diameter influences adhesion under flow. *Biophys J*, **80**, 1733–1743.
[10] Champion, J.A., Walker, A., Mitragotri, S. (2008) Role of particle size in phagocytosis of polymeric microspheres. *Pharmaceutical Research*, **25**, 1815–1821.
[11] Gratton, S.E., Ropp, P.A., Pohlhaus, P.D., *et al.* (2008) The effect of particle design on cellular internalization pathways. *Proc Natl Acad Sci USA*, **105**, 11613–11618.
[12] He, C., Hu, Y., Yin, L., *et al.* (2010) Effects of particle size and surface charge on cellular uptake and biodistribution of polymeric nanoparticles. *Biomaterials*, **31**, 3657–3666.
[13] Choi, H.S., Liu, W., Misra, P., *et al.* (2007) Renal clearance of quantum dots. *Nat Biotechnol*, **25**, 1165–1170.
[14] Blanco, E., Shen, H., Ferrari, M. (2015) Principles of nanoparticle design for overcoming biological barriers to drug delivery. *Nat Biotechnol*, **33**, 941–951.
[15] Foged, C., Brodin, B., Frokjaer, S., Sundblad, A. (2005) Particle size and surface charge affect particle uptake by human dendritic cells in an in vitro model. *Int J Pharm*, **298**, 315–322.
[16] Takeuchi, I., Nobata, S., Oiri, N., *et al.* (2017) Biodistribution and excretion of colloidal gold nanoparticles after intravenous injection: Effects of particle size. *Biomed Mater Eng*, **28**, 315–323.
[17] Balasubramanian, S.K., Poh, K.W., Ong, C.N., *et al.* (2013) The effect of primary particle size on biodistribution of inhaled gold nano-agglomerates. *Biomaterials*, **34**, 5439–5452.
[18] Semmler-Behnke, M., Lipka, J., Wenk, A., *et al.* (2014) Size dependent translocation and fetal accumulation of gold nanoparticles from maternal blood in the rat. *Part Fibre Toxicol*, **11**, 33.
[19] Muoth, C., Grossgarten, M., Karst, U., *et al.* (2017) Impact of particle size and surface modification on gold nanoparticle penetration into human placental microtissues. *Nanomedicine*, **12**, 1119–1133.

[20] Gaillet, S., Rouanet, J.M. (2015) Silver nanoparticles: their potential toxic effects after oral exposure and underlying mechanisms – a review. *Food Chem Toxicol*, **77**, 58–63.
[21] Kim, Y.S., Kim, J.S., Cho, H.S., *et al.* (2008) Twenty-eight-day oral toxicity, genotoxicity, and gender-related tissue distribution of silver nanoparticles in Sprague-Dawley rats. *Inhal Toxicol*, **20**, 575–583.
[22] Lankveld, D.P., Oomen, A.G., Krystek, P., *et al.* (2010) The kinetics of the tissue distribution of silver nanoparticles of different sizes. *Biomaterials*, **31**, 8350–8361.
[23] Braakhuis, H.M., Gosens, I., Krystek, P., *et al.* (2014) Particle size dependent deposition and pulmonary inflammation after short-term inhalation of silver nanoparticles. *Part Fibre Toxicol*, **11**, 49.
[24] Swierczewska, M., Han, H.S., Kim, K., *et al.* (2016) Polysaccharide-based nanoparticles for theranostic nanomedicine. *Adv Drug Deliv Rev*, **99**, 70–84.
[25] Schipper, M.L., Iyer, G., Koh, A.L., *et al.* (2009) Particle size, surface coating, and PEGylation influence the biodistribution of quantum dots in living mice. *Small*, **5**, 126–134.
[26] Verma, A., Stellacci, F. (2010) Effect of surface properties on nanoparticle-cell interactions. *Small*, **6**, 12–21.
[27] Alexis, F., Pridgen, E., Molnar, L.K., Farokhzad, O.C. (2008) Factors affecting the clearance and biodistribution of polymeric nanoparticles. *Mol Pharm*, **5**, 505–515.
[28] Jiang, S.Y., Cao, Z.Q. (2010) Ultralow-fouling, functionalizable, and hydrolyzable zwitterionic materials and their derivatives for biological applications. *Advanced Materials*, **22**, 920–932.
[29] Xiao, K., Li, Y., Luo, J., *et al.* (2011) The effect of surface charge on in vivo biodistribution of PEG-oligocholic acid based micellar nanoparticles. *Biomaterials*, **32**, 3435–3446.
[30] Levchenko, T.S., Rammohan, R., Lukyanov, A.N., *et al.* (2002) Liposome clearance in mice: the effect of a separate and combined presence of surface charge and polymer coating. *International Journal of Pharmaceutics*, **240** (1–2), 95–102. doi:10.1016/s0378-5173(02)00129-1
[31] Zhang, J.S., Liu, F., Huang, L. (2005) Implications of pharmacokinetic behavior of lipoplex for its inflammatory toxicity. *Adv Drug Deli Rev*, **57**, 689–698.
[32] Petros, R.A., DeSimone, J.M. (2010) Strategies in the design of nanoparticles for therapeutic applications. *Nat Rev Drug Discov*, **9**, 615–627.
[33] Duan, X., Li, Y. (2013) Physicochemical characteristics of nanoparticles affect circulation, biodistribution, cellular internalization, and trafficking. *Small*, **9**, 1521–1532.
[34] Muro, S., Garnacho, C., Champion, J.A., *et al.* (2008) Control of endothelial targeting and intracellular delivery of therapeutic enzymes by modulating the size and shape of ICAM-1-targeted carriers. *Mol Ther*, **16**, 1450–1458.
[35] Geng, Y., Dalhaimer, P., Cai, S., *et al.* (2007) Shape effects of filaments versus spherical particles in flow and drug delivery. *Nat Nanotechnol*, **2**, 249–255.
[36] Park, J.H., von Maltzahn, G., Zhang, L., *et al.* (2008) Magnetic iron oxide nanoworms for tumor targeting and imaging. *Adv Mater*, **20**, 1630–1635.
[37] Champion, J.A., Mitragotri, S. (2006) Role of target geometry in phagocytosis. *Proc Natl Acad Sci USA*, **103**, 4930–4934.
[38] Hutter, E., Boridy, S., Labrecque, S., *et al.* (2010) Microglial response to gold nanoparticles. *ACS Nano*, **4**, 2595–2606.
[39] Champion, J.A., Katare, Y.K., Mitragotri, S. (2007) Particle shape: a new design parameter for micro- and nanoscale drug delivery carriers. *J Control Release*, **121**, 3–9.
[40] Arnida, M.M., Janat-Amsbury, A., Ray, C.M., *et al.* (2011) Geometry and surface characteristics of gold nanoparticles influence their biodistribution and uptake by macrophages. *Eur J Pharm Biopharm*, **77**, 417–423.
[41] Decuzzi, P., Ferrari, M. (2006) The adhesive strength of non-spherical particles mediated by specific interactions. *Biomaterials*, **27**, 5307–5314.

[42] Doshi, N., Prabhakarpandian, B., Rea-Ramsey, A., *et al.* (2010) Flow and adhesion of drug carriers in blood vessels depend on their shape: a study using model synthetic microvascular networks. *J Control Release*, **146**, 196–200.
[43] Gentile, F., Chiappini, C., Fine, D., *et al.* (2008) The effect of shape on the margination dynamics of non-neutrally buoyant particles in two-dimensional shear flows. *J Biomech*, **41**, 2312–2318.
[44] Decuzzi, P., Godin, B., Tanaka, T., *et al.* (2010) Size and shape effects in the biodistribution of intravascularly injected particles. *J Control Release*, **141**, 320–327.
[45] Anselmo, A.C., Kumar, S., Gupta, V., *et al.* (2015) Exploiting shape, cellular-hitchhiking and antibodies to target nanoparticles to lung endothelium: Synergy between physical, chemical and biological approaches. *Biomaterials*, **68**, 1–8.
[46] Sun, Y.N., Wang, C.D., Zhang, X.M., *et al.* (2011) Shape dependence of gold nanoparticles on in vivo acute toxicological effects and biodistribution. *J Nanosci Nanotechnol*, **11**, 1210–1216.
[47] Kolhar, P., Anselmo, A.C., Gupta, V., *et al.* (2013) Using shape effects to target antibody-coated nanoparticles to lung and brain endothelium. *Proc Natl Acad Sci USA*, **110**, 10753–10758.
[48] Li, L., Liu, T., Fu, C., *et al.* (2015) Biodistribution, excretion, and toxicity of mesoporous silica nanoparticles after oral administration depend on their shape. *Nanomedicine*, **11**, 1915–1924.
[49] Monge-Fuentes, V., Garcia, M.P., Tavares, M.C., *et al.* (2011) Biodistribution and biocompatibility of DMSA-stabilized maghemite magnetic nanoparticles in nonhuman primates (Cebus spp.). *Nanomedicine (Lond)*, **6**, 1529–1544.
[50] Mejias, R., Perez-Yague, S., Gutierrez, L., *et al.* (2011) Dimercaptosuccinic acid-coated magnetite nanoparticles for magnetically guided in vivo delivery of interferon gamma for cancer immunotherapy. *Biomaterials*, **32**, 2938–2952.
[51] Di Bona, K.R., Xu, Y., Ramirez, P.A., *et al.* (2014) Surface charge and dosage dependent potential developmental toxicity and biodistribution of iron oxide nanoparticles in pregnant CD-1 mice. *Reprod Toxicol*, **50**, 36–42.
[52] Hirst, S.M., Karakoti, A., Singh, S., *et al.* (2013) Bio-distribution and in vivo antioxidant effects of cerium oxide nanoparticles in mice. *Environ Toxicol*, **28**, 107–118.
[53] Tang, J., Xiong, L., Wang, S., *et al.* (2009) Distribution, translocation and accumulation of silver nanoparticles in rats. *J Nanosci Nanotechnol*, **9**, 4924–4932.
[54] Ma, L., Liu, J., Li, N., *et al.* (2010) Oxidative stress in the brain of mice caused by translocated nanoparticulate TiO_2 delivered to the abdominal cavity. *Biomaterials*, **31**, 99–105.
[55] Chen, Y.S., Hung, Y.C., Liau, I., Huang, G.S. (2009) Assessment of the in vivo toxicity of gold nanoparticles. *Nanoscale Res Lett*, **4**, 858–864.
[56] De Jong, W.H., Hagens, W.I., Krystek, P., *et al.* (2008) Particle size-dependent organ distribution of gold nanoparticles after intravenous administration. *Biomaterials*, **29**, 1912–1919.
[57] Cho, W.S., Cho, M., Jeong, J., *et al.* (2009) Acute toxicity and pharmacokinetics of 13 nm-sized PEG-coated gold nanoparticles. *Toxicol Appl Pharmacol*, **236**, 16–24.
[58] Cho, W.S., Cho, M., Jeong, J., *et al.* (2010) Size-dependent tissue kinetics of PEG-coated gold nanoparticles. *Toxicology and Applied Pharmacology*, **245**, 116–123.
[59] Terentyuk, G.S., Maslyakova, G.N., Suleymanova, L.V., *et al.* (2009) Circulation and distribution of gold nanoparticles and induced alterations of tissue morphology at intravenous particle delivery. *Journal of Biophotonics*, **2**, 292–302.
[60] van Furth, R., Cohn, Z.A., Hirsch, J.G., *et al.* (1972) The mononuclear phagocyte system: a new classification of macrophages, monocytes, and their precursor cells. *Bulletin of the World Health Organization*, **46**, 845–852.
[61] Hume, D.A. (2006) The mononuclear phagocyte system. *Curr Opin Immunol*, **18**, 49–53.

[62] Laginha, K.M., Verwoert, S., Charrois, G.J.R., Allen, T.M. (2005) Determination of doxorubicin levels in whole tumor and tumor nuclei in murine breast cancer tumors. *Clinical Cancer Research*, **11**, 6944–6949.

[63] Suk, J.S., Xu, Q., Kim, N., *et al.* (2016) PEGylation as a strategy for improving nanoparticle-based drug and gene delivery. *Adv Drug Deliv Rev*, **99**, 28–51.

[64] Working, P.K., Newman, M.S., Huang, S.K., *et al.* (1994) Pharmacokinetics, biodistribution and therapeutic efficacy of doxorubicin encapsulated in Stealth® liposomes (Doxil®). *Journal of Liposome Research*, **4**, 667–687.

[65] Harper, G.R., Davies, M.C., Davis, S.S., *et al.* (1991) Sterie stabilization of microspheres with grafted polyethylene oxide reduces phagocytosis by rat Kupffer cells in vitro. *Biomaterials*, **12**, 695–700.

[66] Bazile, D., Prud'homme, C., Bassoullet, M.T., *et al.* (1995) Stealth Me.PEG-PLA nanoparticles avoid uptake by the mononuclear phagocytes system. *J Pharm Sci*, **84**, 493–498.

[67] Owens, D.E., Peppas, N.A. (2006) Opsonization, biodistribution, and pharmacokinetics of polymeric nanoparticles. *Int J Pharm*, **307**, 93–102.

[68] Oliveira, H., Thevenot, J., Garanger, E., *et al.* (2014) Nano-encapsulation of plitidepsin: in vivo pharmacokinetics, biodistribution, and efficacy in a renal xenograft tumor model. *Pharm Res*, **31**, 983–991.

[69] Redhead, H.M., Davis, S.S., Illum, L. (2001) Drug delivery in poly(lactide-co-glycolide) nanoparticles surface modified with poloxamer 407 and poloxamine 908: in vitro characterisation and in vivo evaluation. *J Control Release*, **70**, 353–363.

[70] Sadzuka, Y., Hirotsu, S., Hirota, S. (1998) Effect of liposomalization on the antitumor activity, side-effects and tissue distribution of CPT-11. *Cancer Letters*, **127**, 99–106.

[71] Miteva, M., Kirkbride, K.C., Kilchrist, K.V., *et al.* (2015) Tuning PEGylation of mixed micelles to overcome intracellular and systemic siRNA delivery barriers. *Biomaterials*, **38**, 97–107.

[72] Perrault, S.D., Walkey, C., Jennings, T., *et al.* (2009) Mediating tumor targeting efficiency of nanoparticles through design. *Nano Letters*, **9**, 1909–1915.

[73] Lipka, J., Semmler-Behnke, M., Sperling, R.A., *et al.* (2010) Biodistribution of PEG-modified gold nanoparticles following intratracheal instillation and intravenous injection. *Biomaterials*, **31**, 6574–6581.

[74] Braeckmans, K., Buyens, K., Bouquet, W., *et al.* (2010) Sizing nanomatter in biological fluids by fluorescence single particle tracking. *Nano Letters*, **10**, 4435–4442.

[75] Khalid, M.N., Simard, P., Hoarau, D., *et al.* (2006) Long circulating poly(ethylene glycol)-decorated lipid nanocapsules deliver docetaxel to solid tumors. *Pharm Res*, **23**, 752–758.

[76] Mosqueira, V.C., Legrand, P., Morgat, J.L., *et al.* (2001) Biodistribution of long-circulating PEG-grafted nanocapsules in mice: effects of PEG chain length and density. *Pharm Res*, **18**, 1411–1419.

[77] Yang, Q., Jones, S.W., Parker, C.L., *et al.* (2014) Evading immune cell uptake and clearance requires PEG grafting at densities substantially exceeding the minimum for brush conformation. *Mol Pharm*, **11**, 1250–1258.

[78] Walkey, C.D., Olsen, J.B., Guo, H., *et al.* (2012) Nanoparticle size and surface chemistry determine serum protein adsorption and macrophage uptake. *Journal of the American Chemical Society*, **134**, 2139–2147.

[79] Chanan-Khan, A., Szebeni, J., Savay, S., *et al.* (2003) Complement activation following first exposure to pegylated liposomal doxorubicin (Doxil): possible role in hypersensitivity reactions. *Ann Oncol*, **14**, 1430–1437.

[80] Judge, A., McClintock, K., Phelps, J.R., Maclachlan, I. (2006) Hypersensitivity and loss of disease site targeting caused by antibody responses to PEGylated liposomes. *Mol Ther*, **13**, 328–337.

[81] Nascimento, A.V., Gattacceca, F., Singh, A., *et al.* (2016) Biodistribution and pharmacokinetics of Mad2 siRNA-loaded EGFR-targeted chitosan nanoparticles in cisplatin sensitive and resistant lung cancer models. *Nanomedicine (Lond)*, **11**, 767–781.

[82] Beck, M., Mandal, T., Buske, C., Linden, M. (2017) Serum protein adsorption enhances active leukemia stem cell targeting of mesoporous silica nanoparticles. *ACS Appl Mater Interfaces*, **9**, 18566–18574.

9

Opportunities and Challenges of Silicon-based Nanoparticles for Drug Delivery and Imaging

Didem Şen Karaman[1], Martti Kaasalainen[2,3], Helene Kettiger[1] and Jessica M. Rosenholm[1]

[1] Pharmaceutical Sciences Laboratory, Faculty of Science and Engineering, Åbo Akademi University, Turku, Finland
[2] Laboratory of Industrial Physics, University of Turku, Finland
[3] Medicortex Finland Oy

Silicon-based materials, especially porous ones, have been established as flexible platforms in drug delivery applications since the first report on sol-gel derived silica as a drug carrier appeared in 1983. Later, mesoporous materials made their entrance into the drug delivery field, and nowadays both mesoporous silica and silicon are widely exploited especially in nanomedical research. Despite exhibiting corresponding properties and application areas, mesoporous silica nanoparticles (MSNs) and porous silicon (PSi) are fabricated via two distinctively different approaches. While PSi is fabricated via a top-down approach, MSNs (or mesoporous silica in general) are synthesized via a bottom-up technique, generally based on sol-gel methodology. Sol-gel techniques are also predominantly used to prepare nonporous silica nanoparticles, especially via the so-called Stöber process.

The pores in mesoporous silica are formed via a molecular templating process (meaning very precise architectures can be produced), while for PSi, electrochemical etching under controlled conditions causes the porosity. This endows both material types with high drug-loading properties [1]. Apart from the general challenges associated with the particles

Characterization of Pharmaceutical Nano- and Microsystems, First Edition. Edited by Leena Peltonen.

being on the nanoscale, especially the porosity, with pore sizes approaching the molecular scale, this presents characterization and handling requirements with distinct demands. Mesoporous materials can be prepared in the form of films (1D), wires/fibers (2D) and particles (3D nanostructures), with the pores accounting for another nanoscaled dimension; the pore structures are also usually classified as 1D, 2D or 3D mesostructures. In addition, hollow, rattle-like and core@shell structures can be prepared with certain advantages, such as in drug delivery applications.

Despite being inorganic materials, both silica and silicon are bioerodible materials that degrade via hydrolysis in the presence of water. This is an aspect that has been largely ignored even by the research community itself, but this property becomes very prominent especially for nanoscale materials. Contact with residual water or even air humidity should be accounted for in each processing step, making sure via proper characterization that the structure of the material has been retained after each step if water is present. It is not unusual to find in the literature examples where, for example, drug loading has been conducted for prolonged times in aqueous solvent, with characterization only of the starting material shown, or silica dissolution studies being conducted far above the solubility limit of silica, consequently leading to false claims that silica is not biodegradable. Sophisticated drug release mechanisms may also become challenging to translate to *in vivo* conditions if the carrier material itself starts degrading, and thus pore gating functions may be released before the drug is scheduled to do so.

Since it is evident that different characterization methods and associated considerations may vary throughout the processing of the silicon-based materials towards their quest in being applied as drug carriers, we have divided the chapter according to the stage for which the method may be most relevant. For in-depth studies on any of the presented topics, the interested reader is referred to the cited references for more detailed information.

9.1 Synthesis and Characteristics of Silica-based Nanoparticles

Silica nanoparticles are a type of colloidal amorphous metal oxide, which can be synthesized by a bottom-up approach via the sol-gel process. In the sol-gel system, the formation of oxide networks through polycondensation reactions of the molecular precursor in liquid occurs; that is, monomeric silicon alkoxide precursors are hydrolyzed and subsequently condensate with the additional acid or base catalyst. Synthesis protocols denoted the Stöber process, or reverse microemulsion mediated sol-gel approaches, are commonly used for the preparation of silica nanoparticles to be used in drug delivery and bioimaging applications. Numerous studies have demonstrated that the performance of silica nanoparticles in biomedical applications is strongly related to morphology, structure, and the composition of the fabricated particles [2]. In the following sections, different architectures of silica nanoparticles prepared using the aforementioned methods are presented, with case studies of drug delivery and imaging from the literature.

9.1.1 Nonporous Silica NPs

The Stöber method is one of the most used chemical approaches employed to prepare nonporous silica materials (Figure 9.1). The process was discovered in 1968 by Werner Stöber

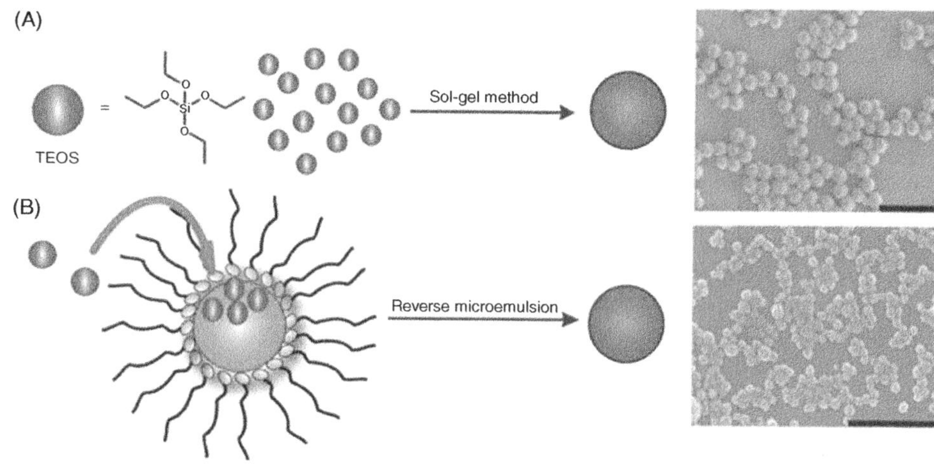

Figure 9.1 *The synthesis of nonporous silica nanoparticles by the Stöber method (A), in which the hydrolysis and condensation of TEOS is facilitated by a base in ethanol and water, and via reverse microemulsion (B), in which TEOS is hydrolyzed at the micellar interface and enters the aqueous droplet to form a silica nanoparticle. The scale bars represent 1000 nm and 500 nm, respectively. Adapted and reprinted with permission from [3].*

et al. [4], building on the earlier work of G. Kolbe. In the Kolbe 1956 study, synthesis of monodispersed silica particles based on the hydrolysis and subsequent condensation of silicon alkoxides in ethanol was investigated [5]. Later on, Stöber *et al.* systematically worked on the experimental conditions of this reaction and investigated the controlled growth of spherical silica particles, which is now the well-known Stöber process. Later on, the parameters of reaction were thoroughly investigated by Bogush and Van Blaaderen to control the size of the ensuing particles [6, 7]. Although there have been many studies related to size-controlled synthesis of nonporous silica nanoparticles, only a limited number of studies exist on the preparation of differently shaped nonporous silica nanoparticles. For the shape control of nonporous silica nanoparticles, templates are generally used [8, 9]. Surface functionalization can be carried out either via chemical conjugation or physical adsorption of the functional moieties. Depending on the functional groups, the surface charge can be tuned to positive, negative, or zwitterionic.

The reverse microemulsion method is the other method used for the production of nonporous silica nanoparticles (Figure 9.1). The resultant silica particles usually yield within a diameter range of tens to a few hundred nanometers. In the reverse microemulsion method, the silica nanoparticles form in inverse micelles compartmentalized by a suitable surfactant in a nonpolar organic solvent. In 1999, Arriagada *et al.* synthesized ultrafine silica nanoparticles within the optimized non-ionic water-in-oil microemulsion system [10]. In their study, they investigated the relation between the created reverse microemulsion system and the characteristics of the obtained nanoparticles. In the reverse microemulsion mediated silica nanoparticle synthesis, the size of the particles can be altered by changing the kinetics of hydrolysis and condensation processes [11] and the ratios among the contents of the microemulsion [12] (i.e., continuous phases, co-solvent amounts and surfactant [13]).

Compared with the reverse microemulsion mediated method, the Stöber method provides much simpler synthesis options to control the size of the silica nanoparticles [14]. However, the reverse microemulsion mediated synthesis method has been successfully applied for the coating of nanoparticles with different functional groups for various applications [15].

There are three strategies for incorporating active (cargo) molecules into nonporous silica nanoparticles, as presented in Figure 9.2. The first is the physical incorporation of the cargo molecules during synthesis (type 1). In this approach, the cargo molecules (i.e., small drug molecules or dyes) are encapsulated into the nanoparticles during the sol-gel process by controlling the amount of silica source. In such designs, the release of cargo from the interior of the nonporous silica matrix is difficult, which may cause insufficient efficacy of the administration and considerable side-effects. Therefore, this

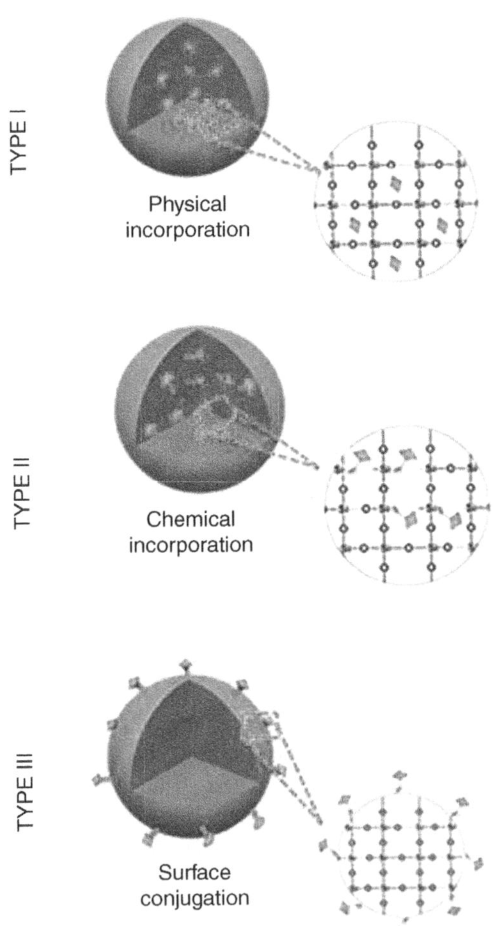

Figure 9.2 Loading of cargos on nonporous silica NPs and the proposed mechanisms of cargo release. Type 1: physical incorporation; type 2: chemical incorporation by modified silane chemistry; type 3: surface conjugation. Adapted and reprinted with permission from [16].

type of silica-cargo composite nanoparticles is usually used in bioimaging applications by incorporating the small dye molecules into the synthesis. Fluorophores such as the tris(2,2′-bipyridyl)dichlororuthenium(II) complex $(Ru(bpy3))^{2+}$ can be doped into the nanoparticles during the synthesis to impart luminescent properties [17, 18]. Literature studies reveal that entrapped dye molecules exhibit a higher quantum yield and enhanced photostability than the free dye molecules [19].

The other commonly employed strategy is the chemical conjugation of the cargo molecule within the silica matrix (type 2 in Figure 9.2) (i.e., co-condensation simultaneously during the synthesis), by which the modified silane chemistry leads to a change of internal composition of the silica nanoparticles. In type 2, the cargo molecules have to possess sufficient chemical groups to be functionalized and take part in the chemical reaction, which sometimes leads to difficulties in the synthesis. However, the incorporation yield of the dye into the silica matrix under non-covalent bonding is poor and dependent on the adsorption force between the dye itself and the silica precursor. To the best of our knowledge, the majority of the reports incorporate dyes or fluorophores through covalent bonds into colloidal silica nanoparticles, as covalent conjugation yields higher incorporation efficiency, especially via the Stöber method [19–23].

In the type 3 preparation, post-synthesis treatments are carried out with the functionalized dye molecules. Typically, cargo molecules are modified with silanes (often aminosilanes) and then conjugated onto the surface of already prepared silica nanoparticles. In this approach, the integrity of the conjugation needs to be well maintained to prevent the cargo from detaching from the silica nanoparticles during applications. If the chemical bond is stimuli-responsive, the loaded cargo can be released in a controlled fashion from the surface of the carrier. Although surface conjugation is relatively easier than incorporation into the silica matrix, the long-term stability in the physiological environment is considered as the limitation for the application.

9.1.2 Mesoporous Silica NPs

The preparation of mesoporous silica materials was first reported by Yanagisawa *et al.* in 1990, using a different synthesis approach than the more well-known studies by Mobil Oil scientists from 1992 where the synthesis of silica/alumina ordered mesoporous molecular sieves, designated as M41S, was reported [24–26]. The preparation of these mesostructured silicates is based on self-assembly principles in concentrated surfactant (C_nTMA) aqueous solution and their reactions with the monomeric silicates. Differently named mesophase silicates (MCM-41, -48, -50) were prepared with the same approach as presented in Figure 9.3. These materials have regular cylindrical pores with tunable pore sizes between 2 and 30 nm and a large surface area. For the actual particle growth, the synthesis of mesoporous silica nanoparticles (MSNs) employs the same principles as the Stöber process. MCM-41 is the initial and the most widely studied member of the M41S family, and consists of hexagonally ordered channels having amorphous silica walls based on the hexagonally packed rod-shaped micelle structures (Figure 9.3). Conventional MCM-41 pore sizes are within the range of 3 to 4 nm, which is the commonly used type for drug delivery applications. A few years after the reports on the M41S family, a new set of families of mesoporous silica materials (e.g., MSU, FSU, SBA) have been developed with unique porosities and shapes

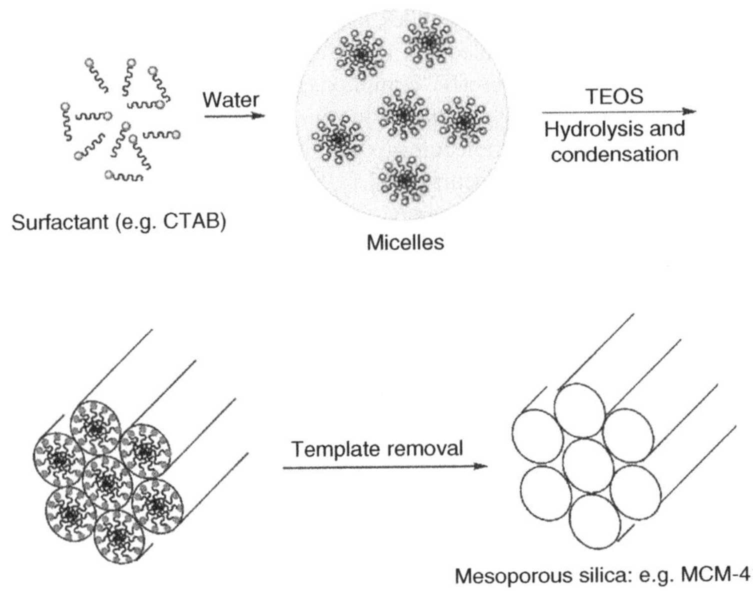

Figure 9.3 *Schematic illustration of the synthesis of mesoporous silica materials. Reprinted with permission from [27].*

[28–30]. Mobile Composition of Matter-41 (MCM-41) and Santa Barbara Amorphous-15 (SBA-15), both with a hexagonal arrangement of the mesopores, and MCM-48 with a cubic arrangement of the mesopores, are the best-known mesoporous silica materials [31].

The preparation of mesoporous silica materials using small-molecule surfactants such as $C_{16}TAB$ leads to ordered mesoporous silica materials with a pore size less than 5 nm, i.e., MCM-41. This may limit their application as a carrier for large guest cargos such as proteins. Therefore, it was necessary to develop large-pore mesoporous silica materials for accommodating large-sized cargos and to promote drug dissolution rates. In one of the recent reviews by Maleki *et al.*, the impact of crucial parameters (e.g., surfactant type, hydrothermal temperature, and introducing organic swelling agents) on controlling the pore size was discussed [27]. As emphasized, while aiming for pore size increments, it is also critical to be able to control the synthetic approaches to preparing appropriate mesophases of the silica materials for a rational development of mesoporous silica-based drug delivery systems.

Modulating the size (<1 µm in diameter) of the mesoporous silica particles has attracted keen interest in the field of drug delivery, with the merits of better dispersibility and mass transport [26]. The size of MSNs can be tuned by adjusting the composition of reactants, temperature and the pH of the synthesis, which are the known parameters in the sol-gel approach that affect the hydrolysis rate of silane and siloxane bond condensation, which leads to the alteration of particulate size. Engineering the shape of MSNs (e.g., spherical, rod, ellipsoid, film, platelet, and cube) has been revealed as an important physicochemical parameter that determines the biological fate of the materials [31]. Shape alteration of

MSNs can be achieved by disturbing the self-assembly process in the particle synthesis or by employing defined templating structures [31].

In the literature, MSNs are functionalized to be employed as multifunctional nanomedicines via different modification strategies. The characteristics of mesoporous silica nanoparticles determine their extensive functionalization capabilities. *In situ*-/post-synthesis modifications are the strategies employed for the modification of MSNs. The *in situ*-synthesis modifications involve either electrostatic interaction or covalent conjugation of organic molecules into the silica matrix during the condensation step of the sol-gel reaction, which yields functional groups on both the exterior surface and the pore walls of the MSNs. The surface properties play vital roles in the interactions of particles with the physiological environment and target cells/tissue. Therefore, the surface modification strategies should be decided in advance by considering the aimed application; for instance, when aiming at targeted delivery applications, additional functionalities, such as targeting ligands, have to be introduced to the outer particle surface by post-synthesis approaches [32]. In the case of preparing fluorescent nanoparticles, uniformly dispersed and segregated fluorophores in the silica network can be obtained via *in situ* chemical conjugation and, accordingly, self-quenching and premature release of the fluorophores can be avoided to a large extent such that enhancement in fluorescence detection can be provided [33].

9.1.3 Core@Shell Materials

Silica is also, due to its relatively facile fabrication processes (especially the sol-gel process that is devoid of both high temperature and high pressure, in contrast to many other inorganic materials synthesis), the most common coating material for other inorganic nanoparticles. Coating with nonporous silica is commonly applied to enhance the dispersability of nanoparticles, especially in aqueous solvent, in which most other inorganic nanoparticles tend to be quite hydrophobic. A silica layer also provides for easy further functionalization of the particle system, which is often needed to carefully tune or design the biological applicability (interactions with a biological system). With the increased hydrophilicity and tunable surface chemistry, increased biocompatibility is also introduced; this is why this strategy is often implemented for the coating of iron oxides, for example, for use as MRI contrast agents. The same strategy can be utilized to create dual or even multimodal imaging agents [34] based on iron oxide nanoparticles by physically separating the magnetic modality from the optical one (e.g., a fluorescent dye), whereby direct contact would otherwise lead to fluorescence quenching. Another advantage is protecting the underlying core from water, for example, which could otherwise lead to quenching (in the case of upconverting nanoparticles) or pH changes, which would lead to core dissolution (e.g., in the case of iron oxide and low pH). Silica itself, however, dissolves rapidly at high pH (>9).

Modified Stöber processes can readily be used to synthesize nonporous silica shells, even in a layer-by-layer fashion to control the thickness. Another approach is the microemulsion technique, where cores are added into the microemulsion droplets in which the silica forms through sol-gel reactions. Porous silica shells can be synthesized by modified mesoporous silica syntheses, with the vital control of the ratio of solvents [35]. For smaller cores, a direct porous silica coating is often deposited (hydrophilic vs. hydrophobic), while for larger cores (microbeads) a middle layer of nonporous silica is often deposited first to render the surface more attractive for mesoporous silica deposition. For iron oxide beads, the nonporous silica coating is also needed for protection of the core material from futher processing (i.e., mesoporous silica coating), which takes place under either acidic or basic conditions where the acid or base is needed to catalyze silica hydrolysis and condensation. This approach can also conversely be exploited for the synthesis of hollow mesoporous silica particles, where the hollow interior space is templated by nanoscaled cores of a material that can easily be leached out after the synthesis of a porous shell (and in the absence of a nonporous shell). This approach is referred to as "hard-templating" where a nanoparticle core functions as the void-forming agent, while surfactants function as pore-forming agents, as in any mesoporous silica synthesis. Consequently, this approach is sometimes referred to as the "dual templating approach". So-called "soft-templating" methods have also been developed, where the hollow void is formed via controlled etching of the less-condensed center of mesoporous silica particles [36]. In addition, approaches for direct coating of metal oxide cores with mesoporous shells under neutral conditions have been developed in order to preserve the core material during shell synthesis [37].

9.1.4 Hollow Silica Nanoparticles

The unique void and the solid/porous shell structure of silica nanoparticles are denoted as hollow silica nanoparticles. Hollow silica nanoparticles have attracted extensive attention in both fundamental and practical studies due to their low density, high cargo-loading capacity, as well as their potential application in therapeutics delivery, biomedical imaging, and catalysis [38]. Hollow-type mesoporous silica nanoparticles (HMSNs) are able to achieve a super-high drug-loading capacity using its void in the core of particles to load drugs, typically >1 g drug/1 g of silica [39].

Preparation techniques for hollow silica nanoparticles include template-assisted synthesis, self-templating methods, and spray-drying methods (Figure 9.4). The template assistance method is known to be more effective compared with the template-free method. The template (organic or inorganic) is the determining factor for the synthesis of different shapes/sizes of hollow silica nanoparticles.

Intensive research has been carried out in order to investigate the variation conditions required to tailor the morphology, size and pore structure for hollow silica particles [38]. As reviewed by Bao *et al.*, hollow silica with diverse morphologies has been successfully prepared to be used in different applications (Table 9.1).

HMSNs stand out as promising nanocarriers to address the challenges of combination multidrug therapy, whereby hydrophobic and hydrophilic drugs can be carried by the same particles, thus obtaining the benefit of the void core as extra reservoirs [41, 42]. The same structure of silica nanoparticles can also be designed to achieve stimuli-responsive controlled release of drug molecules and to prevent premature release of the loaded cargos [43, 44]. The drug loading capacity and release profile from HMSNs can be tuned by altering its structural parameters, such as shell thickness. By increasing the shell thickness of HMSNs, the release rate of the drug from the carrier can be increased and sustained release can be achieved.

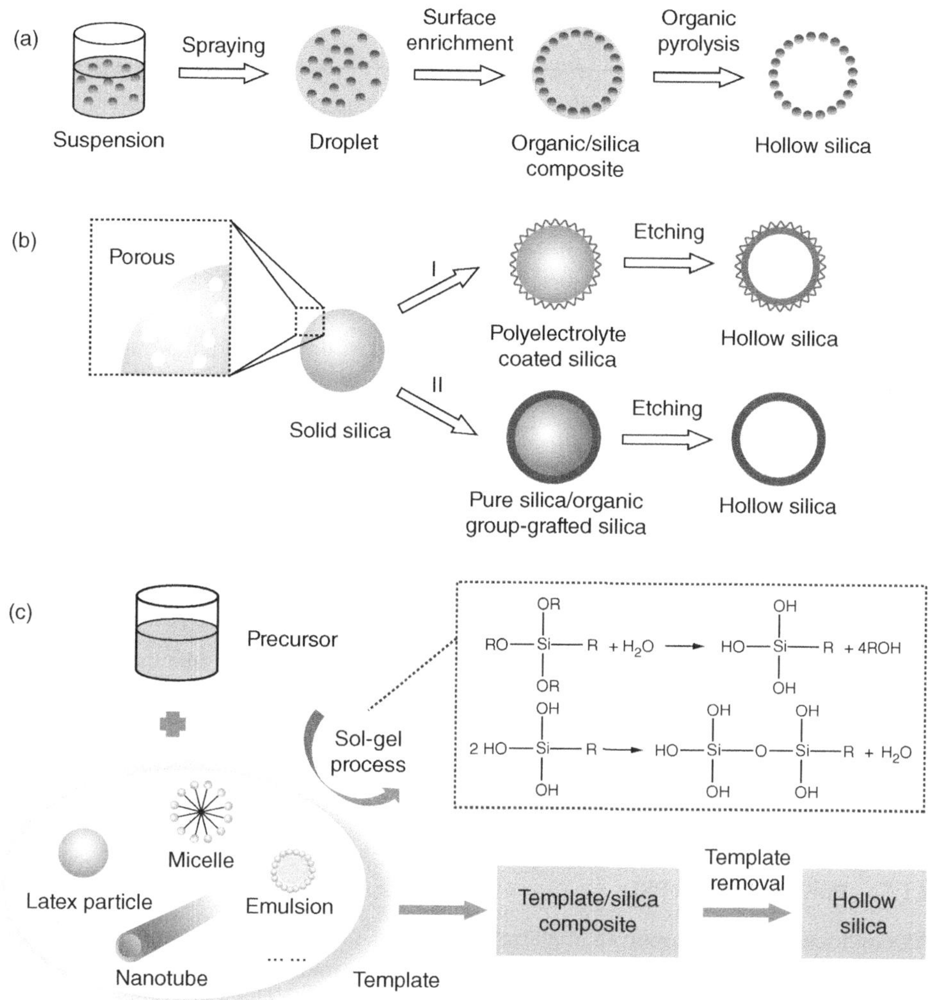

Figure 9.4 *Schematic illustration for: (a) spray drying, (b) the self-templating method, and (c) template synthesis. Reprinted with permission from [40].*

Over the last decade, HMSNs have been engineered to be used in biomedical imaging applications such as optical, PET, MRI, and ultrasound [45]. Engineering of HMSNs to integrate two or more imaging modalities into the same design is a highly useful strategy for providing complementary and more accurate information about the pharmacokinetics, biodistribution, clearance, and so on, of the silica material *in vivo*, which is an exciting new area of cancer theranostics [34]. With the presence of large cavities for loading therapeutic agents (e.g., small anti-cancer or antibiotics), and more importantly, the capability to integrate a vast array of inorganic functional nanocrystals inside or at their surfaces,

Table 9.1 Common hollow silica morphologies and their special performance. Reprinted with permission from [40]

	Hollow silica spheres with disordered pores	Excellent adsorption and penetration properties
	Hollow silica spheres with well-ordered cylindrical channels	High specific surface area
	Mono-porous hollow silica spheres	Enhanced rates of encapsulation and higher effective diffusivity into or out from hollow silica
	Nonporous hollow silica spheres	Avoid mass transfer and improve the surface smoothness and dispersion uniformity
	Multi-shelled hollow silica spheres	Multiply the specific area of hollow silica spheres and potential to enhance their optical properties
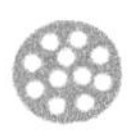	Silica spheres with multi-hollow cores	Promising for sustained release
	One-dimensional hollow silica nanostructures	Large high-aspect-ratio, surface area and accessible both inside and outside surfaces
	Hollow silica cubes	Higher specific surface area than hollow spheres of the same volume

HMSNs are seen to be one of the most promising nanoplatforms for future image-guided drug delivery [45].

9.1.5 Porous Silicon (PSi)

As our focus is on mesoporous materials made out of silicon, we would like to present another way to fabricate a similar, and yet so different, material that fits into these frames. PSi has been studied as a biomaterial since 1995 when Leigh Canham found the first hints of the biocompatible nature of the material [46]. He found that hydroxyapatite, the major constituent of bone, was growing on the surface of PSi in phosphate-buffered saline (PBS) solution. In addition, PSi showed signs of degradation in physiological conditions. Soon after this, the ability of cells to adhere to the surface of PSi was recognized [47].

Porous silicon (PSi) is fabricated via a top-down method from crystalline silicon that is mainly used in the microelectronics industry. The most common method for fabrication is so-called electrochemical etching or anodization [48]. During electrochemical etching, mesopores are formed in the silicon surface due to the highly localized electrochemical dissolution of crystalline silicon. In the simplest fabrication setup, only a hydrofluoric acid (HF) compatible plastic vial, electrolyte, platinum counter-electrode, anode connector, and power supply are needed, although in most laboratories a dedicated etching cell is used (Figure 9.5) [50]. Normal crystalline silicon is stable in acidic aqueous solution because a native oxide passivates the silicon surface. If the solution contains HF, slow isotropic dissolution occurs. Electric potential, applied over the wafer, facilitates the diffusion of

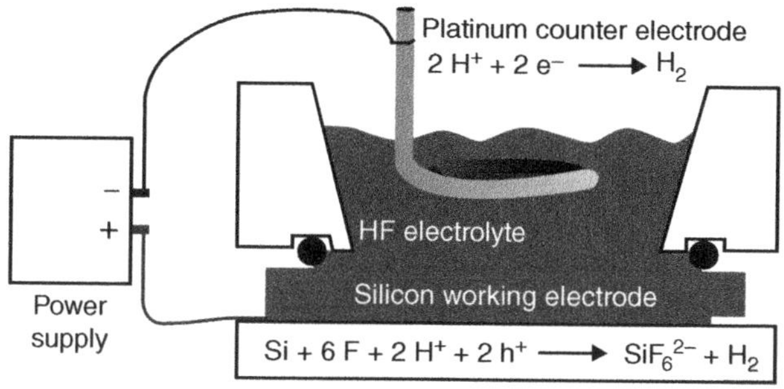

Figure 9.5 *PSi anodization cell. Reprinted with permission from [49].*

semiconductor charge carriers towards the counter-electrode (Figure 9.5). Typically, these charge carriers are positively charged (holes), which facilitate the disturbance of the ionic balance of the terminal Si–H bond. This causes a nucleophilic attack of HF^{-}_{2} ion to the Si atom, which starts a chain reaction that ends with the release of an Si atom to the solution. Because the dissolution rate is typically dictated by the amount of holes, the current density can be used for the fine-tuning of PSi properties, such as pore size and pore volume [51]. This also facilitates the fabrication of multilayered structures [52] by altering current density which can be used as a material for nanoparticle fabrication [53, 54].

As mentioned, surface chemistry is one of the most important properties when it comes to biological applications. The etching process leaves the surface of PSi hydrogen terminated (Figure 9.6), which is prone to oxidation. The great benefit of a top-down nanomaterial such as PSi, which can be easily handled (before milling into nanoparticles) in a dry state, is the possibility to use thermal gas phase treatments for comprehensive stabilization of the surface.

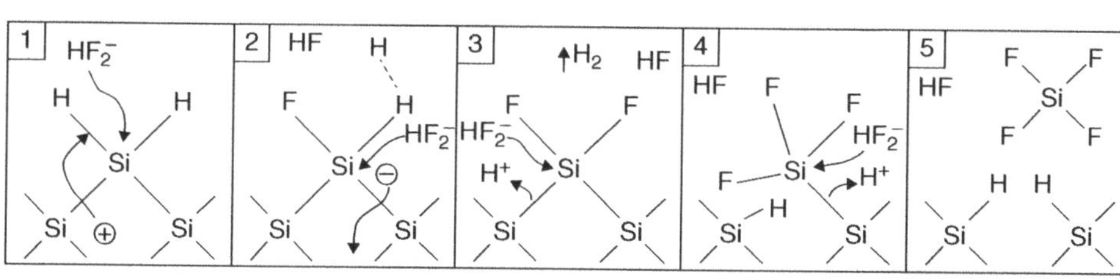

Figure 9.6 *Anodic divalent dissolution of silicon in HF. Reprinted with permission from [55].*

Oxidation is the most common way to stabilize the silicon surface, owing to its simplicity and the variety of chemical entities that can be attached to the surface via siloxane bonds [54, 56]. In thermal oxidation, the used atmosphere, temperature, and treatment time play a major role. The so-called "back-bone" oxidation takes place in ambient air at temperatures above 250°C [57]. The siloxane bridges are formed between silicon atoms at the surface and at the second atom layer (Si–O–Si–H) during this process [58].

There are a lot of benefits to oxidizing passivated surfaces, but for some applications, such as sensors [59] and drug delivery [60], its stability against dissolution may not be sufficient. Alternative gas phase treatment is thermal carbonization. Thermal carbonization is carried out by pre-exposing the freshly etched PSi film to a mixture of an inert carrier gas and carbon-containing gas. A typical combination is acetylene and nitrogen [51, 61]. Acetylene adsorbs to the surface, heating breaks the double bond, and carbon reacts with the silicon. The treatment temperature has a significant impact on the hygroscopic properties of the formed surface [62], which is attributed to changes in surface chemistry [63, 64]. When the treatment temperature is kept below 700°C, the thermal energy cannot break all of the C–H bonds, and the surface stays covered with hydrophobic hydrocarbons. When the temperature is raised above 700°C, all the hydrogen desorbs from the surface and carbon diffuses deeper into the structure [65]. This yields a very thin, non-stoichiometric silicon carbide structure which will become terminated with a hydrophilic thin oxide layer after the films are taken back into ambient air [66]. In order to distinguish these significantly different surface chemistries, thermal carbonization (TC) refers to acetylene treatment at higher temperature, and thermal hydrocarbonization (THC) refers to acetylene treatment at lower temperature (Figure 9.7).

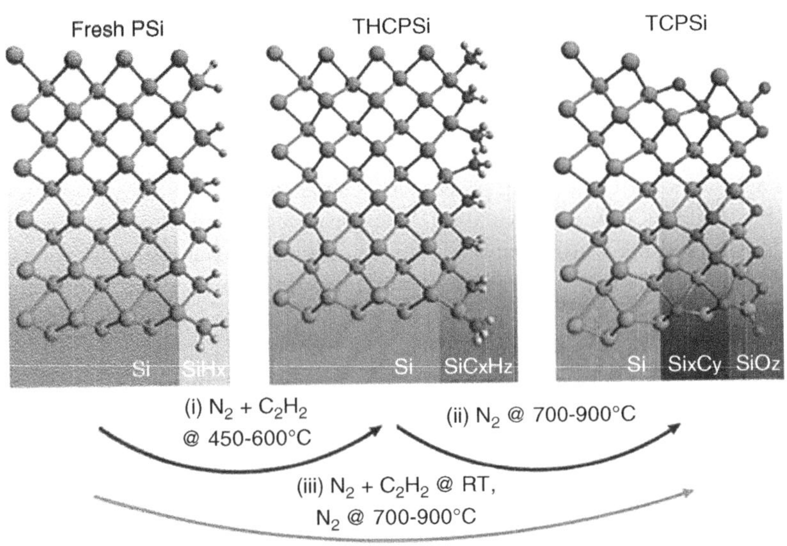

Figure 9.7 *Surface treatments of PSi yielding different surface layers.*

Various organosilanes can be attached to the silicon oxide surface via rather stable siloxane (–Si–O–Si–R) linkages [67, 68]. A frequently used organosilane is APTES (3-aminopropyltriethoxysilane), which creates an amine termination and positive charge [69] and is widely used to attach biomolecules to silica surfaces. Silane coupling chemistry can also be used with TC surfaces because carbon penetrates deeper into the structure and silicon oxide covers the surface. This can be changed to a hydroxyl termination simply by dipping the sample in a HF-containing solution. Thermal carbonization makes the structure of PSi durable against dissolution and can also increase the stability of amine treatment when compared to TOPSi [70].

The situation is different when the functionalization needs to be made for a hydrophobic THC PSi surface. Now the hydrocarbon groups terminate the surface and silane chemistry is not conveniently applied. Nevertheless, THC PSi surface provides superior stability against aqueous degradation, which is a desired property in some applications. In recent studies, functionalization of a THC PSi surface has been managed using hydrosilylation, resembling the thermal reaction of unsaturated carboxylic acid (undecylenic acid) [71]. The strong C–C bond between the carbon on the PSi surface and hydrocarbon chains shows excellent stability compared with many other surface chemistries [72].

9.2 Solid-state Characterization

9.2.1 Porosity and Morphology on the Nanoscale

The complex and variable structures of mesoporous nanoparticles provide a variety of physicochemical properties that may be utilized in applications, but at the same time, are challenging to characterize. Actually, the ability to measure these nanoscale properties has been the prerequisite for the development of nanoscience, and one of the most influential techniques has been electron microscopy (EM). The main advantage of EM measurements is the direct characterization of particle morphology at the micro- and nanoscale [73]. The scanning electron microscope (SEM) has been especially important. Knowledge of the impact of fabrication parameters on the final structure has significantly increased, along with the development of better imaging techniques, such as SEM and TEM (transmission electron microscopy) (Figures 9.8 and 9.10).

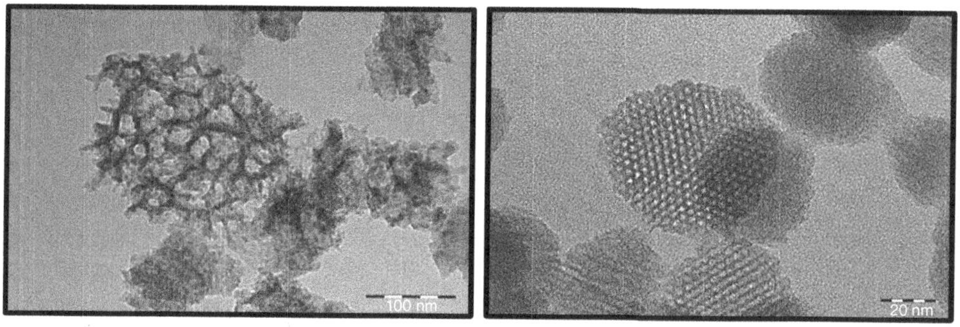

Figure 9.8 *TEM images of porous nanoparticles.*

The drawback of electron microscopes is the low statistical significance. To investigate a sample's porosity in a typical cross-sectional image, there might be 50 pores in a 1 μm row, but when the whole sample is macroscopic, 1 cm^2 for example, in total there may be 250 billion pores.

The porosity of particles in the meso-range (2–50 nm) can be determined with direct and indirect methods such as imbibition, buoyancy, gas expansion, gas adsorption, optical and electron microcopy, core-scale NMR, atomic force microcopy, and imaging methods [74] (Figure 9.9).

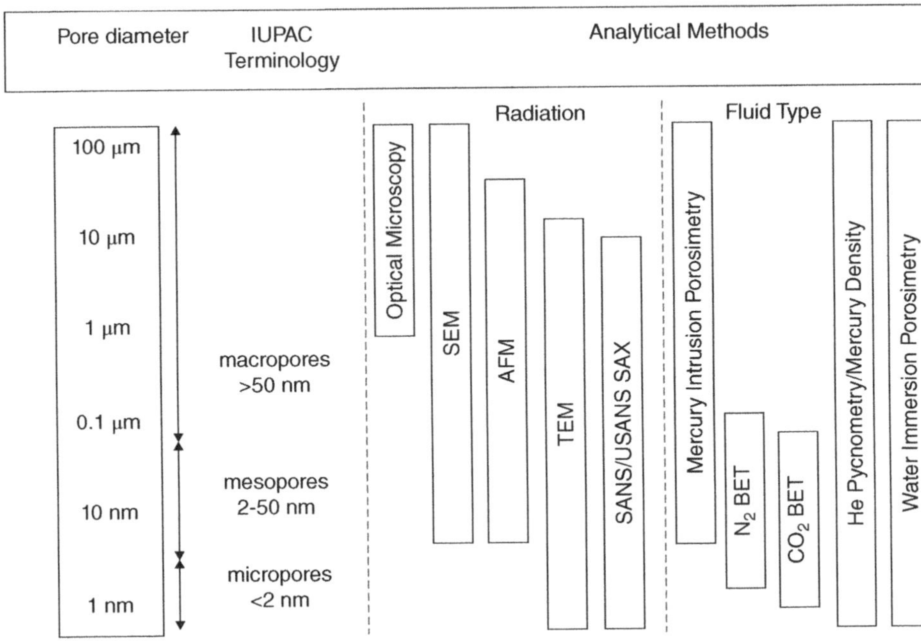

Figure 9.9 *Different methods used to determine porosity and pore size distribution [74].*

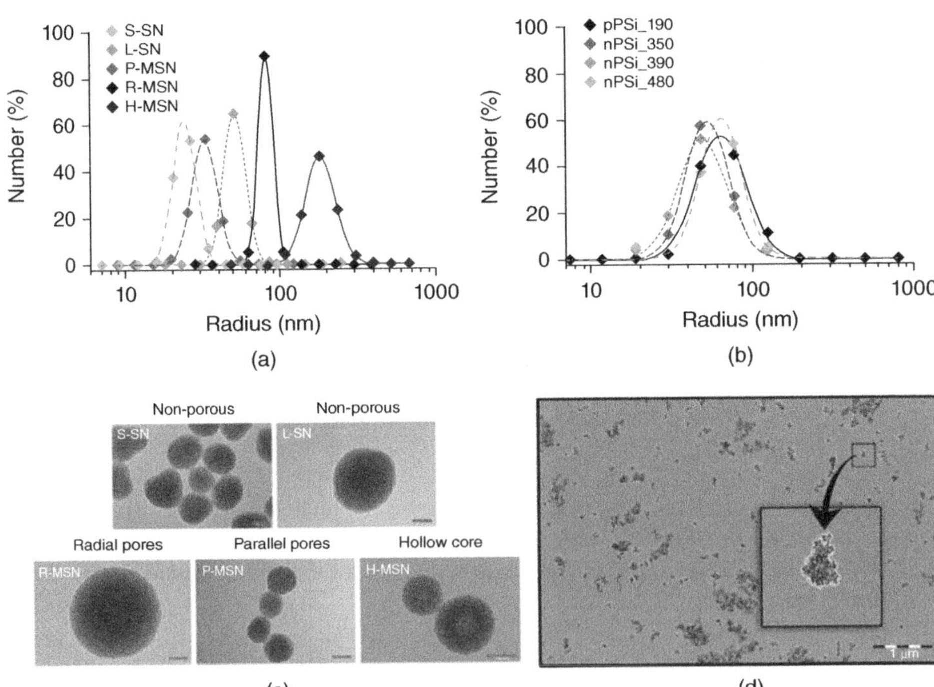

Figure 9.10 *TEM in size measurement, and difference between mesoporous silica and silicon nanoparticles. Particle size distribution of studied (a) silica nanoparticles and (b) PSi nanoparticles. (c, d) Example pictures of TEM results in respective order. The red scale bar in (c) represents 20 nm (S-SN, L-SN and R-MSN), 50 nm (P-MSN) and 200 nm (H-MSN) [106].*

Among those methods, the most commonly used one for the determination of pore size distribution and pore volume of mesoporous silica/silicon nanoparticles is gas sorption. If the above-mentioned sample is 10 μm thick and the pore diameter is 15 nm, the amount of surface area in the pore walls is more than 1000 times bigger than the apparent 1 cm^2 surface area of the sample. With the gas sorption method, the actual surface area in a sample is determined by utilizing the affinity of gas molecules (typically nitrogen) to the material surface. By precisely measuring the volume of injected gas and pressure on the test tube, adsorption of the gas molecules with a well-defined size can be observed, and with predictions from the Brunauer-Emmet-Teller (BET) theory, the surface area can be calculated [75]. By continuing the gas injection into the sample tube near ambient pressure (and near the condensation temperature of nitrogen), the capillary effect eventually fills the pores with liquid nitrogen. When the gas pressure is decreased during the so-called desorption branch, the capillary effect causes a hysteresis loop to form in the adsorption–desorption isotherm. From the shape of this loop, the pore morphology as well as pore size can be determined [76]. One of the most used theories for the determination of pore size distribution is the so-called Barret–Joyner–Halenda (BJH) theory [77] which has also been subject to critique, and is thus sometimes replaced by other methods such as DFT [76].

It is crucial to carry out proper outgassing of the material to be measured prior to N_2 sorption analysis, to remove the humidity in the pores. In the case of inner/exterior pore wall surface modifications, the outgassing temperature should be decided depending on the properties of organic compositions on the silica matrix to avoid decomposition during the outgassing process. The N_2 sorption method can be employed to investigate the change of pore sizes and pore volumes when pore wall modifications are carried out [78]. The hydrophobicity and hydrophilicity of the ensuing silica surfaces can be predicted based on the calculated *C*-value from the N_2 sorption analysis. The *C*-value is related to the affinity of the solid with the adsorbate (i.e., N_2 molecules), and thus to the heat of adsorption. When the *C*-value gets lower, the hydrophobicity of the surface is increased, which can also help to reveal the nature of the existing surface modification.

9.2.2 Structural Analysis

Small-angle X-ray powder diffraction (XRPD) is a commonly used technique to investigate the fine structure of ordered mesoporous silica/silicon materials. From the XRPD pattern of the analyzed material, structural symmetry and the distances between the neighboring pores can be determined. XRPD data provide direct structural information without any semi-empirical parameters or assumptions [79]. The structure analysis of ordered mesoporous materials is carried out based on the peak positions on the X-ray diffractograms. The XRPD pattern simulations of model structures for ordered mesoporous materials (e.g., MCM-41) can be obtained with support from molecular dynamic (MD) simulations [80]. However, MD simulations are not suitable for routine analysis because they require knowledge of the molecular calculations and computational skills for the simulations.

Employing recently developed calculation and refinement methods, techniques such as continuous density function (CDF) and derivative difference minimization (DDM) can be performed with the XRPD data to obtain detailed structural characterization of mesoporous materials [81]. Recently, Ishii *et al.* revised the XRPD pattern calculation by considering interferences for the analysis (i.e., crystallite size effect) more precisely with the Laue-type interference function [82]. They have presented that the theoretically calculated XRPD patterns agree well with experimentally observed patterns. By analyzing the XRPD peak intensities carefully, the mesoporous diameter and unit cell lattice spacing can be obtained even for water-immersed mesoporous silicates.

As reported by Şen Karaman *et al.*, the newly developed and poorly understood structural properties of mesoporous materials can be investigated by performing X-ray powder diffraction analysis as an *in situ* method without complementary TEM or physisorption methods, since they require pretreatments and special conditions [83]. The obtained X-ray diffractograms can also be used to predict the rigidity of modified mesoporous structures by considering biologically relevant environments [84].

9.2.3 Methods for Determination of Surface Functionalization

With the increasing research interest in silicon-based nanopharmaceuticals, it is desirable to identify their surface characteristics, particularly to define the interactions at the bio–nano interface, as well as to overcome the challenges associated with their application and to maintain desired properties during handling and processing due to the dynamic nature of these materials on the nanoscale. It is equally valuable to understand how surface characterization approaches can help to investigate how (and how quickly) materials and properties change in different environments. Both qualitative and quantitative applications of common surface analysis methods are used to address these needs [85].

For these analysis methods, there are critical steps that start with particle preparation and processing. For instance, surface coatings on the nanoparticles might be affected by the presence of some ions in the solution in which they are suspended, or post-synthesis treatments [86]. One of the major challenges associated with using surface analysis methods involves the appropriate preparation of specimens for analysis. The environmental sensitivity of silicon-based nanomaterials and the need to remove materials from solution without introducing artefacts should be properly carried out. It is crucial to investigate whether moving a sample from the "native" environment to the environment of the analysis (vacuum, in many cases) alters the sample and compromises the ability to collect the desired information.

To understand the interaction between the Si/SiOx interface and organic molecules adsorbed on the surface of the silicon particles, a complete knowledge of the electronic (or chemical) structure of the first atomic (or molecular) layers of the interface is required. For this purpose, combined analysis methods are usually used, as in the study of Beyanad *et al.* where a detailed chemical bond analysis based on XPS measurements combined with Fourier transform infrared spectrophotometry (FTIR) and Raman micro-spectroscopy (micro-Raman) were combined to deduce information about the reactivity of the silicon particles [87].

In order to identify the direct chemical bonding between the silanol groups and chemical entities, ^{29}Si solid-state nuclear magnetic resonance (NMR) is used [88]. The quantification of ligand binding, release, ligand exchange on nanoparticles, together with the structural motifs in bound ligands, can be investigated by using solution-phase NMR approaches [89]. These approaches provide understanding of binding and release of supramolecular-bound ligands in an aqueous environment on silica surfaces, and suggests the possibility of full chemical release from a silica matrix [90].

However, in many studies the researchers utilize more convenient and less costly methods such as ζ-potential measurements or electron microscopy imaging to confirm the success of surface functionalization [91–93]. In the determination of high amounts of organic surface coatings on silica nanoparticles, the extent of the organic portion is usually determined by thermogravimetric analysis (TGA) methods. In such cases, the analyst has to carefully distinguish the portion to be analyzed and the other contents (such as adsorbed water).

The interactions of the nanoparticle surface with its environment usually lead to an alteration in their behavior during the application. Therefore, knowledge of the rate of any changes on the surface of the nanoparticles in relevant environments is important, in order to assess their reliability and to predict their environmental impact in the application.

9.3 Medium-dependent Characterization

The problem with all previously mentioned techniques is that they are not compatible with wet samples. When nanoparticles are placed in an aqueous medium, the system changes from a static material studied in a vacuum without "context" to a dynamic part of a bigger system. Constant thermal motion of the medium keeps particles in motion in such a vigorous way that colloids with small particle size (and a few other limitations, as will be seen) can be considered stable. In this context, stability is considered stability against gravitation, sedimentation, or in some cases, against agglomeration.

The world of nanoparticles was studied long before the term nanoscience was even known. Many different disciplines – physics, chemistry, mathematics, and biology – have contributed to building our models of the nanoscale world. Unfortunately, the discussion between these fields has turned out to be challenging at times, and many overlapping definitions are confusing the field. These models concern different kinds of forces, which are manifested in different ways depending on the length scale. Forces can be divided into long-range and short-range categories (Table 9.2), where the latter typically means forces affecting less than 1 nm distance from the surface. In the case of nanoparticles, especially when stability issues are considered, the long-range forces are more important (Table 9.2).

One of the earlier attempts to shine a light on the world of nanoparticles or colloidal particles was made in the field of colloidal sciences. These models are especially beneficial for the studies of nanoparticles because they mostly consider the long-range forces. Many of the theories are well known, but they are not usually included in the curriculum. Nevertheless, these concepts may be particularly useful for grasping the main phenomena dominating the surroundings and interactions of nanoparticles. Introductions to these concepts can be obtained from various textbooks considering surface charging and zeta potential [95–97], stability against agglomeration and long-range colloidal forces [94, 95], Brownian motion [98], light scattering [99, 100], and isotonic solutions [101].

9.3.1 Hydrodynamic Size

Light scattering-based size characterization methods make use of the tendency of light to reflect or refract in a complex fashion when travelling through interfaces with nanoscale

Table 9.2 *Interaction forces between nanoparticles*

Type of interaction	Other names (subclasses)	Attractive or repulsive	A body or surface force
Van der Waals	Dispersion, dipole-induced-dipole, charge-fluctuation, ion-correlation[b], Debye, Keesom, Casimir	Either: attractive	Weak but ubiquitous body force. Force can change sign at some finite distance.
Electrostatic	Coulomb, ionic, salt-bridge, dipolar, hydrogen-bonding, charge-transfer, double-layer[b]	Either: repulsive, attractive between electro-neutral	Strong, long-ranged force arising in polar solvents. Usually a surface force.
Solvation	Structural, epitacial, hydration (in water), depletion[b], hydrophobic[c]	Either, incuding oscillatory	Surface force that modifies the local liquid structure.
Entropic	Steric, osmotic, thermal fluctuation, polymer bridging, depletion[b], double-layer[b]	Usually repulsive	Surface force that arises from the confinement of solvent or solute molecules between the surfaces.
Short-range physical	Adhesive, cohesive, surface tension, wetting, capillary	Usually attractive	Can be either a body or surface force.
Short-range chemical	Covalent, quantum mechanical, metallic exchange, steric, hard core, Born	Either: ultimately repulsive at small distances	Strong short-ranged surface forces, largely independent of the suspending liquid medium.
Specific	Complementary (electrostatic or geometric), host-guest, lock-and-key, ligand-receptor, antibody-antigen	Attractive (specific but not necessarily strong)	Single or multiple noncovalent bond arising from perfect fit of ion or molecule into host pocket or lattice site.
Nonequilibrium	Viscous (drag), friction, shear, lubrication, hydrodynamic, energy dissipating, hysteretic	Either (depends on externally applied forces)	Involves continuous or transient motion of molecules or particles. Heat generating.

[a]Unless specifically mentioned, the interactions refer to the pair-potentials of like particles.
[b]Forces that are difficult to classify unambiguously, or that are made up of two equally important contributors.
[c]Classification still unknown or controversial.
*Reprinted with permission from Elsevier, copyright 2011 [94].

features. When the measurement is made at a certain angle, light intensity either increases or decreases depending on the position of adjacent scattering objects. If particles are in constant movement, which is the case in colloidal suspensions, very rapid intensity fluctuations occur and the phenomenon is called dynamic light scattering (DLS). These intensity fluctuations are very short (less than a few ms), and another way to calculate the size of the nanoparticle under laser illumination is to observe the time-average light intensity as a function of scattering angle. This is called static light scattering (SLS). With SLS, the angular pattern of time-averaged scattered light is collected, and from the results the particle shape and size can be evaluated. Both of these light-scattering techniques, DLS and SLS, give a different kind of approximation for particle size, and comparing them may be beneficial.

9.3.1.1 Dynamic Light Scattering

When studying pollen particles under the microscope, one can clearly see a phenomenon dictating the wet world of nanoparticles. Particles are in constant motion. This was observed by botanist Robert Brown in 1827, after whom the phenomenon was named as Brownian motion. Later the motion was explained by the constant random impacts caused by thermal motion of molecules in solution. When a particle is suspended in a medium, viscosity causes a movement-opposing force and the translational diffusion coefficient D_t can be described with the Stokes-Einstein equation:

$$D_t = (k_B\, T)/(6\pi\, \eta\, r_h) \quad \text{(X)}$$

where k_B is Bolzmann's constant, T is temperature, η is the medium viscosity, and r_h is the hydrodynamic radius of the moving object. This way, by observing the movement of particles in solution, with known temperature and viscosity, it is possible to calculate the particle's translational diffusion coefficient and obtain the value for the hydrodynamic diameter. Since the intensity fluctuations of DLS arise because of the Brownian motion of nanoparticles, this facilitates the convenient determination of the translational diffusion coefficient, and the hydrodynamic radius or even distribution of radii can be determined.

Because the hydrodynamic radius is calculated from the movement of the particle, it is only a close approximation to the actual particle size. This is an analogue to zeta potential and surface potential. In the Stokes-Einstein equation, a few approximations are made and one of them is that the particle is assumed to be spherical and rigid. In other words, the hydrodynamic diameter describes the diameter of a spherical particle with the same diffusion coefficient as the studied particle.

9.3.2 Surface Charge and Zeta Potential

A polar medium, such as water, leads to surface charging of a particle, even if the surface is neutral in the dry state. This is typically caused by the ionization of the surface groups, uneven ion adsorption from the medium, or uneven ion dissolution on a particle surface. The most relevant surface charging mechanism, from our point of view, is ionization, which in many cases is pH-dependent. The pH point where the average surface charge is neutral is called the point of zero charge (PZC). This is very easy to confuse with the isoelectric point, which actually means almost the same thing for proteins, but in the case of nanoparticles, the situation is different. The charge on proteins is mostly caused by amine (e.g. $-NH^{3+}$) or carboxyl groups (–COOH), and the point where the total charge is zero is called the isoelectric point (pI). In the case of nanoparticles, the isoelectric point (IEP) is defined as the pH value where zeta potential equals zero. Luckily, the abbreviations for these physical quantities are typically different and hereinafter, the (calculated) isoelectric point of proteins or peptides is abbreviated as pI and (measured) isoelectric point in the case of particles as IEP.

Zeta potential is a quantity that is defined as movement of the particle under an applied electric field in a (electrolyte) solution. As stated above, all particles have surface charge (σ_0), which means that they also have surface potential (Ψ_0). These properties cause their movement under applied electric fields. In the presence of a polar medium and ions, the so-called electrical double layer (EDL) is formed around the particle as a consequence of uneven distribution of free charges around the charged particle in solution (according to the Gouy-Chapman model). In addition, some counter-ions are adsorbed transiently, or specifically, on the surface, which causes the formation of a Stern or Helmholtz layer. Potential on the surface, right next to the adsorbed layer, is called the Stern potential (Ψ_d). From a technical point of view, the actual surface charge behind the layer of tightly attached adsorbed ions cannot be distinguished when the measurement is electrokinetic, that is, by measuring the movement of the particle under the applied electric field. Stern potential is typically very close to this so-called slipping plane, but in order to distinguish between these planes, the electrokinetically obtained quantity is called the zeta potential (ζ). Because of the zeta potential's sensitivity to adsorption, it is a compelling research method for the characterization of charged molecules such as peptides and proteins [102].

Zeta potential measurements can also be used for *in situ* aqueous (chemical) stability studies (Figure 9.11). In this example, the loss of amine termination was studied with two

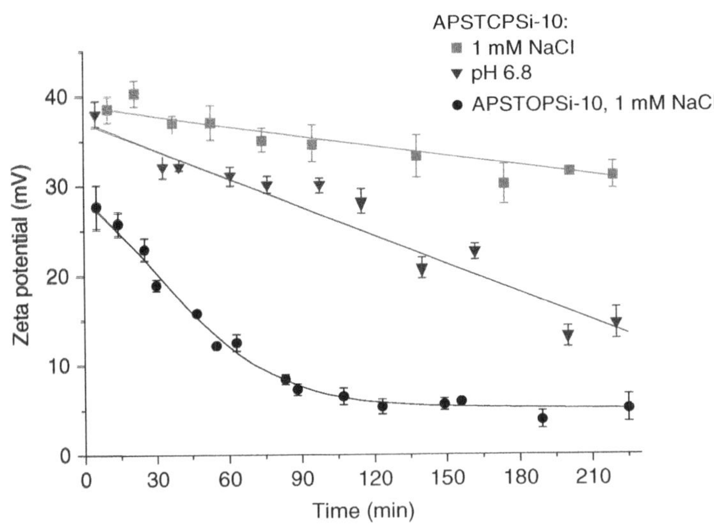

Figure 9.11 Aqueous stability of zeta potential of APSTCPSi-10 in distilled water and in pH 6.8 phosphate buffer and APSTOPSi-10 in distilled water as a function of time. Reprinted with permission from [70].

different surface chemistries, APSTCPSi and APSTOPSi; the former surface is thermally carbonized and latter thermally oxidized before the silanization. The zeta potential of all surfaces decreased as a function of time, which was attributed to the detachment of the positively charged amine groups on the surface. The APSTCPSi surface was found to be more stable against the degradation as the detachment of the amine groups lasted several hours in the aqueous electrolyte solution. The zeta potential of the APSTCPSi-10 decreased by 20% from its initial value in approximately 4 hours. Conversely, the APSTOPSi-10 appeared to undergo a fast hydrolysis, and in the same timeframe, zeta potential was reduced by 70%. In a pH 6.8 buffer solution, the hydrolysis rate of APSTCPSi increased compared with the electrolyte solution. The PSi particles typically cause an electrolyte pH shift in a slightly acidic direction, and in the buffer solution, pH 6.8 is significantly higher. Zeta potential decrease appeared to be slower than with the oxidized sample, falling approximately 50% from the initial value in 4 hours, suggesting that the surface should be stable enough in aqueous conditions to be used, for example, for intravenous drug delivery applications.

9.3.3 Colloidal Stability

The long-range interactions, van der Waals interaction, and double-layer interaction (or electrostatic force) dictate the stability of colloids according to theory, described by Derjaguin, Landau, Verwey and Overbeek [103]. This so-called DLVO theory explains the stabilizing effect of usually repulsive double-layer interactions against the attractive van der Waals interactions (Figure 9.12). The net DLVO interaction forms a primary minimum near the surface, and if the charge is high enough, an energy barrier is formed right

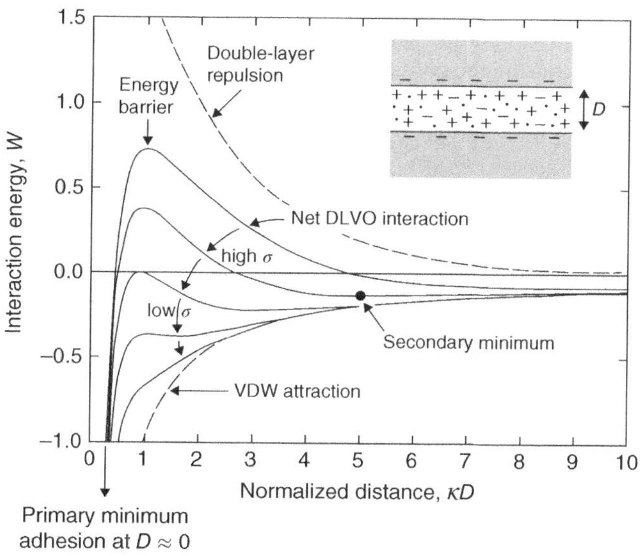

Figure 9.12 *Interaction energy of two approaching surfaces (inset) as a function of distance as modeled by DLVO theory. Reprinted from [94] with permission from Elsevier, copyright 2011.*

outside the minimum. In certain cases, a secondary minimum can also appear farther away from the surface. The secondary minimum typically causes the agglomeration of the particles. Nichols *et al.* proposed that agglomeration means the attachment of the particles in a reversible and easily dispersible manner, as is the case with this secondary minimum [105]. Aggregation is instead a strong (and even chemical) interaction, which is impossible to break without extreme force. Other terminology is also used, such as flocculation to denote interactions at the primary minimum. It is important to note that for pharmaceutical formulations, phenomena such as sedimentation over time due to flocculation, perhaps in combination with gravitational forces, often occurs, but is easily restored by simple shaking before use. Likewise, "eternal" colloidal stability or nanoparticulate systems may not always be required to render them suitable for use after long-term storage.

Colloidal stability is often "proven" via DLS measurements, but static DLS measurements obviously cannot reflect the situation the nanoparticle will encounter after being intravenously injected into the bloodstream, either in terms of the surrounding media or experienced forces caused by blood pressure and flow. In simple solvents and electrolytes, the DLVO theory satisfactorily describes the stability of colloidal solutions, but in a more complex solution, if there are longer polymers on the surface, or if the surface is hydrophobic, the stability is not so straightforward a phenomenon. In this case, the most relevant group of interactions affecting stability are solvation forces. These forces depend on the properties of the surface, but also the medium. In this context, the most important, well studied, and unknown phenomenon is hydrophobicity, the strongly repulsive force between certain kind of surfaces and water molecules. Hydrophobicity originally meant the low solubility and unexpectedly strong interaction between hydrocarbons in water. Hydrophobicity was extended to surfaces when it was discovered that the mutual interaction between two surfaces could be greater than van der Waals forces predict, and water droplets have a large contact angle on hydrophobic surfaces. The latter gives an easy way to estimate the hydrophobicity, by measuring the contact angle between surfaces and water droplets. When the angle is high enough (>90°) the surface is generally considered hydrophobic, although different definitions can be found depending on the research field.

The hydrophobic effect and insufficient electrostatic repulsion can both lead to unstable colloidal suspension in water, which can be avoided with so-called steric stabilization. This typically means that the attachment of hydrophilic polymers or proteins to the particle surface changes the surface interaction with the surroundings in a more solvent-favoring direction. Hydrophilic large molecules form a "hairy" layer around the particle, and when these layers are overlapping while two particles are approaching each other, they cause a repulsive force.

9.3.4 Challenges in Particularly Porous Nanoparticle Characterization

While DLS is the go-to technique in determining whether a nanosuspension is obtained or not, in the case of nanoparticle characterization, the main challenge is that with pure DLS measurement it might be very hard to judge whether the obtained result reflects the size of the aggregate (or agglomerate) or the size of the primary particle [106]. Primary particle size is fairly easy to measure (one day job) by taking a statistically significant amount of TEM images and analyzing them with special software. But it is very hard to distinguish whether particles have agglomerated during drying on the TEM grid or not. This means that both methods (DLS and EM) are needed in order to find the answer. For some reason, nanoparticle characterization with SLS and more detailed multi-angle DLS has not become as popular as back-scattering DLS. This is probably due to the more complicated sample preparation; the sample concentration is crucial and it must be filtered. Despite this, there is a lot of unused characterization power in multi-angle DLS measurement. When comparisons between the DLS and TEM radii are made, it must be noted that these techniques and particle size distributions are fundamentally different. The TEM size distribution is based on the number of the particles, whereas the DLS distribution is based on the intensity of scattered light, which weights the distribution towards larger sizes. Some error is therefore expected.

Puzzled by the huge difference between TEM and DLS radii, Kaasalainen *et al.* studied mesoporous silicon-based nanoparticles with multi-angle methods [106]. They noticed that SLS scattering of agglomerated silica nanoparticles (porous and nonporous) and non-agglomerated PSi nanoparticles resembled each other and the radius of gyration (R_g) was calculated with the Debye-Bueche fit. The scattering of non-agglomerated mesoporous silica nanoparticles resembled the scattering of spherical nanoparticles, enabling the use of a Guinier plot (Figure 9.13). When the radius of gyration and hydrodynamic radius were calculated, the morphology of particles in solution was analyzed by R_g/R_h. This revealed a clear correlation between DLS overestimation (R_h/R_e) and structure factor. Also noteworthy is the result that R_g/R_h did not correlate with the porosity of silica nanoparticles, so the mesoporous structure of silica nanoparticles does not affect this structure factor so much that it could be measured, especially when some of the samples were agglomerated. Actually, this has been seen before, but for some reason nanoparticle research has not adapted the idea [107, 108]. The common assumption is that the hydrodynamic radius and polydispersity index, obtained from the one angle DLS measurement with CUMULANTS-analysis, could reveal whether agglomerates are present.

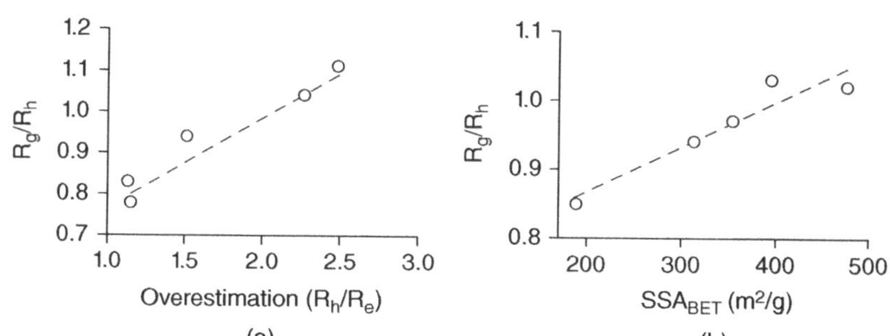

Figure 9.13 *Correlation of structure parameter R_g/R_h compared with (a) overestimation in the hydrodynamic diameter for silica nanoparticles, and (b) to specific surface area SSA_{BET} with PSi nanoparticles [106].*

In Figure 9.13, differences between the TEM radius and hydrodynamic radius were less pronounced in the case of PSi nanoparticles, but still clear. No signs of agglomeration were found, but another interesting behavior arose from the results. Unexpectedly, a strong correlation between R_g/R_h and specific surface area, SSA_{BET} (Figure 9.13b) was found in PSi nanoparticles. The difference could arise from the distribution and size of scattering centers in the PSi matrix, as similarly observed with silica aggregates [108]. This was a very interesting result and needs further study, because it could give a way to characterize the porosity of PSi nanoparticles in solution. It was concluded that both radii, R_g and R_h, seem to overestimate the nanoparticle size compared with the TEM radius, but it may be possible to overcome the issue by a rigorous light scattering study of structure factor R_g/R_h.

9.4 Incorporation of Active Molecules

9.4.1 Drug Loading

Trends in combinatorial chemistry and new approaches in drug design have resulted in more lipophilic potential drug compounds [27]. Depending on the source, up to 70–90% of new chemical entities (NCEs) have been reported to be poorly soluble. As a means of dissolution enhancement, mesoporous silica (whether in bulk or nanoparticulate form) can partially overcome this challenge [27]. The key advantage is the high specific surface area and the pore diameter, where poorly water-soluble drugs can be loaded to remain in their amorphous form due to the small pore size that prevents recrystallization of the drug. To be able to be loaded into the nanoscaled pores, the drug first needs to be dissolved into the molecular state before it can fit into the pores. Consequently, the most common way of loading drugs into mesoporous matrixes is from a solution phase into which the drug has been dissolved. Here, the affinity of the drug to be loaded and the pore surface can be boosted either by choice of solvent, or tuning of the surface chemistry of the drug carrier matrix [109]. If maximum loading degrees are aimed for, a poor solvent for the drug compound can be chosen in order for the interactions between drug and carrier to be dominant, instead of drug–solvent interactions. If the solvent is also a poor dispersant for the silica carrier, the carrier–solvent interactions are simultaneously minimized. To properly drive (or even force) the dissolution and dispersion in these cases, ultrasonication techniques are usually needed as assistance. If the drug is not fully soluble in the loading solvent, the solubility equilibrium should be shifted once the drug is being adsorbed to the carrier material, leading to more drug being able to dissolve in the solvent phase. In these cases, the residual drug in the loading solvent after drug adsorption and separation of the drug-loaded carrier can quite easily be determined via spectrophotometric methods, including high-performance liquid chromatography (HPLC). Care has to be taken that the spectral range (i.e., cut-off wavelengths) for the solvent in question is compatible with the absorbance of the drug compound. This is especially critical since the loading solvent used is typically organic; both because water as loading solvent would partially destroy the carrier structure due to silica dissolution, considering long loading times (hours to days) are usually used, and because the drug to be loaded would typically be a poorly soluble one, in accordance with the above.

The adsorption process can usually be described via adsorption theories such as Langmuir, Henry or Freundlich, once the adsorption isotherm has been determined. Typically, a Langmuir-like adsorption behavior is observed if the adsorption conditions have been chosen correctly. The shape of the isotherm depends on the affinity between carrier and drug, where the surface chemistry of the carrier can be tuned by proper surface functionalization to boost the affinity for a specific drug molecule. Silica is a known adsorbent, and hence an inherent affinity to a range of different drug compounds should be quite sufficient without the need for surface functionalization. The role of surface functionalization becomes more prominent if the loading has to be carried out in aqueous solvent, either due to very high water solubility of the compound to be loaded (i.e., not at all soluble in other solvents), or the drug being incompatible with organic solvents, such as for instance in the case of most biomolecules (enzymes becoming inactivated, other proteins denatured, etc.). In those cases, the adsorption behavior may be very different to that of the organic solvent. Most often the curve will be pH-dependent, due not only to the fact that most drug compounds are weak acids or bases, but also because the same is true for the surface groups of both native and surface functionalized silica. For maximized adsorption, the pH conditions need to be carefully matched between carrier and drug (Figure 9.14a).

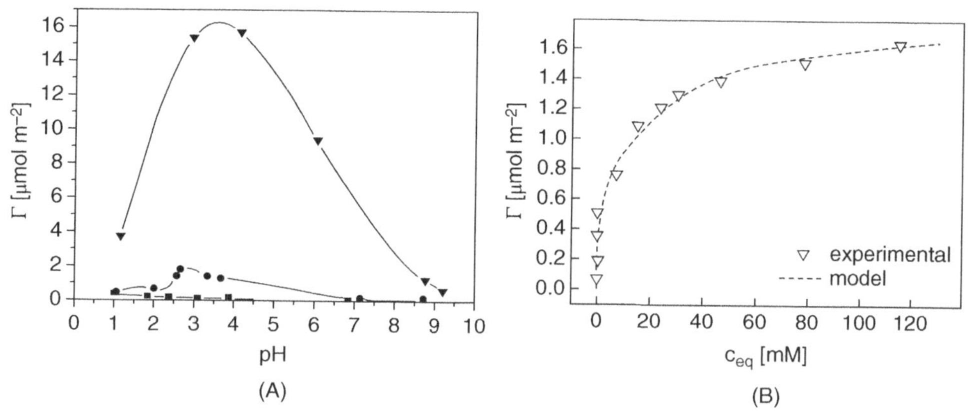

Figure 9.14 *(A) Adsorption of water-soluble salicylic acid from aqueous solvent as a function of pH to PEI-functionalized mesoporous silica (▼), aminosilane-functionalized mesoporous silica (•) and native mesoporous silica (■). Lines are guides for the eye. Reprinted with permission from [109]. B) Adsorption of ibuprofen to native mesoporous silica from organic solvent, as modeled by a hybrid Langmuir model. Reprinted with permission from [110].*

Adsorption isotherms can, consequently, reveal important insights as to the adsorption process and thus how to tune the loading degree; whereby the loading degree may have direct implications for the release rate [111]. When the loading degrees are sufficiently high, bulk methods such as thermogravimetry can also be used for quantification of drug loading degree. Since the mechanism for drug loading in the case of inorganic carriers is adsorption of drug molecules to the surface of the carrier (inside pores/outside particle surface), it is important to pay attention to using the correct term for describing the loading degree, either as a mass fraction ($m_{drug}/m_{carrier}$) or weight percent, whereas the concept of "encapsulation efficiency" bears no relevance in this context. By determining the adsorption isotherm, one can easily pick the optimal loading conditions that would correspond to 100% loading efficiency (only by staying under the plateau value), which only means that a low concentration of initial drug was chosen, rather than that the carrier is exceptionally efficient in terms of drug-carrying capacity. For perfect Langmuir adsorption behavior, the "encapsulation efficiency" is 100% until the plateau value is reached, corresponding to monolayer formation on the surface, after which any more drug added will be excess. In practice, multilayer adsorption most likely also occurs in these cases, but limited by the small pore size that is already in the molecular size range. Thus, the adsorption process is dependent on the surface area and limited by the pore size; whereas the pore volume is not directly controlling the drug-loading process (contrary to what is usually claimed), but is certainly an enabling factor. Recently it was reported that a deviation from "ideal" Langmuir adsorption behavior is not necessarily due to multilayer formation, but to different affinities of the drug molecule to different silanol groups on the carrier surface [112]. This observation was supported by molecular modeling calculations, which is a method that could be utilized more as a complement to experimental observations in describing and predicting drug delivery system behavior. Computational approaches for the optimization of nanocarrier design have indeed been suggested to be able to predict cellular uptake and intracellular distribution, and *in vivo* nanoparticle fate, drug delivery efficiency, prediction of pharmacokinetics, toxicity, as well as biodistribution [113].

Regardless of loading method, given the mechanism of loading and carrier structure, drug elution is perhaps the most accurate method for determining drug loading degree. Here, the drug-loaded dried carrier material is weighed and the drug eluted in a very good solvent for the drug molecule, with subsequent spectrophotometric quantification of eluted drug. Again, but conversely as to the loading process, here the solvent chosen should be good enough for the drug to transfer fully to the solvent phase and not remain adsorbed to the carrier. One method of quantification should preferentially be complemented with another (e.g., drug elution + residual drug in supernatant).

Correctly conducted, adsorption-based loading can yield loading degrees equaling the mass of the carrier, i.e. 100 wt.% if $m_{drug}/m_{carrier}$ is considered, or 50 wt.% if $m_{drug}/(m_{carrier} + m_{drug})$ is considered. The active agent can also be covalently coupled to the carrier via different means, but these approaches usually lead to significantly lower loading degrees. For such approaches, suitable functional groups on the active molecules are also required which can be used for conjugation and subsequently cleaved at the site of action without losing their activity.

For PSi, a common strategy to achieve optimal carrier properties is to keep the pore size and particle size constant and adjust electrostatic and hydrophobic interactions between the surface and the drug molecules [114, 115]. Godin *et al.* studied the loading of so-called secondary state nanoparticles into PSi microparticles, and observed improvement in loading when zeta potentials of microparticles and nanoparticles were opposite [116]. The same observation was made by Huotari *et al.*, when the loading degree of negatively charged GLP-1 peptide in PSi microparticles was higher with a positively charged surface than with a negatively charged surface [117]. In addition, the loading degree decreased from 45% to 24% when the pH was increased to 9.0, where the surface of originally positively charged PSi turns negative. By keeping the concentration of the loading solution low, electrostatic interactions between the PSi surface and the drug molecule might also enable slower, degradation-driven release of payload [118].

Surface adsorption-based loading is a tempting approach, especially with biological cargoes when the molecules may be expensive, and efficient use of the carrying material is not that important an issue. Charged and hydrophobic sections in biomolecules typically enable efficient loading, even though the immersion method is used. As examples, Rytkönen *et al.* achieved a sufficient 14% loading degree with oligonucleotides, and Kovalainen *et al.* were able to load PYY with a 7% loading degree, while both of these loadings were made with a 100% loading efficiency without waste of drug [71, 119, 120]. More basic research on the role of surface chemistry on protein adsorption onto PSi surfaces has been done by Jarvis *et al.* [121]. They noticed that the hydrophobicity of native porous silicon surfaces increased the adsorption of studied proteins (albumin, lysozyme, and papain) but also caused structural rearrangements. A hydrophilic oxide surface instead decreased the adsorbed amount of proteins, but the protein structure remained undisturbed.

9.4.2 Labeling with Imaging Agents

The physical properties, surface modifications, and composition of porous and nonporous silicon-based nanoparticles can be tuned in order to make them powerful tools for imaging. The size and morphology of the particles can be tuned to enhance their interaction with cells and their permeability across biological barriers; reactive groups can be added on to the particle surface for the attachment of different organic moieties, biomolecules, or nanoscopic imaging agents; and the incorporation of MRI, optical, or nuclear imaging agents and therapeutic payloads inside the carrier structure can be carried out.

Among all the molecular imaging modalities, optical imaging (OI) has been a versatile and easy-to-use approach in terms of available imaging agents to be used and relatively low cost of instrumentation, and it can also be employed as complementary to other modalities [122]. In the case of studying silicon-based nanoparticles in optical bioimaging, one of the main concerns is to investigate their efficiencies prior to their use as close as possible to the biological settings for which they are aimed. For instance, in fluorophore attached/incorporated MSNs, fluorescein isothiocyanate (FITC) is one of the commonly used fluorophores within the silicon-base matrix for OI. In their characterization and their efficiency investigations, the unique property of fluorescein as a pH probe should be considered; this essentially means that the fluorescence intensity will depend on the surrounding pH during detection, and also the local environment leads to change in their fluorescence properties [84, 123, 124]. Depending on the preparation conditions, the MSN-conjugated fluorophore can be exposed to conditions (solvents, pH) that induce one of the non-fluorescent forms of FITC, which are also among the existing chemical forms of FITC for certain pH and solvents. Consequently, correlation between fluorescence intensity and amount of fluorophore, frequently used to represent the amount of particles, is highly prone to uncertainties, except under tightly controlled conditions.

For the particles to be traceable in a biological/physiological environment, MSNs are typically tagged with fluorescent dyes. On the one hand, when attempting to quantify particle uptake/biodistribution based on dye-labeled MSNs, it should be kept in mind that the inherent properties of the dye itself, surrounding conditions, surface coating, and silica dissolution consequently leading to leaching of attached labels, may have a tremendous impact on the quantification. On the other hand, the incorporation of the dye into the MSN matrix and further coating of the MSNs may even have beneficiary influences on the dye in terms of photostability enhancement, providing a protective environment, and elimination of pH-induced effects.

When the major portion of the fluorophore is located in a confined space, such as the mesopores in MSNs, the local environment such as pH inside the pores does not necessarily match with the surrounding conditions, due to the acidity of silanols always present on a silica surface. The complexity in investigating the fluorescence properties of MSNs usually increases when adding basic groups, such as aminosilanes used to conjugate the fluorophore. Surface coatings with acid/base properties certainly increase the level of complexity even more, when it comes to the inner versus the outer surface combinations. For instance, poly(ethylene imine), PEI, is a commonly employed polycationic polymer (including primary, secondary, and tertiary amines in its structure) in nanomedical contexts that also possesses enormous buffering capacity due the abundance of amines, which leads the attached molecules to experience this buffering effect when PEI is coated on MSNs [84]. In Figure 9.15, the fluorescence spectra of FITC-conjugated MSNs (F-MSN) and PEI-coated F-MSNs (PEI-F-MSN) in weakly acidic and neutral pH buffer solutions as well

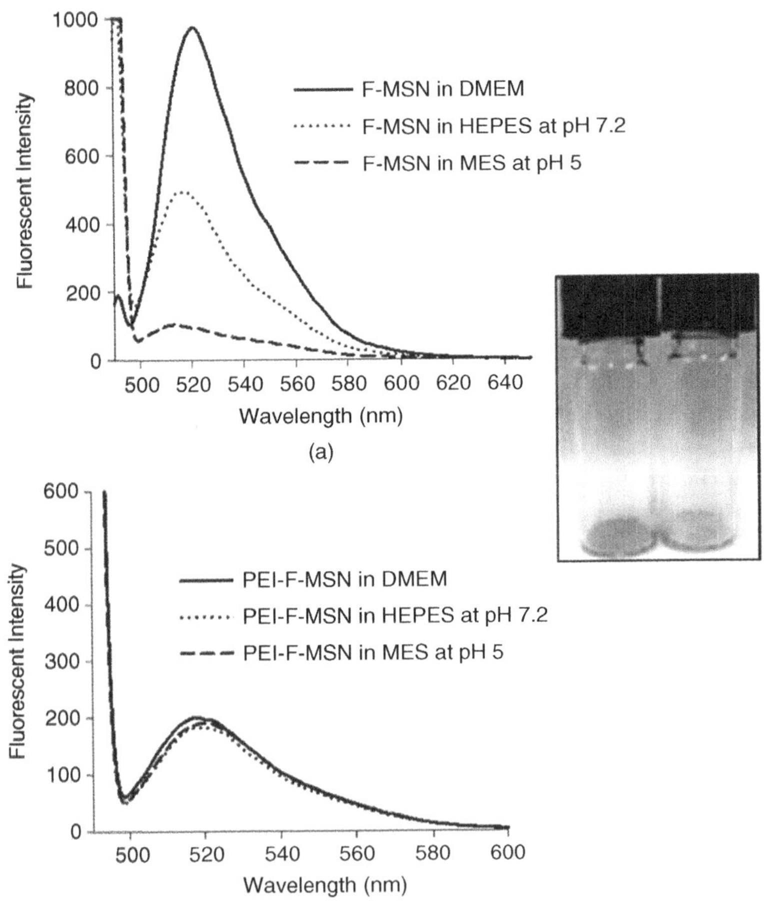

Figure 9.15 *Fluorescence spectra of (a) F-MSN and (b) PEI-F-MSN at neutral and weakly acidic pH buffer solutions, as well as in cell media (DMEM) (λ_{ex} = 490 nm). Inset: dry powder forms of F-MSN and PEI-F-MSN. Adapted and reprinted with permission from [84].*

as in cell media (DMEM) are presented. The figure plots indeed support the notion that the observed fluorescence intensity for the FITC-incorporated MSNs is highly dependent on the experienced pH values, that is, the "local pH" experienced by the incorporated FITC molecules in the MSN matrix.

If water is present in the media used for fluorophore incorporation (where usually buffers are used), it needs to be ensured that the structure of the material is retained after the reaction step [84]. This also applies to any other preparation step, like surface coating. Figure 9.16 shows the structural order of MSNs after several hours of soaking in HEPES buffer (pH 7.2) versus ethanol. No structural deformation is observed in ethanol solution, while the structure is completely destroyed (as determined by XRD) after only 2 h in neutral buffer (Figure 9.16a). This suggests that aqueous solvent should be exchanged to organic whenever possible, for example in fluorophore conjugation reactions but also surface coating (in Figure 16d a PEI coating is shown as an example). In some cases, as reported by Şen Karaman *et al.* [83] for the incorporation of paramagnetic Gd into MSNs, even the incorporation of the imaging agents during synthesis can lead to the destruction of the morphology of the particles, which may cause difficulties in the further applications of the material.

9.5 Biorelevant Physicochemical Characterization

The balance between the toxicity and functionality of nanoparticles is a delicate matter in drug delivery. A good example of this is that, on the one hand, passively targeted nanoparticles are trying to avoid the immune system, and on the other hand, adjuvants for vaccine therapies are trying to activate it [125]. This highlights the importance of understanding the effects of particle properties on the physiological response. Walkey and Chan described the design of nanoparticles using two different causal relationships: synthetic identity determines the biological identity, which in turn determines the physiological response (Figure 9.17) [126]. Synthetic identity is defined by dry state properties like size, shape, and chemical composition, and biological identity by wet state properties like hydrodynamic size, aggregation, and protein adsorption.

It has long been known that a proper material characterization is crucial prior to any *in vitro* or *in vivo* experiment. A long list of methods is available to study distinct physicochemical features of nanoparticles, and in most cases, two methods with different measuring principles are required to exclude a bias [127]. As a "basic set" of characterization, it is commonly accepted to measure size, surface charge (i.e., zeta potential, although this cannot be used as synonym), and porosity, if the material is porous. However, for some (medical) approaches, other characteristics like metal solubility or toxic trace-metal detection from the synthesis might also need to be included in the basic set. "Zetasizers" are relatively easily accessible instruments and can measure both hydrodynamic diameters and surface charge using the same instrument. But how reliable are these results when we want to measure the size in a complex biological media like cell culture media or even blood? If EM is used to image individual particles, to check their shape, size, and maybe porosity, can we still use this method to study the shape, size, and porosity of the particles in cells? The as-characterized nanoparticles in (mostly) buffer will behave completely differently in biological media, and hence the need for other instruments. Size, which is usually measured with TEM or DLS, could still be measured with the two methods, but with some caution

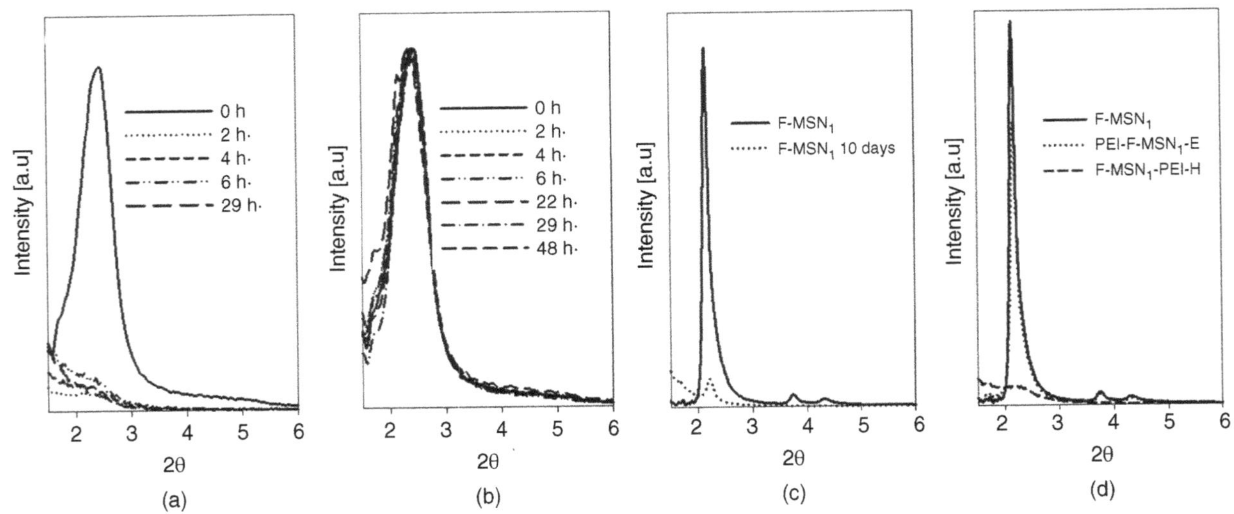

Figure 9.16 *MSNs incubated in (a) HEPES buffer and (b) ethanol; (c) after 10 days incubation in HEPES buffer; and (d) surface adsorption of PEI carried out in HEPES buffer and in ethanol [84].*

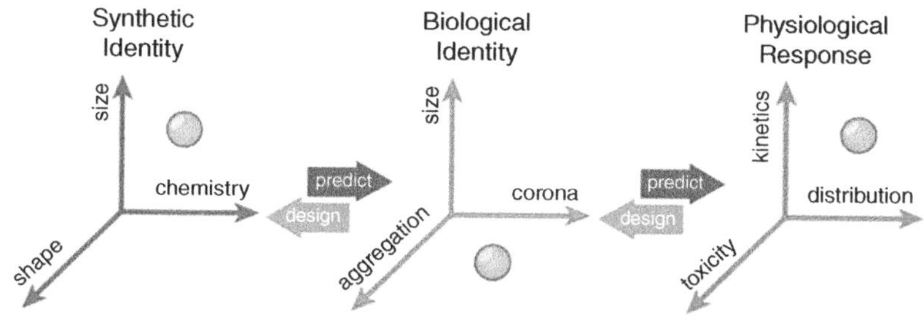

Figure 9.17 Design and prediction cycle as proposed by Walkey and Chan. Reprinted with permission from [126].

in interpreting the results. DLS instruments often have a size discrimination between two populations that is approximately ×4 in magnitude. Hence if the hydrodynamic diameter is measured in fetal bovine serum supplemented by cell culture medium, and the particle size is about 40–50 nm, it will not be possible to distinguish the scattering signal coming from the plasma proteins, because their signal results in an intensity peak at 10 nm. As mentioned above, it has additionally been shown that DLS is not a valid method to truly determine the hydrodynamic size of a mesoporous particle [106]. Likewise, it is often said that DLS measurements over time can provide an insight into the colloidal stability in cell culture media. However, the measurements are intensity based, and even if a major fraction of the particles have settled over time (due to agglomeration, maybe even aggregation), this fraction will not appear in the DLS measurements, as the laser beam is "looking" for the most concentrated spot in the cuvette. Similar precautions must be taken when using electron microscopy, where it is also sometimes claimed that TEM can "show" that particles do agglomerate in biological media, when it could also be a simple drying artefact (contraction of a water drop during evaporation). Other methods that circumvent these obstacles are described and proposed, but most likely not very accessible to nanomedicine labs yet. Hence, it is of utmost importance to know the detection mechanism, the analytical limits, and the corresponding interpretation of the raw data.

9.5.1 Biodegradation/Dissolution of Silica

Biodegradation of a material is a hard-to-define term in general, but is commonly defined as the disintegration of a material's biological means like fungi or bacteria [128]. Recently, the International Union of Pure and Applied Chemistry (IUPAC) has specified the term biodegradation as the "degradation caused by enzymatic processes resulting from the action of cells", and distinguished it from other terms that include isolated enzymatic degradation or so-called bioalteration [129]. The term biodegradable is hence often misused, but since it is a common term used in publications about degradability of silicon-based nanoparticles, the forthcoming section will stick to this very same term. "Bioerosion" would most likely be a more proper term to use in the case of silica.

The main aim of targeted drug carriers, like MSNs, is to safely and specifically deliver the drug to its site of action. The degradation mechanism, or the tuning of the degradation mechanism, could further enhance the potency of the drug, or even act as a depot for sustained release of the drug. Whether the terminal use is a simple vehicle for genes, a probe for photodynamic therapy, or simply an *in vitro* fluorescent probe, the chemical stability of silica nanoparticles in an aqueous environment is critical (if too water-soluble, the cargo may be released prematurely; and if too stable, accumulation and possible chronic inflammation or toxicity). Most of the silica nano- or microparticle degradation studies *in vitro* include observing a change in size, soluble material, porosity, and the specific surface area over time in (mostly) aqueous environments. As a summary, virtually everything from intrinsic physicochemical properties of the nanoparticles to extrinsic factors like ions or proteins in the environment will affect the degradation of mesoporous silica nanoparticles [128].

The dissolution of silica nanoparticles consists of three major steps. Firstly, water is adsorbed onto the surface; secondly, the siloxane bonds are hydrolyzed to silanols (the density of siloxane and silanol bonds on the surface is strongly dependent on the thermal history of the particles); and lastly, silicilic acid is released due to nucleophilic attack [130–132]. The solubility of mono- and polymeric silicilic acid is hence the key step in the particle degradation. If the concentration is kept below the solubility product of silica, the dissolution rate depends on particle size, surface functionalization, and the degree of silica condensation [133–136]. The solubility product of silica is 40 mg/l or 120 ppm in water at neutral pH and room temperature. In acellular media, the degradation is more pronounced for particles with a high specific surface area, as shown by Iller [137], and this is even more evident when comparing the degradation of monodisperse MSNs versus agglomerated MSNs, as the accessible surface area is decreased in the latter. Interestingly, the condensation degree has a pronounced effect on degradation, where siloxane-rich particles degrade much more slowly. This is evident from the degradation mechanism of silica described above, as the second step of the degradation consists of the hydrolysis of siloxanes to silanols. A siloxane-rich surface is achieved by calcination as post-synthetic treatment [138]. It is even postulated that the degree of degradation can be used to fine-tune drug release from silica nanoparticles [139, 140]. Polyethylene glycol (PEG) is often used to enhance biocompatibility and to prolong systemic circulation, due to its so-called "stealthiness". From this perspective, studies have investigated how different PEG chain lengths and different surface PEG densities affect the degradation [141]. They found that denser brushes and longer PEG chains reduced the matrix degradation. Other surface groups like amine, phenols, and others affect the degradation process by generally accelerating the degradation compared with bare silica. However, it is very difficult to establish general rules, as some studies used too high concentrations, or different solvents, which strongly affect the degradation rate, and hence a straightforward comparison of literature is difficult. Techniques that are commonly used to study degradation processes *in vitro* are electron microscopy, absorbance (REF silicomolybdic assay), nitrogen sorption, and small-angle X-ray scattering to investigate the integrity/structure of the pores. Most of these methods rely on "clean" samples and are suitable for simulated degradation processes *in vitro*, and mostly acellular media only. However, they can give valid insight into the degradation mechanism.

If moving to cell-based experiments or even *in vivo*, the translation of mechanistic studies into a completely new environment (different compartments with lesser volumes of water/ions, different pH in lysosomes/endosomes, proteins that immediately adsorb onto the particle surface, and others) is not straightforward, and most of the degradation data may not be valid in the different setting. However, it is possible to measure intracellular degradation, and it was found that 200 nm mesoporous silica nanoparticles degrade over a time period of 48 h, as measured by electron microscopy and inductively coupled plasma mass spectroscopy [142]. Another important point of degradation is that the degradatation products themselves are inherently nontoxic, which has been shown by several researchers (and can even favor wound healing) [143]. *In vivo* experiments have shown that silica nanoparticles are mainly found in the liver and spleen when injected intravenously [144, 145]. MSNs in particular are cleared from the organism within 48 h to a few days. Si can mainly be found in the urine (roughly 70–80%), whereas the other 20–30% is excreted via feces within 96 h [146, 147]. One should be very wary, though, in assigning very accurate values when it comes to determining biodistribution and excretion levels of silicon-based materials; silicon is ubiquitous in both food and tissue, and hence the Si background levels may be even higher than the miniscule amount, of particles administered. Further, the inter-individual variation can also be quite distinct, meaning valid controls cannot be set [148]. Thus, the investigation of silicon-based NP biodistribution and excretion is often indirectly conducted via optical means by the detection of labeled fluorescent dyes, instead of determining Si levels; both of these have still indicated that the major part of MSNs can be renally excreted, which essentially means the material must first be biodegraded [149]. The upper size limit for nanoparticles that can be excreted via urine is commonly set at 5 nm, as this is the size limit in the kidneys – anything larger than 5 nm is retained in the body [150]. Recently, the Cornell silica dots received USFDA approval for the first in-human clinical trial. The Cornell dots are used for imaging and are targeted with a RGD peptide against integrin receptors. Due to their small size (diameter of about 6 nm), they were cleared rapidly via the kidney and excreted via urine, and systemic accumulation is highly unlikely [151].

9.5.2 Biocompatibility and Nano–Bio Interactions

Biocompatibility is defined by IUPAC as the "ability to be in contact with a living system without producing an adverse effect" (IUPAC definition, [129]). When speaking of biocompatibility and nanomaterials, a plethora of tests is needed to deem a nanomaterial as biocompatible. Mostly, biocompatibility testing is time-consuming and expensive. Recently, the USFDA has incorporated ISO 10993-1 (mostly used for medical devices) as guidance for industry to test biocompatibility. The document describes biocompatibility as including cytotoxicity, sensitization, genotoxicity, and hemocompatibility [152]. There is still no guideline on how to test biocompatibility for nanomaterials, and according to the USFDA, industry is advised to "consult relevant literature and standards during the development of test protocols." It is thus evident, that only a minority, if any, of the published studies describe comprehensive biocompatibility as such. However, the majority of studies suggest that silica nanoparticles are well tolerated in *in vitro* biocompatibility assays. Again, different physicochemical properties lead to a plethora of data, and interlaboratory comparisons between results are often hampered by a lack of control experiments or by missing data in the synthesis route or characterization. The ultimate aim is to connect physicochemical properties to the biocompatibility of a specific type of nanoparticles, or to toxicity, or to whatever outcome researchers expect – similarly to the Lipinski rule of five, which gives guidance on the oral bioavailability of small molecules. If we take a step back to look at the larger picture of data, or the overall collection of results, it is possible to summarize a trend about biocompatibility, and to spot where we have an absence of or little data compared with where we have a lot of data. What renders silica nanoparticles bio-incompatible so far? It is commonly accepted that crystalline silica leads to silicosis when inhaled, but how many drug carriers, or how many silica nanoparticles, contain a crystalline structure in their lattices? Virtually none, because the syntheses employed lead to amorphous matter. This was then further investigated, due to the so-called fiber paradigm, whereby mesoporous silica rods with a high aspect ratio could trigger an inflammatory response *in vitro* and *in vivo* [153]. It was shown that for bare mesoporous silica rods up to an aspect ratio of 8, no significant cell viability loss was demonstrated in non-phagocytic A549 cells, whereas viability declined, no matter whether the aspect ratio was 1 or 8, to the same extent at concentrations above 100 ug/ml. Ha *et al.* found no adverse events in preosteoblasts over 14 days [154]. They also observed that the silicon level inside the cells did not decrease significantly, as measured by ICP-AES, because the intracellular, mainly lysosomal acidic compartment slowed down dissolution of silica. Although persistent, no bio-incompatibility could be detected over a time-span of 14 days. Of course, these results are just hints, and long-term testing is needed to confirm whether the shape or surface groups render silica "less biodegradable". When speaking about intravenously injected drug delivery systems, hemocompatibility is of utmost importance. Without specifying any details on how to test hemocompatibility, it has been shown that the inclusion or exclusion of plasma proteins significantly influences the hemolytic potential of silica nanoparticles; plain, nonporous, silanol-rich particles induced hemolysis, whereas hemolysis reduced to nearly 0% when plasma proteins were included in the experimental setup [104, 155]. Although there are a lot of hints that silica is biocompatible for intravenous use, newer and more sensitive methods could give insights into accumulating low-dose effects. *In vivo*, the biodistribution and accumulation have also showed that mainly those organs with residing macrophages like the liver, lung, and spleen are found to accumulate silica nanoparticles, although with a rapid decline over 72 h [156]. A study in Balb/C mice found no adverse effect after administration of silica nanoparticles [157]. However, it must be noted that long-term data, and data after several administrations (as the end user would be exposed to), are still missing. Nevertheless, due to the high solubility of mesoporous silica in an aqueous environment, and hence rapid clearance from the body compared with low water-soluble materials like gold, cerium or others, an accumulation and subsequent chronic inflammation by the particles *per se* is rather unlikely.

9.5.3 Drug Release

As mentioned in Section 9.4.1 (Drug Loading), the most fundamental idea with loading of poorly soluble drugs into MSNs lies in the assumption that this results in the amorphous form of the drug compound being loaded, which, in turn, results in a faster dissolution rate for the drug. The time-dependent drug release can be studied directly by UV or fluorescence measurement *in vitro*, or indirectly via a pharmacological action in cellular systems (e.g., viability decrease, reactive oxygen scavenging, and others) or *in vivo* (e.g., tumor growth suppression) in animal models. While dissolution testing in accordance with pharmacopoeias (e.g., Ph.Eur.) is relatively straightforward, especially in the case of nanoparticulate systems, the volume of release medium required may be a serious limitation. Usually nanoformulations are most advantageous for challenging drug molecules that are available in small quantities and very costly. This restricts the amount of available sample, along with MSN syntheses typically yielding much smaller batches than bulk mesoporous silica which can easily be synthesized at the gram-scale. Being in nanoparticulate form, the sample is often captured in a dialysis bag for release testing; the pore size and material should be compatible with the sample so as not to block the pores with the nanoparticles, but still allow unrestricted diffusion of released drug. In other cases the nanoparticles are simply dispersed into the release media, and separated, for example, by centrifugation for aliquoting liquid for quantification of released drug. Depending on the state of degradation of the carrier material, it may be uncertain whether the sample can be separated in full. Some miniaturized *in vitro* release methods have also been developed that should be especially suitable for nanoformulations [158].

Another concern with regard to the limited release medium volume is the liquid-to-solid ratio, in this case of silica. Since silica dissolves under aqueous conditions, leading to structural distortion (see Section 9.4.2 and Figure 9.16), and the release is dependent on the carrier structure, carrier degradation will inevitably have an impact on the drug release rate [159]. Whilst certainly dependent on the final drug release site in terms of accessible liquid (water), the release mechanism should essentially be studied at conditions under the solubility saturation limit of silica (120 ppm at neutral pH). Once the saturation limit in the release media is reached, the release behavior will change [159]. The often-observed burst release from mesoporous materials is not only connected to the pore geometry and size, but also the hydrolytic stability of the carrier material. This is easily understood as the silica dissolution typically also exhibits burst "release" (dissolution of silica). For poorly soluble drugs, the release could even be chiefly dominated by the silica dissolution (erosion).

The solubility of the drug compound in the release media is, together with the dissolution of silica, another often-disregarded factor. Most drugs are weak acids or bases, and hence their solubility in aqueous media is pH-dependent. (On a minor note, this also means their UV absorbance is pH-dependent, which should be taken into account in the quantification of release.) Quite frequently, claims about "pH-dependent release" are made, which could only be attributed to the solubility of the drug compound in different release media, i.e., with different pH, and not an in-built function of the carrier system. If the fluorescence of the drug molecule is used as means of quantification, it should be taken into account that fluorescence is only semi-quantitative, as the intensity varies significantly depending on the surrounding conditions (see Section 9.4.2).

Depending on the site of delivery, the release media could also be chosen based on the foreseen drug release environment. If the intended site of delivery is intracellular, the main task of the drug delivery system is to retain the cargo during the delivery process, transit the cargo across the cell membrane, and only then release the cargo intracellularly. In these cases, release tests in buffer solution will not shed a lot of light on the release mechanism, as the carrier system is intended to prevent premature release under such conditions. For mimicking intracellular conditions, cell lysate or, in a simpler case, an aqueous solution containing micelles of a mixture of different choline-linked phospholipids typically found in the cell membrane, can be used (Figure 9.18) [160]. As seen in Figure 9.18, drastically different release behaviour is observed under the different conditions; the complete burst release actually corresponds to the release behavior in cells after 24–48 h incubation.

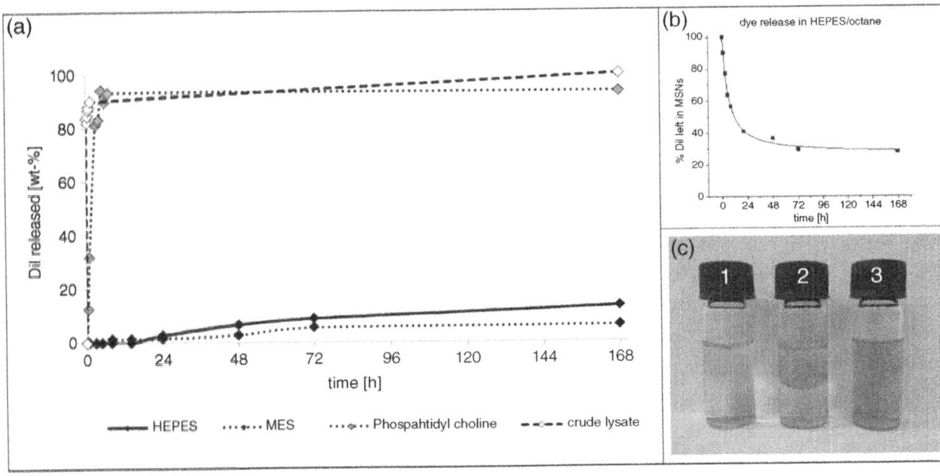

Figure 9.18 *Model drug (DiI dye) release: (a) in different media plotted as wt.% of dye released over time; and (b) in a two-phase HEPES/octane (50/50:V/V) solvent system plotted as wt.% decrease of DiI released from nanoparticles in the HEPES phase into the octane phase. (c) Images of MSN-PEI/DiI in 1: HEPES – poor release, DiI is retained in the MSNs; 2: HEPES/octane – release of dye at the interface between the two solvents while MSNs remain in the water phase; 3: methanol – complete release. Adapted and reprinted with permission from [160].*

The utilization of fluorescent dyes as model drugs is quite convenient for studying the release process in cellular or even animal models using different biomedical imaging techniques. Fluorescence recovery after photobleaching (FRAP) has been used to study the intracellular release of cargo from MSNs in real-time [160]. Fluorescence resonance energy transfer (FRET) has also been used to semi-quantitatively follow the time-course of the intracellular release of a redox-cleavable model drug from a MSN system [161]. The measurement of a biological response (drug action) is always a secondary read-out, and may be more relevant to the onset of action of the drug and/or biological response, than the release kinetics. Free drug is thus always used as control, but may (usually is, if intracellular delivery vehicles are required) be subject to limited intracellular uptake. Thus, it may not be the release kinetics that is (solely) responsible for the time-dependence. If the cargo molecules are large enough to be labeled with a fluorescent dye with the assumption that its properties will not change too much, like in the case of biomolecules, this strategy can also be utilized to study intracellular release over time in live cells [162]. Obviously, the exact means of release as well as kinetics will be determined by the properties of the actual drug molecule to be delivered, but such studies can still reveal a great deal about the (anticipated) function of the carrier system.

9.5.4 Label-free (Imaging) Technologies

Given the challenges in the utilization of molecular imaging agents for the labelling of MSNs discussed in Section 9.4.2, making use of label-free technologies poses an attractive alternative. Therefore, it is a bit surprising to note that there are very few reports yet describing methods for studying the real-time kinetics of label-free NP–cell interactions [163]. One of the drawbacks with labeling agents is that the chemical modification introduced by the fluorescent probe can either affect normal cell behavior, change the surface characteristics of the NPs, or both [164]. Consequently, label-free methodologies capable of tracking NP–cell interactions in real time would be highly desirable in order to rule out possible interferences by labeling agents. Several label-free techniques have been developed for whole cell sensing, including optical waveguide spectroscopy [165], resonant waveguide grating [166], reflectrometric interference spectroscopy [167], surface plasmon resonance [168], and non-optical methods such as impedance spectroscopy [169] and quartz crystal microbalance [170], but these studies have not focused on real-time monitoring of NP–cell interactions. Recently, the utilization of surface plasmon resonance (SPR) for measuring real-time kinetics of NP–cell interactions without the use of any labeling agents was demonstrated, where MSNs in parallel to branched polyethyleneimine-DNA polyplexes were used as model NPs [163]. The approach makes use of a confluent cell monolayer grown on the SPR gold sensor surface, onto which the NPs are fed, while simultaneously measuring the complete SPR angular spectra as a function of time. This allows the determination of the kinetic rate constants of NP–cell interactions at different temperatures, and consequently determination of the activation energy of these interactions by Arrhenius plots (Figure 9.19). The signal responses recorded can further be utilized for calculating the activation energy of endocytosis and determining the phase transition temperature of cell membranes. The SPR-based approach should thus hold great potential for becoming a powerful platform for mechanistic studies of cell uptake and cell interactions of various types of NPs, including elucidation of the real-time kinetics at the very early stages of NP interactions with cells, a phenomenon that has not previously been monitored much due to the lack of suitable label-free methodologies.

Another technique for investigating the NP–cell interactions without the use of labels is hyperspectral imaging technology. Hyperspectral imaging (HSI) combines spectrophotometry and imaging, using advanced optics and algorithms to capture a spectrum from 400 to 1000 nm at each pixel in an enhanced dark-field microscopic (EDFM) image [171]. The integration of enhanced dark-field microscopy and hyperspectral imaging (HSI-EDFM) in particular enables rapid optical observation and spectral confirmation to detect, locate and characterize unlabeled NPs interacting with cells and tissue, by confirming the identity of the materials of interest in a biological sample and generation of an image "mapping" their presence and location in the sample. Another "mapping" technique is coherent anti-Stokes Raman scattering (CARS) microscopy, which has advantages over other analytical approaches used to date and is non-destructive, rapid, label-free, confocal, and has chemical and physical specificity [172]. In the mesoporous silica context, CARS have been used to study both the 3D distribution of loaded drugs in the carrier matrix of microparticles, as well as the intracellular distribution of fluorescently labeled MSNs in combination with two-photon imaging [173]. While the optical transparency may hamper the visualization of MSNs using optical methods, both linear and nonlinear optical imaging techniques can readily be used to study the localization of silicon nanoparticles in cells [174].

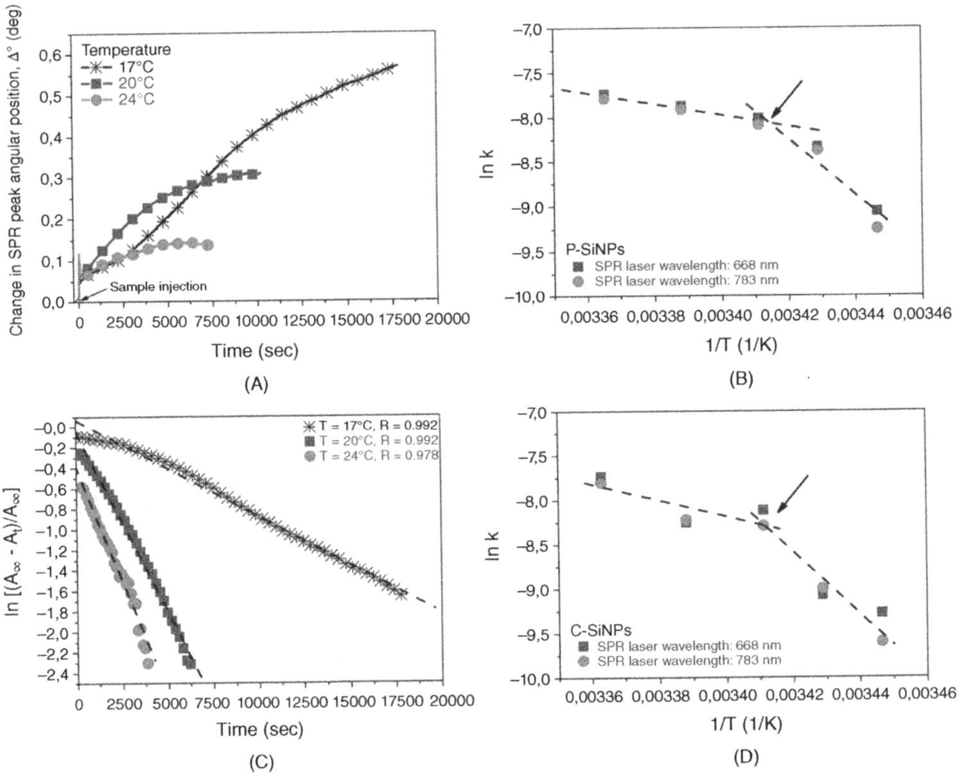

Figure 9.19 *(A) Real-time SPR signal response measured with an SPR wavelength of 668 nm during plain MSNs (P-SiNPs) interaction with HeLa cells at three different temperatures: 17°C, 20°C and 24°C. (B) First-order kinetic fit of P-SiNP cell interactions at three different temperatures: 17°C (cross), 20°C (square) and 24°C (dot). Dashed lines are linear fits to data points:* $R^2 = 0.992$ *(17°C), 0.992 (20°C) and 0.978 (24°C). Arrhenius plots of (C) P-SiNP and (D) PEG-PEI coated MSNs (C-SiNP) for cell interactions measured at five different temperatures; 17, 18.5, 20, 22 and 24°C. Measurements were carried out with two different SPR wavelengths of 668 nm (square) and 783 nm (dot). Dashed lines are linear fits to data points below and above the phase transition temperature (arrow) of the HeLa cell membrane. Reprinted with permission from [163].*

9.6 Conclusions

Due to their versatile synthesis and easy-to-modify surfaces, the possibility of tuning pore size, and other physicochemical parameters, silicon-based nanoparticles offer a huge potential to serve as drug delivery systems or imaging agents. Their physicochemical characteristics have been used as tools to design more effective and personalized nanomedicines for theranostic applications. When aiming for these kinds of applications, proper care must be taken at each step to ensure a rational development. In this chapter, we have tried to point out considerations to be taken into account in their design and characterization, to help promote these developments.

References

[1] Shen, S., Wu, Y., Liu, Y., Wu, D. (2017) High drug-loading nanomedicines: progress, current status, and prospects. *Int. J. Nanomedicine*, **12**, 4085–4109.

[2] Berbel Manaia, E., Paiva Abuçafy, M., Chiari-Andréo, B.G., *et al.* (2017) Physicochemical characterization of drug nanocarriers. *Int. J. Nanomedicine*, **12**, 4991–5011.

[3] Vivero-Escoto, J.L., Huxford-Phillips, R.C., Lin, W. (2012) Silica-based nanoprobes for biomedical imaging and theranostic applications. *Chem. Soc. Rev.*, **41** (7), 2673.

[4] Stöber, W., Fink, A., Bohn, E. (1968) Controlled growth of monodisperse silica spheres in the micron size range. *J. Colloid Interface Sci.*, **26** (1), 62–69.

[5] Kolbe, G. (1956) *Das komplexchemische Verhalten der Kieselsäiure*. Dissertation.

[6] Bogush, G., Zukoski, C. (1991) Studies of the kinetics of the precipitation of uniform silica particles through the hydrolysis and condensation of silicon alkoxides. *J. Colloid Interface Sci.*, **142** (1), 1–18.

[7] Van Blaaderen, A., Van Geest, J., Vrij, A. (1992) Monodisperse colloidal silica spheres from tetraalkoxysilanes: Particle formation and growth mechanism. *Journal of Colloid and Interface Science*, **154** (2), 481–501.

[8] Kwon, N.H., Beaux, M.F., Ebert, C., *et al.* (2007) Nanowire-based delivery of Escherichia coli O157 Shiga Toxin 1 A Subunit into human and bovine cells. *Nano Lett.*, **7** (9), 2718–2723.

[9] Deng, Z., Zhen, Z., Hu, X., *et al.* (2011) Hollow chitosan–silica nanospheres as pH-sensitive targeted delivery carriers in breast cancer therapy. *Biomaterials*, **32** (21), 4976–4986.

[10] Arriagada, F.J., Osseo-Asare, K. (1999) Synthesis of nanosize silica in a nonionic water-in-oil microemulsion: Effects of the water/surfactant molar ratio and ammonia concentration. *J Colloid Interface Sci.*, **211** (2), 210–220.

[11] Finnie, K.S., Bartlett, J.R., Barbé, C.J.A., Kong, L. (2007) Formation of silica nanoparticles in microemulsions. *Langmuir*, **23** (6), 3017–3024.

[12] Richard, B., Lemyre, J.-L., Ritcey, A.M. (2017) Nanoparticle size control in microemulsion synthesis. *Langmuir*, **33** (19), 4748–4757.

[13] Lin, C.-H., Chang, J.-H., Yeh, Y.-Q., *et al.* (2015) Formation of hollow silica nanospheres by reverse microemulsion. *Nanoscale*, **7** (21), 9614–9626.

[14] Song, X., Li, F., Ma, J., *et al.* (2011) Synthesis of fluorescent silica nanoparticles and their applications as fluorescence probes. *J. Fluoresc.*, **21** (3), 1205–1212.

[15] Rahman, I.A., Padavettan, V. (2012) Synthesis of silica nanoparticles by sol-gel: size-dependent properties, surface modification, and applications in silica-polymer nanocomposites – A review. *J. Nanomater.*, **2012**, 1–15.

[16] Xu, Z., Ma, X., Gao, Y.-E., *et al.* (2017) Multifunctional silica nanoparticles as a promising theranostic platform for biomedical applications. *Mater. Chem. Front.*, **1** (7), 1257–1272.

[17] Bagwe, R.P., Yang, C., Hilliard, L.R., Tan, W. (2004) Optimization of dye-doped silica nanoparticles prepared using a reverse microemulsion method. *Langmuir*, **20** (19), 8336–8342.

[18] Rossi, L.M., Shi, L., Rosenzweig, N., Rosenzweig, Z. (2006) Fluorescent silica nanospheres for digital counting bioassay of the breast cancer marker HER2/nue. *Biosens. Bioelectron.*, **21** (10), 1900–1906.

[19] Auger, A., Samuel, J., Poncelet, O., Raccurt, O. (2011) A comparative study of non-covalent encapsulation methods for organic dyes into silica nanoparticles. *Nanoscale Res. Lett.*, **6** (1), 328.

[20] van Blaaderen, A., Vrij, A. (1993) Synthesis and characterization of monodisperse colloidal organo-silica spheres. *J. Colloid Interface Sci.*, **156** (1), 1–18.

[21] Santra, S., Wang, K., Tapec, R., Tan, W. (2001) Development of novel dye-doped silica nanoparticles for biomarker application. *J. Biomed. Opt.*, **6** (2), 160.

[22] Santra, S., Zhang, P., Wang, K., *et al.* (2001) Conjugation of biomolecules with luminophore-doped silica nanoparticles for photostable biomarkers. *Anal. Chem.*, **73** (20), 4988–4993.

[23] Lian, W., Litherland, S.A., Badrane, H., *et al.* (2004) Ultrasensitive detection of biomolecules with fluorescent dye-doped nanoparticles. *Anal. Biochem.*, **334** (1), 135–144.

[24] Beck, J.S., Vartuli, J.C., Roth, W.J., *et al.* (1992) A new family of mesoporous molecular sieves prepared with liquid crystal templates. *J. Am. Chem. Soc.*, **114** (27), 10834–10843.

[25] Kresge, C.T., Leonowicz, M.E., Roth, W.J., *et al.* (1992) Ordered mesoporous molecular sieves synthesized by a liquid-crystal template mechanism. *Nature*, **359** (6397), 710–712.

[26] Beck, J.S., Vartuli, J.C., Kennedy, G.J., *et al.* (1994) Molecular or supramolecular templating: Defining the role of surfactant chemistry in the formation of microporous and mesoporous molecular sieves. *Chem. Mater.*, **6** (10), 1816–1821.

[27] Maleki, A., Kettiger, H., Schoubben, A., *et al.* (2017) Mesoporous silica materials: From physico-chemical properties to enhanced dissolution of poorly water-soluble drugs. *J. Controlled Release*, **262**, 329–347.

[28] Bagshaw, S.A., Prouzet, E., Pinnavaia, T.J. (1995) Templating of mesoporous molecular sieves by nonionic polyethylene oxide surfactants. *Science*, **269** (5228), 1242–1244.

[29] Inagaki, S., Fukushima, Y., Kuroda, K. (1993) Synthesis of highly ordered mesoporous materials from a layered polysilicate. *J. Chem. Soc. Chem. Commun.*, (8), 680.

[30] Zhao, D. (1998) Triblock copolymer syntheses of mesoporous silica with periodic 50 to 300 angstrom pores. *Science*, **279** (5350), 548–552.

[31] Hao, N., Li, L., Tang, F. (2016) Shape matters when engineering mesoporous silica-based nanomedicines. *Biomater. Sci.*, **4** (4), 575–591.

[32] Mai, W.X., Meng, H. (2013) Mesoporous silica nanoparticles: A multifunctional nano therapeutic system. *Integr Biol*, **5** (1), 19–28.

[33] Şen Karaman, D. (2016) *Physicochemical Characteristics of Silicananoparticles Tailored for Nanomedicine*. Åbo Akademi University, Finland. https://www.doria.fi/handle/10024/124250

[34] Şen Karaman, D., Sarparanta, M.P., Rosenholm, J.M., Airaksinen, A.J. (2018) Multimodality imaging of silica and silicon materials in vivo. *Advanced Materials*. DOI: 10.1002/adma.201703651

[35] Rosenholm, J.M., Vlasov, I.I., Burikov, S.A., *et al.* (2015) Nanodiamond-based composite structures for biomedical imaging and drug delivery. *J. Nanosci. Nanotechnol.*, **15** (2), 959–971.

[36] Şen Karaman, D., Gulin-Sarfraz, T., Zhang, J., Rosenholm, J.M. (2015) One-pot synthesis of pore-expanded hollow mesoporous silica particles. *Mater. Lett.*, **143**, 140–143.

[37] Rosenholm, J.M., Zhang, J., Sun, W., Gu, H. (2011) Large-pore mesoporous silica-coated magnetite core-shell nanocomposites and their relevance for biomedical applications. *Microporous Mesoporous Mater.*, **145** (1–3), 14–20.

[38] Fuji, M., Takai, C., Imabeppu, H., Xu, X. (2015) Synthesis and shell structure design of hollow silica nanoparticles using polyelectrolyte as template. *J. Phys. Conf. Ser.*, **596**, 12007.

[39] Bharti, C., Gulati, N., Nagaich, U., Pal, A. (2015) Mesoporous silica nanoparticles in target drug delivery system: A review. *Int. J. Pharm. Investig.*, **5** (3), 124.

[40] Bao, Y., Shi, C., Wang, T., *et al.* (2016) Recent progress in hollow silica: Template synthesis, morphologies and applications. *Microporous Mesoporous Mater.*, **227**, 121–136.

[41] Jiao, Y., Shen, S., Sun, Y., *et al.* (2015) A functionalized hollow mesoporous silica nanoparticles-based controlled dual-drug delivery system for improved tumor cell cytotoxicity. *Part. Part. Syst. Charact.*, **32** (2), 222–233.

[42] Palanikumar, L., Jeena, M.T., Kim, K., *et al.* (2017) Spatiotemporally and sequentially-controlled drug release from polymer gatekeeper–hollow silica nanoparticles. *Sci. Rep.*, **7**, 46540.

[43] Zhu, Y., Shi, J., Shen, W., *et al.* (2005) Stimuli-responsive controlled drug release from a hollow mesoporous silica sphere/polyelectrolyte multilayer core-shell structure. *Angew. Chem.*, **117** (32), 5213–5217.

[44] Tian, Y., Guo, R., Jiao, Y., *et al.* (2016) Redox stimuli-responsive hollow mesoporous silica nanocarriers for targeted drug delivery in cancer therapy. *Nanoscale Horiz.*, **1** (6), 480–487.

[45] Shi, S., Chen, F., Cai, W. (2013) Biomedical applications of functionalized hollow mesoporous silica nanoparticles: focusing on molecular imaging. *Nanomed.*, **8** (12), 2027–2039.

[46] Canham, L.T. (1995) Bioactive silicon structure fabrication through nanoetching techniques. *Adv. Mater.*, **7** (12), 1033–1037.

[47] Bayliss, S., Buckberry, L., Harris, P., Rousseau, C. (1997) Nanostructured semiconductors: compatibility with biomaterials. *Thin Solid Films*, **297** (1–2), 308–310.

[48] Canham, L. (2014) *Handbook of Porous Silicon*, Springer, New York.

[49] Sailor, M.J. (2011) *Porous Silicon in Practice: Preparation, Characterization and Applications*, Wiley-VCH Verlag GmbH & Co. KGaA, Weinheim, Germany.

[50] Santos, H.A., Mäkilä, E., Bimbo, L.M., *et al.* (2013) *Porous silicon nanoparticles*, in *Fundamentals of Pharmaceutical Nanoscience*, Springer Science+Business Media, New York, pp. 235–275.

[51] Salonen, J., Lehto, V.-P., Björkqvist, M., *et al.* (2000) Studies of thermally-carbonized porous silicon surfaces. *Phys. Status Solidi A*, **182** (1), 123–126.

[52] Bychto, L., Makushok, Y., Chirvony, V., Matveeva, E. (2008) Pulse electrochemical method for porosification of silicon and preparation of porSi powder with controllable particles size distribution. *Phys. Status Solidi C*, **5** (12), 3789–3793.

[53] Bimbo, L.M., Sarparanta, M., Santos, H.A., *et al.* (2010) Biocompatibility of thermally hydrocarbonized porous silicon nanoparticles and their biodistribution in rats. *ACS Nano*, **4** (6), 3023–3032.

[54] Nissinen, T., Ikonen, T., Lama, M., *et al.* (2016) Improved production efficiency of mesoporous silicon nanoparticles by pulsed electrochemical etching. *Powder Technol.*, **288**, 360–365.

[55] Lehmann, V. (2002) *Electrochemistry of Silicon*, Wiley-VCH Verlag GmbH, Weinheim.

[56] Riikonen, J., Salomäki, M., van Wonderen, J., *et al.* (2012) Surface chemistry, reactivity, and pore structure of porous silicon oxidized by various methods. *Langmuir*, **28** (28), 10573–10583.

[57] Salonen, J., Lehto, V.-P., Laine, E. (1997) Thermal oxidation of free-standing porous silicon films. *Appl. Phys. Lett.*, **70** (5), 637–639.

[58] Mawhinney, D.B., Glass, J.A., Yates, J.T. (1997) FTIR study of the oxidation of porous silicon. *J. Phys. Chem. B*, **101** (7), 1202–1206.

[59] Jalkanen, T., Torres-Costa, V., Mäkilä, E., *et al.* (2014) Selective optical response of hydrolytically stable stratified Si rugate mirrors to liquid infiltration. *ACS Appl. Mater. Interfaces*, **6** (4), 2884–2892.

[60] Godin, B., Gu, J., Serda, R.E., *et al.* (2010) Tailoring the degradation kinetics of mesoporous silicon structures through PEGylation. *J. Biomed. Mater. Res. A*, **94** (4), 1236–1243.

[61] Seo, Y.J., Cheon, H.J., Choi, D.J. (1998) Enhancement of the thermal stability of photoluminescence by the carbonization of porous silicon. *Journal of Materials Science Letters*, **17** (4), 313–315.

[62] Paski, J., Björkqvist, M., Salonen, J., Lehto, V.-P. (2005) Effects of treatment temperature on thermally-carbonized porous silicon hygroscopicity. *Phys. Status Solidi C*, **2** (9), 3379–3383.

[63] Salonen, J., Laine, E., Niinistö, L. (2002) Thermal carbonization of porous silicon surface by acetylene. *J. Appl. Phys.*, **91** (1), 456.

[64] Salonen, J., Björkqvist, M., Laine, E., Niinistö, L. (2004) Stabilization of porous silicon surface by thermal decomposition of acetylene. *Appl. Surf. Sci.*, **225** (1–4), 389–394.

[65] Cheng, C.C., Taylor, P.A., Wallace, R.M., *et al.* (1993) Hydrocarbon surface chemistry on Si(100). *Thin Solid Films*, **225** (1–2), 196–202.

[66] Sarparanta, M., Mäkilä, E., Heikkilä, T., *et al.* (2011) ^{18}F-Labeled modified porous silicon particles for investigation of drug delivery carrier distribution in vivo with positron emission tomography. *Mol. Pharm.*, **8** (5), 1799–1806.

[67] Janshoff, A., Dancil, K.-P.S., Steinem, C., *et al.* (1998) Macroporous p-type silicon Fabry−Perot layers. Fabrication, characterization, and applications in biosensing. *J. Am. Chem. Soc.*, **120** (46), 12108–12116.

[68] Tinsley-Bown, A.M., Canham, L.T., Hollings, M., *et al.* (2000) Tuning the pore size and surface chemistry of porous silicon for immunoassays. *Phys. Status Solidi A*, **182** (1), 547–553.

[69] Arroyohernandez, M. (2003) Biofunctionalization of surfaces of nanostructured porous silicon. *Mater. Sci. Eng. C*, **23** (6–8), 697–701.

[70] Mäkilä, E., Bimbo, L.M., Kaasalainen, M., *et al.* (2012) Amine modification of thermally carbonized porous silicon with silane coupling chemistry. *Langmuir*, **28** (39), 14045–14054.

[71] Kovalainen, M., Mönkäre, J., Mäkilä, E., *et al.* (2012) Mesoporous silicon (PSi) for sustained peptide delivery: Effect of PSi microparticle surface chemistry on peptide YY3-36 release. *Pharm. Res.*, **29** (3), 837–846.

[72] Jalkanen, T., Mäkilä, E., Suzuki, Y.-I., *et al.* (2012) Studies on chemical modification of porous silicon-based graded-index optical microcavities for improved stability under alkaline conditions. *Adv. Funct. Mater.*, **22** (18), 3890–3898.

[73] Martin-Palma, R.J., Torres-Costa, V. (2014) *Microscopy of porous silicon*, in *Handbook of Porous Silicon*, Springer International Publishing Switzerland, Cham, pp. 413–421.

[74] Anovitz, L.M., Cole, D.R. (2015) Characterization and analysis of porosity and pore structures. *Rev. Mineral. Geochem.*, **80** (1), 61–164.

[75] Brunauer, S., Emmett, P.H., Teller, E. (1938) Adsorption of gases in multimolecular layers. *J. Am. Chem. Soc.*, **60** (2), 309–319.

[76] Rouquerol, F., Rouquerol, J., Sing, K.S.W. (1999) *Adsorption by Powders and Porous Solids: Principles, Methodology, and Applications*, Academic Press, San Diego.

[77] Barrett, E.P., Joyner, L.G., Halenda, P.P. (1951) The determination of pore volume and area distributions in porous substances. *I. Computations from nitrogen isotherms. J. Am. Chem. Soc.*, **73** (1), 373–380.

[78] Rosenholm, J.M., Penninkangas, A., Lindén, M. (2006) Amino-functionalization of large-pore mesoscopically ordered silica by a one-step hyperbranching polymerization of a surface-grown polyethyleneimine. *Chem Commun, Oct.* **1**, 3909–3911.

[79] Ishii, Y., Nishiwaki, Y., Al-Zubaidi, A., Kawasaki, S. (2013) Pore size determination in ordered mesoporous materials using powder X-ray diffraction. *J. Phys. Chem. C*, **117** (35), 18120–18130.

[80] Feuston, B.P., Higgins, J.B. (1994) Model structures for MCM-41 materials: A molecular dynamics simulation. *J. Phys. Chem.*, **98** (16), 4459–4462.

[81] Solovyov, L.A., Belousov, O.V., Dinnebier, R.E., *et al.* (2005) X-ray diffraction structure analysis of MCM-48 mesoporous silica. *J. Phys. Chem. B*, **109** (8), 3233–3237.

[82] Impéror-Clerc, M., Davidson, P., Davidson, A. (2000) Existence of a microporous corona around the mesopores of silica-based SBA-15 materials templated by triblock copolymers. *J. Am. Chem. Soc.*, **122** (48), 11925–11933.

[83] Şen Karaman, D., Desai, D., Zhang, J., *et al.* (2016) Modulation of the structural properties of mesoporous silica nanoparticles to enhance the T1-weighted MR imaging capability. *J. Mater. Chem. B*, **4** (9), 1720–1732.

[84] Desai, D., Karaman, D.S., Prabhakar, N., *et al.* (2014) Design considerations for mesoporous silica nanoparticulate systems in facilitating biomedical applications. *Mesoporous Biomater.*, **1** (1).

[85] Baer, D.R., Engelhard, M.H., Johnson, G.E., *et al.* (2013) Surface characterization of nanomaterials and nanoparticles: Important needs and challenging opportunities. *J. Vac. Sci. Technol. Vac. Surf. Films*, **31** (5), 50820.

[86] Kamiya, H., Iijima, M. (2010) Surface modification and characterization for dispersion stability of inorganic nanometer-scaled particles in liquid media. *Sci. Technol. Adv. Mater.*, **11** (4), 44304.

[87] Benayad, A., Hajjaji, H., Coustier, F., *et al.* (2016) Surface chemical-bonds analysis of silicon particles from diamond-wire cutting of crystalline silicon. *J. Appl. Phys.*, **120** (23), 235308.

[88] Mijatovic, J., Binder, W.H., Gruber, H. (2000) Characterization of surface modified silica nanoparticles by ^{29}Si solid state NMR spectroscopy. *Microchim. Acta*, **133** (1–4), 175–181.

[89] Moreels, I., Martins, J., Hens, Z. (2007) Solution NMR techniques for investigating colloidal nanocrystal ligands: A case study on trioctylphosphine oxide at InP quantum dots. *Sens. Actuators B Chem.*, **126** (1), 283–288.

[90] Lehman, S.E., Tataurova, Y., Mueller, P.S., *et al.* (2014) Ligand characterization of covalently functionalized mesoporous silica nanoparticles: An NMR toolbox approach. *J. Phys. Chem. C*, **118** (51), 29943–29951.

[91] Helle, M., Rampazzo, E., Monchanin, M., *et al.* (2013) Surface chemistry architecture of silica nanoparticles determine the efficiency of in vivo fluorescence lymph node mapping. *ACS Nano*, **7** (10), 8645–8657.

[92] Zheng, X., Zhang, J., Wang, J., *et al.* (2015) Polydopamine coatings in confined nanopore space: Toward improved retention and release of hydrophilic cargo. *J. Phys. Chem. C*, **119** (43), 24512–24521.

[93] Zhang, J., Desai, D., Rosenholm, J.M. (2014) Tethered lipid bilayer gates: Toward extended retention of hydrophilic cargo in porous nanocarriers. *Adv. Funct. Mater.*, **24** (16), 2352–2360.

[94] Israelachvili, J.N. (2011) *Intermolecular and Surface Forces*, Elsevier, Acad. Press, Amsterdam.

[95] Hunter, R.J. (2001) *Foundations of Colloid Science*, Oxford University Press, Oxford.

[96] Kosmulski, M. (ed.) (2009) *Surface Charging and Points of Zero Charge*, CRC Press, Boca Raton, FL.

[97] Shaw, D.J. (1992) *Introduction to Colloid and Surface Chemistry*, Butterworth-Heinemann, Oxford.

[98] Jones, R.A.L. (2002) *Soft Condensed Matter*, Oxford University Press, Oxford.

[99] Hassan, P.A., Rana, S., Verma, G. (2015) Making sense of Brownian motion: Colloid characterization by dynamic light scattering. *Langmuir*, **31** (1), 3–12.

[100] Xu, R.L. (2000) *Particle Characterization: Light Scattering Methods*, Springer, Dordrecht.

[101] Florence, A.T., Attwood, D. (2007) *Physicochemical Principles of Pharmacy*, Pharmaceutical Press, London.

[102] Zadaka, D., Radian, A., Mishael, Y.G. (2010) Applying zeta potential measurements to characterize the adsorption on montmorillonite of organic cations as monomers, micelles, or polymers. *J. Colloid Interface Sci.*, **352** (1), 171–177.

[103] Verwey, E.J.W. (1947) Theory of the stability of lyophobic colloids. *J. Phys. Colloid Chem.*, **51** (3), 631–636.

[104] Martinez, D.S.T., Paula, A.J., Fonseca, L.C., *et al.* (2015) Monitoring the hemolytic effect of mesoporous silica nanoparticles after human blood protein corona formation: Hemolytic effect of mesoporous silica nanoparticles. *Eur. J. Inorg. Chem.*, **2015** (27), 4595–4602.

[105] Nichols, G., Byard, S., Bloxham, M.J., *et al.* (2002) A review of the terms agglomerate and aggregate with a recommendation for nomenclature used in powder and particle characterization. *J. Pharm. Sci.*, **91** (10), 2103–2109.

[106] Kaasalainen, M., Aseyev, V., von Haartman, E., *et al.* (2017) Size, stability, and porosity of mesoporous nanoparticles characterized with light scattering. *Nanoscale Res. Lett.*, **12** (1).

[107] Kätzel, U., Bedrich, R., Stintz, M., *et al.* (2008) Dynamic light scattering for the characterization of polydisperse fractal systems: I. Simulation of the diffusional behavior. *Part. Part. Syst. Charact.*, **25** (1), 9–18.

[108] Kätzel, U., Vorbau, M., Stintz, M., *et al.* (2008) Dynamic light scattering for the characterization of polydisperse fractal systems: II. Relation between structure and DLS results. *Part. Part. Syst. Charact.*, **25** (1), 19–30.

[109] Rosenholm, J.M., Lindén, M. (2008) Towards establishing structure–activity relationships for mesoporous silica in drug delivery applications. *J. Controlled Release*, **128** (2), 157–164.

[110] Sanderg, T., Weinberger, C., Sen Karaman, D., Rosenholm, J.M. (2017) Modeling of a hybrid Langmuir adsorption isotherm for describing interactions between drug molecules and silica surfaces. *Journal of Pharmaceutical Sciences*, **107** (5), 1392–1397.

[111] Rosenholm, J. (2002) *Synthesis and characterization of ordered micro- and mesoporous silica materials prepared by supramolecular templating and their application as carrier matrixes for controlled drug release*. MSc dissertation, Åbo Akademi University.

[112] Sandberg, T., Rosenholm, J., Hotokka, M. (2008) The molecular structure of disulfiram and its complexation with silica. A quantum chemical study. *J. Mol. Struct. THEOCHEM*, **861** (1–3), 57–61.

[113] Henriksen-Lacey, M., Carregal-Romero, S., Liz-Marzán, L.M. (2017) Current challenges toward in vitro cellular validation of inorganic nanoparticles. *Bioconjug. Chem.*, **28** (1), 212–221.

[114] Nieto, A., Colilla, M., Balas, F., Vallet-Regí, M. (2010) Surface electrochemistry of mesoporous silicas as a key factor in the design of tailored delivery devices. *Langmuir*, **26** (7), 5038–5049.

[115] Anglin, E., Cheng, L., Freeman, W., Sailor, M. (2008) Porous silicon in drug delivery devices and materials. *Adv. Drug Deliv. Rev.*, **60** (11), 1266–1277.

[116] Godin, B., Tasciotti, E., Liu, X., *et al.* (2011) Multistage nanovectors: From concept to novel imaging contrast agents and therapeutics. *Acc. Chem. Res.*, **44** (10), 979–989.

[117] Huotari, A., Xu, W., Mönkäre, J., *et al.* (2013) Effect of surface chemistry of porous silicon microparticles on glucagon-like peptide-1 (GLP-1) loading, release and biological activity. *Int. J. Pharm.*, **454** (1), 67–73.

[118] Park, J.-H., Gu, L., von Maltzahn, G., *et al.* (2009) Biodegradable luminescent porous silicon nanoparticles for in vivo applications. *Nat. Mater.*, **8** (4), 331–336.

[119] Rytkönen, J., Arukuusk, P., Xu, W., *et al.* (2014) Porous silicon–cell penetrating peptide hybrid nanocarrier for intracellular delivery of oligonucleotides. *Mol. Pharm.*, **11** (2), 382–390.

[120] Kovalainen, M., Monkare, J., Riikonen, J., *et al.* (2015) Novel delivery systems for improving the clinical use of peptides. *Pharmacol. Rev.*, **67** (3), 541–561.

[121] Jarvis, K.L., Barnes, T.J., Prestidge, C.A. (2010) Thermal oxidation for controlling protein interactions with porous silicon. *Langmuir*, **26** (17), 14316–14322.

[122] Lee, C.-H., Cheng, S.-H., Wang, Y.-J., *et al.* (2009) Near-infrared mesoporous silica nanoparticles for optical imaging: Characterization and in vivo biodistribution. *Adv. Funct. Mater.*, **19** (2), 215–222.

[123] Sjöback, R., Nygren, J., Kubista, M. (1995) Absorption and fluorescence properties of fluorescein. *Spectrochim. Acta. A. Mol. Biomol. Spectrosc.*, **51** (6), L7–L21.

[124] Zhu, H., Derksen, R.C., Krause, C.R., *et al.* (2005) Fluorescent intensity of dye solutions under different pH conditions. *Journal of ASTM International*, **2**, 1–7.

[125] Moyano, D.F., Goldsmith, M., Solfiell, D.J., *et al.* (2012) Nanoparticle hydrophobicity dictates immune response. *J. Am. Chem. Soc.*, **134** (9), 3965–3967.
[126] Walkey, C.D., Chan, W.C.W. (2012) Understanding and controlling the interaction of nanomaterials with proteins in a physiological environment. *Chem Soc Rev*, **41** (7), 2780–2799.
[127] Krug, H.F., Wick, P. (2011) Nanotoxicology: An interdisciplinary challenge. *Angew. Chem. Int. Ed.*, **50** (6), 1260–1278.
[128] Croissant, J.G., Fatieiev, Y., Khashab, N.M. (2017) Degradability and clearance of silicon, organosilica, silsesquioxane, silica mixed oxide, and mesoporous silica nanoparticles. *Adv. Mater.*, **29** (9), 1604634.
[129] Vert, M., Doi, Y., Hellwich, K.-H., *et al.* (2012) Terminology for biorelated polymers and applications (IUPAC Recommendations 2012). *Pure Appl. Chem.*, **84** (2).
[130] Gin, S., Jollivet, P., Fournier, M., *et al.* (2015) Origin and consequences of silicate glass passivation by surface layers. *Nat. Commun.*, **6**, 6360.
[131] Bunker, B.C. (1994) Molecular mechanisms for corrosion of silica and silicate glasses. *J. Non-Cryst. Solids*, **179**, 300–308.
[132] Dove, P.M., Han, N., Wallace, A.F., De Yoreo, J.J. (2008) Kinetics of amorphous silica dissolution and the paradox of the silica polymorphs. *Proc. Natl. Acad. Sci.*, **105** (29), 9903–9908.
[133] Cauda, V., Schlossbauer, A., Bein, T. (2010) Bio-degradation study of colloidal mesoporous silica nanoparticles: Effect of surface functionalization with organo-silanes and poly(ethylene glycol). *Microporous Mesoporous Mater.*, **132** (1–2), 60–71.
[134] Etienne, M. (2003) Analytical investigation of the chemical reactivity and stability of aminopropyl-grafted silica in aqueous medium. *Talanta*, **59** (6), 1173–1188.
[135] Chen, K., Zhang, J., Gu, H. (2012) Dissolution from inside: a unique degradation behaviour of core–shell magnetic mesoporous silica nanoparticles and the effect of polyethyleneimine coating. *J. Mater. Chem.*, **22** (41), 22005.
[136] Choi, Y., Lee, J.E., Lee, J.H., *et al.* (2015) A biodegradation study of SBA-15 microparticles in simulated body fluid and *in vivo*. *Langmuir*, **31** (23), 6457–6462.
[137] Iler, R.K. (1979) *The Chemistry of Silica: Solubility, Polymerization, Colloid and Surface Properties, and Biochemistry*, Wiley, New York.
[138] Cypryk, M., Apeloig, Y. (2002) Mechanism of the acid-catalyzed Si–O bond cleavage in siloxanes and siloxanols. A theoretical study. *Organometallics*, **21** (11), 2165–2175.
[139] Maggini, L., Cabrera, I., Ruiz-Carretero, A., *et al.* (2016) Breakable mesoporous silica nanoparticles for targeted drug delivery. *Nanoscale*, **8** (13), 7240–7247.
[140] Zhou, X., Chen, L., Wang, W., *et al.* (2015) Electrospun nanofibers incorporating self-decomposable silica nanoparticles as carriers for controlled delivery of anticancer drug. *RSC Adv.*, **5** (81), 65897–65904.
[141] Lin, Y.-S., Abadeer, N., Hurley, K.R., Haynes, C.L. (2011) Ultrastable, redispersible, small, and highly organomodified mesoporous silica nanotherapeutics. *J. Am. Chem. Soc.*, **133** (50), 20444–20457.
[142] Chen, G., Teng, Z., Su, X., *et al.* (2015) Unique biological degradation behavior of Stöber mesoporous silica nanoparticles from their interiors to their exteriors. *J. Biomed. Nanotechnol.*, **11** (4), 722–729.
[143] Kettiger, H., Sen Karaman, D., Schiesser, L., *et al.* (2015) Comparative safety evaluation of silica-based particles. *Toxicol. In Vitro*, **30** (1), 355–363.
[144] Xie, G., Sun, J., Zhong, G., *et al.* (2010) Biodistribution and toxicity of intravenously administered silica nanoparticles in mice. *Arch. Toxicol.*, **84** (3), 183–190.
[145] Mamaeva, V., Sahlgren, C., Lindén, M. (2013) Mesoporous silica nanoparticles in medicine –Recent advances. *Adv. Drug Deliv. Rev.*, **65** (5), 689–702.

[146] Lu, J., Liong, M., Li, Z., *et al.* (2010) Biocompatibility, biodistribution, and drug-delivery efficiency of mesoporous silica nanoparticles for cancer therapy in animals. *Small*, **6** (16), 1794–1805.

[147] Finnie, K.S., Waller, D.J., Perret, F.L., *et al.* (2009) Biodegradability of sol-gel silica microparticles for drug delivery. *J. Sol-Gel Sci. Technol.*, **49** (1), 12–18.

[148] Rosenholm, J., Desai, D., Prabhakar, N., *et al.* (2016) Targeted modulation of cell differentiation in distinct regions of the gastrointestinal tract via oral administration of differently PEG-PEI functionalized mesoporous silica nanoparticles. *Int. J. Nanomedicine*, 299.

[149] Mamaeva, V., Rosenholm, J.M., Bate-Eya, L.T., *et al.* (2011) Mesoporous silica nanoparticles as drug delivery systems for targeted inhibition of notch signaling in cancer. *Mol. Ther.*, **19** (8), 1538–1546.

[150] Soo Choi, H., Liu, W., Misra, P., *et al.* (2007) Renal clearance of quantum dots. *Nat. Biotechnol.*, **25** (10), 1165–1170.

[151] Benezra, M., Penate-Medina, O., Zanzonico, P.B., *et al.* (2011) Multimodal silica nanoparticles are effective cancer-targeted probes in a model of human melanoma. *J. Clin. Invest.*, **121** (7), 2768–2780.

[152] International Organization for Standardization (2018) *ISO 10993-1:2018 Biological evaluation of medical devices - Part 1: Evaluation and testing within a risk management process.* https://www.iso.org/standard/68936.html

[153] Yu, T., Malugin, A., Ghandehari, H. (2011) Impact of silica nanoparticle design on cellular toxicity and hemolytic activity. *ACS Nano*, **5** (7), 5717–5728.

[154] Ha, S.-W., Camalier, C.E., Weitzmann, M.N., *et al.* (2013) Long-term monitoring of the physicochemical properties of silica-based nanoparticles on the rate of endocytosis and exocytosis and consequences of cell division. *Soft Mater.*, **11** (2), 195–203.

[155] Schmoldt, A., Benthe, H.F., Haberland, G. (1975) Digitoxin metabolism by rat liver microsomes. *Biochem. Pharmacol.*, **24** (17), 1639–1641.

[156] Yu, T., Hubbard, D., Ray, A., Ghandehari, H. (2012) In vivo biodistribution and pharmacokinetics of silica nanoparticles as a function of geometry, porosity and surface characteristics. *J. Controlled Release*, **163** (1), 46–54.

[157] Chan, W.-T., Liu, C.-C., Chiang Chiau, J.-S., *et al.* (2017) In vivo toxicologic study of larger silica nanoparticles in mice. *Int. J. Nanomedicine*, **12**, 3421–3432.

[158] Ahnfelt, E., Sjögren, E., Axén, N., Lennernäs, H. (2015) A miniaturized in vitro release method for investigating drug-release mechanisms. *Int. J. Pharm.*, **486** (1–2), 339–349.

[159] Andersson, J., Rosenholm, J., Areva, S., Lindén, M. (2004) Influences of material characteristics on ibuprofen drug loading and release profiles from ordered micro- and mesoporous silica matrices. *Chem. Mater.*, **16** (21), 4160–4167.

[160] von Haartman, E., Lindberg, D., Prabhakar, N., Rosenholm, J.M. (2016) On the intracellular release mechanism of hydrophobic cargo and its relation to the biodegradation behavior of mesoporous silica nanocarriers. *Eur. J. Pharm. Sci.*, **95**, 17–27.

[161] Gulin-Sarfraz, T., Sarfraz, J., Didem Şen Karaman, D.Ş.K., *et al.* (2014) FRET-reporter nanoparticles to monitor redox-induced intracellular delivery of active compounds. *RSC Adv*, **4** (32), 16429–16437.

[162] Prabhakar, N., Zhang, J., Desai, D., *et al.* (2016) Stimuli-responsive hybrid nanocarriers developed by controllable integration of hyperbranched PEI with mesoporous silica nanoparticles for sustained intracellular siRNA delivery. *Int. J. Nanomedicine*, **11**, 6591–6608.

[163] Suutari, T., Silen, T., Sen Karaman, D., *et al.* (2016) Real-time label-free monitoring of nanoparticle cell uptake. *Small*, **12** (45), 6289–6300.

[164] Bergman, L., Rosenholm, J., Öst, A.-B., *et al.* (2008) On the complexity of electrostatic suspension stabilization of functionalized silica nanoparticles for biotargeting and imaging applications. *J. Nanomater.*, **2008**, 1–9.

[165] Hug, T.S., Prenosil, J.E., Morbidelli, M. (2001) Optical waveguide lightmode spectroscopy as a new method to study adhesion of anchorage-dependent cells as an indicator of metabolic state. *Biosens. Bioelectron.*, **16** (9–12), 865–874.

[166] Fang, Y., Ferrie, A.M., Fontaine, N.H., *et al.* (2006) Resonant waveguide grating biosensor for living cell sensing. *Biophys. J.*, **91** (5), 1925–1940.

[167] Möhrle, B.P., Köhler, K., Jaehrling, J., *et al.* (2005) Label-free characterization of cell adhesion using reflectometric interference spectroscopy (RIfS). *Anal. Bioanal. Chem.*, **384** (2), 407–413.

[168] Hide, M., Tsutsui, T., Sato, H., *et al.* (2002) Real-time analysis of ligand-induced cell surface and intracellular reactions of living mast cells using a surface plasmon resonance-based biosensor. *Anal. Biochem.*, **302** (1), 28–37.

[169] Xiao, C., Lachance, B., Sunahara, G., Luong, J.H.T. (2002) An in-depth analysis of electric cell–substrate impedance sensing to study the attachment and spreading of mammalian cells. *Anal. Chem.*, **74** (6), 1333–1339.

[170] Fredriksson, C., Kihlman, S., Rodahl, M., Kasemo, B. (1998) The piezoelectric quartz crystal mass and dissipation sensor: A means of studying cell adhesion. *Langmuir*, **14** (2), 248–251.

[171] Peña, M.D.P.S., Gottipati, A., Tahiliani, S., *et al.* (2016) Hyperspectral imaging of nanoparticles in biological samples: Simultaneous visualization and elemental identification: Hyperspectral mapping in biological samples. *Microsc. Res. Tech.*, **79** (5), 349–358.

[172] Fussell, A.L., Mah, P.T., Offerhaus, H., *et al.* (2014) Coherent anti-Stokes Raman scattering microscopy driving the future of loaded mesoporous silica imaging. *Acta Biomater.*, **10** (11), 4870–4877.

[173] Tian, Y., Kong, Y., Li, X., *et al.* (2015) Light- and pH-activated intracellular drug release from polymeric mesoporous silica nanoparticles. *Colloids Surf. B Biointerfaces*, **134**, 147–155.

[174] Tolstik, E., Osminkina, L., Akimov, D., *et al.* (2016) Linear and non-linear optical imaging of cancer cells with silicon nanoparticles. *Int. J. Mol. Sci.*, **17** (9), 1536.

10

Statistical Analysis and Multidimensional Modeling in Research

Osmo Antikainen

Faculty of Pharmacy, University of Helsinki, Finland

10.1 Measurement in Research

Measurement is an integral part of modern science, as well as of engineering and commerce. We usually measure how variables (factors) affect the properties of the system or process we are investigating. We receive knowledge about nature by doing measurements and then generalize the data by making models. In science, we always make repeatable measurements. One-time measurements do not belong to science. All measurements always incorporate some variability. This is not only due to the limits of technology in measuring devices. Even the most careful scientific measurement cannot yield an exact value. Repeating a measurement yields a scatter of measurements distributed around some mean value. This scatter is not only caused by small errors in the measurement device, but also by natural variability. Measurements themselves, independent of any human or instrument inaccuracies, always exhibit scatter.

10.2 Mean and Sample Mean

"Mean" usually refers to the population mean. This is the mean of the entire population of a set. Quite often in scientific research, it is not, however, possible to measure every

Characterization of Pharmaceutical Nano- and Microsystems, First Edition. Edited by Leena Peltonen.

individual member of a set. It is only possible to measure a smaller sample from the set. The mean of the sample group is called the sample mean.

If we want to measure some quality attribute of our product, we need to take note that every measurement is influenced by random noise. That is why the sample mean is expressed in the form of:

$$\text{Estimate of the mean} \pm \text{estimate of the error} \tag{10.1}$$

Case 1

We have compressed a batch of 100 000 vitamin D3 tablets.

We want to analyze the average content of vitamin D3 in those tablets. We first take a sample of ten tablets (Table 10.1) and analyze the vitamin content.

Table 10.1 *Vitamin content of ten randomly selected tablets*

Tablet number	Vitamin D3 content (μg)
1	18.7
2	21.1
3	21.5
4	22.6
5	21.1
6	20.8
7	20.0
8	17.6
9	19.7
10	20.6

Every tablet in the batch is slightly different. If we want to know the real mean value of the vitamin content of tablets in this batch, we would need to analyze every tablet. In this example we get an estimate of the average $\mu = 20.4$ μg and an estimate of the standard deviation $\sigma = 1.43$ μg. If we had measured one extra tablet, the estimate of the average and the estimate of the standard deviation would have changed a little bit. How accurate then is our estimate? We can calculate the 95% confidence interval for the average $\overline{X}$ by using Equation (10.2):

$$\overline{X} = \mu \pm z\frac{\sigma}{\sqrt{n}} \tag{10.2}$$

where n is the sample size and $z = 1.96$, when the 95% confidence interval is calculated ($z = 2.58$ for 99% confidence interval, and $z = 3.29$ for 99.9% confidence interval) [1].

The real average tablet weight in this example batch is, with 95% certainty, between 19.5–21.3 mg. If we need to know the real average with 1 μg accuracy (meaning ±0.5 μg), how many tablets would we need to measure?

To get the answer, n needs to be calculated based on Equation (10.2):

$$n = \left(\frac{1.96 \cdot \sigma}{0.5\ \mu g} \right)^2 = \left(\frac{1.96 \cdot 1.43\ \mu g}{0.5\ \mu g} \right)^2 = 55.9$$

Accordingly, we would need to measure 56 tablets, with the assumption that the standard deviation of the sample is 1.43 μg and the content of vitamin D3 in tablets is normally distributed.

Using the equation above, it is possible to estimate how many times measurements need to be repeated in order to get a result with the desired accuracy.

10.3 Correlation

Correlation is a commonly used statistical technique that can show whether and how strongly two measured variables are related. For example, tablet breaking force and tablet compression pressure are related; tablets compressed with higher compression pressure tend to have higher crushing strengths than tablets compressed with low compression pressures. Or, in nanomilling, a longer milling duration tends to lead to smaller particle sizes [2]. However, the relationship is not perfect. Crushing strengths of tablets compressed with the same pressure vary to some extent. Nonetheless, the average crushing strength is higher with tablets compressed with higher compression pressure [3]. Correlation can tell you just how much of the variation in compression pressure is related to the variation of crushing strengths of tablets. If two variables are correlated, it is possible to estimate the value of other variable just by knowing the value of the first variable. The better the two variables correlate, the more accurately it is possible to estimate the value of the other variable.

Pearson's correlation coefficient (r) is a measure of the linear correlation between two variables x and y. It can have values between +1 and −1. A correlation coefficient value of 1 means a total positive linear correlation; a value of 0 or close to zero that there is no correlation between variables; and a value of −1 that a total negative linear correlation exists. Pearson's correlation coefficient is widely used in the sciences.

Correlations are commonly illustrated using scatterplots. In Figure 10.1 four different scatterplots are presented. In all the four plots there is the same correlation coefficient ($r = 0.84$) between the two variables. Only in Figure 10.1a is Pearson's correlation coefficient used correctly. In Figure 10.1b there is one outlier in the dataset, which is easy to see from the scatterplot. The real correlation between variables is for that reason higher than $r = 0.84$. In Figure 10.1c, variables are correlated nonlinearly. However, Pearson's correlation coefficient can describe only linear correlation. In Figure 10.1d the data are not distributed evenly over whole detection area. The correlation coefficient gives too high values, if data is settled down in two groups. It is always possible to draw a straight line precisely between two points.

When the correlation coefficient between two variables is high enough, can we say that the correlation is real (statistically significant) and not coincidence? We can test that by calculating Student's t-value with $N - 2$ degrees of freedom (N is the number of measurement

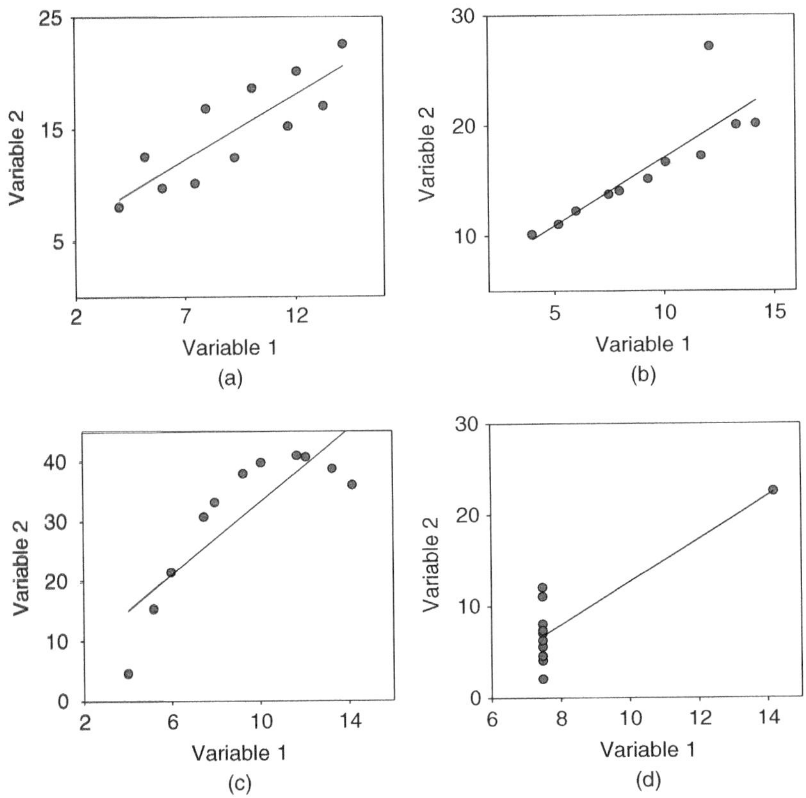

Figure 10.1 *Four different scatterplots between two variables. Pearson's correlation coefficient between the two variables is the same in all the pictures, r = 0.84, but it is only correct in (a), as explained in the text.*

points) using Equation (10.3) [4].

$$t_{N-2} = \frac{|r \cdot \sqrt{N-2}|}{\sqrt{1-r^2}} \tag{10.3}$$

Case 2

Is there a statistically significant correlation between dissolution time (time to release 80% of active substance) and *in vivo* test results (time to concentration peak value, C_{max}) (Table 10.2)?

Pearson's correlation coefficient value, r, is 0.77. You can calculate it in Microsoft Excel using command PEARSON(array1, array2). Student's t-value with $N-2$ degrees of freedom can be calculated based on Equation (10.3) to be:

$$t_{N-2} = \frac{|r \cdot \sqrt{N-2}|}{\sqrt{1-r^2}} = \frac{|0.77 \cdot \sqrt{10-2}|}{\sqrt{1-0.77}} = 4.6$$

Table 10.2 Results of dissolution tests (time to 80% drug release) and C_{max} values from in vivo tests of ten different products: positive correlation could be found

Dissolution (min)	C_{max} *in vivo* (h)
5	1.0
8	1.6
4	0.9
9	1.4
20	2.0
8	1.2
3	0.8
6	0.4
14	1.6
9	0.7

Next, we take the two-tailed inverse of Student's t-distribution. You can get it from statistics texts [4] or Excel using command TINV(probability,deg_freedom). In this case, TINV(0.05, 10-2) gives the value 2.3.

Because this value is smaller than the calculated t-value (4.6), we can say that the correlation is statistically significant with 95% certainty. If we require 99.9% certainty we would get TINV(0.001, 8) value of 5.0, which is higher than t-value 4.6. In this case, we would conclude that the correlation is not significant at the 99.9% certainty level.

It is possible that the correlation coefficient is low, but correlation is still significant. For small samples, strong correlations can be produced by chance and it is important to calculate significance of the correlation in those cases to avoid incorrect conclusions. It is good to remember that correlation does not imply that there is a cause-and-effect relationship between datasets that correlate. For example, the rate of deaths by drowning and ice cream consumption are significantly correlated, but there is no causation between them. Correlation is a good tool to screen possible causation between different variables, but alone it does not reveal whether there is a real cause-and-effect relationship between correlating variables. However, if there is real cause-and-effect relationship between two variables (that is, when we understand its mechanism), there is always correlation [5].

10.4 Modeling Relationships Between Series of Observations

When two series of observations have been made, and we want to test whether there is a relationship between them, it is good first to construct a scatterplot from the data. Usually we put the observed classification of the dataset on the x-axis. This can be, for example, time, pressure, etc., and it is called the "factor" or "independent variable". The other dataset consists of experimental results, and it is usual to put them on the vertical axis (y-axis). These represent what is called the "dependent variable" or "response". The relationship between x and y can be represented by a simple equation called the regression equation:

$$y = f(x) \tag{10.4}$$

This means that the value of y is a "function" of x, that is, it changes with x.

The regression equation, representing how much y changes with any given change of x, can be used to construct a regression equation on a scatter diagram, and in the simplest case this is a straight line given by Equation (10.5):

$$y = \alpha \cdot x + \text{constant} \tag{10.5}$$

Given that the association between factor and response is well described by a straight line, we must define the two numerical constants of the line equation if we are to place the straight line in the best possible way on the scatterplot. The first of these is a slope (α) and the second one is a *constant* that determines distance above the baseline. The direction in which the line slopes depends on whether the correlation is positive or negative. When the two sets of observations increase or decrease together (positive) the line slopes upwards from left to right; when one set decreases as the other increases, the line slopes downwards from left to right [6–8]. As the equation is a straight line, it will probably pass through only a few, if any, of the data points in the scatterplot. To fit the line over the scatterplot in the best possible way, the method of least squares is commonly used [9]. In this method the parameters of a model function are adjusted to give the best fit to the dataset.

10.5 Quality of a Model

The most common way to estimate the "quality" of a model is to calculate the squared correlation coefficient, or R^2 value (also written r^2). This is a numerical value between 0 and 1 and it indicates the proportion of the variation in the dependent variable that is explained by the regression equation.

10.5.1 The Meaning of R^2 in Linear Regression

As mentioned before, R^2 (or r^2) is a fraction between 0.0 and 1.0, and it has no unit. An r^2 value of 0.0 means that knowing x does not help you predict y. There is no linear relationship between x and y, and the best-fit line is a horizontal line going through the mean of all y values (Figure 10.2). When r^2 equals 1.0, all points lie exactly on a straight line with no

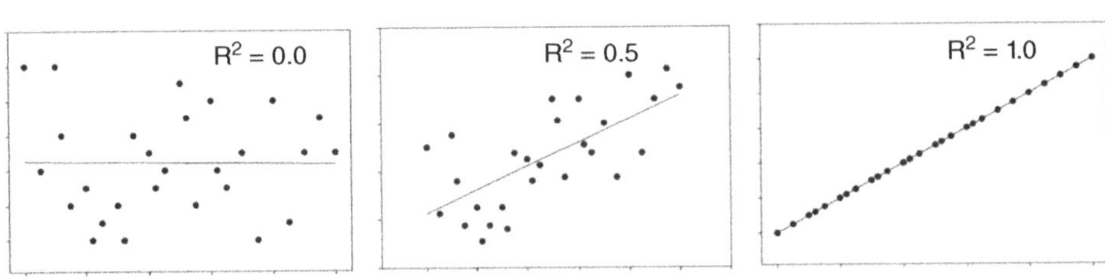

Figure 10.2 Examples of different R^2 values in linear regression.

scatter. Knowing x lets you predict y perfectly. Unfortunately, a high R^2 value does not guarantee that the model can predict response values well in all cases.

10.5.2 Cross-validation

Cross-validation methods provide a way to try to overcome some of the problems inherent in the use of the R^2 value alone.

Cross-validation involves the removal of some of the values from the dataset; the model is then fitted with the remaining data. This model is then used to predict the values of the data that have been removed. The simplest form of cross-validation is the "leave-one-out" approach, where just a single data value is removed. Repeating this process for every value in the dataset leads to a cross-validated Q^2 value [10]:

$$Q^2 = 1 - \frac{\sum_{i=1}^{N} (y_i - y_{\text{pred,i}})^2}{\sum_{i=1}^{N} (y_i - \langle y \rangle)^2} \tag{10.6}$$

where y is responses, y_{pred} the predicted values and $\langle y \rangle$ is the mean value.

R^2 measures goodness of fit. Q^2 measures goodness of prediction. Always $R^2 > Q^2$, but for a well-behaved model R^2 should not be significantly larger than Q^2. If there is large discrepancy, then it is likely the data have been overfitted and the predictive ability of the model equation will be suspect [11].

Case 3

In next case we model how compression pressure affects the crushing strength of tablets (Table 10.3).

Table 10.3 *Results of test where the effect of compression pressure on the crushing strength of tablets was measured*

Compression pressure (MPa)	Crushing strength (N)
29	30
35	37
38	45
55	55
61	60
92	77
106	82
129	100
165	110
200	115

It is beneficial first to draw a scatterplot of the data (Figure 10.3) to get a good overview how it is distributed over the research region.

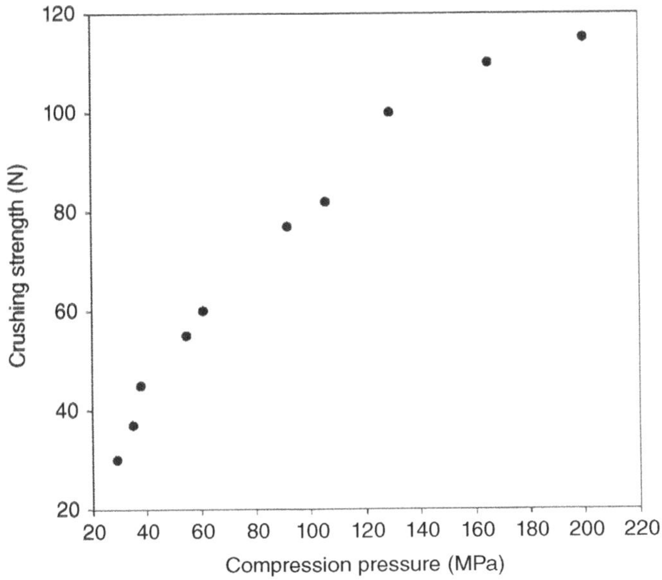

Figure 10.3 *Scatterplot of data from Table 10.3.*

Next, we fit to the data using the most simple model, a straight line. This regression equation is a simple model. Often it is more useful than the correlation coefficient. Using this model, we can predict the value of y from such x values we have not measured, and this model gives us a better summary of the relationship between the two variables. We get best data fit in this case (minimum square error of residuals) when according to Equation (10.4) the slope, α, is 0.506 and the *constant* is 25.1 (Figure 10.4). We have now developed a simple model of how crushing strength increases as a function of compression pressure.

It is now possible to predict what the crushing strength of a tablet would be with, for example, a 150 MPa compression pressure (which we did not measure earlier). Based on Equation (10.5), it can be calculated to be:

$$y = 0.506 \cdot x + 25.1 = 0.506 \cdot 150 + 25.1 = 101$$

Using this model, the prediction for crushing strength is 101 N, the R^2 value of the model is 0.951, and the prediction ability Q^2 is 0.892. Measured data points seem to behave nonlinearly as the compression pressure increases. Accordingly, we can test whether a nonlinear second-order polynomial model would explain the measured results better than a simple line (first-order polynomial).

A second-order polynomial model can be expressed using the following regression equation:

$$y = \alpha \cdot x + \beta \cdot x^2 + constant \qquad (10.7)$$

We get the best data fit in this case using these numerical values: $\alpha = 0.987$, $\beta = -0.00219$ and $constant = 6.37$ (Equation (10.7)). Because this second-order

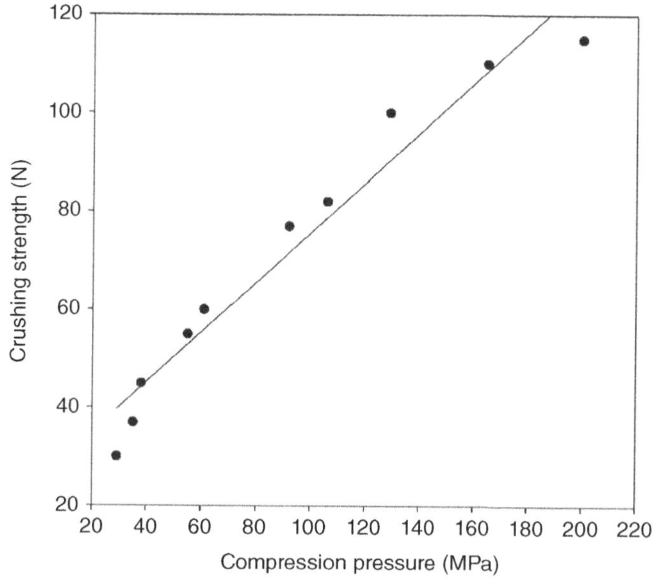

Figure 10.4 *Scatterplot of the data from Table 10.3 and modeled by a first-order equation (straight line).*

polynomial model is more complicated, it can adapt to the measured result better than a simple straight line (Figure 10.5). In this case, the R^2 value is 0.992 and Q^2 is 0.984: these values are higher for this model, which indicate that this is a more useful model.

$$y = 0.987 \cdot x - 0.00219 \cdot x^2 + 6.37 \tag{10.8}$$

Using this new model (Equation (10.8)) we can predict what the crushing strength of the tablet would be for this same 150 MPa compression pressure:

$$y = 0.987 \cdot 150 - 0.00219 \cdot 150^2 + 6.37 = 105$$

The predicted result (105 N) has increased a little compared with the result predicted by the simpler model, the straight line. Is this second-order model proper to describe this phenomenon, or would we still need a more complicated model? A third-order polynomial model (Equation (10.9)) would be more complicated, and we can test how well it will fit to the measured dataset.

$$y = \alpha \cdot x + \beta \cdot x^2 + \gamma x^3 + constant \tag{10.9}$$

We get best data fit in this case using the following numerical values: $\alpha = 0.986$, $\beta = -0.00219$, $\gamma = -4.97 \cdot 10^{-8}$ and $constant = 6.42$ (Equation (10.10)). Because this third-order polynomial model is more complicated than the second-order model, it can adopt to the measured result better than the second-order model (Figure 10.6). The R^2 value increases slightly (not seen when using only 3 significant figures, $R^2 = 0.992$). The value of Q^2 in turn decreases ($Q^2 = 0.967$), which tells us that we have overfitted

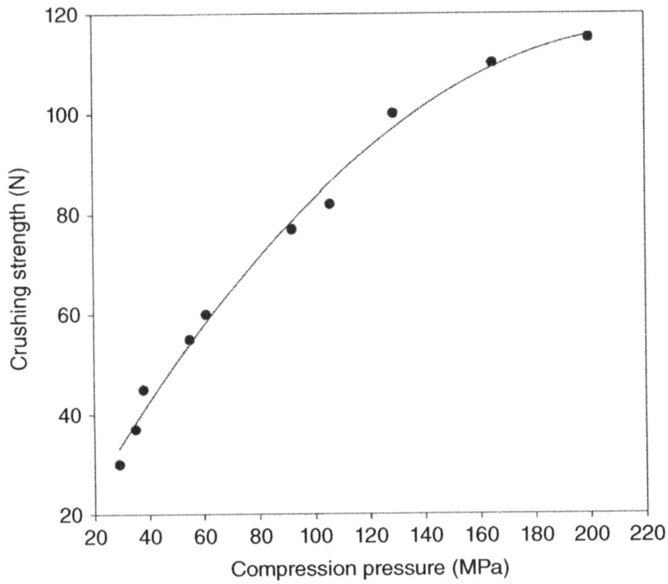

Figure 10.5 *Data from Table 10.3 modeled by a second-order equation (solid line).*

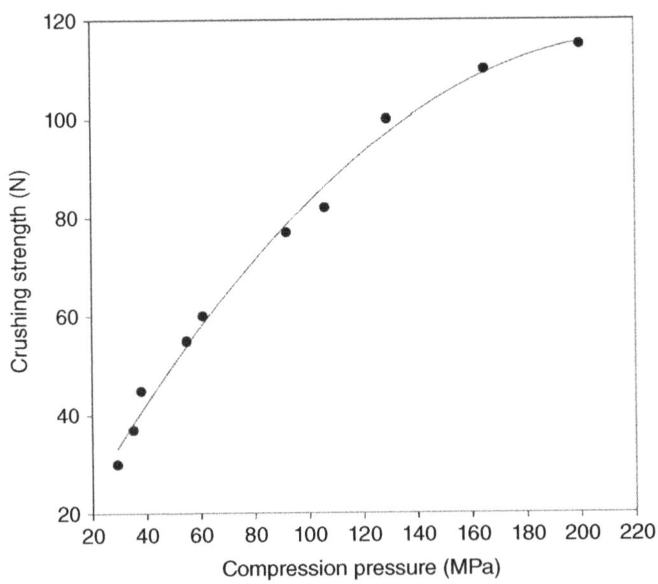

Figure 10.6 *Data from Table 10.3 modeled by third-order equation (solid line).*

our model and the second-order polynomial model is the best model to describe and predict measured results in Table 10.3.

$$y = 0.986x - 0.00219 \cdot x^2 - 4.97 \cdot 10^{-8} \cdot x^3 + 6.42 \quad (10.10)$$

In this case, the predicted crushing strength for a tablet compressed with 150 MPa compression pressure is practically the same (105 N) for the third-order polynomial as with the second-order polynomial equation.

$$y = 0.986 \cdot 150 - 0.00219 \cdot 150^2 - 4.97 \cdot 10^{-8} \cdot 150^3 + 6.42 = 105$$

Finally, in Figure 10.7, a 9th-order polynomial fitting to the same data points is shown. Because the model is so complicated, the R^2 value is 1.0 and the curve goes exactly through all experimental points. However, its ability to predict results is extremely poor. For a 150 MPa compression pressure, the model predicts a negative crushing strength value, which is of course impossible.

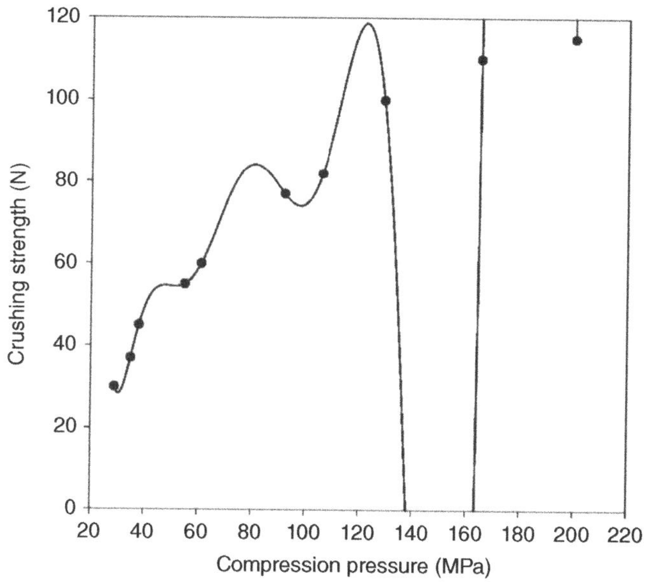

Figure 10.7 Data from Table 10.3 modeled by a 9th-order equation.

It is important to remember that the R^2 value alone does not reveal a good model, and a more complicated model is not always a better model, as described in Case 3. In Figure 10.8 the horizontal axis depicts the model complexity (number of terms, number of latent variables, etc.). The vertical axis corresponds to the amount of explained R^2 and predicted Q^2 variation of measured data. The more complex model we use, the nearer to the limit value 1.0 model's R^2 values increases. When we exceed a certain model complexity, we start to

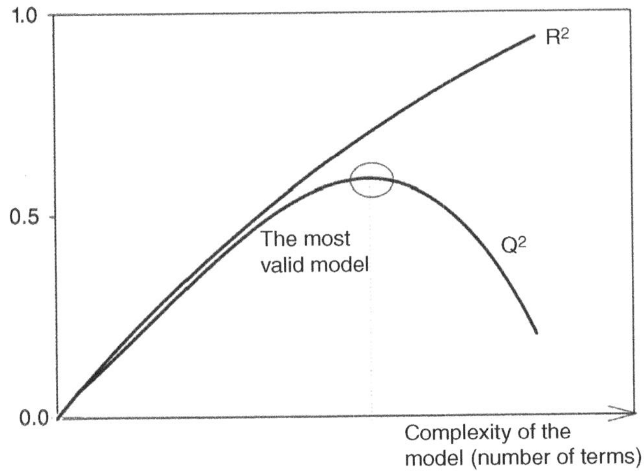

Figure 10.8 The trade-off between the goodness of fit, R^2, and goodness of prediction, Q^2. Modified from [10].

overfit our model and the prediction ability of the model decreases. The most valid model is usually achieved when Q^2 get its highest value.

10.6 Multivariate Data

The data we collect in science and technology are often multivariate. We need in those cases two or more variables to explain variation in our responses. To understand and use purposefully a multivariable dataset we have measured from the system we investigate, it is not enough just to view the measured data table. The measured data must be analyzed such that the information it includes is expressed in an understandable way. This can be achieved by graphs or numerical parameters. Using these graphs and parameters, the result is related to the objectives of the investigation and the scientific context. It is not affordable to make experiments in a trial-and-error fashion. A far more effective method is to use a computer-enhanced, systematic approach that considers all factors simultaneously. This approach is called design of experiments (DOE). When we use DOE we get information about the possible interactions of factors. This is not obtainable through testing one factor at a time (OFAT) while holding the rest of the factors constant [12]. In OFAT, we obtain in the beginning the "best" value for the first factor. This best value is then fixed, and we examine the best value for the next factor and then fix it, continuing on to the last factor. Because each experimental run considers only one factor, many runs are needed to get sufficient information about the optimum settings of the factors. This consumes a lot of time and money. Another drawback is that factors usually interact, and it is not possible to find the optimum solution by examining a single isolated factor at a time. The DOE approach provides information about the interactions of the variables and the way in which the total experimental system works; this is not usually attainable through the OFAT approach. In DOE a carefully selected set of experiments (usually 12–27 runs) are constructed, whereby all the important factors are simultaneously varied systematically [13].

Over the past decades, scientists manually applied DOE principles by using algorithms found in statistical handbooks. During the last 20 years, the development of DOE software packages has made DOE more accessible for researchers having not programming skills and deep statistical experience. Today's software is mostly menu-driven, and programming skills are not needed. These software packages ease the complexity of DOE by increasing the emphasis on graphically displayed results over numerical tables. The software packages are relatively inexpensive, ranging from €100 to €2000. With three or four days of training, people can integrate DOE into their experimentation strategies.

The first step in using DOE is to define the objective of the study and to find the right factors and responses. It is important to choose the proper range (minimum and maximum values) between which each factor varies. The range is usually chosen to be as broad as possible. Factors must in any case be in a sensible area of operation. To find a rational area of operation for factors, preliminary tests are often done. If the range chosen for a certain factor is too narrow, the result of the experiment might be that the factor has no influence on the responses. It is also important to check that all the factors are independent. The independent variable is "independent" because its variation does not depend on the variation of another variable in the study design. The independent variable is controlled or changed only by the researcher. For example, if you analyze the effect of temperature of the granulating mass in a fluidized bed granulator *and* the temperature of the fluidizing air, they are not independent factors. This is due to the fact that, if you change the temperature of the fluidizing air, the temperature of the granulating mass will also change on some timescale. Thus it is not possible to change the levels of those two factors independently. You must be careful when you choose factors for your study design that they are not related to each other in any way. Otherwise, interpretation of the results is impossible; you cannot identify the reason for the change in response, or which of the factors was most influential.

The next step is to choose a proper experimental design. If we have many factors and we do not have prior knowledge of which of these are the most important influences on the responses, it is good to first choose a screening experiment design. Screening designs are used to explore many factors and to reveal which of these have most influence on the responses. Screening designs are economical experimental plans (the number of experiments is low) that can determine the relative significance of many factors. If we have only 2–4 factors we can usually perform directly a full factorial design, and a screening design is not needed [14].

10.6.1 Screening Designs

The primary purpose of a screening design is to identify which of the factors are the most important, that is, which factors most affect the response. After that we can make a more detailed design using those most important factors, for example, a full factorial design in three levels. If we use as a so-called two-level fractional factorial design, or Plackett-Burman design, which involves a carefully chosen subset of the experimental conditions in a complete factorial design, we cannot find out all possible interactions between factors [15]. This is because these interactions are confounded. Always, when we reduce the number of experiments in the study design compared with the full factorial design, we lose some information about the phenomena we study. The more compact the design, the more we lose information. Design resolutions describe how much we lose information and in what way. Most commonly used resolutions in screening designs are resolution III, IV and V designs. In resolution III designs, all the two-factor interactions are confounded. Resolution III designs are excellent for the initial screening to find out the most important factors from many potential factors. If researchers believe that there is a risk that two or more factors interact, this kind of design is not recommended. In resolution IV designs, some, but not all, two-factor interactions are confounded. If two factors are suspected to interact, they can be assigned in the resolution IV study design in such a way that possible interaction between those two factors can anyway be detected. In resolution V designs, no two-factor interactions are confounded. Three- or higher-factor interactions are confounded also in resolution V designs, but these kinds of interactions are extremely rare. Resolution III–V designs are commonly used when the number of factors is six or more. Because there are only two levels for each factor in screening designs, we can *only model linear relationships* between factors and responses.

Case 4: Number of Experiments

If we have six factors in our study design and we change them in two levels, we would need to make $2^6 = 64$ different experiments in full factorial design. With resolution V design the number of experiments would be 32, with resolution IV 16 experiments, and with resolution III only 8 experiments.

10.6.2 Full Factorial Designs

10.6.2.1 Full Factorial Designs in Two Levels

In this design, all input factors are set at two levels. These levels are called "high" and "low" or "+1" and "−1", respectively. A design with all possible high/low combinations of all the input factors is called a full factorial design in two levels. The number of experiments in full factorial designs in two levels is 2^n, where n is the number of factors. The 2^n design is useful in the early stages of experimental work. It is most commonly used when the number of factors is from 2 to 5. With 6 or higher number of factors, the number of experiments is 64 or more, which is usually too high (see Table 10.4).

Table 10.4 *Number of experiments in full factorial designs in two levels (2^n)*

Number of factors (n)	Number of experiments (2^n)
1	2
2	4
3	8
4	16
5	32
6	64

These designs are widely used in factor screening experiments to find the main effects and possible interactions between factors. Because there are only two levels for each factor, we can model only linear relationship between factors and responses.

Case 5

We want to find out the effects of the amount of magnesium stearate in the powder we compress to tablets and the compression pressure on the quality of tablets. Responses are: friability, crushing strength (*CrSt*), weight variation, dissolution rate, roughness of the tablets, etc.

In full factorial design in two levels (2^2) we need to make four experiments (see Table 10.5 and Figure 10.9).

Table 10.5 *Design of 2^2 experiments*

Experiment	Amount of magnesium stearate (Mg)	Compression pressure (CP)
1	"High" or "+1" 0.7%	"High" or "+1" 100 MPa
2	"Low" or "−1" 0.1%	"High" or "+1" 100 MPa
3	"High" or "+1" 0.7%	"Low" or "−1" 50 MPa
4	"Low" or "−1" 0.1%	"Low" or "−1" 50 MPa

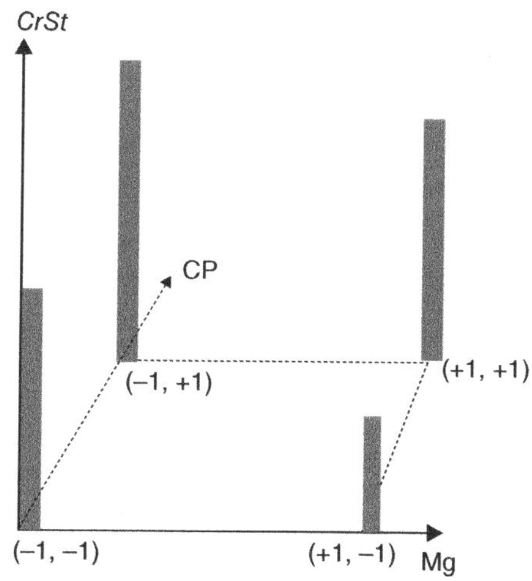

Figure 10.9 *Design of experiments and response (bars) for crushing strength (CrSt). The higher the bar, the higher the response.*

Because with 2^n design we can use only a first-order model (Equation (10.11)), the fitted response surface model is a plane (Figure 10.10).

$$CrSt = a_1 \cdot \mathrm{Mg} + a_2 \cdot \mathrm{CP} + a_3 \cdot \mathrm{Mg} \cdot \mathrm{CP} + constant \qquad (10.11)$$

where CrSt is the crushing strength, Mg is the amount of magnesium stearate and CP is the compression pressure.

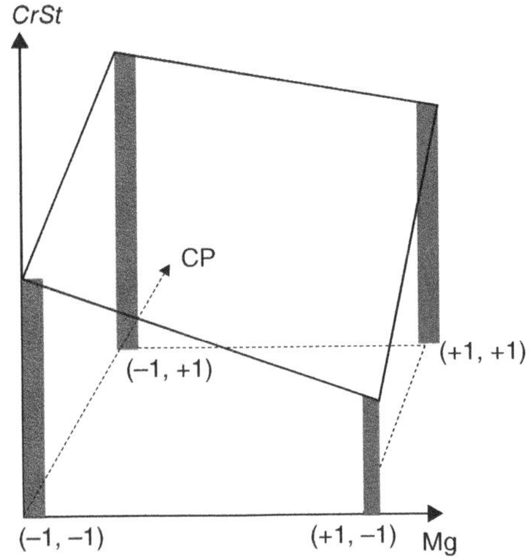

Figure 10.10 *First-order model (plane) fitted to the measured results.*

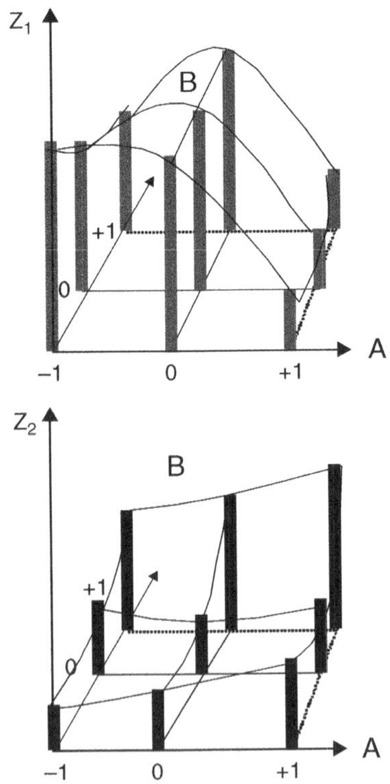

Figure 10.11 *Second-order model (curved plane) fitted to the measured results of responses Z_1 and Z_2.*

10.6.2.2 Full Factorial Designs in Three Levels (3^n Design)

In 3^n factorial design, n factors are at three levels. There are several different notations used to represent these factor levels. One commonly used method is to represent the factor levels by the digits: −1 (low), 0 (intermediate), and +1 (high) (see Figure 10.11). Because factors are used in three levels, it is possible to draw a parabolic response surface (Figure 10.11). A regression model relating the response Z_2 to factors A and B that is supported by 3^2 design is:

$$Z^2 = a_1 \cdot A + a_2 \cdot B + a_3 \cdot A \times B + a_4 \cdot A^2 + a_5 \cdot B^2 + constant \quad (10.12)$$

3^n designs are a good choice for a researcher who is concerned about curvature in the response function.

Case 6

We need to know how compression pressure (CP) in the tablet machine, and the amount of a disintegrating agent in tablet formulation (Di), affect the crushing strength and disintegration time of tablets. We choose a 3^2 study design, and both factors are varied in three different levels. Compression pressure is varied between 100–300 MPa. Amount of disintegrating agent is varied between 0–10%. The total number of different experiments is 9 plus three replicated tests in the middle of the study design. Accordingly, the total number of tests is then 12 (Table 10.6 and Figure 10.12).

Table 10.6 *3^2 study design*

Compression pressure (MPa)	Amount of disintegrating agent (%)
100	0
200	0
300	0
100	5
200	5
300	5
100	10
200	10
300	10
200	5
200	5
200	5

Factors in this study design are compression pressure and amount of disintegrating agent, and the responses are crushing strength and disintegration time (Table 10.7). In the first column a randomized run order is shown. Because experiments are usually done in some time order sequence, there is a risk that measured results are affected by some

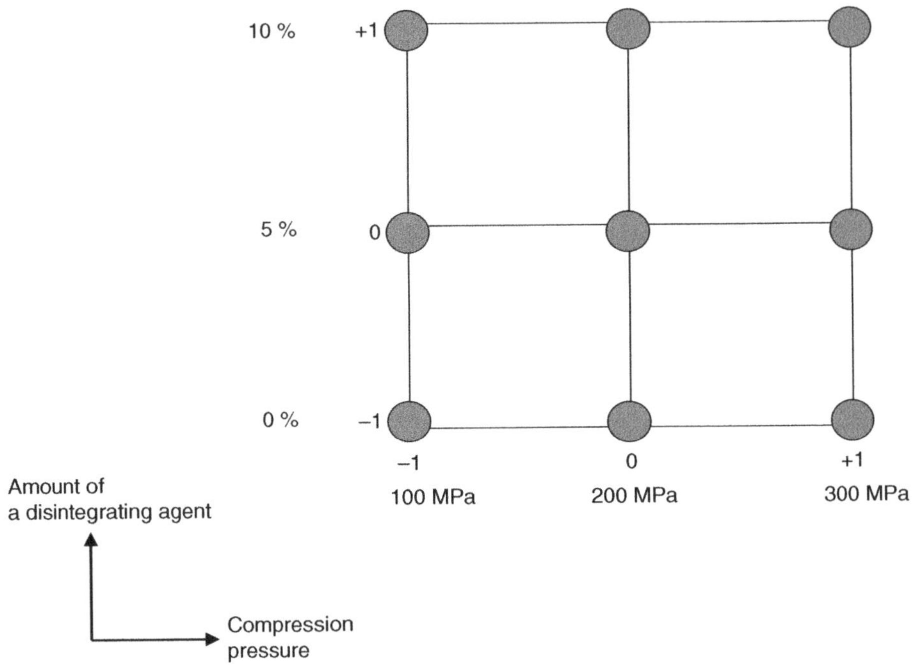

Figure 10.12 *Picture of the 3^2 study design.*

Table 10.7 *Study design and measured responses*

	Factors		Responses	
Run order	Compression pressure (MPa)	Amount of disintegrating agent (%)	Crushing strength of tablets (N)	Disintegration time of tablets (s)
6	100	0	114	500
1	200	0	193	680
2	300	0	232	1020
5	100	5	81	220
10	200	5	148	400
12	300	5	164	720
7	100	10	20	70
9	200	10	82	80
3	300	10	118	270
4	200	5	145	410
8	200	5	140	405
11	200	5	149	390

changes in environment, for example, temperature and humidity may change during seasons. Operators may change, or they may learn to make experiments more efficiently, which might have some effect on the results. These kinds of error sources do not reflect in the results so much if measurements in a study design are made in a random order.

Results are usually modeled by the following second-order polynomial model. Why do we not model results with third-order or even higher order models? The second-order polynomial model is the highest we can use because we have tested our factors in three different levels. If we used a higher-order model, we would not get just one solution, but an infinite number of solutions. We could use a third-order model if we had factors in four different levels in our study design. This would, however, increase greatly the number of necessary experiments in the study design. The maximum order of the model polynomial is always one lower than the number of different levels we have used in our study design. Modeling is usually done using coded factors (−1, 0 ,+1) [16]. By using coded factors, it is possible to compare the size of the coefficients and determine which factor has the largest impact on the response. We fit first crushing strength using this equation:

$$Y = a_1 \cdot CP + a_2 \cdot Di + a_3 \cdot CP \cdot Di + a_4 \cdot CP^2 + a_5 \cdot Di^2 + constant \qquad (10.13)$$

Fitting is usually done by using some modeling program, like Modde, Matlab, SAS, Minitab, and so on.

In Table 10.8, numerical coefficients (coded factors) for the fitted model are presented. In this example, crushing strength is the response parameter.

Table 10.8 *Numerical coefficients for the fitted model*

Crushing strength	Coefficients	*P*
CP	*49.83*	2.46×10^{-6}
Di	*−53.17*	1.68×10^{-6}
CP·Di	−5.000	*0.208*
CP^2	−19.75	0.00390
Di^2	−4.750	*0.316*
Constant	*144.4*	8.50×10^{-9}

For this model, R^2 is 0.991 and Q^2 is 0.923. Column *P* gives the probability that we do not need a certain term in our model. In this experiment, Di^2 has the highest *P* value ($P = 0.316$). The effect of this term on the response is not statistically significant (in italics in Table 10.8). The limit value for *P* to be statistically significant is less than or equal to 0.05. The term *Di* has $P = 0.00000168$. It is the smallest of all the terms in this polynomial model and its effect on the crushing strength is most strong, and also statistically very significant. There is a minus sign before this coefficient. This means that when the amount of disintegrating agent in the tablet formulation increases, the tablet crushing strength decreases.

The next stage is to reduce the model by removing the least significant term (Di^2) from the model polynomial and fit the new reduced model to the data. Because the model has changed, all the coefficients and P values change, as shown in Table 10.9. The R^2 value has decreased because model is more simple and it is not able to adapt as well to the measured data points. Q^2 has increased and the model has better prediction ability for factor levels we have not used in the study design. The mixture term $CP \cdot Di$ now has the highest P value ($P = 0.207$). We will reduce our model further until Q^2 has reached its highest value by removing the $CP \cdot Di$ term from the model.

Table 10.9 *Numerical coefficients for the reduced model*

Crushing strength	Coefficients	P
CP	*49.83*	6.04×10^{-7}
Di	*−53.17*	3.87×10^{-7}
$CP \cdot Di$	−5.000	*0.207*
CP^2	−21.34	0.00134
Di^2	–	–
Constant	*142.8*	4.06×10^{-10}

$R^2 = 0.989$, $Q^2 = 0.947$.

After the above-mentioned recalculations, all the terms are now statistically significant in the final model (Table 10.10), the R^2 value is 0.986 and Q^2 has increased to a value of 0.962.

Table 10.10 *Numerical coefficients for the final model*

Crushing strength	Coefficients	P
CP	*49.83*	2.26×10^{-7}
Di	*−53.17*	1.37×10^{-7}
$CP \cdot Di$	–	–
CP^2	−21.33	0.00125
Di^2	–	–
Constant	*142.8*	5.47×10^{-11}

If we would further reduce the model by removing the term CP^2, the model would become too simple, R^2 would be 0.946, and also Q^2 would decrease to a value of 0.894.

Our final model for crushing strength is thus:

$$Cr = 49.8 \cdot Cp - 53.2 \cdot Di - 21.3 \cdot Cr^2 + 142.8 \tag{10.14}$$

We can use this model for coded factors only. If we calculate the model value for the compression pressure of 200 MPa (0 in coded factors) and 5% of the amount of disintegrating agent (0 in coded factors), we would get the following prediction (142.8 N) for the crushing strength:

$$Cr = 49.8 \cdot 0 - 53.2 \cdot 0 - 21.3 \cdot 0^2 + 142.8 = 142.8$$

Usually we want to use original factor levels, and modeling programs can also transform coefficients to unscaled form. In our example, the equation would become:

$$Cr = 1.35 \cdot Cp - 10.6 \cdot Di - 0.00213 \cdot Cr^2 + 11.0 \quad (10.15)$$

With this model we can use original factor levels. For 200 MPa compression pressure and for 5% amount of disintegrating agent, we would get the same result as earlier:

$$Cr = 1.35 \cdot 200 - 10.6 \cdot 5 - 0.00213 \cdot 200^2 + 11.0 = 142.8 = 143$$

Model validation is an important step in the model building sequence. It reveals how well the final model fits to the measured data points. There are many statistical tools for model validation, but the most used is residual analysis. The residuals from a fitted model (Table 10.11) are the differences between the measured responses and the corresponding prediction of the response computed using the regression equation. If the residuals are normally distributed (Gaussian distribution) around zero, it tells us that our model fits the data well. From residuals we can also check whether our dataset contains outliers. An outlier is an observation point that is distant from other observations (has very high absolute residual value). An outlier may be due to variability in the measurement or it may also indicate experimental error. If you suspect that your data contains outliers, you should first check whether it is generated due to a misprint or some other kind of human-generated error. It is not acceptable to delete outliers from a data table without finding the real reason for the erroneous values. One way to deal the situation is to carry out additional experiments in that point of study design, if it is still possible.

Table 10.11 *The residuals of the fitted final model*

Factors		Responses		
Compression pressure (Mpa)	Amount of disintegration agent (%)	Measured crushing strength (N)	Predicted crushing strength (N)	Residual value (N)
100	0	114	125	−11
200	0	193	196	−3
300	0	232	225	7
100	5	81	72	9
200	5	148	143	5
300	5	164	171	−7
100	10	20	19	1
200	10	82	90	−8
300	10	118	118	0
200	5	145	143	2
200	5	140	143	−3
200	5	149	143	6

Visualising the fitted model: It is difficult to understand the relationship between factors and response just by looking at the model equation. By using a contour plot, it is much easier to understand the dependence between a fitted response and two continuous

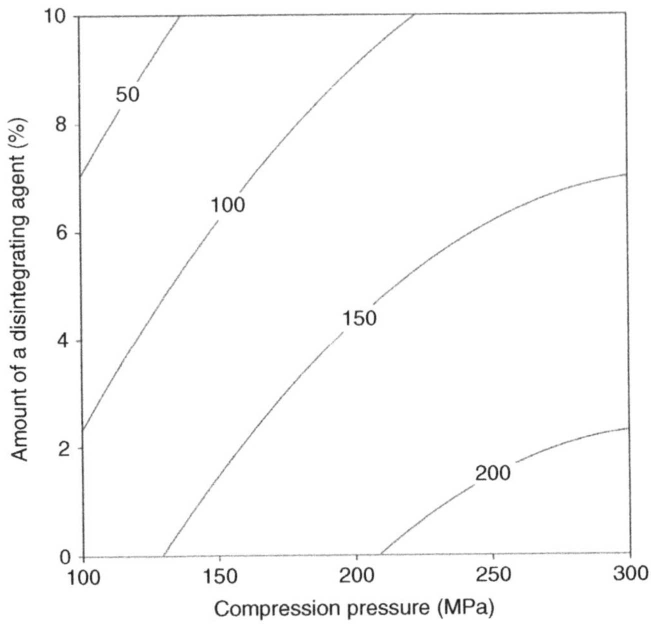

Figure 10.13 *Contour plot of the crushing strength model.*

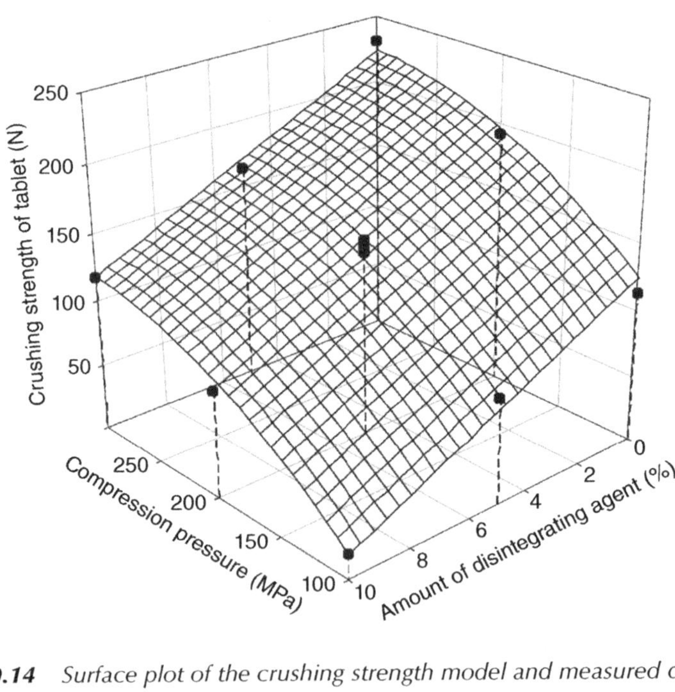

Figure 10.14 *Surface plot of the crushing strength model and measured data points.*

variables. A contour plot (Figure 10.13) displays a two-dimensional view in which points that have the same response value are connected with a contour line. The contour plot is an alternative to a surface plot (Figure 10.14). From the surface plot, we can see that our model is a curved plane that fits as well as possible to the measured data points.

In a second response, the disintegration time of the tablets (*DT*), was modeled with the same principles as the crushing strength and the final reduced model equation is:

$$DT = -0.0667 \cdot -27.3 \cdot Di + 0.00725 \cdot Cr^2 - 0.160 \cdot Cr \cdot Di + 414$$

Figure 10.15 shows the contour plot of this second model.

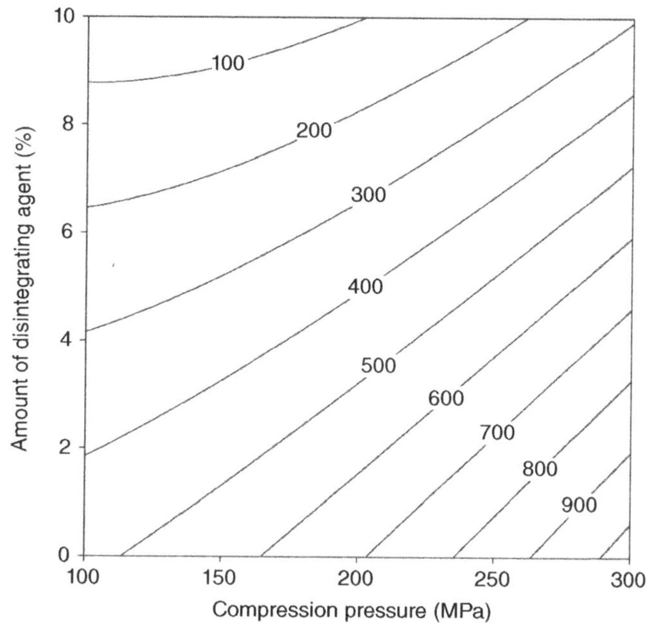

Figure 10.15 *Contour plot of the disintegration time.*

Optimizing process conditions: We can use these two models and contour plots for optimizing. If we would like, for example, to compress tablets having crushing strength over 60 N but not greater than 100 N and disintegration time below 200 seconds, we can use these two contour plots to find a solution. We can combine these two contour plots and find the area where both requirements are fulfilled. In Figure 10.16, critical contour lines from the contour plots of both responses are combined.

We can see from the graph that in the coloured area of the contour plot, both requirements are fulfilled. If we choose tabletting conditions in the middle of this area (compression pressure 151 MPa and the amount of disintegrating agent 8.7%), we most probably will get tablets that meet our quality requirements.

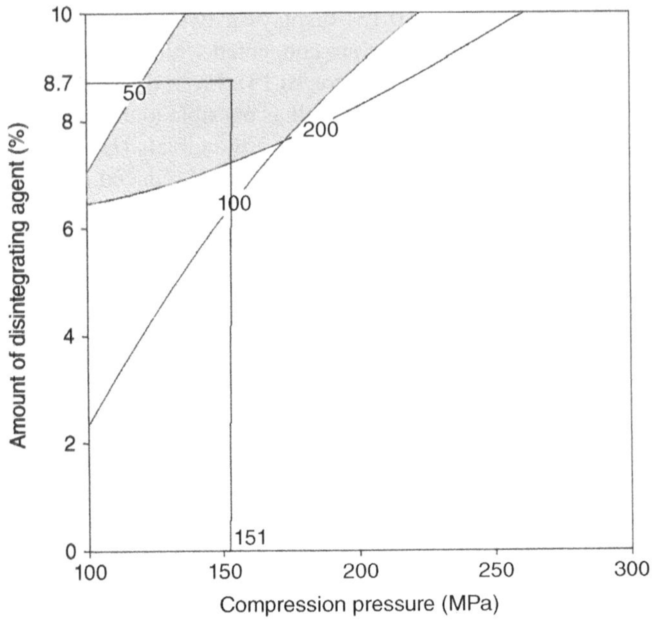

Figure 10.16 Contour plot used to optimize tabletting conditions.

10.7 Principal Component Analysis (PCA)

PCA can be used to reduce the dimensions of a dataset [17]. Dimension reduction is analogous to being philosophically reductionist: it reduces the data down into its basic components, stripping away any unnecessary parts. PCA is based on a decomposition of the original data matrix (X) into a loadings matrix (L) and a score matrix (S) (Equation (10.16)). The loadings are the weights for each original variable when calculating the principal component. The scores are the original data in a rotated coordinate system.

$$X = S \cdot L^T \tag{10.16}$$

If we multiply the score matrix with the transpose of the loadings matrix (L^T), we get the original data matrix as a result.

PCA is a good tool for visualizing spectral and all kinds of multivariate data, where there are lots of variables describing a measurement result [18, 19]. Principal components can reveal an underlying structure in the multivariate data table. Principal components (new coordinate axis) points to the directions where there is the most variance in the data – where the data is most spread out. In this new coordinate system, the first axis corresponds to the first principal component, which is the component that explains the greatest amount of the variance in the original data. PCA transforms the data into a new, lower-dimensional coordinate system where there are usually only two or three new coordinate axes. This new lower-dimensional coordinate system can, in many cases, describe most of the information the original data table contained, but it is much easier to understand and visualize.

Case 7

In this example [20], crystallization of amorphous indomethacin on the tablet surface is investigated. Tablets were compressed from amorphous indomethacin and stored for 22 days at 30°C and in 23% relative humidity. Amorphous indomethacin crystallizes on the surface of the tablet either to gamma or alpha form, depending on the storage temperature and relative humidity. In Figure 10.17, Fourier transform infrared (FTIR) spectroscopy measurements of amorphous indomethacin, tablet surface measurements during storage, and pure gamma and alpha forms of crystalline indomethacin are presented. FTIR spectra are measured with wavenumbers between 650–1801 cm^{-1}.

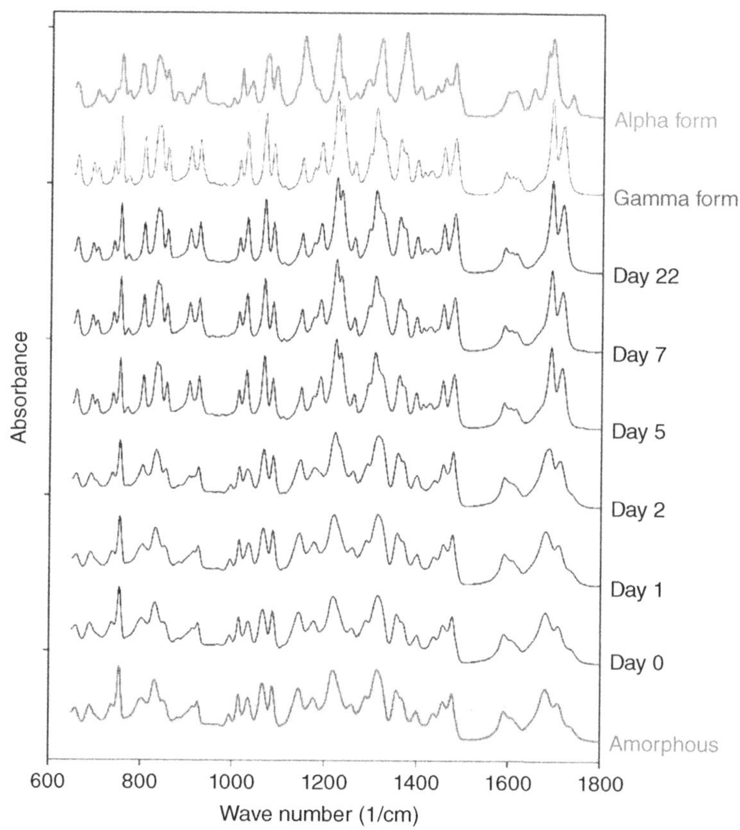

Figure 10.17 Fourier transform infrared (FTIR) spectroscopy measurements of alpha and gamma forms of indomethacin, amorphous indomethacin and freshly prepared indomethacin tablets, and the tablets after 1, 2, 5, 7 and 22 days of storage. Figure is based on data taken from [20].

One measurement (spectra) consists of 598 individual measurement points or variables. When we perform PCA on the data, we reduce the number of variables usually to two or three. In Table 10.12, we can see the spectral data reduced from 598 variables to only three variables (scores). These thee variables cannot represent all information content of the original measurement data table. First principal component (PC1) can explain 65.0%, second principal component (PC2) 29.6%, and third principal component (PC3) 3.4% of the variation in the original data table. Together these first three principal components can explain 98.0% of the information content in original measurements.

In Figure 10.18, the two first score values (Score 1 versus Score 2) of all the FTIR spectroscopy measurements are presented. Because only two variables are now used,

Table 10.12 Score values

Measurement	Score 1	Score 2	Score 3
Alpha form	−7.16	11.1	−0.351
Gamma form	8.51	1.21	0.199
Day 22	7.20	0.895	1.52
Day 7	5.73	−0.697	−2.02
Day 5	6.26	−0.200	−0.756
Day 2	−2.40	−1.95	2.10
Day 1	−5.50	−2.89	1.56
Day 0	−6.56	−4.15	−2.25
Amorphous	−6.08	−3.31	−0.0120

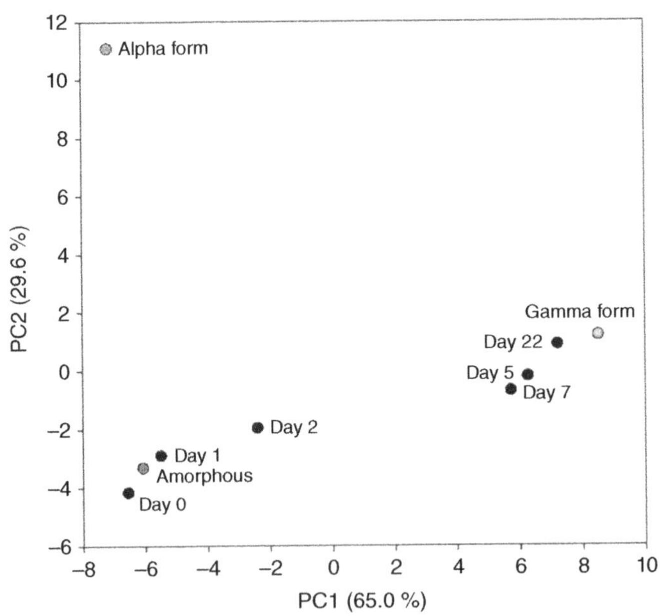

Figure 10.18 Score plot.

we can represent the whole FTIR spectrum as a single point in this score plot. We can see that the FTIR spectrum of amorphous indomethacin is very similar to FTIR spectra of Day 0 and Day 1 from the surface of the tablet. This is because all those three measurements have quite similar first two score values. If two FTIR spectra in the original data table are identical, then their score values will also be identical, and those spectra would be projected over each other on the score plot. The more similar original spectra are, the closer together they will be projected on the score plot. As the storage time increases, the spectra gradually change towards the spectrum of the gamma form.

In Figure 10.19, the first component loadings are shown as a function of wavenumber.

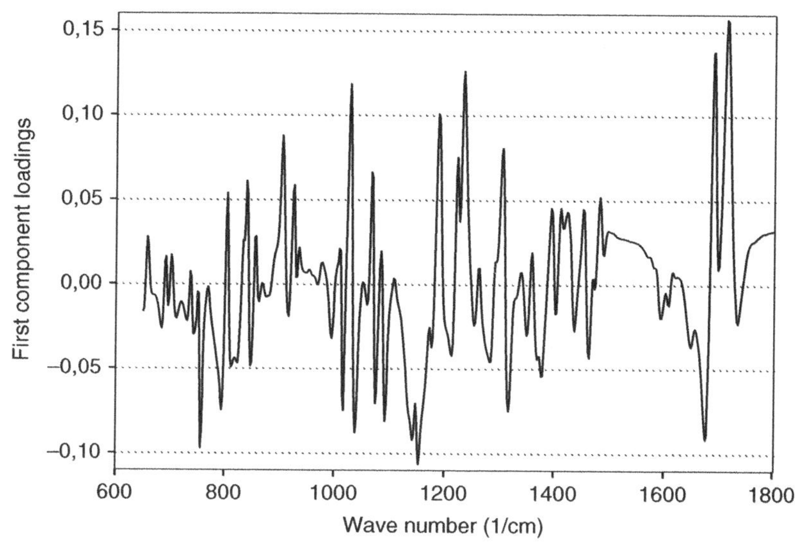

Figure 10.19 *Loadings plot of the first principal component.*

Loadings have the highest values at wavenumbers 1715 cm^{-1} and 1717 cm^{-1}. Those areas in the original spectra explain most of the movement of the score plots in the direction of the first principal component. The higher the absolute value of loadings is in Figure 10.19, the more that area explains the most variation in original spectra as the surface of the tablet crystallizes toward gamma form. If loadings have values near zero, those areas of the original spectra do not explain the crystallization of the tablet surface in this example. Loadings values of the second principal component (Figure 10.20) explain where in the original spectra is the variation leading to score plot movement in the direction of the second principal component. Loadings of the second principal component largely explain variation in the spectrum of alpha form. Loadings have the highest value in this case at wave number 1370 cm^{-1}.

Related methods for PCA are factor analysis [21] and self-organizing maps [22]. All of these three methods are able to organize the original data matrix to a much more visual and understandable form.

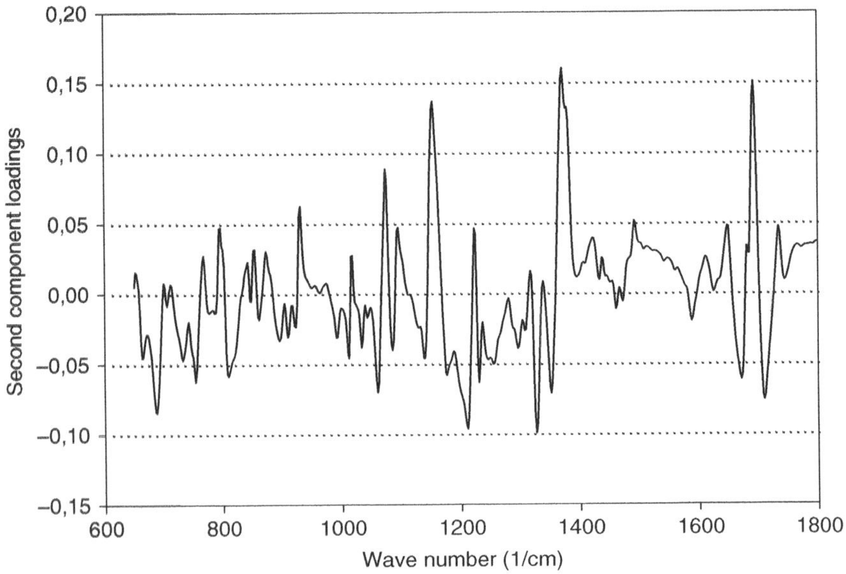

Figure 10.20 *Loadings plot of the second principal component.*

10.8 Conclusions

The continuous development of measuring instruments brings an exponentially increasing amount of new measurement data to researchers. The challenge here is how to convert the measured data and the information contained therein into an understandable form. This can only be done if the data can be analyzed and interpreted correctly, while managing modern multivariate modeling methods and understanding the inherent constraints. These include understanding the importance of repeatable measurements in scientific research, noting that correlation does not in every case imply causality, avoiding overfitting, and identifying the possible interdependencies between the factors, and thus choosing the right predictors for the model.

References

[1] Apte, D.P. (2008) *Statistical Tools for Managers: Using MS Excel*, Excel Books, pp. 269–276.

[2] Peltonen, L. (2018) Design space and QbD approach for production of drug nanocrystals by wet media milling techniques. *Pharmaceutics*, **10** (3), 104.

[3] Fell, J.T., *et al.* (1970) Determination of tablet strength by the diametral compression test. *Journal of Pharmaceutical Sciences*, **59**, 688–691.

[4] Bolton, S., *et al.* (2009) *Pharmaceutical Statistics: Practical and Clinical Applications*, CRC Press, p. 173.

[5] Rodgers, J. (1988) Thirteen ways to look at the correlation coefficient. *The American Statistician*, **42**, 59–66.

[6] Quirk, T. (2015) *Excel 2013 for Biological and Life Sciences Statistics: A Guide to Solving Practical Problems*, Springer, p. 109–112.

[7] Mah, P., *et al.* (2015) Differential scanning calorimetry predicts the critical quality attributes of amorphous glibenclamide. *European Journal of Pharmaceutical Sciences*, **80**, 74–81.

[8] Colombo, M., *et al.* (2017) *In situ determination of the saturation solubility of nanocrystals of poorly soluble drugs for dermal application. International Journal of Pharmaceutics*, **521**, 156–166.

[9] Wolberg, J. (2006) *Data Analysis Using the Method of Least Squares: Extracting the Most Information from Experiments*, Springer.

[10] Eriksson, L. (2013) *Multi- and Megavariate Data Analysis: Basic Principles and Applications*, Umetrics Academy, pp. 83–462.

[11] Merkku, P., *et al.* (1993) Use of 3^3 factorial design and multilinear stepwise regression analysis in studying the fluidized bed granulation process, Part II. *European Journal of Pharmacy and Biopharmacy*, **39**, 112–116.

[12] Montgomery, D. (2003) *Design and Analysis of Experiments*, John Wiley & Sons, pp. 1–21.

[13] Politis, S., *et al.* (2017) Design of experiments (DoE) in pharmaceutical development. *Drug Development and Industrial Pharmacy*, **43**, 889–901.

[14] Teja, S. (2018) 2^3 full factorial model for particle size optimization of methotrexate loaded chitosan nanocarriers: A design of experiments (DoE) approach. *BioMed Research International*, 1–9.

[15] Gareth, A., *et al.* (1998) *Pharmaceutical Experimental Design*, CRC Press, pp. 42–56.

[16] Lawson, J. (2010) *Design and Analysis of Experiments with SAS*, CRC Press, pp. 200–210.

[17] Jolliffe, I. (2003) *Principal Component Analysis*, Springer, pp. 64–71.

[18] Jørgensen, A., *et al.* (2006) Multivariate data analysis as a fast tool in evaluation of solid state phenomena. *Journal of Pharmaceutical Sciences*, **95**, 906–916.

[19] Palttala, I., *et al.* (2013) Towards more reliable automated multi-dose dispensing: retrospective follow-up study on medication dose errors and product defects. *Drug Development and Industrial Pharmacy*, **39**, 489–498.

[20] Novakovic, D., *et al.* (2017) Multimodal nonlinear optical imaging for sensitive detection of multiple pharmaceutical solid-state forms and surface transformations. *Analytical Chemistry*, **89**, 11460–11467.

[21] Kline, P. (1993) *An Easy Guide to Factor Analysis*, Routledge.

[22] Kohonen, T. (2001) *Self-Organizing Maps*, Springer Science & Business Media.

Index

Characterization of Pharmaceutical Nano- and Microsystems, First Edition. Edited by Leena Peltonen.

Printed and bound by CPI Group (UK) Ltd, Croydon, CR0 4YY
17/09/2026
14964307-0001